Methods in
INSECT SENSORY NEUROSCIENCE

FRONTIERS IN NEUROSCIENCE

Series Editors
Sidney A. Simon, Ph.D.
Miguel A.L. Nicolelis, M.D., Ph.D.

Published Titles

Apoptosis in Neurobiology
Yusuf A. Hannun, M.D., Professor of Biomedical Research and Chairman/Department of Biochemistry and Molecular Biology, Medical University of South Carolina
Rose-Mary Boustany, M.D., tenured Associate Professor of Pediatrics and Neurobiology, Duke University Medical Center

Methods for Neural Ensemble Recordings
Miguel A.L. Nicolelis, M.D., Ph.D., Professor of Neurobiology and Biomedical Engineering, Duke University Medical Center

Methods of Behavioral Analysis in Neuroscience
Jerry J. Buccafusco, Ph.D., Alzheimer's Research Center, Professor of Pharmacology and Toxicology, Professor of Psychiatry and Health Behavior, Medical College of Georgia

Neural Prostheses for Restoration of Sensory and Motor Function
John K. Chapin, Ph.D., Professor of Physiology and Pharmacology, State University of New York Health Science Center
Karen A. Moxon, Ph.D., Assistant Professor/School of Biomedical Engineering, Science, and Health Systems, Drexel University

Computational Neuroscience: Realistic Modeling for Experimentalists
Eric DeSchutter, M.D., Ph.D., Professor/Department of Medicine, University of Antwerp

Methods in Pain Research
Lawrence Kruger, Ph.D., Professor of Neurobiology (Emeritus), UCLA School of Medicine and Brain Research Institute

Motor Neurobiology of the Spinal Cord
Timothy C. Cope, Ph.D., Professor of Physiology, Emory University School of Medicine

Nicotinic Receptors in the Nervous System
Edward D. Levin, Ph.D., Associate Professor/Department of Psychiatry and Pharmacology and Molecular Cancer Biology and Department of Psychiatry and Behavioral Sciences, Duke University School of Medicine

Methods in Genomic Neuroscience
Helmin R. Chin, Ph.D., Genetics Research Branch, NIMH, NIH
Steven O. Moldin, Ph.D, Genetics Research Branch, NIMH, NIH

Methods in Chemosensory Research
Sidney A. Simon, Ph.D., Professor of Neurobiology, Biomedical Engineering, and Anesthesiology, Duke University
Miguel A.L. Nicolelis, M.D., Ph.D., Professor of Neurobiology and Biomedical Engineering, Duke University

Methods in
INSECT SENSORY NEUROSCIENCE

EDITED BY

Thomas A. Christensen
Arizona Research Laboratories
Division of Neurobiology
University of Arizona
Tucson, Arizona

CRC Press is an imprint of the
Taylor & Francis Group, an **informa** business

CRC Press
Taylor & Francis Group
6000 Broken Sound Parkway NW, Suite 300
Boca Raton, FL 33487-2742

First issued in paperback 2019

ISBN-13: 978-0-8493-2024-8 (hbk)
ISBN-13: 978-0-367-39346-5 (pbk)

Library of Congress Cataloging-in-Publication Data

Methods in insect sensory neuroscience / edited by Thomas A. Christensen.
p. cm. — (Methods & new frontiers in neuroscience)
Includes bibliographical references and index.
ISBN 0-8493-2024-0 (alk. paper)
1. Insects—Nervous system–Research–Methodology. I. Christensen, Thomas A., 1956– II. Title. III. Methods & new frontiers in neuroscience series.

QL495 .N49 2004
571.7'157--dc22 2004055218

Library of Congress Card Number 2004055218

Visit the Taylor & Francis Web site at
http://www.taylorandfrancis.com

and the CRC Press Web site at
http://www.crcpress.com

Frontiers in Neuroscience

Our goal in creating the **Frontiers in Neuroscience** series is to present the insights of experts on emerging experimental techniques and theoretical concepts that are, or will be, at the vanguard of neuroscience. Books in the series cover topics ranging from methods to investigate apoptosis, to modern techniques for neural ensemble recordings in behaving animals. The series also covers new and exciting multidisciplinary areas of brain research, such as computational neuroscience and neuroengineering, and describes breakthroughs in classical fields like behavioral neuroscience. We want these books to be the books every neuroscientist will use in order to get acquainted with new methodologies in brain research. These books can be given to graduate students and postdoctoral fellows when they are looking for guidance to start a new line of research.

Each book is edited by an expert and consists of chapters written by the leaders in a particular field. Books are richly illustrated and contain comprehensive bibliographies. Chapters provide substantial background material relevant to the particular subject. Hence, they are not only " methods books," but they also contain detailed "tricks of the trade" and information as to where these methods can be safely applied. In addition, they include information about where to buy equipment and about web sites helpful in solving both practical and theoretical problems

We hope that as the volumes become available, the effort put in by us, by the publisher, by the book editors, and by individual authors will contribute to the further development of brain research. The extent that we achieve this goal will be determined by the utility of these books.

To Ann, Cassie, James & Emily ~
for making my life complete

Preface

With the number of identified species estimated at several million and growing steadily, insects are arguably among the most diverse and adaptable organisms on Earth. They have long been and remain our chief competitors for food, and they represent dangerous disease vectors, responsible for spreading devastating human afflictions such as malaria, encephalitis, and dengue fever, to name only a few. The great success of these animals is due in large part to their well-developed sensory systems, and it is now common knowledge that insects possess a host of novel physiological, biochemical, and behavioral attributes that underlie these remarkable feats of sensory performance. Insects face the formidable task of navigating through a range of different and complex environments in order to locate appropriate food sources, oviposition sites, and even mates. Evolutionary processes over millions of years have shaped the development of highly efficient sensory systems that underlie these sophisticated sensory adaptations. Insects are, therefore, eminently suited as models for sensory research, and it follows that investigating the neural mechanisms that underlie sensory processing in insects will yield important insights into the neurobiology of sensation and perception, as well as aid in the development of new strategies to reduce or eliminate insect-borne disease. Additionally, insect sensory systems have recently been shown to be excellent models for designing practical devices that humans can use to enhance their own sensory experiences or to help them navigate their own complex environments.

This book is the first to showcase the tremendous variety of methods that are currently available to study the amazing sensory capabilities of insects. It covers the complete spectrum of sensory modalities in insects, from vision and audition to chemoreception and multimodal processing. My hope is that it will serve as a useful how-to guide for putting into practice a wide range of exciting techniques, including behavioral observation, brain imaging, single- and multi-unit electrophysiology, computer modeling and signal processing, and robotics to address innumerable questions related to sensory system functioning. The book also helps bridge the gap between experimentalists, engineers, and theorists who have long based their research on sensory systems but, sadly, do not interact often enough. I believe that a truly multidisciplinary synthesis of neurobiological, behavioral, and computational approaches to sensory-information processing is most likely to yield our richest understanding of the fundamental mechanisms that underlie sensation and perception. In that spirit, the book contains chapters by leading neuroethologists, comparative biologists, neuroscientists, computational biologists, geneticists, and bioengineers who have adopted insects as their models and have developed or applied to their own studies these new experimental approaches. Their hard work and dedication in assembling this volume is evident in the quality of every chapter, and I am sure the reader will benefit from their careful attention to detail.

This book is intended for the seasoned neuroscientist who is looking for state-of-the-art information and discussions of the most challenging questions that face sensory neuroscience today. At the other end of the spectrum, it is also intended as a primer for newcomers who are considering new avenues of research by taking advantage of insects as premier models to study sensory mechanisms. In preparing the book, I asked all the authors to be generous in providing their "tricks" and "tips," which seldom appear in mainstream publications. You will notice that many authors took great pains in describing some procedures down to the minute details in order to facilitate their implementation in the laboratory. As you read through the book, you will also find some special formatting that may be helpful. For example, many key terms are shown in italics to emphasize the importance of particular structures, mechanisms, and concepts in the discussion. Also, as a guide to newcomers, references to vendors of particularly critical supplies and equipment are indicated by superscript numerals in the main text. The reader can then turn to the Appendix, where they will find an alphabetized list that includes detailed contact information for the vendors cited throughout the book. Finally, each chapter is capped with a section entitled Conclusions and Outlook to help put everything into perspective.

This book was never intended to cover every possible aspect of this enormous and ever-growing field. Indeed, it would be impossible to document all of the possible ways that insects have contributed to our understanding of sensory neuroscience in a single volume of this size. In other cases, as fate would have it, some chapters that were planned never materialized. In the end, however, 35 authors contributed their expertise to what I believe is a truly unique synopsis of the exciting advancements that are driving this exciting field forward. The first section of the book serves as an overview, which I hope will provide the reader with sufficient background information and references to learn more about the basic organization of the insect brain and the behavioral strategies used by insects to navigate their complex and varied environments. Later sections are designed to provide the reader with more detailed information about specific sensory modalities and the tools that are used to study them.

Finally, this volume would not have been possible without the dedicated people at CRC Press who helped pull it all together. I would like to thank Series Editors Sid Simon and Miguel Nicolelis for having faith in me, Publisher Barbara Norwitz for her much appreciated cooperation, and especially my Project Coordinator Erika Dery, who had pity on a rookie editor and patiently and promptly answered every one of my 1237 e-mail messages.

Last, but not least, I am indebted to John Hildebrand for giving me my first break 21 years ago. Over the past several decades, he has attracted an incredibly talented group of individuals to work with him, and it has been a great pleasure to work side by side with them. I would especially like to acknowledge the current cast of characters — Andrew Dacks, Norm Davis, Pablo Guerenstein, Hong Lei, Erin McKiernan, Carolina Reisenman, Heather Stein, Corinna Thom, Vince Pawlowski, and Caroline Wilson — for their clever ideas, thoughtful discussions, and especially, their friendship.

Assembling these diverse chapters into a coherent whole was a tremendous learning experience for me. It is my sincere hope that the reader is struck with the same sense of astonishment in learning about the unprecedented sensory capabilities of insects and the enormous range of sophisticated tools that we use to study them.

Thomas A. Christensen
The University of Arizona
Tucson, Arizona

About the Editor

Thomas A. Christensen, Ph.D., was born in Queens, New York, in 1956. After his family relocated farther east to Suffolk County, he quickly developed a fascination for the teeming invertebrate life that inhabits the rocky beaches of Long Island. He earned his bachelor's degree in biology from the State University of New York at Stony Brook in 1978 and his Ph.D. in neurobiology from the same institution in 1983. Having acquired a passion for insect neurobiology as an undergraduate, he set out to unravel one of the oldest insect mysteries, and his graduate work led to his discovery of the neurons that control luminescence in the firefly. Before the ink had dried on his diploma, he joined the lab of his longtime colleague John Hildebrand at Columbia University and began work as an NIH Postdoctoral Fellow. At Columbia, he developed his interest in the neurobiology of olfaction and its critical importance in insect behavior. Now a senior research scientist at the University of Arizona, Dr. Christensen is a pioneer in adapting multielectrode electrophysiological tools to probe the tiny insect brain.

Dr. Christensen's awards and honors include a visiting professorship at the University of Trondheim in Norway, and he is the recipient of the 1994 Kenji Nakanishi Research Award in Olfaction, the first insect neurobiologist so honored. He is a member of the Society for Neuroscience, the International Congress of Neuroethology, and the International Brain Research Organization, and he has served on the Executive Committee of the Association for Chemoreception Sciences as Counselor from 1996 to 1998 and as Membership Chair from 2000 to 2002. As an active participant in these societies, he has organized and coorganized numerous national and international symposia and scientific conferences. Dr. Christensen is also a staunch supporter of science literacy in our schools, and he regularly organizes outreach activities using insects as teaching tools. He is the author of over 100 published articles, reviews, and chapters, and his most recent research uses a strongly multidisciplinary approach to decipher the roles of complex spatial and temporal patterning in the brain's representation of olfactory stimuli, particularly in the context of the insect's behavior in its natural learning environment.

Contributors

Jan Benda
Department of Physics
University of Ottawa
Ottawa, Ontario, Canada

John R. Carlson
Department of Molecular, Cellular, and Developmental Biology
Yale University
New Haven, Connecticut, U.S.A.

Thomas A. Christensen
Arizona Research Laboratories Division of Neurobiology
University of Arizona
Tucson, Arizona, U.S.A.

Anupama Dahanukar
Department of Molecular, Cellular, and Developmental Biology
Yale University
New Haven, Connecticut, U.S.A.

Martin Egelhaaf
Lehrstuhl Neurobiologie,
Universität Bielefeld,
Bielefeld, Germany

Lawrence H. Field
School of Biological Science
University of Canterbury
Christchurch, New Zealand

Mark A. Frye
Department of Physiological Science
UCLA
Los Angeles, California, U.S.A.

C. Giovanni Galizia
Department of Entomology
University of California at Riverside
Riverside, California, U.S.A.

Tim Gollisch
Institute for Theoretical Biology
Humboldt-Universität zu Berlin
Berlin, Germany

John R. Gray
Department of Biology
University of Saskatchewan
Canada

Jan Grewe
Lehrstuhl Neurobiologie
Universität Bielefeld
Bielefeld, Germany

Bill Hansson
Department of Crop Science
Swedish University of Agricultural Sciences
Alnarp, Sweden

Andreas V.M. Herz
Institute for Theoretical Biology
Humboldt-Universität zu Berlin
Berlin, Germany

John G. Hildebrand
Arizona Research Laboratories Division of Neurobiology
University of Arizona
Tucson, Arizona, U.S.A.

Uwe Homberg
Fachbereich Biologie, Tierphysiologie
University of Marburg
Marburg, Germany

Rickard Ignell
Department of Crop Science
Swedish University of Agricultural Sciences
Alnarp, Sweden

Katja Karmeier
Lehrstuhl Neurobiologie
Universität Bielefeld
Bielefeld, Germany

Roland Kern
Lehrstuhl Neurobiologie,
Universität Bielefeld
Bielefeld, Germany

Rafael Kurtz
Lehrstuhl Neurobiologie
Universität Bielefeld
Bielefeld, Germany

Mattias C. Larsson
Department of Crop Sciences
Swedish University of Agricultural Sciences
Alnarp, Sweden

Hong Lei
Arizona Research Laboratories
Division of Neurobiology
University of Arizona
Tucson, Arizona, U.S.A.

Christian K. Machens
Cold Spring Harbor Laboratory
Cold Spring Harbor, New York, U.S.A.

Philip L. Newland
Southampton Neuroscience Group
University of Southampton,
Southampton, U.K.

Alan J. Nighorn
Arizona Research Laboratories
University of Arizona
Tucson, Arizona, U.S.A.

Vincent M. Pawlowski
Arizona Research Laboratories
Division of Neurobiology
University of Arizona
Tucson, Arizona, U.S.A.

Jennifer J. Perry
Department of Molecular, Cellular, and Developmental Biology
Yale University
New Haven, Connecticut, U.S.A.

F. Claire Rind
School of Biology
University of Newcastle
Newcastle-upon-Tyne, U.K.

Roland Schaette
Institute for Theoretical Biology
Humboldt-Universität
Berlin, Germany

Hartmut Schütze
Institute for Theoretical Biology
Humboldt-Universität
Berlin, Germany

Martin B. Stemmler
Institute for Theoretical Biology
Humboldt-Universität
Berlin, Germany

Doekele G. Stavenga
Department of Neurobiophysics
University of Groningen
Groningen, The Netherlands

Glenn P. Svensson
Department of Ecology
Lund University
Lund, Sweden

Richard S. Vetter
Department of Entomology
University of California at Riverside
Riverside, California, U.S.A.

Anne-Kathrin Warzecha
Lehrstuhl Neurobiologie
Universität Bielefeld
Bielefeld, Germany

Lawrence J. Zwiebel
Department of Biological Sciences
Center for Molecular Neuroscience
Vanderbilt University
Nashville, Tennessee, U.S.A.

Table of Contents

Section I

Introduction: Multimodal Signal Integration and Behavior

1 Multisensory Processing in the Insect Brain

Uwe Homberg

CONTENTS

1.1 INTRODUCTION

Insects, like all animal species, are equipped with multiple sensory channels that extract information from the environment and convert it to adaptive behavioral responses and memories. Past research has greatly advanced our understanding of how information encoding occurs within single sensory channels (such as vision, hearing, somatosensation, and chemosensation) but has paid relatively little attention to specific interactions between different sensory modalities. However, different sensory channels usually have evolved to work in concert, and parallel stimulation of several sensory channels is often essential for the appropriate execution of a particular behavior (Stein and Meredith, 1993; Stein, 1998; Calvert,

0-8493-2024-0/05/$0.00+$1.50

2001). In recognition of this observation, a surge of interest in multisensory or crossmodal processing in vertebrates has provided several important insights into the biological roles and underlying neural mechanisms of multisensory interactions (Driver and Spence, 2000; Meredith, 2002; Calvert et al., 2004). Much of this work has employed recently refined imaging techniques and is directed toward an understanding of human multisensory integration in cognitive functions of object/event identification and localization (Stein, 1998; Calvert, 2001; Maravita et al., 2003; De Gelder and Bertelson, 2003; Small, 2004). However, crossmodal interactions are probably essential for analysis of the environment in all animals. This is especially true for insects.

The significance of multisensory processing in insects has been demonstrated particularly in three functional contexts:

1. **Object perception:** A rewarding food source providing nectar for an insect may be recognized by the specific combination of color and odor from a particular patch of flowers.
2. **Control of posture and goal-directed movement:** Flight balance requires integration of visual signals from the flow field and proprioceptive input from the body and wings.
3. **Learning and memory:** Classical conditioning involves precisely timed sequences of the conditioned and unconditioned stimuli.

Numerous instances of crossmodal processing have been demonstrated within these contexts at the behavioral level (e.g., Böhm et al., 1991; Srinivasan et al., 1998; Hölldobler, 1999; Menzel, 2001; Raguso and Willis, 2002; Frye et al., 2003; Ye et al., 2003), but the underlying neural mechanisms have been studied in relatively few cases. These include olfactory learning, control of flight balance, and centrifugal modulation of sensory attention (e.g., Reichert, 1993; Bacon et al., 1995; Hammer, 1997; Menzel, 2001), all of which are covered in greater detail in later chapters.

Continued interest and further analysis remain stimulated by the fact that insects offer distinct advantages over vertebrate models, perhaps most strikingly:

- The possibility to identify individual nerve cells and to study their functional properties under various contextual conditions
- The feasibility to study the physiology of identified neurons during the expression of natural behaviors

Crossmodal processing can be studied at several different levels, ranging from anatomical tracing studies in prepared tissues to electrophysiological recording techniques in behaving animals. Owing to space limits, this review will focus primarily on multisensory interactions in the insect brain and the specific neural pathways that mediate them (more detailed information on single-modality pathways can be found in later sections of the book).

1.2 ANATOMICAL TRACING STUDIES: SENSORY PATHWAYS IN THE BRAIN

Large areas of the insect brain are involved in decoding sensory signals, especially in the visual (see chapters in Section III), chemosensory (Section IV and Section V), and mechanosensory (Section II) domains, and transforming them into adaptive commands for motor circuits. In the following paragraphs, I will outline the anatomical organization of visual, chemosensory, and mechanosensory pathways in the insect brain.

In each of these sensory systems, primary afferent axons, as well as second- or higher-order interneurons, can be traced in fiber bundles to specific areas in the midbrain (Figure 1.1). Each of these midbrain projections can largely be assigned a modality-specific function, although matching and integration of various aspects within a particular modality (e.g., color and shape; sound and substrate vibration) may occur at relatively early stages in the pathway, before crossmodal processing takes place (see, e.g., Nebeling, 2000; Nishino et al., 2003). As will be evident, the median protocerebrum is thus supplied with a topographical arrangement of sensory-specific areas. With few exceptions, modality-specific projections cannot be traced beyond these midbrain areas, largely because follower interneurons make diffuse connections with various other heterosensory or multisensory brain areas. This suggests that crossmodal processing is probably the rule rather than the exception in interneurons within the median protocerebrum.

1.2.1 VISUAL PATHWAYS

Processing of visual signals dominates the brains of many insect species, as judged by the large numbers of interneurons involved. Signals from compound eye photoreceptors (see Chapter 6) are processed sequentially in three retinotopically organized neuropil areas of the optic lobe: the lamina, the medulla, and the lobula (see Chapter 7 and Chapter 8). Parallel processing for the analysis of different aspects of the retinal image already originates at the photoreceptor level and includes analysis of large- and small-field motion, form, pattern, color, and polarization of the visual signal (Strausfeld and Lee, 1991). Distinct sets of specialized photoreceptors in or near the compound eye and on the frons or vertex of the head serve special functions in insects. There are up to three single-lens eyes (ocelli) on the head that are involved in flight balance and various other functions (Mizunami, 1994). The small, dorsal rim areas of the compound eyes are devoted to the analysis of the polarization patterns in the sky (Labhart and Meyer, 1999; Chapter 6), and extraretinal photoreceptors in the optic lobe or near the compound eye are involved in photic entrainment of the circadian clock (Fleissner and Fleissner, 2003). The orderly, retinotopic organization within the optic lobe is abandoned at deeper levels of the lobula. Particular sets of projection neurons from the lobula and medulla, running in a variety of fiber tracts and bundles, target many regions in the midbrain or provide interhemispheric connections between the optic lobes (Strausfeld, 1976; Figure 1.1a,b).

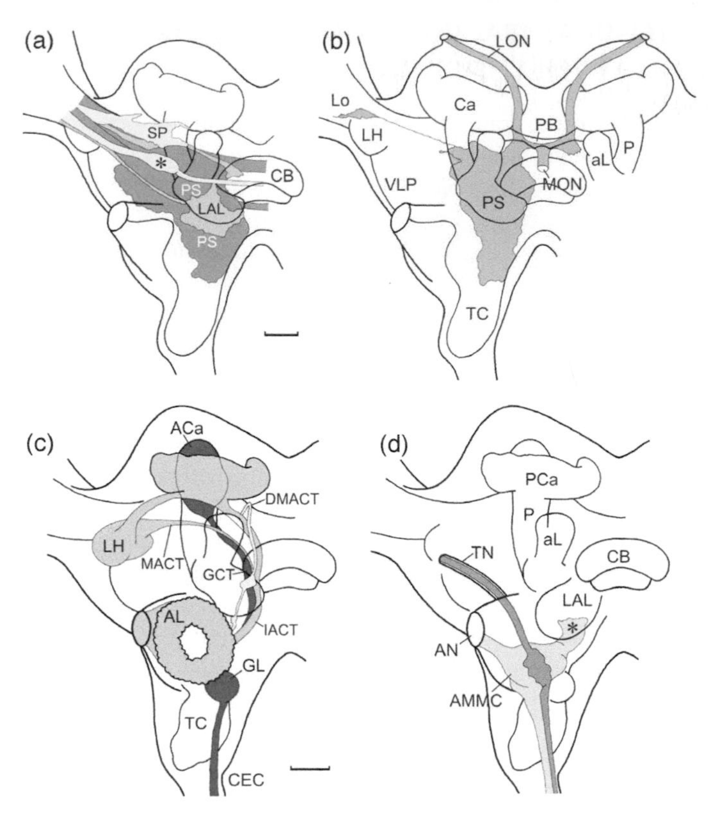

FIGURE 1.1 Summary of sensory pathways in the brain of the locust. **a, b**: Visual pathways from the compound eye (**a**) and ocelli (**b**). Projections are shown from a frontal perspective and in one brain hemisphere only. **a**: Projection neurons from the lobula and medulla invade several anterior foci in the median protocerebrum, such as the anterior optic tubercle (asterisk; light gray), the superior protocerebrum (SP; light gray), and the lateral accessory lobe (LAL, medium gray), but have most prominent arborizations in the deep protocerebrum, especially in the posterior slope area (PS; dark gray) and in median aspects of the ventro-lateral protocerebrum (labeled VLP in [b]; dark gray). **b**: Ocellar interneurons invade large areas in the deep ventro-median protocerebrum (posterior slope, PS), other ocellar tracts, and, more sparsely, the lobula (Lo). **c**: Chemosensory pathways. The antennal lobe (AL; light gray) is the first relay station for olfactory afferent fibers entering the brain, but certain afferents continue via the dorso-median antenno-cerebral tract (DMACT) to the superior median protocerebrum. Major targets of AL outputs are the primary calyx of the mushroom body (labeled PCa in [d]) and the lateral horn area (LH; light gray). Ascending afferents, presumably from contact chemoreceptors on the maxillary palps, are connected to secondary neurons in the glomerular lobe (GL; dark gray), which project to the accessory calyx (ACa; dark gray) of the mushroom body. **d**: Mechanosensory pathways. Mechanosensory fibers from the antenna (light gray) and head capsule (dark gray) enter the antennal mechanosensory and motor center (AMMC) via the antennal and tegumentary nerves (AN, TN). Some AN fibers descend to the ventral nerve cord or continue into the deep ventro-median protocerebrum (asterisk). TN fibers descend to the ventral nerve cord. Other abbreviations: aL, α-lobe; Ca, calyx of the mushroom body; CB, central body; CEC, circumesophageal connective; GCT, glomerular-calycal tract; IACT, MACT, inner and middle antenno-cerebral tract, respectively; LON, MON, lateral and median ocellar nerve; P, pedunculus; PB, protocerebral bridge; TC tritocerebrum. Scale bars: 200 μm.

Areas in the posterior brain, such as the posterior slope area or the posterior optic tubercle, are densely supplied by neurons from the posterior optic tract. Many of these neurons are sensitive to movement stimuli and are involved in optomotor control, motion parallax, and figure–ground discrimination (Strausfeld, 1989; Douglass and Strausfeld, 2003). The posterior optic foci receive convergent visual input from ocellar interneurons (Mizunami, 1994; Figure 1.1b). Many of these projections are directly connected with descending neurons to enable fast optomotor or escape responses (see Chapter 8). The anterior optic tract carries fibers from the lobula and medulla and supplies visual input to the anterior optic tubercle and surrounding areas and, via several fiber fascicles, to the lateral accessory lobes, which are closely associated with the central complex (Homberg et al., 2003a; Figure 1.1a; see Section 1.3.3). Certain neurons of this pathway in the locust are involved in signaling solar azimuth and sky compass information, and perhaps other visual signals involved in spatial orientation (Homberg, 2004). Pathways from the accessory medulla to the superior median protocerebrum provide circadian output from the internal clock (Helfrich-Förster et al., 1998; Helfrich-Förster, 2003).

It is important to note that not all insect sensory systems share the same construction, potentially reflecting different species' behavioral requirements for sensory input (see also Chapter 2). For example, the mushroom-body calyces are free of visual input in flies and are invaded by a few visual fibers in crickets, cockroaches, and locusts, but they receive substantial input from the medulla and lobula in bees, ants, and other Hymenoptera (Strausfeld, 2000). Several other areas in the superior protocerebrum and the ventro-median and ventro-lateral protocerebrum are more sparsely supplied with visual input, and the lateral horn and adjacent areas in the ventro-lateral protocerebrum are essentially free of direct input from the optic lobes (Figure 1.1a).

1.2.2 Chemosensory Pathways

Insect chemosensory organs are located on the antennae, labial and maxillary palps, legs, wings, and other body parts. Antennal chemosensory sensilla largely serve an olfactory function and are specified for long-range perception of volatiles, but gustatory receptors are also present in some species. Mouthpart and tarsal receptors, instead, are largely gustatory. Exceptions are the olfactory CO_2 receptors on labial and maxillary palps of certain species (see chapters in Section IV).

Antennal olfactory pathways are well understood and have been described in detail for a number of species (reviewed by Anton and Homberg, 1999; see also Chapter 12 and Chapter 13). A seemingly universal feature is that all olfactory receptor neurons terminate in the antennal lobe glomeruli of the deutocerebrum. They converge directly or indirectly onto antennal-lobe projection neurons, which provide olfactory input to the protocerebrum via three to five distinct fiber tracts (Figure 1.1c). Because convergence of olfactory and other modalities already exists in the antennal lobe (see Section 1.3.1), at least some of these fiber tracts already carry multisensory signals to the protocerebrum. Targets of antennal-lobe outputs in the protocerebrum are the mushroom-body calyces, the lateral horn, and adjacent areas in the lateral protocerebrum (Figure 1.1a). In addition, there are diffuse projections to regions in the inferior median protocerebrum in different species (Homberg et al., 1989; Stocker et al., 1990; Malun et al., 1993; Anton and Homberg, 1999; Marin et al., 2002; Wong et al., 2002). Mapping studies in moths, cockroaches, and flies showed that projection neurons maintain a chemotopic organization in the calyces and lateral protocerebrum (Homberg et al., 1988; Malun et al., 1993; Wong et al., 2002; Nishino et al., 2003).

Ascending chemosensory pathways from the mouthparts have been mapped in less detail, but some information on brain projections exists in flies and mosquitoes, honeybees, crickets, locusts and cockroaches, and several species of moth (see also Chapter 11). In orthopteromorph insects (locusts, crickets, cockroaches), ascending afferents, apparently from contact chemoreceptors on the maxillary palps, terminate in a glomerular neuropil adjacent to the antennal lobe, the glomerular lobe at the tritocerebrum/deutocerebrum border (Ernst et al., 1977; Ignell et al., 2000; Figure 1.1c). Second-order neurons ascend via the glomerular–calycal tract to the mushroom-body calyces (Ernst et al., 1977; Weiss, 1981; Frambach and Schürmann, 2004; Homberg et al., 2004). In holometabolous insects, a glomerular lobe is not present. Instead, second-order gustatory interneurons in honeybees project directly from the subesophageal ganglion to the calyces (Schröter and Menzel, 2003; Figure 1.2). In Diptera and Lepidoptera, ascending afferents from mouthpart receptors enter specific glomeruli in the antennal lobe that are not targeted by antennal afferents. Many and

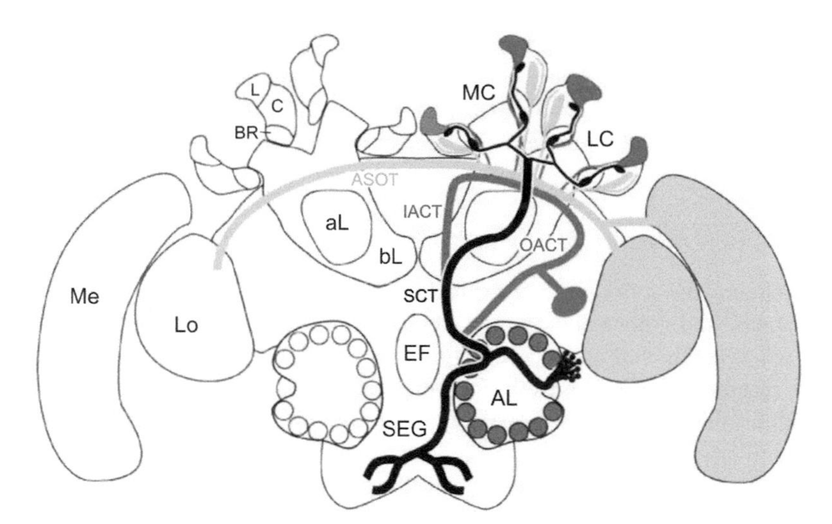

FIGURE 1.2 **(See color insert following page 202.)** Topographic segregation of multisensory input to the calyces of the mushroom body in the honeybee brain. Projection neurons from the antennal lobe (AL) enter the calyces via the inner and outer antenno–cerebral tracts (IACT, OACT) and arborize primarily in the lip area (L). Projection neurons from the subesophageal ganglion (SEG), probably carrying contact chemosensory information, enter via the subesophageal calycal tract (SCT) and ramify in parts of the collar (C) and basal ring (BR) neuropils. Visual input from the lobula (Lo) and medulla (Me) enters the calyces via the anterior superior optic tract (ASOT) and invade collar neuropil. aL, bL, α- and β-lobe of the mushroom body; EF, esophageal foramen; LC, MC, lateral and median calyx of the mushroom body. (Adapted from Schröter, U. and Menzel, R., *J. Comp. Neurol.*, 465: 168–178, 2003. With permission.)

perhaps all of these afferents are sensitive to CO_2 (Anton and Homberg, 1999). In the fruit fly Drosophila and the sphinx moth *Manduca*, antennal-lobe projection neurons from glomeruli that receive maxillary input continue via the inner antenno-cerebral tract to the mushroom-body calyces and lateral horn of the protocerebrum (Stocker et al., 1990; Wong et al., 2002; Guerenstein et al., 2004).

1.2.3 Mechanosensory Pathways

Mechanoreceptors are distributed over the whole body surface and, corresponding to position and cuticular specializations, serve a variety of functions. These include touch, perception of air currents, surface vibrations, airborne vibrations (sound), proprioreception, equilibrium and gravity, temperature, and humidity (see Section II). Only some of these receptors are located on the antennae and head capsule; others are connected indirectly via ascending neurons to the brain.

Axons from tactile and wind receptors on the antennae and head project via the antennal and tegumentary nerves to several mechanosensory foci in the antennal mechanosensory and motor center (Figure 1.1d). Projections from unknown antennal mechanoreceptors extend beyond the deutocerebral domain and, as demonstrated in locusts, crickets, bees, and flies, innervate areas in the ventro-median protocerebrum, which partly overlap with visual projections (Gewecke, 1979; Strausfeld and Bacon, 1983; Maronde, 1991; Figure 1.1d).

Although hygro- and thermoreceptors on the antennal flagellum are specifically modified mechanoreceptor neurons, their central projections, as revealed in the cockroach, extend into the “olfactory” antennal lobe and not to the mechanosensory neuropil of the deutocerebrum (Nishikawa et al., 1995). Three small groups of glomeruli receiving hygro- and thermoreceptive afferents have been identified in the cockroach: DC1 glomeruli receive exclusive input from cold receptors, DC2 glomeruli from moist receptors, and DC3 glomeruli from dry receptors (Nishikawa et al., 1995; Nishino et al., 2003). Projection neurons from these glomeruli innervate the calyces of the mushroom bodies and areas in the lateral horn of the protocerebrum that are distinct from the projection areas of uniglomerular or macroglomerular projection neurons, signaling general odors and female pheromones, respectively (Nishino et al., 2003). Some of the identified hygro- and thermosensory projection neurons are unimodal, whereas others also respond to olfactory stimuli (Zeiner and Tichy, 2000; Nishino et al., 2003).

Auditory organs are located on various body parts, and therefore, auditory signals use species-specific mechanosensory pathways to the brain. In orthopteroid insects, projections from ascending auditory, vibratory, and wind-sensitive interneurons to the brain have been studied in some detail (Eichendorf and Kalmring, 1980; Boyan and Williams, 1982; Boyan, 1983; Shen, 1983; Nebeling, 2000; see also Chapter 3 and Chapter 5). In locusts and bush crickets, these neurons generally extend processes into adjacent areas in the ventro-lateral protocerebrum, whereas ascending auditory interneurons in the cricket project to superior median areas, the optic stalk, and the posterior median brain. In bush crickets, Nebeling (2000) reported ascending vibratory, auditory, and bimodal vibratory–auditory interneurons. Vibration-sensitive and bimodal interneurons have more lateral, and auditory interneurons more medial, projections in the ventro-lateral protocerebrum. Side branches of auditory interneurons extend to mechanosensory neuropil of the deutocerebrum in bush crickets (Nebeling, 2000) and locusts (Boyan, 1983) but are missing in the vibration-sensitive neurons (Nebeling, 2000). Wind-sensitive giant interneurons receiving input from cercal afferents also terminate in the deep ventro-lateral protocerebrum (reviewed by Boyan and Ball, 1990), but the spatial relationship of their projections with those of auditory and vibratory fibers has been documented only poorly.

1.3 MULTISENSORY CONVERGENCE IN HIGHLY STRUCTURED BRAIN AREAS

Anatomical tracings and neuroimaging studies can suggest areas of convergence of different sensory modalities, but whether a synthesis of crossmodal cues actually occurs can only be studied physiologically. Insects offer the possibility to study the organization of crossmodal processing at the systems level (see Section V) and, with intracellular recordings, at the level of single, identifiable cells. Although it is likely that most intrinsic neurons of the protocerebrum are multisensory, the exact properties of intersensory matching and integration have only been studied in a few well-accessible descending neurons (see Section 1.4). The highly organized brain areas of the deutocerebrum, the mushroom body, and the central body, on the other hand, offer the possibility to highlight neuronal architectures that underlie multisensory convergence for specific functional roles of crossmodal processing within these brain areas.

1.3.1 DEUTOCEREBRUM

Although the antenna houses mechanosensory, olfactory, and contact chemoreceptor cells, integration of these modalities within the deutocerebrum has received less attention than analysis of coding within the olfactory domain. Olfactory receptor neurons project to the antennal lobe and mechanosensory afferents (at least those from the antennal scape and pedicel) to the antennal mechanosensory and motor center. Although there is little evidence for the presence of direct connections between both areas, bimodal mechanosensory/olfactory responses were encountered in substantial numbers of antennal-lobe neurons in moths (Kanzaki and Shibuya, 1986; Kanzaki et al., 1989; Anton and Hansson, 1994; Hansson et al., 1994), cockroaches (Waldow, 1975; Ernst and Boeckh, 1983; Zeiner and Tichy, 1998; Strausfeld and Li, 1999; Nishino et al., 2003), and honeybees (Homberg, 1984; Flanagan and Mercer, 1989; Fonta et al., 1991; Iwama et al., 1995; Abel et al., 2001; Müller et al., 2002). Several of these studies took specific care to avoid olfactory contamination during stimulation by using pure air or stimulating by direct mechanical touching of the antenna (e.g., Waldow, 1975; Zeiner and Tichy, 1998). In the honeybee, certain antennal-lobe projection neurons responded both to odors and to gustatory stimulation with sugar water (Homberg, 1984), whereas in the cockroach, thermo- and hygrosensitive projection neurons also were generally sensitive to odors (Zeiner and Tichy, 2000; Nishino et al., 2003). The latter neurons showed characteristic excitatory/inhibitory response profiles to moist air, dry air, cold air, and various natural odors (orange, banana, apple). In all of these studies, neurons most often responded synergistically to two modalities, resulting in an increased response to the combination of two stimuli, such as odor and an air stream (Zeiner and Tichy, 1998). Certain neurons, however, showed antagonistic responses (i.e., inhibition to pure air but strong excitation to an odor) (Kanzaki and Shibuya, 1986; Fonta et al., 1991; Iwama et al., 1995). Although antagonistic responses to two modalities might be interpreted as a mechanism of contrast enhancement between background stimulation (such as wind) and the stimulus of interest (such as an odor), synergistic responses to two stimuli (e.g., a combination of banana odor plus humidity) could suggest that a neuron is contextually tuned to a bimodal stimulus (i.e., banana odor carried in moist air). Further experimentation is needed to clarify the synaptic mechanisms that underlie these multimodal interactions at early stages of processing in the deutocerebrum.

1.3.2 Mushroom Bodies

Mushroom bodies are neuropil structures in the insect midbrain, involved in olfactory learning and various related cognitive functions (Strausfeld, 2000). Each mushroom body typically consists of a pair of cup-shaped calyces, which may be fused in certain species. Both calyces are joined to a stalk or pedunculus which gives rise to two major lobes, termed α-lobe (or vertical lobe) and β-lobe (or horizontal lobe), which may be subdivided further (Strausfeld, 2000). Thousands of interneurons, the Kenyon cells, are arranged in parallel and form the matrix of the mushroom body. From cell bodies in and around the calyces, Kenyon cells send dendritic arborizations into the calyx neuropil. Single axonal fibers run through the pedunculus and bifurcate, with one collateral extending into each of the two lobes.

1.3.2.1 Modality-Specific Topography

In all insect species studied, the calyces receive input from the antennal lobe and, therefore, from olfactory projection neurons (reviewed by Anton and Homberg, 1999). In certain groups of insects (e.g., Diptera and Lepidoptera), antennal-lobe input may be the only input from primary sensory neuropils, but in Orthopteromorpha (locusts, crickets) and Hymenoptera (honeybees), putatively gustatory input is also present and maps to a calycal area (the accessory or secondary calyx in the cricket and locust), which is distinct from the olfactory domain (Figure 1.1c; Ernst et al., 1977; Weiss, 1981; Schröter and Menzel, 2003; Frambach and Schürmann, 2004; Homberg et al., 2004). In Orthopteromorpha, but much more prominently in Hymenoptera, the calyces also receive visual input (Figure 1.2), which again maps to a discrete calycal zone (Gronenberg and Hölldobler, 1999; Gronenberg, 2001; Ehmer and Gronenberg, 2002). Finally, a variety of multisensory interneurons from various brain areas and recurrent neurons from the mushroom-body lobes feed into the calyces (Schildberger, 1984; Nishino and Mizunami, 1998; Strausfeld and Li, 1999).

How and where are these modalities integrated? Intracellular analysis of output elements from the mushroom body lobes, particularly in the cockroach, the cricket, and the honeybee, shows that multisensory processing is the general rule (Schildberger, 1981, 1984; Homberg, 1984; Gronenberg, 1987; Li and Strausfeld, 1997,

1999; Rybak and Menzel, 1998; Grünewald, 1999; Okada et al., 1999). The modality-specific topographic organization of the calyces suggests that integration of different modalities may occur in the calyces and in the lobes, depending on the domains of the dendritic processes of Kenyon cells and extrinsic output neurons. Bimodal responses to antennal touch/wind stimulation and odors are most common, but additional sensitivity to sound (cricket, cockroach), cercal stimulation (cricket, cockroach), sugar water (honeybee), and simple visual (lights on/off) stimulation (cockroach, honeybee, cricket) also occurs. Although bimodal mechanosensory/olfactory response profiles may already be present in the antennal-lobe derived inputs, combination with visual, auditory, gustatory, and cercal mechanosensory input might occur largely at the level of the mushroom body. Because olfactory responses usually dominate, it appears that mushroom-body neurons respond in a specific heterosensory context to particular odors. One possible interpretation of crossmodal processing in the mushroom body, put forward by Strausfeld and Li (1999), is that response thresholds for particular odors (food odors, pheromones) may change depending on specific multisensory conditions. In extreme cases, neurons only responded to a modality when it was presented together with a second, different modality (Li and Strausfeld, 1999). In other cases, the response or its sign (inhibitory, excitatory) to a sensory stimulus depended on the modality of a preceding stimulus (Schildberger, 1981; Gronenberg, 1987; Li and Strausfeld, 1999).

1.3.2.2 Learning-Associated Heterosensory Interactions

Olfactory learning using classical or operant conditioning has been demonstrated in several insect species, including cockroaches, crickets, honeybees, flies, and moths (Hartlieb et al., 1999; Menzel, 2001; Sakura and Mizunami, 2001; Waddell and Quinn, 2001; Matsumoto and Mizunami, 2002). Work in the honeybee and the fruit fly established the essential role of mushroom bodies in olfactory learning, but a neurophysiological analysis of heterosensory interactions during stimulus associations has been performed only in bees. After a single pairing of an odor with sucrose stimulation of the proboscis, honeybees learn to associate the odor with the sucrose reward and show a proboscis extension response that is not present in naive bees. Hammer (1993, 1997) identified a gustatory interneuron in the subesophageal ganglion, termed VUMmx1, which mediates the reinforcing function of the sugar reward during olfactory conditioning. VUMmx1 shows long-lasting excitation to sugar water stimulation of the antennae and proboscis, and forward-pairing of an odor with depolarization of VUMmx1 increased the odor-evoked response of the main proboscis muscle in a later test. Ascending projections of VUMmx1 converge with olfactory pathways in three areas: the antennal lobe, the lateral horn of the protocerebrum, and the calyces of the mushroom bodies (Hammer, 1993, 1997; Menzel, 2001). This suggests that coincidence of VUMmx1-mediated sucrose reward signals and olfactory inputs at these three sites may underlie olfactory memory formation. Interestingly, local injections of octopamine, the likely neurotransmitter of VUMmx1, showed that memory could be established after separate injection into the antennal lobe and calyces, but not after injection into the lateral horn. Furthermore, although injections into the antennal lobe led to a normal acquisition function, memory development occurred in a slower, consolidation-like process after injections into the calyces (Hammer and Menzel, 1998; Menzel, 2001). These data show that gustatory–olfactory interactions in the antennal lobe, calyces, and lateral horn play different roles in memory formation and may lead only in concert to a solid long-term memory trace.

Direct physiological evidence for a role of the antennal lobe–mushroom body pathway in olfactory associative plasticity was provided by intracellular recordings from extrinsic neurons with ramifications in the mushroom-body lobes (Mauelshagen, 1993; Grünewald, 1999). These neurons responded to various odors, to sucrose, and to other modalities. Following a single olfactory conditioning trial (odor stimulus followed by sugar stimulation), the excitatory olfactory response decreased transiently but, following multiple conditioning trials, increased with respect to the naive response. Whether nonolfactory projections to the calyces in Hymenoptera and other insects are also involved in memory formation, as suggested by impaired visual place memory after severance of mushroom bodies in the cockroach (Mizunami et al., 1998), remains to be explored. Interestingly, a positive interaction between visual and olfactory learning in honeybees has been demonstrated (Gerber and Smith, 1998), and the mushroom bodies may well be the sites for this multisensory interaction.

1.3.3 Central Complex

The central complex is a group of neuropils spanning the midline of the brain. Work in the fruit fly and locust points to a role in spatial orientation and navigation, and in the coordination of right–left motor output (Homberg, 1994; Strauss, 2002; Vitzthum et al., 2002). The central complex consists of four subunits: the upper and lower divisions of the central body (also termed *fan-shaped body* and *ellipsoid body*), a pair of ventral noduli, and the protocerebral bridge. The anatomical organization of the central complex is characterized by a quasi-crystalline matrix of layers composed of arrays of exactly 16 columns in all species studied so far. Two categories of neurons predominate: tangential neurons innervate all columns of a particular array, and columnar neurons interconnect single columns of different arrays in a highly stereotypic pattern of interhemispheric connections. Input and output connections are with several areas in the midbrain, but most prominently with the lateral accessory lobes adjacent to the central body. In locusts and honeybees, sensory input from the visual system dominates (Homberg, 1985, 1994; Milde, 1988; Vitzthum et al., 2002), but additional responses to mechanosensory and/or chemosensory stimuli are common (Homberg, 1985).

In locusts, single cell recordings provided the first insights into the specific sensory functions of this brain area. Neurons associated with the lower division of the central body respond to dorsally presented polarized light (see also Chapter 2 and Chapter 6), suggesting a function in sky-compass orientation (Vitzthum et al., 2002). Mechanosensory input, provided through ascending fibers via the lateral accessory lobe (Homberg, 1994), appears to be associated with flight-related sensory feedback, but so far, combinations of polarized light during stationary flight have not been tested. Visual input via the posterior optic tubercle to the protocerebral bridge, studied only anatomically (Homberg and Würden, 1997), originates from the accessory medulla, the circadian pacemaker in the brains of cockroaches and flies (Helfrich-Förster, 2003; Homberg et al., 2003b). This pathway may be involved in time compensation, essential for the proper functioning of an internal sky compass. It is clear from these studies that the central complex plays a cardinal role in spatial orientation of the behaving animal and is probably involved in evaluating external sensory cues (sky polarization pattern, solar azimuth, pheromones) with motion-generated sensory feedback for tasks of path integration and right–left orientation, but how and in which neurons integration of these signals takes place is still unresolved.

1.4 MULTISENSORY INTEGRATION IN DESCENDING NEURONS

Neurons with dendritic ramifications in the brain and axonal processes in the ventral nerve cord (descending neurons) are involved in many aspects of behavioral control. Several of these neurons in a number of insects have been studied extensively and are now classical examples for the integration of multisensory signals at the level of identified neurons (for details, see Sections 1.4.1 and 1.4.2). A morphological character of virtually all of these neurons is their elaborate, highly partitioned dendritic tree, which integrates signals from several brain areas supplied by different sensory modalities in a behaviorally meaningful way. Among the functional roles of descending neurons, it is possible to distinguish neurons mediating escape responses, neurons stabilizing motor activity (such as flight direction and balance), and neurons controlling goal-directed behavior.

1.4.1 Neurons Mediating Escape Responses

Well-studied examples of descending neurons mediating escape responses include the following:

- The giant descending neuron (GDN) and a cluster of similar descending neurons in flies
- The descending contralateral movement detector neuron (DCMD) and related neurons in locusts (see Chapter 8)
- Descending mechanosensory interneurons (DMIs) in the cockroach

In all three species, integration of specific visual and mechanosensory cues is essential for eliciting a response in these interneurons and, consequently, a rapid escape.

The GDN in flies is the most prominent cell in a cluster of similarly organized neurons (the GDNC). It is electrically and chemically coupled to thoracic motor neurons, triggering an escape jump of the fly followed by opening of the wings (reviewed by Strausfeld, 1989). The dendritic tree of the GDN in the brain is highly partitioned, with one arbor extending into antennal mechanosensory neuropil and two dendritic fields extending laterally into optic foci of the protocerebrum (Bacon and Strausfeld, 1986). The GDN rarely spikes to single-modality stimulation; it shows vigorous spiking to corresponding ipsilateral wind and visual stimulation but not to "nonsense" combinations such as ipsilateral wind and contralateral light

(Milde and Strausfeld, 1990). In addition, the neuron receives input from ipsilateral wind-sensitive ascending interneurons from the thoracic ganglia. These results illustrate the specific multisensory filter characteristics of the GDNC cells in the fly brain, which cooperatively initiate the escape jump.

The specific stimulus features that elicit a response have been particularly well studied in the locust DCMD neuron (reviewed by Rowell, 1971; Rind and Simmons, 1999; Rind, 2002; Chapter 8). Like the fly GDN, the DCMD is the most prominent of a group of descending interneurons which, upon activation, prime and release hindleg extension, resulting in a powerful escape jump (Burrows and Rowell, 1973; Burrows, 1996). The DCMD neuron receives input in the posterior lateral protocerebrum via electrical and chemical synapses from a lobula tangential neuron, the lobula giant movement detector neuron (LGMD; Chapter 8). This neuron receives columnar input in the lobula and specifically responds to approaching objects on a collision course with the locust (Rind, 2002). In addition, the LGMD receives auditory input (loud sound) at a second site in the lateral protocerebrum (O'Shea, 1975). Sensitivity of the DCMD to compound eye stimulation is also modulated by ocellar input — reducing light intensity detected by ocellar L-neurons increases excitability to compound eye stimulation (Simmons, 1981). This multisensory response profile of the LGMD–DCMD network is well adapted for mediating an efficient escape response, but more is yet to be learned about how different combinations of sensory stimuli affect its firing properties.

The DMIs of the cockroach initiate a directional escape run in response to direct touch of the antenna (Burdohan and Comer, 1996; Ye and Comer, 1996). There is at present no evidence for sensory input to the DMIs in addition to antennal touch, but behavioral studies have shown that the turning response depends on visually controlled guidance of antennal movements toward the touching object and that visual control also contributes to the run phase (Ye et al., 2003). The two phases of the cockroach escape response are, therefore, controlled differently by visual and mechanosensory input. The underlying neuronal circuitries controlling this behavioral sequence have yet to be identified.

1.4.2 Neurons Controlling Flight Course and Balance

More than a dozen descending interneurons in locusts, collectively termed descending deviation detector neurons (DDNs), illustrate in a classical way multisensory integration at the single-cell level (Rowell, 1988; Hensler, 1992; Reichert, 1993). The neurons respond to aerodynamically relevant combinations of large-field visual input detected through the compound eyes and ocelli, such as changes in pitch or roll position of an artificial horizon, and concurrent wind stimulation of mechanosensitive hairs on the head and antennae (Figure 1.3b through 1.3d). Each DDN, therefore, detects a specific deviation from a straight flight course. The responses of the neurons to single modality stimulation (such as wind) depend strongly on simultaneous panoramic visual stimulation (Figure 1.3d), suggesting that gating mechanisms are involved in crossmodal integration (Rowell and Reichert, 1986). The combinations of sensory input in these neurons are nicely reflected by their dendritic trees, which extend to corresponding domains of ocellar neuropil, compound-eye visual neuropil, and mechanosensory neuropil in the locust midbrain (Figure 1.3a; Griss and Rowell, 1986; Rowell and

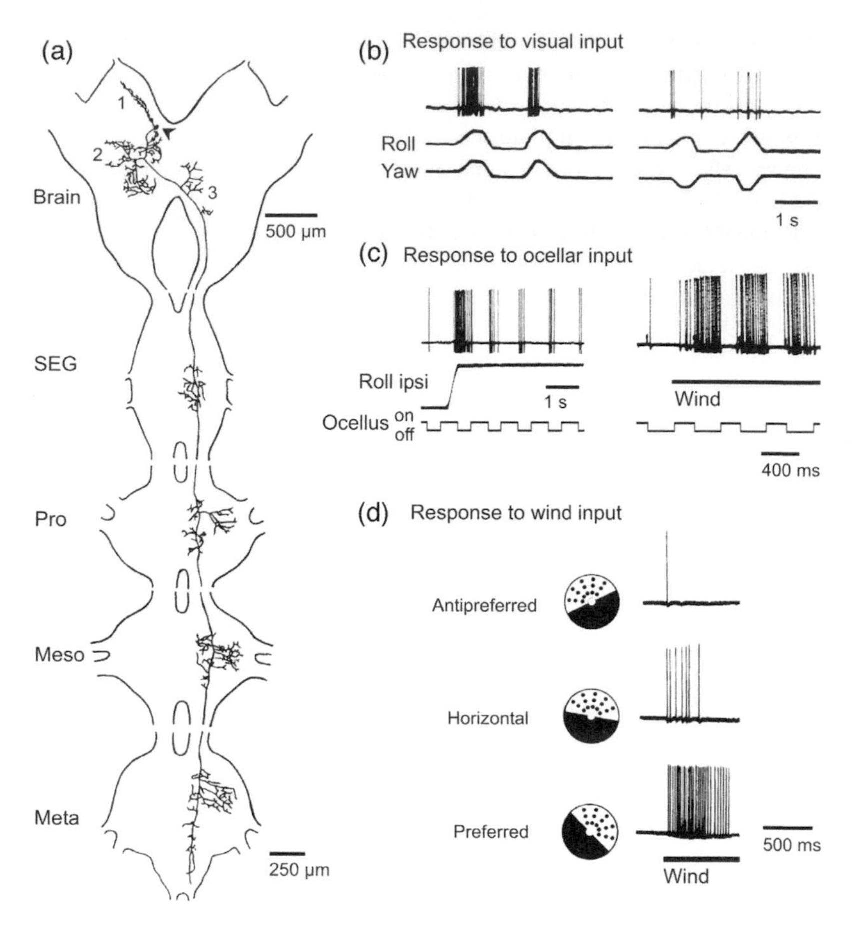

FIGURE 1.3 Morphology and physiology of deviation detector neurons (deviation detector neuron with input from the contralateral ocellus [DNCs]) in the locust *Locusta migratoria*. **a**: Morphology. Dendritic processes in the brain invade ocellar (1), compound eye visual (2), and mechanosensory (3) neuropil. Arrowhead points to cell body. The axon descends through the contralateral circumesophageal connective to the subesophageal ganglion (SEG), pro-, meso-, and metathoracic ganglia (Pro, Meso, Meta). Projections in the brain and ventral nerve cord are shown at different scales. **b–d**: Physiology. The neuron responds to a banked turn of the locust toward the ipsilateral side. **b**: Simulated roll and yaw to the ipsilateral, preferred side produce strong bursts of action potentials (left). Combination of antipreferred yaw and preferred roll strongly reduces the response (right). **c**: Response to ipsilateral ocellar stimulation is modified by rotation of the horizon (left) and by frontal wind stimulation. **d**: The response to frontal wind strongly depends on the position of the visual world. (**a**: Adapted from Griss, C. and Rowell, C.H.F., *J. Comp. Physiol. A* **158**:765–774, 1986. With permission.) (**b–d**: Adapted from Rowell, C.H.F. and Reichert, H., *J. Comp. Physiol. A* **158**:775–794, 1986. With permission.)

Reichert, 1986; Hensler, 1992). The DDNs make direct synaptic contact with premotor thoracic interneurons and with most flight motor neurons but not with interneurons of the flight oscillator. Input to flight motor neurons and premotor interneurons is gated by the thoracic flight oscillator. The net results of these output connections are changes in onset, recruitment, and most importantly, phase of motoneuron activity during the wingbeat cycle and a resultant corrective steering response through changes in wingbeat (Rowell, 1988; Reichert, 1993). Descending neurons with similar properties are present in other insects but have been studied less systematically with respect to multisensory feature extraction (e.g., Olberg, 1981; Gronenberg and Strausfeld, 1990).

1.4.3 Neurons Controlling Goal-Directed Behavior

Integration of multisensory signals plays an eminent role in directional responses of insects walking or flying toward particular targets (e.g., a food source or a mating partner), but the underlying characteristics of crossmodal processing have not yet been analyzed in detail. In several species of moth, best studied in the silk moth *Bombyx mori*, descending neurons with flip-flopping response characteristics carry right–left turning instructions underlying pheromone-mediated anemotaxis toward a female (Olberg, 1983; Olberg and Willis, 1990; Kanzaki et al., 1991, 1994). The neurons switch between low and high firing rates in response to successive pulses of pheromone odor. Some of these neurons show, in addition, directionally selective responses to wind and visual stimuli (a moving grating) in a way suggesting synergistic effects on the turning signal conveyed by the neurons (Olberg, 1983; Kanzaki et al., 1991). In some neuron types, pheromonal responses are enhanced by light (Kanzaki et al., 1994); in others, directionally selective visual responses are amplified by pheromones (Olberg and Willis, 1990). Although the precise behavioral consequences of these interactions are not fully clear in moths, a behavioral study in flies demonstrates the particular importance of vertical visual contrast for successful odor localization (Frye et al., 2003; see Chapter 4).

Processing of bimodal mechanosensory and visual input is a prominent feature of descending interneurons selective for local visual stimuli in flies (Gronenberg and Strausfeld, 1991, 1992), but this has not been found in visual target-selective descending neurons in the dragonfly (Olberg, 1986). It is possible, therefore, that in the latter species, bimodal integration occurs via parallel pathways or beyond the level of descending neurons (at the thoracic level). In female crickets, a variety of descending neurons involved in the control of phonotactive steering toward a calling male are also sensitive to bimodal mechanosensory and optomotor stimuli (Böhm and Schildberger, 1992; Staudacher and Schildberger, 1998; Staudacher, 2001). A particularly important finding was that in certain neurons, the responsiveness to a moving grating or an artificial calling song depended on the behavioral context and was only observed while the animal was walking. Responses to an aversive stimulus (ultrasound leading to negative phonotaxis), however, were not gated. Whether reafferent sensory input or central behavior-related inputs are responsible for these effects remains unclear.

1.5 CENTRIFUGAL MODULATION OF SENSORY PROCESSING

Several studies show that sensory processing in the insect brain, even at early stages of the sensory pathway, is subject to modulation via centrifugal multisensory or heterosensory input. In locusts, a pair of octopaminergic centrifugal neurons, termed

PM4, with dendritic processes in antennal mechanosensory neuropil and in the ventro-lateral protocerebrum and wide axonal projections to the medulla and lobula of the optic lobe respond to various tactile stimuli, sound, and simple light flashes (Stern et al., 1995; Stern, 1999). The neurons are involved in dishabituating the response of the DCMD neuron to moving stimuli and are generally proposed to act as "novelty detectors" leading to an overall arousal in the visual system (Bacon et al., 1995; Stern et al., 1995). In several species, the activity of ocellar interneurons is modulated, apparently at their dendritic processes in the ocellus by centrifugal neurons, which in cockroaches are sensitive to compound-eye visual stimulation, movement of antennae and wings, air puffs to cerci, and vibratory stimuli (Ohyama and Toh, 1986, 1990; Mizunami, 1994). The functional consequences of these inputs for ocellar function remain to be explored.

In the olfactory system of several insect species, a single centrifugal serotonin-immunoreactive neuron to the antennal lobe has been studied intensely. The neuron receives input primarily in the superior protocerebrum, the mushroom bodies, and lateral accessory lobe, and invades all antennal-lobe glomeruli. Judged from the effects of serotonin on antennal-lobe interneurons, activity of this neuron leads to enhanced olfactory responses of antennal-lobe interneurons (Kloppenburg and Hildebrand, 1995; Kloppenburg et al., 1999). The only two successful recordings from this neuron in the silk moth *Bombyx mori* revealed sensitivity to antennal wind stimulation but not to pheromone odors or changes in ambient light intensity (Hill et al., 2002). The role of this input in the modulation of odor representations in the antennal lobe is currently under investigation.

1.6 CONCLUSIONS AND OUTLOOK

Integration of multisensory signals has been demonstrated in numerous instances in the insect brain. Judging from the involvement of multiple brain sites and the complex properties of these responses, multimodal processing serves a variety of functions. These include identification of events or objects through matching of heterosensory inputs, detection of meaningful stimulus combinations for processes of learning and memory, adjustment of body posture and goal-directed movement by integration of self-generated and external sensory input, and adjustments in sensory attention and triggering of escape responses through convergence of various novelty or potentially threatening stimuli. Direct evidence for multisensory processing has usually been provided through intracellular recordings from identified brain interneurons. Most of these studies have demonstrated that a particular neuron responds to several sensory modalities, but how these modalities influence each other (e.g., through additive or multiplicative effects, or through a dependence on the heterosensory background) has been addressed only in very few cases, most notably in the deviation detector neurons of locusts.

In order to achieve further progress in our understanding of the mechanisms and relevance of multisensory processing, focused efforts to characterize the relevant multisensory configurations will be essential, as has been shown with locust DDNs. The unique multisensory layering within the mushroom-body calyces of hymenopterans provokes analysis of the underlying multisensory register or match of this configuration. Is this the neural basis for the unique capability of honeybees to transform visually derived navigational information (distance and direction of food source) into a mechanosensory (gravitational) framework when performing their waggle dance inside the dark hive? Particularly revealing for studies in higher-level brain areas is the work of Staudacher and Schildberger (1998), who demonstrated the dependence of multisensory integration on particular behavioral contexts. Considerable advances in recording neuronal activity during tethered flight (Gray et al., 2002) or even in freely moving insects (Mizunami et al., 1998; Okada et al., 1999) are highly promising developments to take these considerations into account. Beyond the single-cell level, multiunit recordings of Ca^{2+} signals (Chapter 13) or spiking activities (Chapter 14) might reveal insights into multisensory processing at the network level. Finally, a question that has not been addressed concerns the kinds of synaptic mechanisms underlying the different types of multisensory processing. What is the spatial relationship of heterosensory input synapses, and where and how are presynaptic mechanisms involved and configured? Structures such as the mushroom-body calyces, which provide the possibility to label selectively the different neuronal elements for ultrastructural analysis (e.g., Yasuyama et al., 2002), may be particularly suited to this approach.

ACKNOWLEDGMENTS

I wish to thank Monika Stengl for insightful thoughts and comments during the preparation of this manuscript. Research from my laboratory reported in this chapter has been supported by DFG grants HO 950/13 and HO 950/14.

REFERENCES

Abel R, Rybak J, Menzel R (2001) Structure and response patterns of olfactory interneurons in the honeybee, *Apis mellifera*. *J. Comp. Neurol.* **437**:363–383.

Anton S, Hansson BS (1994) Central processing of sex pheromone, host odour, and oviposition deterrent information by interneurons in the antennal lobe of female *Spodoptera littoralis* (Lepidoptera: Noctuidae). *J. Comp. Neurol.* **350**:199–214.

Anton S, Homberg U (1999) Antennal lobe structure. In: *Insect Olfaction*, Hansson BS, Ed., Springer, Berlin, pp. 97–124.

Bacon JP, Strausfeld NJ (1986) The dipteran “giant fibre” pathway: neurons and signals. *J. Comp. Physiol. A* **158**:529–548.

Bacon JP, Thompson KS, Stern M (1995) Identified octopaminergic neurons provide an arousal mechanism in the locust brain. *J. Neurophysiol.* **74**:2739–2743.

Böhm H, Schildberger K (1992) Brain neurones involved in the control of walking in the cricket *Gryllus bimaculatus*. *J. Exp. Biol.* **166**:113–130.

Böhm H, Schildberger K, Huber F (1991) Visual and acoustic course control in the cricket *Gryllus bimaculatus*. *J. Exp. Biol.* **159**:235–248.

Boyan GS (1983) Postembryonic development in the auditory system of the locust. *J. Comp. Physiol.* **151**:499–513.

Boyan GS, Ball EE (1990) Neuronal organization and information processing in the wind-sensitive cercal receptor/giant interneurone system of the locust and other orthopteroid insects. *Prog. Neurobiol.* **35**:217–243.

Boyan GS, Williams JLD (1982) Auditory neurones in the brain of the cricket *Gryllus bimaculatus* (De Geer): ascending interneurones. *J. Insect Physiol.* **28**:493–501.

Burdohan JA, Comer CM (1996) Cellular organization of an antennal mechanosensory pathway in the cockroach, *Periplaneta americana. J. Neurosci.* **16**:5830–5843.

Burrows M (1996) *The Neurobiology of an Insect Brain.* Oxford University Press, Oxford.

Burrows M, Rowell CHF (1973) Connections between descending visual interneurons and metathoracic motoneurons in the locust. *J. Comp. Physiol.* **85**:221–234.

Calvert GA (2001) Crossmodal processing in the human brain: insights from functional neuroimaging studies. *Cereb. Cortex* **11**:1110–1123.

Calvert GA, Spence C, Stein BE (2004) *The Handbook of Multisensory Processes.* MIT Press, Cambridge, MA.

De Gelder B, Bertelson P (2003) Multisensory integration, perception and ecological validity. *Trends Cogn. Sci.* **7**:460–467.

Douglass JK, Strausfeld NJ (2003) Anatomical organization of retinotopic motion-sensitive pathways in the optic lobes of flies. *Microsc. Res. Tech.* **62**:132–150.

Driver J, Spence C (2000) Multisensory perception: beyond modularity and convergence. *Current Biol.* **10**:R731–R735.

Eichendorf A, Kalmring K (1980) Projections of auditory ventral-cord neurons in the supraesophageal ganglion of *Locusta migratoria. Zoomorphology* **94**:133–149.

Ehmer B, Gronenberg W (2002) Segregation of visual input to the mushroom bodies in the honeybee (*Apis mellifera*). *J. Comp. Neurol.* **451**:362–373.

Ernst K-D, Boeckh J (1983) A neuroanatomical study on the organization of the central antennal pathways in insects. III. Neuroanatomical characterization of physiologically defined types of deutocerebral neurons in *Periplaneta americana. Cell Tissue Res.* **229**:1–22.

Ernst K-D, Boeckh J, Boeckh V (1977) A neuroanatomical study on the organization of the central antennal pathways in insects. II. Deutocerebral connections in *Locusta migratoria* and *Periplaneta americana. Cell Tissue Res.* **176**:285–308.

Flanagan D, Mercer A (1989) Morphology and response characteristics of neurons in the deutocerebrum of the brain of the honeybee *Apis mellifera. J. Comp. Physiol. A* **164**:483–494.

Fleissner G, Fleissner G (2003) Nonvisual photoreceptors in arthropods with emphasis on their putative role as receptors of natural zeitgeber stimuli. *Chronobiol. Int.* **20**:593–616.

Fonta C, Sun XJ, Masson C (1991) Cellular analysis of odour integration in the honeybee antennal lobe. In: *Behavior and Physiology of Bees.* Goodman LJ, Fischer RC, Eds., CAB International, Wallingford, pp. 227–241.

Frambach I, Schürmann F-W (2004) Separate distribution of deutocerebral projection neurones in the mushroom bodies of the cricket brain. *Acta Biol. Hung.* **55**:21–29.

Frye MA, Tarsitano M, Dickinson MH (2003) Odor localization requires visual feedback during free flight in *Drosophila melanogaster. J. Exp. Biol.* **206**:843–855.

Gerber B, Smith BH (1998) Visual modulation of olfactory learning in honeybees. *J. Exp. Biol.* **201**:2213–2217.

Gewecke M (1979) Central projections of antennal afferents for the flight motor in *Locusta migratoria* (Orthoptera: Acrididae). *Entomol. Gen.* **5**:317–320.

Gray JR, Pawlowski V, Willis MA (2002) A method for recording behavior and multineuronal CNS activity from tethered insects flying in virtual space. *J. Neurosci. Meth.* **120**:211–223.

Griss C, Rowell CHF (1986) Three descending interneurons reporting deviation from course in the locust. I. Anatomy. *J. Comp. Physiol. A* **158**:765–774.

Gronenberg W (1987) Anatomical and physiological properties of feedback neurons of the mushroom bodies in the bee brain. *Exp. Biol.* **46**:115–125.

Gronenberg W (2001) Subdivisions of hymenopteran mushroom body calyces by their afferent supply. *J. Comp. Neurol.* **436**:474–489.

Gronenberg W, Hölldobler B (1999) Morphologic representation of visual and antennal information in the ant brain. *J. Comp. Neurol.* **412**:229–240.

Gronenberg W, Strausfeld NJ (1990) Descending neurons supplying the neck and flight motor of Diptera: physiological and anatomical characteristics. *J. Comp. Neurol.* **302**:973–991.

Gronenberg W, Strausfeld NJ (1991) Descending pathways connecting the male-specific visual system of flies to the neck and flight motor. *J. Comp. Physiol. A* **169**:413–426.

Gronenberg W, Strausfeld NJ (1992) Premotor descending neurons responding selectively to local visual stimuli in flies. *J. Comp. Neurol.* **316**:87–103.

Grünewald B (1999) Physiological properties and response modulations of mushroom body feedback neurons during olfactory learning in the honeybee, *Apis mellifera. J. Comp. Physiol. A* **185**:565–576.

Guerenstein PG, Christensen TA, Hildebrand JG (in press) Sensory processing of ambient-CO_2 information in the brain of the moth *Manduca sexta. J. Comp. Physiol. A*.

Hammer M (1993) An identified neuron mediates the unconditioned stimulus in associative olfactory learning in honeybees. *Nature* **366**:59–63.

Hammer M (1997) The neural basis of associative reward learning in honeybees. *Trends Neurosci.* **20**:245–252.

Hammer M, Menzel R (1998) Multiple sites of associative odor learning as revealed by local brain microinjection of octopamine in honeybees. *Learn. Mem.* **5**:146–156.

Hansson BS, Anton S, Christensen TA (1994) Structure and function of antennal lobe neurons in the male turnip moth, *Agrotis segetum* (Lepidoptera: Noctuidae). *J. Comp. Physiol. A* **175**:547–562.

Hartlieb E, Anderson P, Hansson BS (1999) Appetitive learning of odours with different behavioural meaning in moths. *Physiol. Behav.* **67**:671–677.

Helfrich-Förster C (2003) The neuroarchitecture of the circadian clock in the brain of *Drosophila melanogaster. Microsc. Res. Tech.* **62**:94–102.

Helfrich-Förster C, Stengl M, Homberg U (1998) Organization of the circadian system in insects. *Chronobiol. Int.* **15**:567–594.

Hensler K (1992) Neuronal co-processing of course deviation and head movements in locusts. I. Descending deviation detectors. *J. Comp. Physiol. A* **171**:257–271.

Hill ES, Iwano M, Gatellier L, Kanzaki R (2002) Morphology and physiology of the serotonin-immunoreactive putative antennal lobe feedback neuron in the male silkmoth *Bombyx mori. Chem. Senses* **27**:475–483.

Hölldobler B (1999) Multimodal signals in ant communication. *J. Comp. Physiol. A* **184**:129–141.

Homberg U (1984) Processing of antennal information in extrinsic mushroom body neurons of the bee brain. *J. Comp. Physiol. A* **154**:825–836.

Homberg U (1985) Interneurones of the central complex in the bee brain (*Apis mellifera, L.*). *J. Insect Physiol.* **31**:251–264.

Homberg U (1994) Flight-correlated activity changes in neurons of the lateral accessory lobes in the brain of the locust *Schistocerca gregaria. J. Comp. Physiol. A* **175**:597–610.

Homberg U (in press) In search of the sky compass in the insect brain. *Naturwissenschaften*.

Homberg U, Würden S (1997) Movement-sensitive, polarization-sensitive, and light-sensitive neurons of the medulla and accessory medulla of the locust, *Schistocerca gregaria. J. Comp. Neurol.* **386**:329–346.

Homberg U, Montague RA, Hildebrand JG (1988) Anatomy of antenno-cerebral pathways in the brain of the sphinx moth *Manduca sexta*. *Cell Tissue Res.* **254**:255–281.

Homberg U, Christensen TA, Hildebrand JG (1989) Structure and function of the deutocerebrum in insects. *Annu. Rev. Entomol.* **34**:477–501.

Homberg U, Hofer S, Pfeiffer K, Gebhardt S (2003a) Organization and neural connections of the anterior optic tubercle in the brain of the locust, *Schistocerca gregaria*. *J. Comp. Neurol.* **462**:415–430.

Homberg U, Reischig T, Stengl M (2003b) Neural organization of the circadian system of the cockroach *Leucophaea maderae*. *Chronobiol. Int.* **20**:577–591.

Homberg U, Brandl C, Clynen E, Schoofs L, Veenstra JA (2004) Mas-allatotropin/Lom-AG myotropin I immunostaining in the brain of the locust, *Schistocerca gregaria*. *Cell Tissue Res.* (submitted).

Ignell R, Anton S, Hansson BS (2000) The maxillary palp sensory pathway of Orthoptera. *Arthropod Struct. Devel.* **29**:295–305.

Iwama A, Sugihara D, Shibuya T (1995) Morphology and physiology of neurons responding to the Nasanov pheromone in the antennal lobe of the honeybee, *Apis mellifera*. *Zool. Sci.* **12**:207–218.

Kanzaki R, Shibuya T (1986) Identification of the deutocerebral neurons responding to the sexual pheromone in the male silkworm moth brain. *Zool. Sci.* **3**:409–418.

Kanzaki R, Arbas EA, Strausfeld NJ, Hildebrand JG (1989) Physiology and morphology of projection neurons in the antennal lobe of the male moth *Manduca sexta*. *J. Comp. Physiol. A* **165**:427–453.

Kanzaki R, Arbas EA, Hildebrand JG (1991) Physiology and morphology of descending neurons in pheromone-processing olfactory pathways in the male moth *Manduca sexta*. *J. Comp. Physiol. A* **169**:1–14.

Kanzaki R, Ikeda A, Shibuya T (1994) Morphological and physiological properties of pheromone-triggered flipflopping descending interneurons of the male silkworm moth, *Bombyx mori*. *J. Comp. Physiol. A* **175**:1–14.

Kloppenburg P, Hildebrand JG (1995) Neuromodulation by 5-hydroxytryptamine in the antennal lobe of the sphinx moth *Manduca sexta*. *J. Exp. Biol.* **198**:603–611.

Kloppenburg P, Ferns D, Mercer AR (1999) Serotonin enhances central olfactory neuron responses to sex pheromone in the male sphinx moth *Manduca sexta*. *J. Neurosci.* **19**:8172–8181.

Labhart T, Meyer EP (1999) Detectors for polarized skylight in insects: a survey of ommatidial specializations in the dorsal rim area of the compound eye. *Microsc. Res. Tech.* **47**:368–379.

Li Y, Strausfeld NJ (1997) Morphology and sensory modality of mushroom body extrinsic neurons in the brain of the cockroach *Periplaneta americana*. *J. Comp. Neurol.* **387**:631–650.

Li Y, Strausfeld NJ (1999) Multimodal efferent and recurrent neurons in the medial lobes of cockroach mushroom bodies. *J. Comp. Neurol.* **409**:647–663.

Malun D, Waldow U, Krauss D, Boeckh J (1993) Connection between the deutocerebrum and the protocerebrum, and neuroanatomy of several classes of deutocerebral projection neurons in the brain of male *Periplaneta americana*. *J. Comp. Neurol.* **329**:143–162.

Maravita A, Spence C, Driver J (2003) Multisensory integration and the body scheme: close to hand and within reach. *Current Biol.* **13**:R531–R539.

Marin EC, Jeffris, GS, Komiyama, T, Zhu H, Luo L (2002) Representation of the glomerular olfactory map in the *Drosophila* brain. *Cell* **109**: 243–255.

Maronde U (1991) Common projection areas of antennal and visual pathways in the honeybee brain, *Apis mellifera. J. Comp. Neurol.* **309**:328–340.

Matsumoto Y, Mizunami M (2002) Lifetime olfactory memory in the cricket *Gryllus bimaculatus. J. Comp. Physiol. A* **188**:295–299.

Mauelshagen J (1993) Neural correlates of olfactory learning paradigms in an identified neuron in the honeybee brain. *J. Neurophysiol.* **69**:609–625.

Menzel R (2001) Searching for the memory trace in a mini-brain, the honeybee. *Learn. Mem.* **8**:53–62.

Meredith MA (2002) On the neuronal basis for multisensory convergence: a brief overview, *Cogn. Brain Res.* **14**:31–40.

Milde JJ (1988) Visual responses of interneurones in the posterior median protocerebrum and the central complex of the honeybee *Apis mellifera. J. Insect Physiol.* **34**:427–436.

Milde JJ, Strausfeld NJ (1990) Cluster organization and response characteristics of the giant fiber pathway of the blowfly *Calliphora erythrocephala. J. Comp. Neurol.* **294**:59–75.

Mizunami M (1994) Information processing in the insect ocellar system: comparative approaches to the evolution of visual processing and neural circuits. *Adv. Insect Physiol.* **25**:151–265.

Mizunami M, Okada R, Li Y, Strausfeld NJ (1998) Mushroom bodies of the cockroach: activity and identities of neurons recorded in freely moving animals. *J. Comp. Neurol.* **402**:501–519.

Mizunami M, Weibrecht JM, Strausfeld NJ (1998) Mushroom bodies of the cockroach: their participation in place memory. *J. Comp. Neurol.* **402**:520–537.

Müller D, Abel R, Brandt R, Zöckler M, Menzel R (2002) Differential parallel processing of olfactory information in the honeybee, *Apis mellifera* L. *J. Comp. Physiol. A* **188**:359–370.

Nebeling B (2000) Morphology and physiology of auditory and vibratory ascending interneurones in bushcrickets. *J. Exp. Zool.* **286**:219–230.

Nishikawa M, Yokohari F, Ishibashi T (1995) Central projections of the antennal cold receptor neurons and hygroreceptor neurons of the cockroach *Periplaneta americana. J. Comp. Neurol.* **361**:165–178.

Nishino H, Mizunami M (1998) Giant input neurons of the mushroom body: intracellular recording and staining in the cockroach. *Neurosci. Lett.* **246**:57–60.

Nishino H, Yamashita S, Yamazaki Y, Nishikawa M, Yokohari F, Mizunami M (2003) Projection neurons originating from thermo- and hygrosensory glomeruli in the antennal lobe of the cockroach. *J. Comp. Neurol.* **455**:40–55.

Ohyama T, Toh Y (1986) Multimodality of ocellar interneurones of the American cockroach. *J. Exp. Biol.* **125**:405–409.

Ohyama T, Toh Y (1990) Morphological and physiological characterization of small multimodal ocellar interneurons in the American cockroach. *J. Comp. Neurol.* **301**:501–510.

Okada R, Ikeda J, Mizunami M (1999) Sensory responses and movement-related activities in extrinsic neurons of the cockroach mushroom body. *J. Comp. Physiol. A* **185**:115–129.

Olberg RM (1981) Parallel encoding of detection of wind, head, abdomen, and visual pattern movement by single interneurons in the dragonfly. *J. Comp. Physiol.* **142**:27–41.

Olberg RM (1983) Pheromone-triggering flip-flopping interneurons in the ventral nerve cord of the silkworm moth, *Bombyx mori. J. Comp. Physiol.* **152**:297–307.

Olberg RM (1986) Identified target-selective visual interneurons descending from the dragonfly brain. *J. Comp. Physiol. A* **159**:827–840.

Olberg RM, Willis MA (1990) Pheromone-mediated optomotor response in male gypsy moths, *Lymantria dispar* L.: directionally selective visual interneurons in the ventral nerve cord. *J. Comp. Physiol. A* **167**:707–714.

O'Shea M (1975) Two sites of axonal spike initiation in a bimodal interneuron. *Brain Res.* **96**:93–98.

Raguso RA, Willis MA (2002) Synergy between visual and olfactory cues in nectar feeding by naïve hawkmoths, *Manduca sexta. Anim. Behav.* **64**:685–695.

Reichert H (1993) Sensory input and flight orientation in locusts. *Comp. Biochem. Physiol.* **104A**: 647–657.

Rind FC (2002) Motion detectors in the locust visual system: from biology to robot sensors. *Microsc. Res. Tech.* **56**:256–269.

Rind FC, Simmons PJ (1999) Seeing what is coming: building collision-sensitive neurones. *Trends Neurosci.* **22**:215–220.

Rowell CHF (1971) The orthopteran descending movement detector (DMD) neurones: a characterization and review. *Z. Vgl. Physiol.* **73**:167–194.

Rowell CHF (1988) Mechanisms of flight steering in locusts. *Experientia* **44**:389–395.

Rowell CHF, Reichert H (1986) Three descending interneurons reporting deviation from course in the locust. II: Physiology. *J. Comp. Physiol. A* **158**:775–794.

Rybak J, Menzel R (1998) Integrative properties of the Pe1 neuron, a unique mushroom body output neuron. *Learn. Mem.* **3**:133–143.

Sakura M, Mizunami M (2001) Olfactory learning and memory in the cockroach *Periplaneta americana. Zool. Sci.* **18**:21–28.

Schildberger K (1981) Some physiological features of mushroom-body linked fibers in the house cricket brain. *Naturwissenschaften* **68**:623–624.

Schildberger K (1984) Multimodal interneurons in the cricket brain: properties of identified extrinsic mushroom body cells. *J. Comp. Physiol. A* **154**:71–79.

Schröter U, Menzel R (2003) A new ascending sensory tract to the calyces of the honeybee mushroom body, the subesophageal-calycal tract. *J. Comp. Neurol.* **465**:168–178.

Shen JX (1983) The cercus-to-giant interneurone system in the bushcricket *Tettigonia cantans*: morphology and response to low-frequency sound. *J. Comp. Physiol. A* **151**:449–459.

Simmons PJ (1981) Ocellar excitation of the DCMD: an identified locust interneurone. *J. Exp. Biol.* **91**:355–359.

Small DM (2004) Crossmodal integration: insights from the chemical senses. *Trends Neurosci.* **27**:120–123.

Srinivasan MV, Zhang SW, Zhu H (1998) Honeybees link sites to smells. *Nature* **396**:637–638.

Staudacher EM (2001) Sensory responses of descending brain neurons in the walking cricket, *Gryllus bimaculatus. J. Comp. Physiol. A* **187**:1–17.

Staudacher E, Schildberger K (1998) Gating of sensory responses of descending brain neurones during walking in crickets. *J. Exp. Biol.* **201**:559–572.

Stein BE (1998) Neural mechanisms for synthesizing sensory information and producing adaptive behaviors. *Exp. Brain Res.* **123**:124–135.

Stein BE, Meredith MA (1993) *Merging of the Senses.* MIT Press, Cambridge, MA.

Stern M (1999) Octopamine in the locust brain: cellular distribution and functional significance in an arousal mechanism. *Microsc. Res. Tech.* **45**:135–141.

Stern M, Thompson KSJ, Zhou P, Watson DG, Midgley JM, Gewecke M, Bacon JP (1995) Octopaminergic neurons in the locust brain: morphological, biochemical and electrophysiological characterization of potential modulators of the visual system. *J. Comp. Physiol. A* **177**:611–625.

Stocker RF, Lienhard MC, Borst A, Fischbach K-F (1990) Neuronal architecture of the antennal lobe in *Drosophila melanogaster. Cell Tissue Res.* **262**:9–34.

Strausfeld NJ (1976) *Atlas of an Insect Brain*. Springer, Heidelberg.

Strausfeld NJ (1989) Beneath the compound eye: neuroanatomical analysis and physiological correlates in the study of insect vision. In: *Facets of Vision*, Stavenga DG, Hardie RC, Eds., Springer, New York, pp. 317–359.

Strausfeld NJ (2000) Insect brain. In: *Brain Evolution and Cognition*, Roth G, Wullimann MF, Eds., Wiley, New York, and Spektrum, Heidelberg, pp. 368–400.

Strausfeld NJ, Bacon JP (1983) Multimodal convergence in the central nervous system of dipterous insects. In: *Fortschritte der Zoologie, Vol 28*, Horn E, Ed., Fischer, Stuttgart, pp. 47–76.

Strausfeld NJ, Lee J-K (1991) Neuronal basis for parallel processing in the brain. *Visual Neurosci.* **7**:13–33.

Strausfeld NJ, Li Y (1999) Organization of olfactory and multimodal afferent neurons supplying the calyx and pedunculus of the cockroach mushroom bodies. *J. Comp. Neurol.* **409**:603–625.

Strauss R (2002) The central complex and the genetic dissection of locomotor behaviour. *Curr. Opin. Neurobiol.* **12**:633–638.

Vitzthum H, Müller M, Homberg U (2002) Neurons of the central complex of the locust *Schistocerca gregaria* are sensitive to polarized light. *J. Neurosci.* **22**:1114–1125.

Waddell S, Quinn WG (2001) What can we teach *Drosophila*? What can they teach us? *Trends Genet.* **17**:719–726.

Waldow U (1975) Multimodale Neurone im Deutocerebrum von *Periplaneta americana*. *J. Comp. Physiol.* **101**:329–341.

Weiss MJ (1981) Structural patterns in the corpora pedunculata of Orthoptera: a reduced silver analysis. *J. Comp. Neurol.* **203**:515–553.

Wong AM, Wang JW, Axel R (2002) Spatial representation of the glomerular map in the *Drosophila* protocerebrum. *Cell* **109**:229–241.

Yasuyama K, Meinertzhagen IA, Schürmann F-W (2002) Synaptic organization of the mushroom body calyx in *Drosophila melanogaster*. *J. Comp. Neurol.* **445**:211–226.

Ye S, Comer CM (1996) Correspondence of escape-turning behavior with activity of descending mechanosensory interneurons in the cockroach, *Periplaneta americana*. *J. Neurosci.* **16**:5844–5853.

Ye S, Leung V, Khan A, Baba Y, Comer CM (2003) The antennal system and cockroach evasive behavior. I. Roles for visual and mechanosensory cues in the response. *J. Comp. Physiol. A* **189**:89–96.

Zeiner R, Tichy H (1998) Combined effects of olfactory and mechanical inputs in antennal lobe neurons of the cockroach. *J. Comp. Physiol. A* **182**:467–473.

Zeiner R, Tichy H (2000) Integration of temperature and olfactory information in cockroach antennal lobe glomeruli. *J. Comp. Physiol. A* **186**:717–727.

2 Methods in Insect Sensory Ecology

Mattias C. Larsson and Glenn P. Svensson

CONTENTS

0-8493-2024-0/05/$0.00+$1.50

2.1 INTRODUCTION

Most people enjoying warm summer nights outdoors will also marvel at the formidable efficiency with which blood-sucking insects such as mosquitoes localize their hosts. This and other impressive insect performances are products of an adaptive use of specific sensory cues inducing a limited set of stereotyped behavioral responses. Due to their great diversity, insects as a group provide a wide range of models for virtually any aspect of sensory interactions. The specialized nature of most insects also makes them a challenge for scientists, because their usefulness as behavioral subjects depends on whether they are provided with proper key stimuli in a relevant context. Some investigators may have noticed a discrepancy between insects, as the supposedly most successful group of organisms on Earth, and their often pathetic performances when expected to yield quantifiable data under controlled conditions. In order to get the most out of their model organisms, scientists aiming to understand the relationship between neural processes and behavior should know as much as possible about the relationship between the model and the natural environment in which it evolved.

Sensory ecology is the study of how organisms use sensory information in their interactions with the environment, including all its biotic and abiotic components. As in most scientific disciplines, this includes a descriptive aspect, aiming to classify different types of sensory interactions and to elucidate the mechanisms by which they occur. In this way, sensory ecology embodies the physics, chemistry, and physiology behind the generation and perception of sensory stimuli. The main focus of sensory ecology, however, is to learn the adaptive uses of sensory stimuli (why organisms use them the way they do) with the understanding that adaptations represent imperfect compromises between conflicting selective pressures working within the physical constraints and evolutionary history of an organism. Here, we will limit our discussion primarily to how insects use sensory information to discriminate among stimulus emitters and to locate and exploit important resources. Further reading on general sensory ecology can be found in Dusenbery (1992) and Barth and Schmidt (2001). Bradbury and Vehrencamp (1998) discuss animal communication in depth, but with a lamentable scarcity of insect models; these, on the other hand, are well covered by Greenfield (2002).

2.2 SENSORY SIGNALS AND ECOLOGY

Sensory information naturally constitutes the basis of all interactions between an organism and its environment. To a great extent, these interactions shape and are shaped by evolutionary events. It has been suggested that the Cambrian explosion, the greatest radiation of animal forms in evolutionary history, is the result of adaptations and counteradaptations to the first appearance of eyes that could form spatial images (Land and Nilsson, 2002). Speciation may thus be driven by increased sensory sophistication, which leads to more diversified niches. Organisms must also specialize in order to compete in a diversified world, which to a large extent may be linked to the need for efficient use of sensory information. Host shifts among insects are also intimately associated with their sensory ecology. These events may occur when alternate hosts resemble the original host in critical ways, causing "mistakes" that eventually lead to the formation of new host races and species through genetic drift or divergent selection (Schoonhoven et al., 1998; Linn et al., 2003).

2.2.1 Detection and Processing of Sensory Signals

A common denominator of sensory stimuli is the transfer of information through interactions between chemical or physical signals and specialized receptors that transduce the information into nervous impulses. Sensory receptors often operate close to their maximum theoretical sensitivity, registering the presence of single photons or odor molecules (Hardie, 2001; Minor and Kaissling, 2003). If receptor sensitivity were the only criterion for the performance of sensory systems, there would be little room for specific adaptations — and thus for sensory ecology. The ability of an organism to detect and extract information from an environmental signal depends mainly on factors other than the absolute sensitivity of sensory receptors, however, and most sensory adaptations reflect solutions to the specific needs of the animal. Some adaptations, such as large ears, olfactory antennae, or pupils, increase sensitivity by increasing the total signal capture. Other adaptations, such as large eyes with many receptors, increase the resolution of the sensory system and lead to a more fine-grained representation of the external world. Another important class of adaptations increases the ability of the nervous system to distinguish sensory signals from various types of internal or external noise (e.g., spontaneous activity of receptors or neurons and signals from other sources interacting with the same receptor) that tend to mask its presence. Limits to detection are ultimately set by the ratio between signal and noise that the sensory system can obtain (Dusenbery, 1992).

Organisms employ a multitude of strategies to improve signal-to-noise ratios in their sensory systems. A common strategy to overcome noise at low signal intensity is *summation*: averaging over many neurons or over extended time periods to give more reliable estimates. Extreme examples of summation over many receptors can be exemplified by the enlarged antennae of some male moths, with tens of thousands of receptors detecting single components of the female-released pheromone blend (Todd and Baker, 1999). In this case, extensive temporal summation of signals is not an option because male moths need to track very fast changes in odor concentration in order to follow the shifting odor plume in the air. Other strategies to improve signal-to-noise ratio may be collectively termed *matched filters* or *feature detectors*, whose role is to detect and amplify primarily those features of the signals that are of particular interest. Such filters include receptors tuned to match specific stimuli, as well as neural networks designed to extract specific information from the signal, such as movement or shape detectors in the visual system (Dusenbery, 1992; Warrant, 2001; see also Chapter 10).

2.2.2 Sensory Systems and the Environment

Sensory systems are adapted to extract meaningful information in specific contexts, some important elements of which are summarized in Figure 2.1. The most important context is perhaps the relation between the source/emitter and the receiver. Emitters

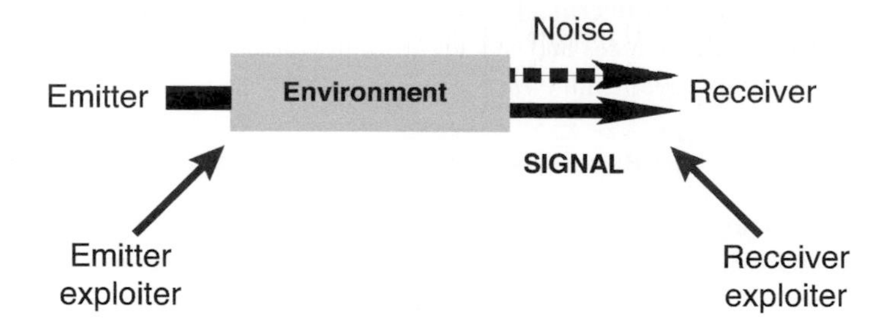

FIGURE 2.1 Basic components of the emitter–receiver system (after Endler, J.A., *Phil. Trans. R. Soc. Lond. B* **340**:215–225, 1993; Bradbury, J.W. and Vehrencamp, S.L., *Principles of Animal Communication*. Sinauer Associates, Sunderland, MS, 1998.) An object, the *emitter*, produces a signal that is transmitted through a medium (air, water, or substrate) and detected by an individual, the *receiver*. The signal can be affected by various biotic and abiotic factors during its transfer through the environment. Two forms of exploitation of the communication channel can take place. First, the emitter can be exploited by enemies that are able to detect and locate the source of the signal (e.g., tachinid fly parasitoids homing in on the male song of their cricket host; Robert et al. 1992). Second, receiver exploiters can either lure the receiver by using signals similar to the "true" one from the emitter (e.g., deceptive *Ophrys* orchids attracting male bees for pollination by releasing a chemical mimic of the female bee sex pheromone; Schiestl et al., 1999), or they can use the signal to intercept receivers on their way to the source.

may be animate or inanimate objects, but here we usually assume the former. Chemical signals may be unique in having specific nomenclature defined by relations between senders and receivers (see Section 2.3.3 of this chapter), but the same type of interactions apply to all signals. If the interaction is beneficial to both participants, true communication often evolves, where the sender and receiver can agree upon a signal and its meaning and strive to optimize signal transmission while excluding unwanted participants (Wyatt, 2003). If the interaction is detrimental, the party should try to opt out; the emitter by reducing or confounding the signal — an adaptation known as *crypsis* — and the receiver by increasing signal processing capacity to better distinguish unfavorable signals from adaptive ones.

The choice of signals that are utilized in any specific context depends on many factors, such as the type of information transmitted, the surrounding medium, timing, and the range over which the signal is active (Table 2.1). In communication, costs in terms of energy expenditure and predation risk are also very important factors. For example, sexual selection has probably shaped the emitter–receiver relationship of mating signals in crickets and moths. Male crickets use auditory signals that are costly in terms of both energy expenditure and predation risk; female moths use "inexpensive" chemical signals that are hard to detect. In either case, the pressure is mostly on the males, independent of the mode of signal transmission (Phelan, 1992; Zuk and Kolluru, 1998). Even if some signals are generally inexpensive to produce, they can be made costly in order to signal mate quality, by being inherently dependent on, for example, dietary resources in limited supply, such as color pigments (Knüttel and Fiedler, 2001) or chemicals used as pheromone precursors (Dussourd et al., 1991).

TABLE 2.1
Some Characteristics of Different Sensory Modes of Communication Covered in This Chapter: Type of Signal

	Acoustic	Olfactory	Visual
Requirements of Medium	**Air or Water**	**Current Flow**	**Ambient Light**
Range	Large	Large	Medium
Speed	High	Low	High
Attenuation	Fast	Slow	Fast
Ability to circumvent obstacles	Good	Good	Poor
Localizability	Medium	Variable	Good
Directional control of emission	Good	Poor	Good
Complexity	High	Low	High
Energetic cost	High	Low	Low
Exploitation risk	Medium	Low	High

Note: Depending on the circumstances, specific signals or combinations of signals may be advantageous to an emitter or a receiver. Characteristics of visual signals presented here are based on reflected light and do not always apply when an emitter (such as a firefly) is able to produce its own light.

Source: After Alcock, J., *Animal Behavior: An Evolutionary Perspective*. 3rd edition. Sinauer Associates, Sunderland, MS, 1984; Endler, J.A., *Phil. Trans. R. Soc. Lond. B* **340**:215–225, 1993; Bradbury, J.W. and Vehrencamp, S.L., *Principles of Animal Communication*. Sinauer Associates, Sunderland, MS, 1998.

It is important to realize that a good match between a trait of a sensory system and some feature of the environment does not necessarily result from a specific adaptation to this feature. Although such correlations are the common currency in sensory ecology, in the absence of further evidence speculations about the adaptive value of a trait are nothing more than just speculations. Chittka and Briscoe (2001) point to the need for greater stringency in analysis and suggest additional criteria, such as phylogenetic comparisons and direct tests of the fitness value of a trait, before drawing any conclusions about its adaptive significance.

2.2.3 Costs and Constraints in Sensory Systems

The benefit of having an effective sensory apparatus is always balanced by the cost of carrying it. For example, there may be a trade-off between higher sensitivity and resolution obtained with a bigger compound eye and relatively higher energy demands associated with such an eye. The cost–benefit balance is probably more pronounced in small animals, such as insects, with relatively heavy and bulky sensory organs in relation to the total mass and size of the body. Laughlin (2001) estimated the metabolic costs associated with the support and electrical coding of the compound eye of the blowfly and showed that these account for 10% of total energy expenditure when at rest and 3% when in flight, significantly contributing to the total metabolic rate.

Quantitative estimates of costs associated with sensory systems are rare, but studies of loss or degeneration of sensory structures under conditions of relaxed selection pressure suggest that there may be considerable costs associated with the maintenance of a fine-tuned sensory system. Moths with ultrasonic hearing that are spatially or temporally isolated from insectivorous bats can serve as models to test hypotheses regarding such costs. Investigations of sympatric species of diurnal and nocturnal moths (Fullard et al., 1997), moth populations that are spatially isolated from bats (Fullard, 1994), and winter moths that experience reduced predation risk from bats (Surlykke and Treat, 1995) have shown evidence for both persistence and reduction of ultrasonic hearing ranging from normal hearing to complete deafness. Rydell et al. (1997) demonstrated sexual dimorphism in ultrasonic hearing ability of geometrid winter moths, showing that these differences reflect sex-specific adaptations rather than general degeneration of the sensory system under relaxed selection pressure (Figure 2.2). Females of these species, which are flightless with reduced wings and thus cryptic to echolocating bats, had strongly reduced ears and were virtually deaf. In contrast, males, which are airborne and therefore potentially

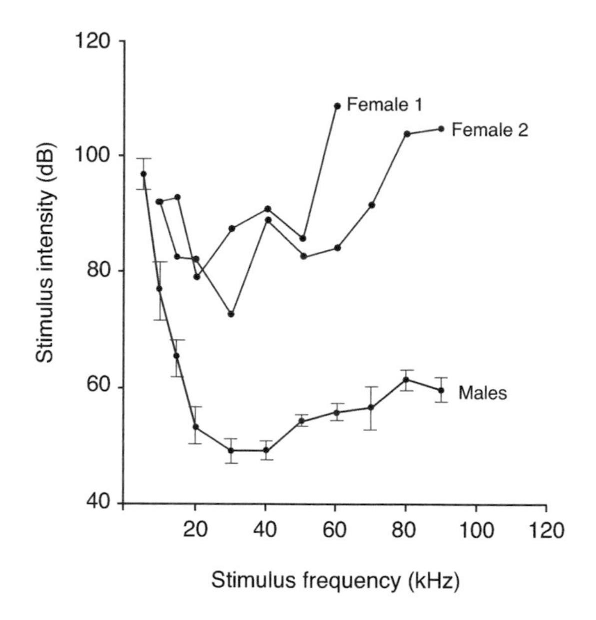

FIGURE 2.2 Audiograms of the geometrid winter moth *Agriopis marginaria*, showing the hearing thresholds for eight males (mean ±SE) and two females at different frequencies of the sound stimulus. Males have good ultrasonic hearing with ears tuned to the frequencies used by echolocating bats. In contrast, the wingless females, which are cryptic to hunting bats, show reduced hearing capacity, presumably due to an absence of the selection pressure maintaining this trait in males. (Modified from Rydell, J. et al., *Proc. R. Soc. Lond. B* **264**:83–88, 1997.)

exposed to bats, had broadly tuned ultrasonic hearing and showed strong evasive reactions when stimulated with ultrasound.

2.2.4 Insects as Signal Specialists

Most insects exhibit a very high degree of niche specialization, exemplified by the vast majority of insect herbivores or parasitoids that feed on single or a few related host species, which correlates with high species diversity among insects (Schoonhoven et al., 1998). Even so-called generalist species (including many hunting predators) usually feed on a limited spectrum of foods or on restricted parts of their hosts. There are many possible explanations for this phenomenon, including physiological adaptations to overcome host defenses, but sensory limitations are likely to play a very important role. Optimizing sensory systems for high efficiency of resource utilization is probably facilitated by selecting a combination of a few predictable stimuli that indicate the essential resource with high certainty (Bernays, 2001). Clearly, insects can modify their behavior and make progressively more informed choices through experience (Menzel and Müller, 1996; Weiss, 1997; McNeely and Singer, 2001), but innate preferences form the basis of specialization in insects.

Another important part of signal specialization may be the need to avoid catastrophic mistakes. Host preferences of ovipositing females are sometimes much narrower than the range of plants on which their larvae can survive, indicating that insects cannot determine the suitability of potential hosts *per se* but utilize indirect cues that have proved reliable in the past (Schoonhoven et al., 1998). Insect orders with a high degree of association to specific hosts (e.g., Coleoptera, Hymenoptera, Diptera, Lepidoptera) typically have juvenile stages that have little opportunity to switch to another host once a decision has been made by the ovipositing female, which should promote conservatism in female choice.

2.2.5 Identifying Sensory Signals

A critical element when studying sensory systems is access to relevant sensory stimuli for experimentation. There are four equally important steps involved in the complete identification and characterization of a sensory signal:

1. The collection of the natural signal
2. The selection of key elements of that signal
3. The synthesis of a new signal based on step 2
4. The release and exposure of the signal to the test animal, under conditions in which the responses to the natural and synthetic signals can be compared

The goal of this process is to reproduce a signal that is indistinguishable in its effects from the natural one. We will not go into great detail here, but instead refer to citations in this chapter and methods described in other chapters of this volume for further information.

Problems involved in the various steps of the identification process differ greatly, depending on the sensory modality and specific situation that is under study. A general problem during the whole process is that our own senses differ from those of insects to an extent that we are often unable to perceive key elements of the signal, which forces us to rely on more or less sophisticated equipment to identify signal parameters (see Eisner et al., 1969 for a classical example). Collection, storage, and release of auditory or visual signals are greatly facilitated by modern techniques for sound and image recording, playback, and synthesis. The modern age could be said to have begun with Johan Regen's use of a telephone in 1913 to demonstrate the attraction of female crickets to male calling song (related in Greenfield, 2002). In comparison, chemical signals are more troublesome to collect, store, and handle than visual and auditory stimuli (Millar and Haynes, 1998). Factors such as degradation and contamination of odor extracts or synthesized substitutes can severely limit the possibility of obtaining reliable physiological and behavioral responses from test animals (Tumlinson et al., 1977; Kozlov et al., 1996). Selection of active elements of a complex stimulus may be aided by comparing active stimulus sources with inactive but otherwise similar ones, for example, comparing males with females, similar species with each other, or dominant individuals with subordinates (Monnin et al., 1998; Sweeney et al., 2003). Later in this chapter, we discuss how to use the insect's own chemosensory system as a detector for relevant stimuli using electrophysiology in combination with gas chromatography (see Section 2.3.3.4).

If several elements of a stimulus have strong synergistic effects, such that their combined presence is necessary for a significant behavioral response, it may be difficult to identify key elements of the signal by testing single elements at a time (sometimes known as a bottom-up approach). This problem is typical for the chemical blend that makes up many insect sex pheromones, but it is also common in social insect communication systems and other forms of multimodal communication (Hölldobler, 1999). In that case, a top-down approach, such as a subtractive assay, may be helpful: one starts with a complete stimulus or mimic able to elicit the full response in the animal, followed by selective elimination of single components of the stimulus. Finally, we must stress the importance of steps 3 and 4 for the identification process. Artificially reproducing or otherwise manipulating signal transfer is the most convincing way to demonstrate a correct identification. In many systems, however, this may be difficult or impossible to achieve. New possibilities for selective manipulation may open up with genetic techniques allowing selective targeting of specific features in emitters and receivers (Savarit et al., 1999; Bradshaw and Schemske, 2003; Bray and Amrein, 2003).

2.3 SENSORY MODALITIES

In spite of their small size and relatively simple nervous systems, insects make use of an impressive variety of sensory stimuli to gather relevant information from the environment. In this chapter, we will introduce the major modes of signal transmission discussed in this book, namely, mechanosensation/audition (Section II), vision (Section III), and chemoreception (Section IV and Section V), and relate their importance for different insects in different contexts. Additional modalities used by insects include thermoreception in mosquitoes and carrion flies (Clements, 1999; Angioy et al., 2004), infrared reception in buprestid beetles ovipositing in fire-damaged trees (Schütz et al., 1997), and magnetoreception (Walker and Bitterman, 1985). Although these modalities represent fascinating specialized sensory adaptations, they are nevertheless of lesser general importance and will not be covered further in this chapter.

2.3.1 Hearing/Mechanosensation

Compared with vision and olfaction, the ability of insects to use auditory information is relatively restricted, although the number of taxonomic groups with documented hearing ability is growing. Insects have evolved auditory systems numerous times for various purposes: to locate and select mates, to detect and evade predators, and to find proper hosts as a food source for their offspring. Insects perceive auditory signals either as airborne sounds or as substrate-borne vibrations, and it is often difficult to draw a line between hearing and mechanosensation (Hoy and Robert, 1996). All sound-registering organs are based on modified mechanoreceptors, which register either pressure waves in air or vibrations in air or substrate. Pressure waves are registered over long distances by highly specialized tympanal organs found in, for example, crickets and moths. Air vibrations are used as auditory mating signals in flies and mosquitoes and are picked up by plumose antennae or antennal aristae and registered by a specialized Johnston's organ in the second antennal segment. They are active over very short distances, from centimeters to a few meters.

Auditory signals can be detected at night, travel quickly over long distances, and penetrate physical barriers such as vegetation (Table 2.1). On the other hand, they attenuate quickly, are costly to produce, and can easily be exploited by enemies. Insects have evolved highly specialized auditory systems to extract relevant information from an acoustically noisy environment. Key features of a signal, such as its frequency, intensity, and temporal structure, provide important information to a receiver about, for example, the species, sex, or mating status of a sender. Sound signals in insects cover a wide frequency spectrum, ranging from a few hundred Hz in the courtship signals of fruit flies (Ewing and Bennet-Clark, 1968) to about 100 kHz in the mating calls of wax moths (Spangler, 1986). Physical and biological factors in the environment, such as attenuation and atmospheric absorption, strongly influence how, when, and where insects use acoustic signals (Römer, 2001).

2.3.1.1 Constraints on Insect Acoustic Communication

Due to their relatively small size, insects face some major limitations in acoustic communication (reviewed by Bennet-Clark, 1998; Gerhardt and Huber, 2002). For example, their small sound radiators and receptors force them to use relatively high-frequency sound signals, and this will in turn reduce the signal range because high-frequency sound attenuates and degrades more rapidly than low-frequency sound. In addition, due to their small size, insects often communicate close to the ground, causing increased attenuation that further constrains their signal range (Römer, 2001). Not surprisingly, insects have evolved various strategies to increase their communication range. Some species have secondary resonators with larger dimensions than their primary ones, the burrow of mole crickets (Daws et al., 1996) and the abdominal air sac of bladder cicadas (Bennet-Clark and Young, 1998) being primary examples. Other species, such as bladder grasshoppers, *Bullacris membracioides*, increase their effective communication range by limiting their calling to times when temperature inversions create a low-attenuation channel close to the ground (van Staaden and Römer, 1997).

Another problem associated with the small size of insects is the narrow separation of the hearing organs, which limits the use of binaural effects (i.e., using the difference between the two "ears" in intensity and arrival time of sound stimuli to determine its direction of origin). This constraint is more pronounced in species using relatively low frequencies in their communication. To produce binaural differences, insects have evolved ears that function as pressure receivers, pressure-difference receivers, or both (Gerhardt and Huber, 2002).

2.3.1.2 The Insect Ear

In contrast to the compound eyes (vision) and the antennae (chemoreception) of insects, which are shared characters, sensory organs used for acoustic communication evolved repeatedly over evolutionary time (Yager, 1999). The tympanate ear, which is the most common type of insect hearing organ, has appeared at least 19 times and today can be found in at least seven orders: Mantodea, Orthoptera, Hemiptera, Coleoptera, Neuroptera, Diptera, and Lepidoptera (Yager, 1999). Although there are great differences in the complexity of tympanal ears among taxonomic groups, they share the same basic features:

- A thin cuticular membrane (the tympanum), which is set into vibration when stimulated with sound
- The associated tracheal sac
- A chordotonal organ

(See Chapter 4 for methods on how to study insect mechanosensory organs.)

Tympanal ears can be found on various parts of the insect body, such as the thorax, abdomen, legs, wings, and mouthparts (Fullard and Yack, 1993). Insects are normally equipped with pairs of ears, but there are exceptions to this rule: praying mantids have only a single hearing organ located on the ventral part of the thorax (Yager and Hoy, 1986). The precursors of tympanal ears were stretch receptors used for recording changes in body segment position (van Staaden and Römer, 1998; Yack et al., 1999; see also Chapter 4). These mechanoreceptors are widely distributed on the insect body, and this preadaptation probably facilitated the repeated evolution of hearing organs on various body parts. The number of chordotonal sensilla innervating each hearing organ also differs greatly among taxonomic groups. Moths of the families Notodontidae and Sphingidae have only a single auditory sensillum per ear (Surlykke, 1984; Göpfert and Wasserthal, 1999), whereas the ears of primitive grasshoppers (family Pneumoridae) house about 2000 sensilla each (van Staaden and Römer, 1998).

Insect ears show enhanced sensitivity to frequencies corresponding to relevant signals (e.g., those used for intraspecific communication). Species using narrow band signals, such as crickets, have ears sharply tuned to the carrier frequency of the song (see Chapter 5). Others that use transient signals, such as grasshoppers, have ears tuned to a broader frequency range. Normally, individual auditory receptors are constrained to cover a relatively narrow frequency range, but by using groups of receptors tuned to different frequencies, insects can cover a much broader frequency range (*range fractionation*) and thereby increase the overall capacity of their auditory system. In addition to the primary uses of ears, some species have secondarily adapted their hearing systems to include supplemental context-dependent signals. Several orthopterans with ears preadapted for recognizing conspecifics have acquired ultrasonic hearing, which enables them to detect echolocating bats. Conversely, some moths with ears preadapted for bat detection today use high-frequency signals for intraspecific communication (Conner, 1999; Nolen and Hoy, 1986). The question arises how the auditory systems in these insects have been adjusted to include antagonistic signals, especially in moths with a limited number of auditory neurons.

2.3.1.3 Hearing in Orthoptera

Auditory organs in Orthoptera evolved mainly for the purpose of detecting intraspecific signals at medium to high frequencies. The complex ears in grasshoppers are positioned between the metathorax and the first abdominal segment (Gerhard and Huber, 2002). The two tympanal membranes are coupled via internal air sacs, so each membrane is influenced by sound stimulation from two sources: the external surface by cues from the outside and the inner surface by cues triggering the contralateral membrane and propagating through the internal air sacs. Depending on the location of the sound source in relation to the insect, the relative strength of the two inputs will result in a net displacement of each tympanic membrane. The elaborate ears in crickets are positioned on the proximal parts of the forelegs (Gerhardt and Huber, 2002). The external surface of the posterior membrane is exposed to the outside, whereas the internal surface is connected to the main leg trachea via a side branch. Additional side branches of the main leg trachea reach the thoracic surface via the acoustic spiracle and the tracheal system of the other leg. Thus, each membrane can receive sensory input from four different sources: the outside, the ipsilateral acoustic spiracle, the contralateral spiracle, and the contralateral tympanic membrane.

2.3.1.4 Hearing in Lepidoptera

The ears of moths are much simpler in morphology and physiology than those of orthopteran insects, and their primary function is to detect the echolocation calls of predatory bats. The best recognized ultrasonic hearing system is that of noctuid moths, examined in numerous elegant studies by Kenneth Roeder and colleagues (Roeder, 1967). The pair of ears in these species is located on the lateral surface of the metathorax. Each organ is innervated by two sensory neurons (A_1 and A_2), which are attached to the tympanum. Both neurons follow the same tuning pattern, with best frequencies at 25 kHz to 50 kHz (thus, perfectly matching the frequencies used by hunting bats), but they differ in intensity threshold, with the A_1 neuron being 20 dB more sensitive than the A_2 neuron. The moth uses this difference in sensitivity between auditory neurons to make appropriate escape maneuvers in relation to the location and distance of an attacking bat. When the intensity of the ultrasound stimulus is low (the bat is far away), only the A_1 neuron responds. By encoding the difference in arrival times of the acoustic signal between ears, the moth can estimate the location of the bat and respond by steering away from the sound source. When the intensity of the sound is high (the bat is close), the A_2 neuron also fires. At these intensities, the moth brain cannot distinguish information derived from the two ears and the evasive reactions become unpredictable, involving dives, loops, and extended zigzagging flight maneuvers. An individual moth may show large variation in the response pattern to the same sound stimulus; this unpredictability would make it harder for the bat predator to learn how the prey would behave when under attack (Roeder, 1967).

2.3.1.5 Mechanosensation

Vibrations transmitted through honeycombs may be an important component of nest mate recruiting through the waggle dance performed by honeybees (Sandeman et al., 1996). Airborne sounds and substrate vibrations are also part of an ongoing arms race between parasitoid wasps and their larval hosts. Larvae can detect parasitoid wasps from wing beat sounds or substrate vibrations produced by foraging parasitoids, which elicit freezing responses or evasive maneuvers (Djemai et al., 2001). Substrate-borne vibrations are also frequently used by parasitoids to mediate their attraction to hosts hidden in plant material or decaying substrates. Pupal parasitoids show remarkable adaptations of their sensory systems to cope with the problem of low detectability of their immobile hosts concealed in the stem, wood, or other part of plants. Females of two hymenopteran families (Ichneumonidae and Orussidae) use a strategy called *vibrational sounding* for host location (Broad and Quicke, 2000). A female induces vibrations by tapping the tips of her antennae against the substrate, and the resonance signals are detected by enlarged subgenual organs on the forelegs (Vilhelmsen et al., 2001). By using this form of "echolocation," a female wasp can determine the exact position of a hidden host (Wäckers et al., 1998; Fischer et al., 2003). No additional chemical or visual cues are necessary for precise host location. The efficiency of the detection system among wasp species seems to be correlated with their ovipositor morphology and the substrate used by their hosts (Fischer et al., 2003).

2.3.2 Vision

Visual abilities of animals range from simple separation of light from darkness to the ability to resolve small, differently colored objects in the field of view with high acuity. Many insects depend on reasonably good visual acuity for their orientation, and their visual systems are sometimes adapted to solve formidable tasks. Insects and most other arthropods use compound eyes as their main visual organs (see Chapter 8 for more on the functional anatomy of insect eyes). Here, we will focus on the design of compound eyes with respect to specific adaptations of insect visual systems. In addition to compound eyes, many insects have single-lens simple eyes or ocelli, which generally do not provide sharp image resolution and whose function may vary among different insects. Observations of their design and behavioral experiments suggest that in some flying insects the eyes could be specifically adapted to function as flight stabilizers by monitoring the position of the horizon in relation to the animal (Stange et al., 2002).

The function of an eye is to form an image of the environment by partitioning incoming light according to the direction of origin. Within this image, individual objects are distinguished through contrast differences at various positions in the visual field. Features of light that eyes can distinguish are intensity, wavelength, and the plane of polarization, which are all important in insects for object identification. The physical principles behind the operation of visual systems are well known, which facilitates comparisons between predictions and experimental results to an extent that must surely be envied by anyone working with other sensory modalities (Land and Nilsson, 2002).

2.3.2.1 Spatial Resolution and the Compound Eye

The level of spatial resolution that an eye can provide depends on the size of the angular differences in incoming light that the eye can resolve. This, in turn, depends on the spacing of its photoreceptive elements, such as the rhabdoms in an insect compound eye (see Chapter 8) or single photoreceptors in a vertebrate retina. The physical optics of the eye also sets an upper limit to resolution, determined mainly by the size of the eye and the aperture diameter of its lenses, both of which conspire to make life difficult for small animals such as insects. In compound eyes, resolution is limited by diffraction due to the small apertures of the ommatidia that make up the eye (see Chapter 8 for details on the construction of compound eyes). This limitation is so severe that overall image resolution in compound eyes is always much lower than that of single-lens eyes of the same size, and very far below that of our own eyes. The world record in resolution of an insect eye is held by the dragonfly. These eyes have a maximum resolution of approximately 4 to 5 ommatidial axes per degree. In comparison, the bee eye has around 1 axis per degree, compared with the human equivalent of approximately 60 per degree (Land and Nilsson, 2002). Consequently, insects have a very coarse impression of the visual world, and the shapes of objects that they can identify are usually limited to rough outlines of, for example, flowers or leaves, which nevertheless may be a very good complement in identifying specific resources (Dafni et al., 1997). Specialized adaptations, exemplified by dragonfly eyes with extremely high resolution and dedicated visual interneurons working as matched filters, allow these insects to detect small objects such as flies at relatively long distances (Warrant, 2001).

Many authors have pointed out the hopeless inefficiency of compound eyes in comparison with single-lens eyes in terms of spatial resolution, and in many ways it remains a mystery why almost all arthropods persist in using them. Either there are other advantages to compound eyes or maximum resolution is not the main feature selected for in arthropod eyes, perhaps because it may be useless without additional investments in neural processing capacity. Another explanation could be that there is no easy transition from compound eyes to single-lens eyes that can be achieved without significant selective disadvantage, meaning that most animals are stuck with their basic ancestral design (Land, 1997; Laughlin, 2001). The low overall image resolution of compound eyes can be compensated for to some degree by using eyes with inhomogeneous construction, containing areas of higher resolution (*acute zones*) that view those parts of the visual environment where it is most needed, while sacrificing resolution in other parts. The localization of zones with high resolution varies in predictable ways according to the lifestyle of many insects, including how they move and where objects of interest are normally located in the visual field (Figure 2.3). The

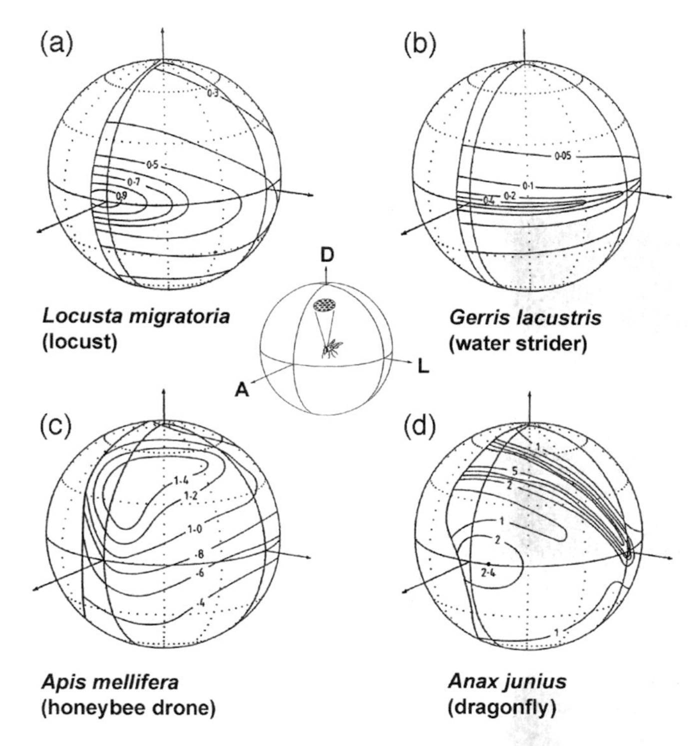

FIGURE 2.3 Distribution of eye resolution, expressed as ommatidial axes per square degree, in four insects with different lifestyles. The image in the center shows the principle behind the projections in **a** through **d**, with ommatidial axes within a conical angle. A = anterior, D = dorsal, L = lateral. **a–d**: Fast-flying insects such as the locust (**a**) typically have highest resolution in the anterior part of the eye, with resolution gradually decreasing toward the lateral and posterior parts, because the maximum resolution that can be obtained decreases with increasing image velocity of the environment viewed by the lateral part of the eye. For many insects, the main objects of interest are found in the equatorial plane of vision, and their eyes correspondingly have relatively high resolution along the equator, with resolution decreasing along a vertical axis toward the dorsal and ventral parts of the eye (**a, b**). Animals living in very flat habitats, such as near water surfaces and in deserts, often have an extremely pronounced horizontal zone of high acuity near the equator of the eye because most objects will be found in the field of view of the horizon (**b**). Insects that chase and capture prey or mates on the wing use an acute zone in the dorsal anterior region of the eye to spot flying objects against the sky (**c, d**), often combined with a forward-pointing zone used for prey catching (**d**). (Drawings courtesy of Michael F. Land.)

foveae of vertebrate single-lens eyes represent an analogous solution to the same problem, and in vertebrates the position and shape of foveae also vary according to the visual environment. In vertebrates, however, limits to overall resolution are set by the optic nerve, which can accommodate enough axons to supply only a fraction of the photoreceptors that could fit in the eye (Land and Nilsson, 2002).

2.3.2.2 Color Vision

The ability to determine the wavelengths of incoming light allows for a whole new dimension in distinguishing contrasts in the visual environment, compared with the intensity differences offered by a monochromatic system (Dafni et al., 1997; Kelber et al., 2003). These abilities may be all the more important in a visual system that is already hard-pressed to distinguish shapes due to relatively low resolution (Figure 2.4).

FIGURE 2.4 (See color insert following page 202.) These four photos of a queen-of-Spain fritillary (*Issoria lathonia*) feeding on flowers of alkanet (*Anchusa officinalis*) illustrate the potential advantages of color vision. Pictures to the right are in grayscale; the lower two pictures have their resolution arbitrarily set to 1/10 of the original, which is still much higher than that of most insect eyes. Even at high resolution, the flowers tend to disappear against the foliage without color, while the butterfly remains visible against the background. Lowering the resolution while maintaining color hardly decreases the detectability of the flowers, and the butterfly can still be distinguished, although the low contrast against this particular background makes it difficult. If the mother-of-pearl spots on the wings (hypothetically) polarize reflected light, they could significantly improve detectability of the butterfly to someone able to see polarized light. The low-resolution grayscale picture to the right contains nothing distinguishable as either a butterfly or flowers. (Photo by Mattias Larsson.)

Although true color vision has been experimentally verified in only a few species (Kelber et al., 2003), the ubiquitous presence of photoreceptors with different wavelength specificities indicates that all or most insects possess it, as do related arthropod groups. Apart from aiding in distinguishing objects through improved contrast, specific colors can in themselves provide important information about resources, such as the flowers or fruit distributed among green foliage. Many insects have strong innate preferences for certain colors related to specific behaviors such as oviposition or flower feeding. Insects can also learn colors and thus quickly specialize on flowers that provide substantial nectar rewards (Schoonhoven et al., 1998; Kelber et al., 2003).

Most insects, including honeybees, most moths, and beetles, have trichromatic color vision with photoreceptors maximally sensitive in the UV, blue, and green parts of the spectra. Some insects, including many butterflies, have an additional photoreceptor type sensitive to red light, and some flies have up to six different photoreceptor types (detailed in Chapter 8). Some species, such as the cockroach *Periplaneta americana,* have only two types of photoreceptors (Briscoe and Chittka, 2001). The tuning curves of specific photoreceptor types also vary among different species of insects. In spite of these great differences in visual systems and great variations in lifestyles among insects, there seems to be surprisingly little correlation between photoreceptor assemblies and visual environment (Vorobyev and Menzel, 1999; Briscoe and Chittka, 2001; Chittka and Briscoe, 2001). The distribution of photoreceptors over the surface of the eye varies extensively in many insects, and some types may be absent or present at higher frequencies in parts viewing specific areas of the visual field, which has been interpreted as adaptations for specific tasks, such as detection of mates, host plants, or flowers (Briscoe and Chittka, 2001; White et al., 2003).

2.3.2.3 Polarization

Direct sunlight is randomly polarized, but atmospheric scattering and reflection from shiny surfaces produces light that is partly polarized at certain angles, providing specific sensory cues to animals able to detect the polarization pattern of light (Wehner, 2001). Insect photoreceptors are inherently sensitive to polarization angle because in the microvillar rhabdomeres of insects and other arthropods, a large fraction of the rhodopsin molecules are oriented in the same plane. Because this could interfere with other aspects of light, such as wavelength discrimination, insect photoreceptors are normally twisted like cork screws to nullify the polarization preference, with polarization sensitivity limited to specific parts of the eye (Wehner and Bernard, 1993; Chapter 8). Polarization caused by reflectance can work like false colors and create patterns that function as oviposition or mate-recognition cues (Kelber, 1999; Sweeney et al., 2003). Aquatic insects can detect pools of water very effectively by the polarized light given off by reflection from the water surface (Schwind, 1991).

Scattering in the atmosphere produces a specific pattern of skylight polarization, with e-vectors concentrically arranged around the sun. This pattern is used by many insects as a compass for orientation and navigation, either by itself or as a guide to the position of the sun when the sky is partially obscured by clouds. This is often associated with homing behavior, the most advanced example of which is perhaps found in the desert ant *Cataglyphus*. These ants live in an environment deprived of visual landmarks and use the polarization compass for continuous path integration during foraging, which allows them to compute a straight path back to their nest on the return trip (Wehner, 2003). Honeybees use the polarization pattern of the sky as a compass for orientation between the hive and foraging sites (Rossel and Wehner, 1984, 1986). Dung beetles use the polarization compass for rolling their dung balls in a straight line from the dung patch away from competitors, and some nocturnal dung beetles can use the polarization pattern from the moon in the same way that diurnal dung beetles use the sun (Dacke et al., 2003).

2.3.2.4 Adaptations to Low Light Intensities

Apart from limiting resolution, the small apertures of apposition eye ommatidia also severely restrict the amount of light that can be captured by every single ommatidium (Warrant, 2001). This is generally not a problem for diurnal insects active in bright light, but at dusk it quickly leads to a catastrophic decline in visual performance. Photon capture can be improved by increasing aperture diameter, but at the cost of decreased resolution. Insects active in dim light have, therefore, usually abandoned the ancestral apposition compound eye in favor of other designs, such as neural superposition eyes (flies) or optical superposition eyes that are many times better at capturing light (most nocturnal insects). The effect of these eye types is to increase light capture without sacrificing image resolution, by increasing the effective surface area over which photons can be collected. Nocturnal insects normally also complement their eye design with a reflecting layer behind the retina (tapetum lucidum, which is standard equipment among nocturnal animals) to increase the probability of photon capture. At very low light levels, spatial or temporal summation mechanisms improve light sensitivity by increasing signal-to-noise ratios. The signals from photoreceptors in adjacent ommatidia can be pooled by lateral interneurons bridging several input channels, at the cost of impaired image resolution. Temporal summation, extending the integration time over which the visual system collects photons, increases sensitivity while decreasing the ability to detect motion. The neural substrates performing these operations are not always identified, but comparisons between empirical data and predictions based on eye optics reveal that insects and other animals use summation to improve visual performance in dim light (Warrant, 1999; Warrant et al., 1996). Combinations of mechanisms mentioned previously allow some insects to use visual cues at very low light intensities. The nocturnal hawkmoth *Deilephila elpenor* uses color vision to detect flowers in dim starlight or at light intensities at which humans are completely color blind (Kelber et al., 2002).

2.3.3 Chemoreception

Chemical signals are probably the most important cues for insects in determining the suitability of feeding and oviposition sites. In terrestrial habitats, chemoreception can be divided into olfaction and taste, depending on whether the chemical stimulus is airborne or detected through contact. Olfactory signals in insects are detected mainly by the antennae and, to a lesser extent, by the maxillary palps, whereas taste receptors are located on various body parts, including palps and other mouthparts, such as the proboscis and tarsi. Here, we will concentrate mostly (but not exclusively) on olfaction and how insects utilize odors to navigate their varied environments.

Chemoreception presents specific problems that are a challenge for scientific investigators as well as for the organisms under study. Each type of odor molecule is unique in that it cannot be described in relation to other molecules in any single dimension along a spectrum, such as the distribution of wavelengths in the case of light or sound. The number of potential odorants or other chemical stimuli that organisms may encounter is enormous and, for all practical purposes, infinite. These factors make it difficult to classify the odor universe according to physical or chemical parameters and to match chemical ligands to sensory receptors. However, as we will discuss, defined chemical stimuli comprising single compounds (or more often blends of several compounds) do convey specific information about odor sources. These stimuli are often collectively classified as *semiochemicals* or *infochemicals*. From an ecological perspective, it makes sense to group chemical stimuli according to their roles in the interaction between sender and receiver, as well as by the type of information they transmit.

2.3.3.1 Classification of Semiochemicals

An excellent discussion about the broad range of different semiochemicals, their roles, and problems associated with their definition can be found in Wyatt (2003); we will give only a brief overview here. Excluding *hormones*, which transmit signals within an individual, other semiochemicals transmit information between individuals. These are primarily divided into *pheromones,* which transmit signals between conspecifics, and *allelochemicals,* which transmit signals between species (Nordlund, 1981). Pheromones are generally assumed to be mutually beneficial to both emitter and receiver. Depending on the interaction they mediate, they can be subdivided into self-explanatory types: sex pheromones, aggregation pheromones, alarm pheromones, etc. Pheromones may exert their effects on different time scales: *releaser* pheromones mediate immediate behavioral responses, whereas *primer* pheromones cause long-term developmental or physiological responses. Unlike pheromones, allelochemicals are subdivided depending on who benefits from the interaction that they mediate: *kairomones* are signals beneficial only to the receiver (as when predators or herbivores localize prey or hosts by their smell); *allomones* are beneficial only to the emitter, typically involved in various cases of deception and other cases of exploitation of the receiver; *synomones* are beneficial to both emitter and receiver, as when flowers advertise nectar rewards in return for pollination. Note that the same chemical signal can be simultaneously classified as more than one type of semiochemical if more than two actors are involved. A pheromone can, for example, also work as a kairomone if it is used by a predator to locate prey.

2.3.3.2 Orientation to Chemical Stimuli

Many olfactory interactions involve orientation toward attractive odor sources, which is difficult to understand because the odor signal itself rarely contains any directional information. If the odor is accompanied by conspicuous visual landmarks, which is often the case with flowers, orientation may be entirely guided by visual cues. Occasionally, volatile substances are deposited by animals moving along a substrate, generating a trail that is comparatively easy to follow. Social ground-dwelling insects such as ants and termites often deposit specific trail pheromones guiding nest mates to resources outside the nest (Hölldobler, 1999). Usually, however, animals must orient toward odor sources by other means.

Odor molecules are dispersed from a source through the combined action of diffusion and current flow of the medium. Diffusion is a slow process, creating an expanding volume of gradually decreasing concentration away from the odor source. Diffusion dominates in situations with little mass flow, in sheltered conditions close to the odor source, or in boundary layers with little airflow close to substrates (Dusenbery, 1992; Greenfield, 2002). In these situations, orientation toward odor sources can be accomplished by following odor gradients if animals are able to monitor odor concentration over time, changing direction in response to changes in odor concentration. In most situations, however, the effects of air or water flow will dominate over diffusion; this can generate an odor plume that meanders downwind from the source. Due to turbulence, the odor plume will be broken up into packages of high odor concentration interspersed with clean air, resulting in a dynamic stimulus with odor concentrations fluctuating rapidly over time (Murlis et al., 1992). Animals navigating in odor plumes usually employ *anemotaxis* (i.e., moving upwind as long as they are in contact with the odor plume). If they lose contact with the plume, they switch to sideways movements in an attempt to reestablish contact. In flying animals, this usually manifests as wide, casting movements perpendicular to the wind direction. Walking animals can keep track of the wind direction by means of mechanosensory input; flying animals employ optomotor responses, using visual input to compute their displacement relative to their surroundings (Mafra-Neto and Cardé, 1994; Vickers and Baker, 1994).

Odor plumes from various sources will be mixed to various degrees in the air, and the integrity of the odor information depends on how efficiently animals can separate odor packages originating from different plumes. Insects flying upwind will encounter odor packages from different sources in sequence, potentially enabling them to distinguish partially intermixed plumes based on the timing of different odor packages' arrival. Experiments with intermixed plumes have shown that moths can differentiate odor sources very well — in some cases, even when separated by only a few millimeters (Baker et al., 1998).

2.3.3.3 Adaptive Use of Odor Information

Insects utilize chemical information during any phase of their orientation, often in a series of stepwise decisions ranging from finding the right habitat, to the identification of a resource, to final evaluation of its suitability through contact chemoreception. Finding suitable resources is generally assumed to involve the interplay between positive signals (attractants and stimulants) and specific negative signals (inhibitors and repellents) that steer the animal away from maladaptive choices (Schoonhoven et al., 1998; Schlyter and Birgersson, 1999). Although the world is full of potential odorants, most insects appear to make use of only a small fraction, by basing their decisions on a very limited number of key compounds.

Any single odorant could potentially serve as an identifier of its source, provided that it is released exclusively from this source alone. In some cases, specific chemicals or classes of chemicals can be reliably associated with certain sources, such as allyl isothiocyanates released from cruciferous plants, certain sulfides released from onions (Visser, 1986), and other sulfides released from putrefying proteins (Stensmyr et al., 2002). Not surprisingly, these compounds often make potent stimuli for specialists exploiting them whereas most other species avoid them. Many flower volatiles apparently are also sufficiently reliable to elicit significant attraction. In reality, this level of specificity is rarely found, and insects use other strategies, involving combinations of rare and very common chemicals, to identify resources. These chemicals, sometimes released in narrowly defined ratios, provide an organism with specific information about the source and mediate appropriate behavior, such as searching, avoidance, or aggression. Species-specific sex-pheromone blends released from female insects guide searching males to conspecific females; other odors give information about kin, social or reproductive status, or whether an individual is a nest mate (Wyatt, 2003). Chemicals released from breath and skin, such as carbon dioxide, 1-octen-3-ol, and lactic acid, guide mosquitoes to their vertebrate hosts and mediate some degree of host specificity in, for example, anthropophilic mosquitoes (Takken and Knols, 1999; Dekker et al., 2002). Many parasitoid wasps would likely have a hard time detecting the smell of prey when hidden on or even inside a plant but exploit volatiles released by host plants in response to herbivore attack (Dicke and van Loon, 2000). Specific blends of plant volatiles released from hawthorn fruits and apples guide each of the two newly diverged host races of the apple maggot fly *Rhagoletis pomonella* to its host and may thus be promoting ongoing sympatric speciation (Linn et al., 2003).

Many chemicals could potentially convey information that an odor source is, in fact, unsuitable as a resource, and these chemicals thus mediate avoidance or inhibit attraction (Schoonhoven et al., 1998; Wyatt, 2003). Insects often avoid pheromone blends that contain pheromone components of sympatric species (Leal, 1996; Cossé et al., 1998; Larsson et al., 2002), as well as "antiaggregation" pheromones and oviposition marking pheromones that show that a resource is already occupied (Wyatt, 2003). Bark beetles specializing on coniferous trees rich in terpenoids avoid odors released primarily from deciduous trees, such as green-leaf volatiles (GLVs): short-chain alcohols, aldehydes, and acetates released from green plant tissue (Zhang and Schlyter, 2004). Specific alkaloids and other secondary plant compounds that are usually assumed to function in host-plant defense often act as insect deterrents at very low doses, except among specialized herbivores that are physiologically adapted to avoid their toxicity, for which they often act as strong attractants or phagostimulants (Schoonhoven et al., 1998).

When we say that certain compounds form a chemosensory signal, we mean that these compounds are sufficient to reproduce the effect of natural stimuli (see Section 2.2.5 on identification of sensory signals). To what extent insects use information from other chemicals released from the same or other sources, or what compounds they can perceive at naturally occurring concentrations, is not clear. Ultimately, the size and complexity of the chemosensory world of any given organism is determined by the scope of compounds that its chemosensory system can detect and distinguish; insects have about an order of magnitude fewer types of olfactory receptor neurons than mammals (cf. Pilpel and Lancet, 1999). Available evidence suggests that olfactory and gustatory receptor assemblies of insects are, to a great extent, specialized to detect those semiochemicals that act in the process of host or mate identification (Schoonhoven et al., 1998; Wyatt, 2003). Insect pheromone receptor neurons are generally considered to be very selective for single pheromone components, and many studies have also demonstrated narrow tuning in receptor neurons for nonpheromonal odors (Hansson et al., 1999; Jönsson and Anderson, 1999; Shields and Hildebrand, 2001).

On the other hand, although the range of odors that insects are able to perceive and discriminate has never been extensively explored for any species, available evidence suggests that this range is rather broad. The relatively limited chemosensory system of insects can thus accommodate both considerable breadth and high discriminatory ability along with high innate specificity — not a small feat! Related insect species often appear to have similar receptor assemblies (Larsson et al., 2001; Stensmyr et al., 2003a; Stranden et al., 2003; see also Figure 2.5 and Figure 2.6). This is true even for species with different host preferences, suggesting that changes in host preference may be initiated primarily by changes in higher-order processing, whereas changes in receptor assemblies are later adaptations. Conversely, some host shifts may be mediated specifically by changes in chemosensory receptors, such as widening of host spectra after loss of receptors mediating deterrence (Schoonhoven et al., 1998).

2.3.3.4 Antennae as Odor Detectors: GC–Electrophysiology

The complexity of chemosensory stimuli can make life difficult for sensory physiologists searching for potent ligands for chemosensory receptor neurons, as well as for chemical ecologists hoping to identify specific semiochemicals. In either case, it is recommended that one start with compounds emanating from sources that have documented behavioral effects on the species under study, such as a preferred host plant eliciting attraction and oviposition, or an unfortunate test subject especially favored by hematophagous insects. However, solvent extraction or collection of headspace volatiles often yields complex extracts, sometimes containing hundreds of compounds. Determining which may be relevant through chemical identification and testing of every compound in different combinations would be an arduous task, especially because specific blends of compounds may be necessary to achieve significant effects.

One popular method to circumvent this problem is to use the insect chemosensory system itself as a detector for relevant compounds. Electroantennographic

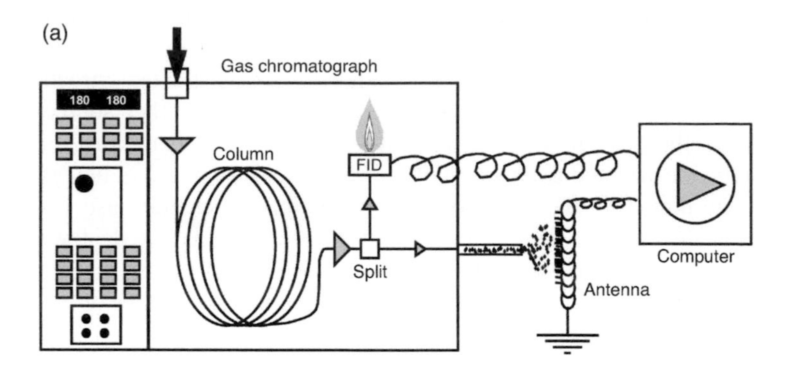

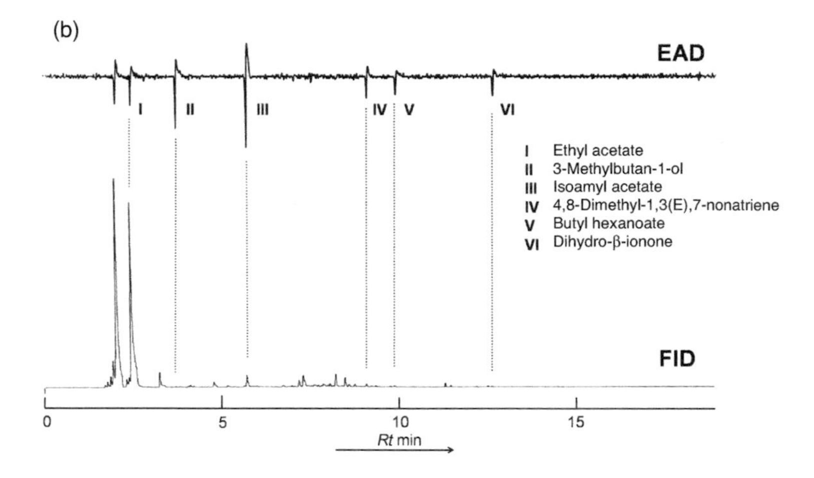

FIGURE 2.5 Combined gas chromatography and electroantennographic detection (GC–EAD). **a**: The principle behind GC–EAD. An extract to be analyzed is injected into the column of a gas chromatograph (arrow). After separation, half of the material is sent to the flame ionization detector (FID) of the GC, while half is simultaneously sent to an antennal preparation. Signals from the FID and antenna are recorded in parallel, and the peaks from the GC can be monitored by responses from the antenna. **b**: GC–EAD recordings showing responses to volatiles from hawthorn fruit (below) using the antenna of an apple maggot fly *Rhagoletis pomonella* (above). Note that this antenna is from an apple race fly, showing that apple race flies can detect the same compounds that are used in host location by hawthorn race flies. (From Nojima, S. et al., *J. Chem. Ecol.* **29**:321–336, 2003. With permission.)

(EAG) responses from whole insect antennae (or palps) (Figure 2.5) or single-cell/single-sensillum (SC) recordings (Figure 2.6) can be used to monitor responses to single synthetic compounds or fractionated parts of extracts to determine their biological activity. First, organic extracts can be obtained from host plants, insects, or other biological material. Then, a gas chromatograph (GC) is used to separate the extract into its single constituents. The individual volatiles can then be used as

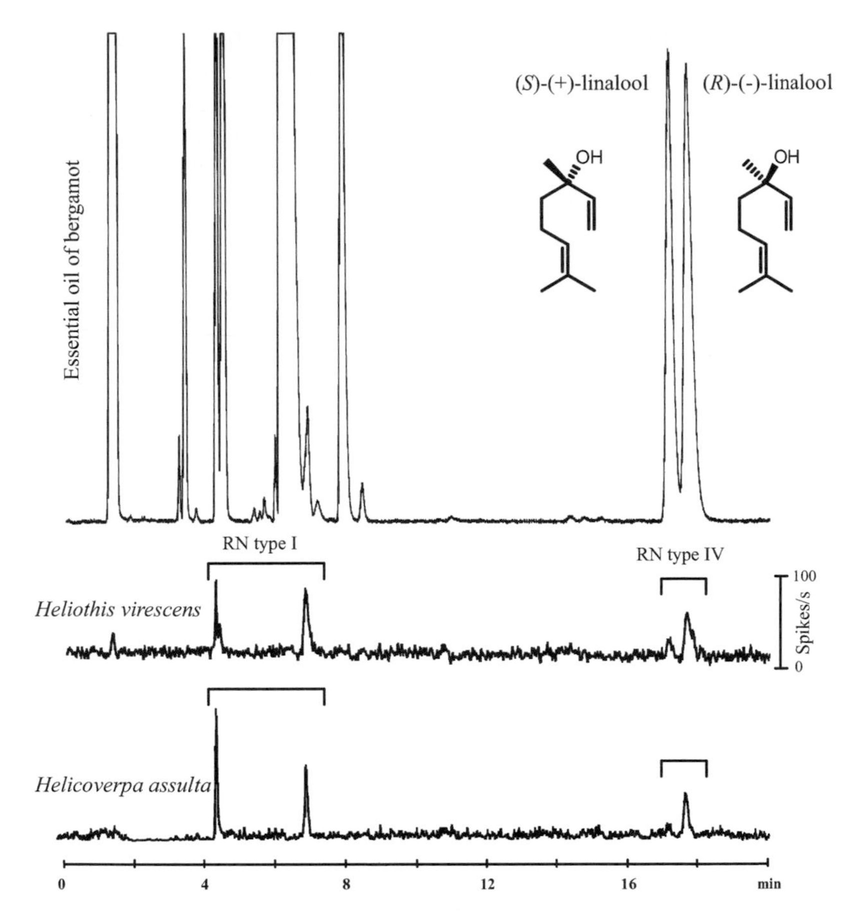

FIGURE 2.6 Monitoring of volatiles from essential oil of bergamot (*Citrus bergamia*) separated in a chiral GC column (upper trace) coupled with the recording of activity from single neurons colocalized in the sensilla of two heliothine moth species (lower traces). Both species have combinations of neurons with identical response spectra: type I cells responding to the two terpenoids β-myrcene and *E*-β-ocimene, and type IV cells with a preference for the *(R)*-enantiomer of linalool. (From Stranden, M. et al., *Chemoecology* **13**:143–154, 2003. With permission).

stimuli to test the responses of a biological detector (antenna/palp/sensillum) in real time as the volatiles elute from the GC column. The specific methods used for electrophysiological recording are described in Chapter 10 and Chapter 12. The extract to be analyzed is separated in a GC with a split installed at the end of the column (Figure 2.5a). In this way, as the single components of the extract elute one after the other, half of the material eluting from the column is sent to the GC detector, while the other half is sent out of the GC and into a glass tube with a continuous air stream flushing over the antennal prep[28]. The signal from the GC detector and the electrophysiological signal are recorded simultaneously, so that the physiological response to each compound can be matched with great precision to the corresponding peaks in the gas chromatogram (Figure 2.5, Figure 2.6).

Monitoring of biological activity with insect antennae has the advantage that no prior assumptions must be made regarding any constituent of an extract; the method rests on the assumption that chemosensory neurons of insects are specialized to detect those compounds that are most important for the species (see previous discussion). The validity of this assumption and the efficiency of the method have been confirmed beyond doubt for specialized pheromone detection systems of many insects, where often a majority of olfactory receptor neurons is dedicated specifically to detecting a few pheromone components (Todd and Baker, 1999). Gas chromatography and electroantennographic detection (GC–EAD) methods are a major factor behind the success story of insect pheromone identifications. To identify the components of a moth sex pheromone, only an extract from a single female and an antenna from a single male may be needed today, compared with the 500,000 female pheromone glands used in the first pheromone identification, in the silk moth *Bombyx mori* (Butenandt et al., 1959). Chemical communication in mammals is generally more complicated than in insects, and the study of mammal semiochemicals and their sensory perception could potentially benefit immensely from similar techniques (Luo et al., 2003). For semiochemicals other than sex pheromones, the usefulness of GC–EAD recordings has been more variable than for pheromones, as some other types of receptor neurons are not numerous enough to give responses with acceptable signal-to-noise ratios from the antenna. Nevertheless, it is a standard tool applied in many successful identifications of nonpheromonal semiochemicals (Light et al., 2001; Stensmyr et al., 2002; Nojima et al., 2003).

GC–SC recordings can be used as an alternative or complementary method to GC–EAD (Blight et al., 1995; Wibe et al., 1997; Stensmyr et al., 2001, 2003b); their advantage is that they may detect neurons present in low numbers, and they allow a characterization of response spectra at the level of single receptor neurons instead of combined responses from the antenna as a whole. In some cases, they may be used when the morphology or size of the antenna makes GC–EAD recordings problematic. On the other hand, GC–SC is a laborious and time-consuming method that requires probing large numbers of sensilla over most of the antennal surface to find a significant proportion of all types of receptor neurons. Thus, it is rarely suitable for high-throughput screening of candidate compounds.

2.3.4 Multimodal Stimuli

Although insects often show extreme specialization to a specific sensory modality, such as detection and processing of sex pheromone signals by male moths, they are normally exposed to a variety of sensory stimuli originating from different modalities (see Chapter 1). This information must be integrated in the nervous system before the insect can perform critical tasks, such as locating a mate or avoiding predators. For the most part, sensory ecologists tend to use a single-modality approach when investigating sensory-mediated behaviors in insects, but more studies are beginning to take a multisensory approach. A trend toward multimodal stimuli should be the focus of future studies.

In insects, the process of searching for and selecting a reliable host or prey often includes multisensory strategies. Multiple modalities can work simultaneously, giving rise to synergistic effects, or sequentially in a fixed sequence of responses to a predicted succession of sensory cues originating from different modalities (van Loon and Dicke, 2001). The number of sensory modalities increases as a phytophagous insect approaches and establishes contact with the host. During the long-distance search phase, vision, olfaction, or a combination of the systems is used to guide the insect to the resource. At contact with a plant, mechanosensation and gustation take over as primary modalities for host evaluation and acceptance, although olfactory information is still important. (For an outline of modalities used during host-plant selection by phytophagous insects, see Schoonhoven et al., 1998.)

Multimodal signaling is frequently used by plants to facilitate the attraction of insects for pollination. Night-flowering plants apply this strategy to attract pollinators (such as the nocturnal hawk moth *Manduca sexta*), where both visual and olfactory cues are required to elicit proboscis extension and feeding (Raguso and Willis, 2002). Multimodal signaling seems to be a common theme in elaborate resource mimicry systems, where insects are fooled into pollinating a plant without obtaining any obvious reward in return. For example, sexually deceptive orchids of the genus *Ophrys* attract male bees for pollination by mimicking both the shape and sex pheromone emission of the female bee (Schiestl et al., 1999). Another case of sensory manipulation includes the dead-horse arum, a Mediterranean species belonging to the Araceae family. This plant attracts carrion blowflies for pollination by copying several features of the rectum of a carcass, which is an important oviposition site for these insects. The plant uses a multisensory strategy, including thermogeny, scent, and possibly also visual and tactile cues, to create an irresistible fly attractant (Angioy et al., 2004; Stensmyr et al., 2002).

Multisensory signals can also convey conflicting information, which may alter the behavioral output of an insect in a specific context. Several studies have recently shown how echolocation calls from predatory bats influence the mating and foraging behaviors of moths. Male *Agrotis segetum* (Noctuidae) and *Plodia interpunctella* (Pyralidae) were exposed to the attack call from a bat while they were orienting in sex pheromone plumes of different quality, in terms of blend composition and release rates (Svensson et al., 2004). Little evasive behavior was observed when moths were flying toward optimal pheromone sources, whereas strong evasive behavior was observed when moths were flying toward suboptimal ones (Figure 2.7). In a similar study, Skals et al. (2003) studied how the noctuid moth *Autographa gamma* responded to bat sound when orienting to context-dependent odor sources. Although both sexes changed their behavior in response to bat sounds, there was no difference in evasive behaviors depending on whether moths were orienting to a flower odor (both sexes) or the female-produced sex pheromone (males only). The results obtained from these studies suggest a dynamic trade-off between reproduction/foraging and predator avoidance in moths with ultrasonic hearing.

2.4 CONCLUSIONS AND OUTLOOK

This chapter was written in the hope that it would help to promote increased communication among scientists in the fields of sensory neuroscience and ecology.

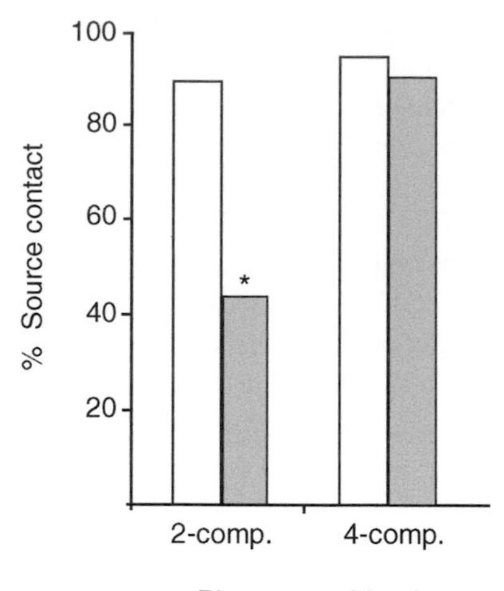

FIGURE 2.7 Percentage of unexposed (white bars) and ultrasound-exposed (gray bars) males of the noctuid moth *Agrotis segetum* reaching sex pheromone sources in a flight tunnel. Evasive reactions to the sound cue mimicking the attack call of a predatory bat are much less pronounced when males orient toward the complete pheromone blend (4-comp.) than toward a blend lacking two components (2-comp.), suggesting a trade-off between mate attraction and predator avoidance in this moth species.* $P<0.01$. (From Svensson, G.P. et al., *Oikos* **104**:91–97, 2004. With permission.)

In order to continue to make significant advances in these fields, we believe that this communication is to their mutual advantage (which, as we mentioned earlier, is a necessary condition for true communication). Sensory neuroscience is confronting increasingly complex questions, including the study of higher brain centers, and progress toward understanding what are commonly called "cognitive" functions is made possible by recent advances in methodology. As studies near the critical point(s) where sensory stimuli are converted into decisions about behavioral responses, the experimental approach should expand to include more "meaningful" and often multimodal sensory stimuli.

In reality, there is no clear boundary between sensory neuroscience and sensory ecology: any differences between them are defined more by the outlook of their practitioners than by specific methodology. According to our own experience, insights from either field often promote the generation and testing of hypotheses in the other. In addition, we feel that our own interest in sensory neuroscience is fueled to a great extent by our fascination with the complex sensory interactions governing the lives of insects.

ACKNOWLEDGMENTS

We are grateful to (in alphabetical order) Tom Christensen, Andreas Keller, Eva Palmqvist, Rob Raguso, and Silke Sachse, for reading and commenting on the manuscript.

REFERENCES

Alcock J (1984) *Animal Behavior: An Evolutionary Perspective*. 3rd edition. Sinauer Associates, Sunderland, MS.

Angioy AM, Stensmyr MC, Urru I, Puliafito M, Collu I, Hansson BS (2004) Function of the heater: the dead horse arum revisited. *Proc. R. Soc. Lond. B* **271**: S13–S15.

Baker TC, Fadamiro HY, Cosse AA (1998) Moths use fine tuning for odour resolution. *Nature* **393**:530.

Barth FG, Schmidt A (2001) *Ecology of Sensing*. Springer Verlag, Berlin.

Bennet-Clark HC (1998) Size and scale effects as constraints in insect sound communication. *Phil. Trans. R. Soc. Lond. B* **353**:407–419.

Bennet-Clark HC, Young D (1998) Sound radiation by the bladder cicada *Cystosoma saundersii*. *J. Exp. Biol.* **201**:701–715.

Bernays AE (2001) Neural limitations in phytophagous insects: implications for diet breadth and evolution of host affiliation. *Annu. Rev. Entomol.* **46**:703–727.

Blight MM, Pickett JA, Wadhams LJ, Woodcock CM (1995) Antennal perception of oilseed rape, *Brassica napus* (Brassicaceae), volatiles by the cabbage seed weevil *Ceutorhynchus assimilis* (Coleoptera, Curculionidae). *J. Chem. Ecol.* **21**:1649–1664.

Bradbury JW, Vehrencamp SL (1998) *Principles of Animal Communication*. Sinauer Associates, Sunderland, MS.

Bradshaw HD, Schemske DW (2003) Allele substitution at a flower colour locus produces a pollinator shift in monkeyflowers. *Nature* **426**:176–178.

Bray S, Amrein H (2003) A putative *Drosophila* pheromone receptor expressed in male-specific taste neurons is required for efficient courtship. *Neuron* **39**:1019–1029.

Briscoe AD, Chittka L (2001) The evolution of color vision in insects. *Annu. Rev. Entomol.* **46**:471–510.

Broad GR, Quicke DLJ (2000) The adaptive significance of host location by vibrational sounding in parasitoid wasps. *Proc. R. Soc. Lond. B* **267**:2403–2409.

Butenandt A, Beckmann R, Stamm D, Hecker E (1959) Über den Sexual-Lockstoff des Seidenspinners *Bombyx mori*. Reindarstellung und Konstitution. *Z. Naturforsch.* **14b**:283–284.

Chittka L, Briscoe A (2001) Why sensory ecology needs to become more evolutionary: insect color vision as a case in point. In: *Ecology of Sensing*, Barth FG and Schmidt A, Eds., Springer Verlag, Berlin, pp. 19–37.

Clements AN (1999) *The Biology of Mosquitoes Vol. 2: Sensory Reception and Behaviour.* CABI Publishing, Wallingford, U.K.

Conner WE (1999) "Un chant d'appel amoureux": acoustic communication in moths. *J. Exp. Biol.* **202**:1711–1723.

Cossé AA, Todd JL, Baker TC (1998) Neurons discovered in male Helicoverpa zea antennae that correlate with pheromone-mediated attraction and interspecific antagonism. *J. Comp. Physiol. A* **182**:585–594.

Dacke M, Nilsson DE, Scholtz CH, Byrne M, Warrant EJ (2003) Insect orientation to polarized moonlight. *Nature* **424**:33.

Daws AG, Bennet-Clark HC, Fletcher NH (1996) The mechanism of tuning of the mole cricket singing burrow. *Bioacoustics* **7**:81–117.

Dafni A, Lehrer M, Kevan PG (1997) Spatial flower parameters and insect spatial vision. *Biol. Rev.* **72**:239–282.

Dekker T, Steib B, Cardé RT, Geier M (2002) L-lactic acid: a human-signifying host cue for the antropophilic mosquito *Anopheles gambiae*. *Med. Vet. Entomol.* **16**:1–8.

Dicke M, van Loon JJA (2000) Multitrophic effects of herbivore-induced plant volatiles in an evolutionary context. *Entomol. Exp. Appl.* **97**:237–249.

Djemai I, Casas J, Magal C (2001) Matching host reactions to parasitoid wasp vibrations. *Proc. Roy. Soc. Lond. B* **268**:2403–2408.

Dusenbery DB (1992) *Sensory Ecology.* WH Freeman and Company, New York.

Dussourd DE, Harvis CA, Meinwald J, Eisner, T (1991) Pheromonal advertisement of a nuptial gift by a male moth (*Utetheisa ornatrix*). *Proc. Natl. Acad. Sci. USA* **88**:9224–9227.

Eisner T, Silberglied RE, Aneshansley D, Carrel JE, Howland HC (1969) Ultraviolet video-viewing: the television camera as an insect eye. *Science* **166**:1172–1174.

Endler JA (1993) Some general comments on the evolution and design of animal communication systems. *Phil. Trans. R. Soc. Lond. B* **340**:215–225.

Ewing AW, Bennet-Clark HC (1968) The courtship songs of *Drosophila. Behaviour* **31**:288–301.

Fischer S, Samietz J, Dorn S (2003) Efficiency of vibrational sounding in parasitoid host location depends on substrate density. *J. Comp. Physiol. A* **189**:723–730.

Fullard JH (1994) Auditory changes in noctuid moths endemic to a bat-free habitat. *J. Evol. Biol.* **7**:435–445.

Fullard JH, Yack JE (1993) The evolutionary biology of insect hearing. *Trends Ecol. Evol.* 8:248–252.

Fullard JH, Dawson JW, Otero LD, Surlykke A (1997) Bat-deafness in day-flying moths (Lepidoptera, Notodontidae, Dioptinae). *J. Comp. Physiol. A* **181**:477–483.

Gerhardt HC, Huber F (2002) *Acoustic Communication in Insects and Anurans: Common Problems and Diverse Solutions.* The University of Chicago Press, Chicago.

Göpfert MC, Wasserthal LT (1999) Auditory sensory cells in hawkmoths: identification, physiology and structure. *J. Exp. Biol.* **202**:1579–1587.

Greenfield MD (2002) *Signalers and Receivers: Mechanisms and Evolution of Arthropod Communication.* Oxford University Press, Oxford, U.K.

Hansson BS, Larsson MC, Leal WS (1999) Green leaf volatile-detecting olfactory receptor neurones display very high sensitivity and specificity in a scarab beetle. *Physiol. Entomol.* **24**:121–126.

Hardie RC (2001) Phototransduction in *Drosophila melanogaster. J. Exp. Biol.* **204**: 3403–3409.

Hölldobler B (1999) Multimodal signals in ant communication. *J. Comp. Physiol. A* **184**:129–141.

Hoy RR, Robert D (1996) Tympanal hearing in insects. *Annu. Rev. Entomol.* **41**: 433–450.

Jönsson M, Anderson P (1999) Electrophysiological response to herbivore-induced host plant volatiles in the moth *Spodoptera littoralis. Physiol. Entomol.* **24**:377–385.

Kelber A (1999) Why 'false' colours are seen by butterflies. *Nature* **402**:251.

Kelber A, Balkenius A, Warrant EJ (2002) Scotopic colour vision in nocturnal hawkmoths. *Nature* **419**: 922–925.

Kelber A, Vorobyev M, Osorio D (2003) Animal colour vision – behavioural tests and physiological concepts. *Biol. Rev.* **78**:81–118.

Knüttel H, Fiedler K (2001) Host-plant-derived variation in ultraviolet patterns influences mate selection by male butterflies. *J. Exp. Biol.* **204**: 2447–2459.

Kozlov MV, Zhu J, Philipp P, Francke W, Zvereva EL, Hansson BS, Löfstedt C (1996) Pheromone specificity in *Eriocrania semipurpurella* (Stephens) and *E. sangii* (Wood) (Lepidoptera: Eriocraniidae) based on chirality of semiochemicals. *J. Chem. Ecol.* **22**:431–454.

Land MF (1997) Visual acuity in insects. *Annu. Rev. Entomol.* **42**:147–177.

Land MF, Nilsson D-E (2002) *Animal Eyes.* Oxford University Press. Oxford, U.K.

Larsson MC, Hallberg E, Kozlov MV, Francke W, Hansson BS, Löfstedt C (2002) Specialized olfactory receptor neurons mediating intra- and interspecific chemical communication in leafminer moths *Eriocrania* spp. (Lepidoptera: Eriocraniidae). *J. Exp. Biol.* **205**:989–998.

Larsson MC, Leal WS, Hansson BS (2001) Olfactory receptor neurons detecting plant odours and male volatiles in *Anomala cuprea* beetles (Coleoptera: Scarabaeidae) *J. Insect Physiol.* **47**:1065–1076.

Laughlin S (2001) The metabolic cost of information: a fundamental factor in visual ecology. In: *Ecology of Sensing*, Barth FG and Schmidt A, Eds., Springer-Verlag, Berlin, pp. 169–186.

Leal WS (1996) Chemical communication in scarab beetles: Reciprocal behavioral agonist–antagonist activities of chiral pheromones. *Proc. Natl. Acad. Sci. USA* **93**:12112–12115.

Light DM, Knight AL, Henrick CA, Rajapaska D, Lingren B, Dickens JC, Reynolds KM, Buttery RG, Merrill G, Roitman J, Campbell BC (2001) A pear-derived kairomone with pheromonal potency that attracts male and female codling moth, *Cydia pomonella* (L.) *Naturwissenschaften* **88**:333–338.

Linn Jr. CJ, Feder JL, Nojima S, Dambroski HR, Berlocher SH, Roelofs W (2003) Fruit odor discrimination and sympatric host race formation in *Rhagoletis*. *Proc. Natl. Acad. Sci. USA* **100**:11490–11493.

Luo M, Fee MS, Katz L (2003) Encoding pheromonal signals in the accessory olfactory bulb of behaving mice. *Science* **299**:1196–1201.

Mafra-Neto A, Cardé RT (1994) Fine structure of pheromone plumes modulate upwind orientation of flying moths. *Nature* **369**: 142–144.

McNeely C, Singer MC (2001) Contrasting the roles of learning in butterflies foraging for nectar and oviposition sites. *Anim. Behav.* **61**: 847–852.

Menzel R, Müller U (1996) Learning and memory in honeybees: from behavior to neural substrates. *Annu. Rev. Neurosci.* **19**: 379–404.

Millar JG, Haynes KF (1998) *Methods in Chemical Ecology I: Chemical Methods*. Chapman & Hall, New York.

Minor AV, Kaissling K-E (2003) Cell responses to single pheromone molecules may reflect the activation kinetics of olfactory receptor molecules. *J. Comp. Physiol. A* **189**: 221–230.

Monnin T, Malosse C, Peeters C (1998) Solid-phase microextraction and cuticular hydrocarbon differences related to reproductive activity in a queenless ant. *J. Chem. Ecol.* **24**:473–490.

Murlis J, Elkinton JS, Cardé RT (1992) Odor plumes and how insects use them. *Annu. Rev. Entomol.* **37**:505–532.

Nojima S, Linn Jr C, Morris B, Zhang A, Roelofs W (2003) Identification of host fruit volatiles from hawthorn (*Crataegus spp.*) attractive to hawthorn-origin *Rhagoletis pomonella* flies. *J. Chem. Ecol.* **29**:321–336.

Nolen TG, Hoy RR (1986) Phonotaxis in flying crickets. I. Attraction to the calling song and avoidance of bat-like ultrasound are discrete behaviors. *J. Comp. Physiol.* **159**:423–439.

Nordlund DA (1981) Semiochemicals: a review of the terminology. In: *Semiochemicals: Their Role in Pest Control*, Nordlund DA, Jones RL, and Lewis WJ, Eds., Wiley, New York, pp. 13–28.

Phelan PL (1992) Evolution of sex pheromones and the role of asymmetric tracking. In: *Insect Chemical Ecology: An Evolutionary Approach*, Roitberg BD and Isman MB, Eds., Chapman & Hall, New York, pp. 265–314.

Pilpel Y, Lancet D (1999) Good reception in fruitfly antennae. *Nature* **398**:285–287.

Raguso RA, Willis MA (2002) Synergy between visual and olfactory cues in nectar feeding by naïve hawkmoths, *Manduca sexta. Anim. Behav.* **64**:685–695.

Robert D, Amoroso J, Hoy RR (1992) The evolutionary convergence of hearing in a parasitoid fly and its cricket host. *Science* **258**:1135–1137.

Roeder KD (1967) *Nerve Cells and Insect Behavior.* Harvard University Press, Cambridge, MA.

Römer H (2001) Ecological constraints for sound communication: from grasshoppers to elephants. In: *Ecology of Sensing*, Barth FG and Schmidt A, Eds., Springer-Verlag, Berlin, pp. 59–77.

Rossel S, Wehner R (1984) How bees analyze the polarization patterns in the sky: experiments and model. *J. Comp. Physiol. A* **154**: 607–615.

Rossel S, Wehner R (1986) Polarization vision in bees. *Nature* **323**:128–131.

Rydell J, Skals N, Surlykke A, Svensson M (1997) Hearing and bat defence in geometrid wintermoths. *Proc. R. Soc. Lond. B* **264**:83–88.

Sandeman DC, Tautz J, Lindauer M (1996) Transmission of vibration across honeycombs and its detection by bee leg receptors. *J. Exp. Biol.* **199**:2585–2594.

Savarit F, Sureau G, Cobb M, Ferveur J-F (1999) Genetic elimination of known pheromones reveals the fundamental chemical bases of mating and isolation in *Drosophila. Proc. Natl. Acad. Sci. USA* **96**:9015–9020.

Schiestl FP, Ayasse M, Paulus HF, Löfstedt C, Hansson BS, Ibarra F, Francke W (1999) Orchid pollination by sexual swindle. *Nature* **399**:421–422.

Schlyter F, Birgersson GA (1999) Forest beetles. In: Hardie J, Minks AK, Eds., *Pheromones of Non-Lepidopteran Insects Associated with Agricultural Plants*, CABI Publishing, Wallingford, U.K., pp. 113–148.

Schoonhoven LM, Jermy T, van Loon JJA (1998) *Insect–Plant Biology.* Chapman & Hall, London.

Schütz S, Bleckmann H, Mürtz M (1997) Infrared detection in a beetle. *Nature* **386**:773–774.

Schwind R (1991) Polarization vision in water insects and insects living on a moist substrate. *J. Comp. Physiol.* A **169**, 531–540.

Shields VDC, Hildebrand JG (2001) Responses of a population of antennal olfactory receptor cells in the female moth *Manduca sexta* to plant-associated volatile organic compounds. *J. Comp. Physiol.* A **186**:1135–1151.

Skals N, Plepys D, Löfstedt C (2003) Foraging and mate-finding in the silver Y moth, *Autographa gamma* L. (Lepidoptera: Noctuidae) under the risk of predation. *Oikos* **102**:351–357.

Spangler HG (1986) Functional and temporal analysis of sound production in *Galleria mellonella* L. Lepidoptera Pyralidae. *J. Comp. Physiol. A* **159**:751–756.

Stange G, Stowe S, Chahl JS, Massaro A (2002) Anisotropic imaging in the dragonfly median ocellus: a matched filter for horizon detection. *J. Comp. Physiol. A* **188**: 455–467.

Stensmyr MC, Larsson MC, Bice S, Hansson BS (2001) Detection of fruit- and flower-emitted volatiles by olfactory receptor neurons in the polyphagous fruit chafer *Pachnoda marginata* (Coleoptera: Cetoniinae). *J. Comp. Physiol. A* **187**:509–519.

Stensmyr MC, Urru I, Collu I, Celander M, Hansson BS, Angioy AM (2002) Rotting smell of dead-horse arum florets. *Nature* **420**:625–626.

Stensmyr MC, Dekker T, Hansson BS (2003a) Evolution of the olfactory code in the *Drosophila melanogaster* subgroup. *Proc. R. Soc. Lond. B* **270**:2333–2340.

Stensmyr MC, Giordano E, Balloi A, Angioy A-M, Hansson BS (2003b) Novel natural ligands for *Drosophila* olfactory neurones. *J. Exp. Biol.* **206**:715–724.

Stranden M, Røstelien T, Liblikas I, Almaas TJ, Borg-Karlsson A-K, Mustaparta H (2003) Receptor neurones in three heliothine moths responding to floral and inducible plant volatiles. *Chemoecology* **13**:143–154.

Surlykke A (1984) Hearing in Notodontid moths: a tympanic organ with a single auditory neuron. *J. Exp. Biol.* **113**:323–336.

Surlykke A, Treat AE (1995) Hearing in winter moths. *Naturwissenschaften* **82**:382–384.

Svensson GP, Löfstedt C, Skals N (2004) The odour makes the difference: male moths attracted by sex pheromones ignore the threat from predatory bats. *Oikos* **104**:91–97.

Sweeney A, Jiggins C, Johnsen S (2003) Polarized light as a butterfly mating signal. *Nature* **423**:31–32.

Takken W, Knols BGJ (1999) Odor-mediated behaviour of afrotropical malaria mosquitoes. *Annu. Rev. Entomol.* **44**:131–157.

Todd JL, Baker TC (1999) Function of peripheral olfactory organs. In: *Insect Olfaction*, Hansson BS, Ed., Springer-Verlag, Berlin, pp. 67–96.

Tumlinson JH, Klein MG, Doolittle RE, Ladd TL, Proveaux AT (1977) Identification of the female Japanese beetle sex pheromone: inhibition of male response by an enantiomer. *Science* **197:**789–792.

van Loon JJA, Dicke M (2001) Sensory ecology of arthropods utilizing plant infochemicals. In: *Ecology of Sensing*, Barth FG and Schmidt A, Eds., Springer-Verlag, Berlin, pp. 253–270.

van Staaden MJ, Römer H (1997) Sexual signalling in bladder grasshoppers: tactical design for maximizing calling range. *J. Exp. Biol.* **200**:2597–2608.

van Staaden MJ, Römer H (1998) Evolutionary transition from stretch to hearing organs in ancient grasshoppers. *Nature* **394**:773–776.

Vickers NJ, Baker TC (1994) Reiterative responses to single strands of odor promote sustained upwind flight and odor source location by moths. *Proc. Natl. Acad. Sci. USA* **91**:5756–5760.

Vilhelmsen L, Isidoro N, Romani R, Basibuyuck HH, Quicke DLJ (2001). Host location and oviposition in a basal group of parasitic wasps: the subgenual organ, ovipositor apparatus and associated structures in the Orussidae (Hymenoptera, Insecta). *Zoomorphology* **121**:63–84.

Visser JH (1986) Host odor perception in phytophagous insects. *Annu. Rev. Entomol.* **31**:121–144.

Vorobyev M, Menzel R (1999) Flower advertisement for insects: bees, a case study. In: Archer SN, Djamgoz MB, Loew E, Partridge JC and Vallerga S, Eds., *Adaptive Mechanisms in the Ecology of Vision*, Kluwer Academic Publishers, Dordrecht, the Netherlands, pp. 537–553.

Wäckers FL, Mitter E, Dorn S (1998) Vibrational sounding by the pupal parasitoid *Pimpla (Coccygomimus) turionellae*: an additional solution to the reliability–detectability problem. *Biol. Contr.* **11**:141–146.

Walker MM, Bitterman ME (1985) Conditioned responding to magnetic fields by honeybees *Apis mellifera*. *J. Comp. Physiol. A* **157**:67–72.

Warrant EJ (1999) Seeing better at night: life style, eye design and the optimum strategy of spatial and temporal summation. *Vision Res.* **39**:1611–1630.

Warrant EJ (2001) The design of compound eyes and the illumination of natural habitats. In: *Ecology of Sensing*, Barth FG and Schmidt A, Eds., Springer-Verlag, Berlin, pp. 187–214.

Warrant EJ, Porombka T, Kirchner WT (1996) Neural image enhancement allows honeybees to see at night. *Proc. R. Soc. Lond. B* **263**:1521–1526.

Wehner R (2001) Polarization vision: a uniform sensory capacity? *J. Exp. Biol.* **204**: 2589–2596.

Wehner R (2003) Desert ant navigation: how miniature brains solve complex tasks. *J. Comp. Physiol. A* **189**: 579–588.

Wehner R, Bernard GD (1993) Photoreceptor twist: A solution to the false-color problem. *Proc. Natl. Acad. Sci. USA* **90**:4132–4135.

Weiss MR (1997) Innate colour preferences and flexible colour learning in the pipevine swallowtail. *Anim. Behav.* **53**: 1043–1052.

White RH, Xu H, Münch TA, Bennett RR, Grable EA (2003) The retina of *Manduca sexta*: rhodopsin expression, the mosaic of green-, blue-, and UV-sensitive photoreceptors, and regional specialization. *J. Exp. Biol.* **206**:3337–3348.

Wibe A, Borg-Karlsson A-K, Norin T, Mustaparta H (1997) Identification of plant volatiles activating single receptor neurons in the pine weevil (*Hylobius abietis*). *J. Comp. Physiol. A* **180**:585–595.

Wyatt TD (2003) *Pheromones and Animal Behaviour: Communication by Smell and Taste.* Cambridge University Press, Cambridge, U.K.

Yack JE, Scudder GGE, Fullard JH (1999) Evolution of the metathoracic tympanal ear and its mesothoracic homologue in the Macrolepidoptera (Insecta). *Zoomorphology* **119**:93–103.

Yager DD (1999) Structure, development, and evolution of insect auditory systems. *Microsc. Res. Tech.* **47**:380–400.

Yager DD, Hoy RR (1986) The cyclopean ear: a new sense for the praying mantids. *Science* **231**:727–729.

Zhang Q-H, Schlyter F (2004) Olfactory recognition and behavioural avoidance of angiosperm nonhost volatiles by conifer-inhabiting bark beetles. *Agr. Forest Entomol.* **6**:1–19.

Zuk M, Kolluru GR (1998) Exploitation of sexual signals by predators and parasitoids. *Quart. Rev. Biol.* **73**: 415–438.

Section II

Mechanosensation and Audition

3 The Chordotonal Organ: A Uniquely Invertebrate Mechanoreceptor

Lawrence H. Field

CONTENTS

0-8493-2024-0/05/$0.00+$1.50

3.1 INTRODUCTION

Chordotonal organs are internal proprioceptors that have evolved only in insects and crustaceans. Although they serve as versatile stretch receptor organs in these classes, their sensory role is greater, owing to their extraordinary sensitivity to micromechanical displacement (Figure 3.1). These sense organs still contain unsolved mysteries of operation: unknown mechanisms of transduction, unusual and little-known spike potentials and unitary receptor "quantum bumps," extraordinary partitioning of stimulus specificity within single organs, and the possibility of several unusual molecular mechanical events involved in transduction and sensitivity (Field and Matheson, 1998). Such features contribute to their scientific appeal. Earlier reviews provided limited coverage of chordotonal organs under the topics of insect or arthropod mechanoreception (Finlayson, 1968; Rice, 1975; Moulins, 1976; Wales, 1976; Wright, 1976; McIvor, 1985). A number of these reviews formed chapters in the excellent book by Mill (1976), which gives a comprehensive illustrative coverage of chordotonal organ anatomy, histology, ultrastructure, and physiology. This mate-

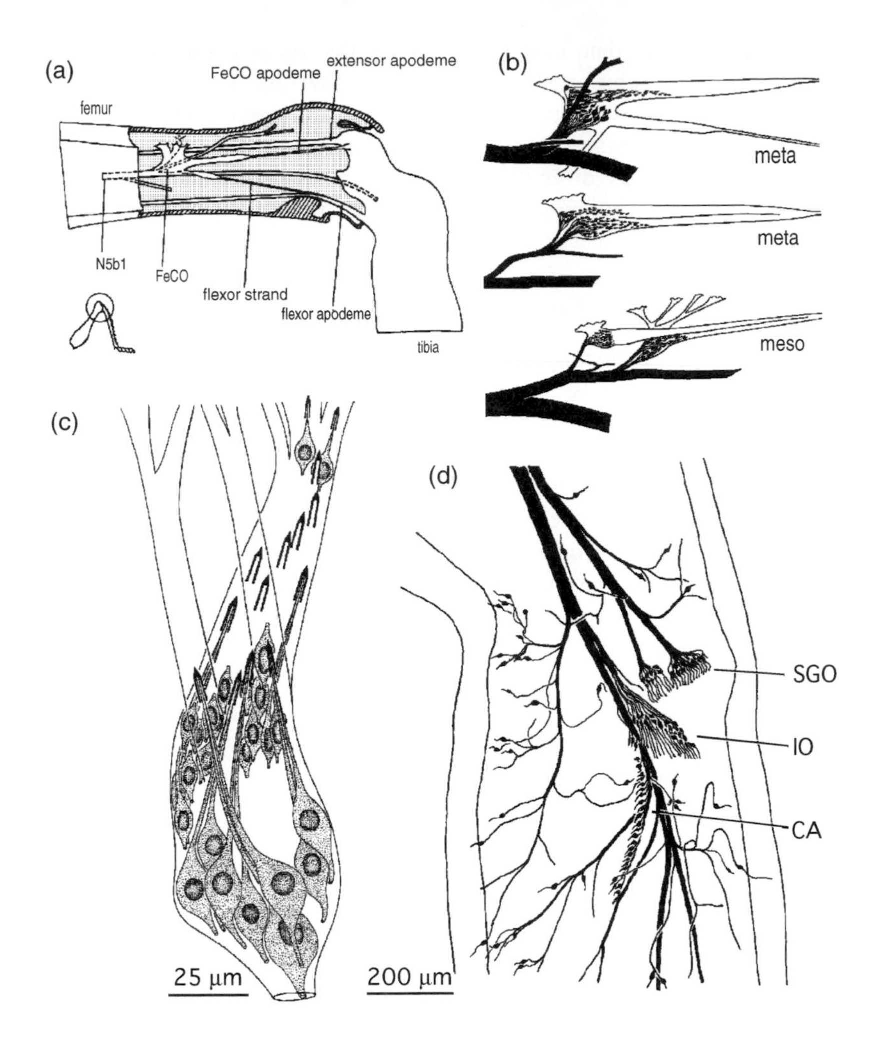

FIGURE 3.1 Connective and nonconnective chordotonal organ gross morphology. **a–c**: Connective chordotonal organs; **d**: Nonconnective chordotonal organ. **a**: Femoral chordotonal organ (FeCO) in distal femur (see inset) of locust hind leg, showing attachment across joint via ligament connected to FeCO apodeme, and additional flexor strand containing a single strand organ. **b**: Variations in fusion of scoloparia (sensory neuronal subgroups) and connective ligaments of joint chordotonal organs (FeCO) from different legs and orthopteran species: upper, locust; middle, New Zealand tree weta (Anostostomatidae); lower, locust. **c**: Tibio-tarsal chordotonal organ of cockroach showing subgroups of scolopidial neurons embedded in connective tissue strands that span the tibio-tarsal joint. **d**: Nonconnective chordotonal organs in proximal region of hind tibia of New Zealand tree weta. These scoloparia, composing the complex tibial organ, include the subgenual organ (SGO), the intermediate organ (IO), and the crista acustica (CA). They lack long connective tissue ligaments and instead are intimately associated with tracheae (CA, IO) or a septate mass occluding the hemocoel (SGO). (**a, b** from Field, L.H. and Matheson, T., *Adv. Insect Physiol.* **27**:1–228, 1998. With permission. **c** from Young, D., *Phil. Trans. Roy. Soc. Lond. B* **256**:401–426, 1970. With permission. **d** from unpublished data of author.)

rial was brought up to date in a recent review that focused on insect chordotonal organs (Field and Matheson, 1998). Extensive coverage was provided for many aspects of these sense organs, including newly developed genetic approaches to studying chordotonal organ function and structure. Here, a descriptive overview of anatomy and terminology, diversity, and function will be presented to lay the groundwork for discussing techniques used to study the structure and sensory physiology of chordotonal organs.

3.1.1 Anatomy of the Chordotonal Organ

Chordotonal organs are distinguished from external cuticular *sensilla* (hairs, bristles, campaniform sensilla) by the presence of a subcuticular *scolopidium,* as well as by their embryonic derivation (for further discussion, see Field and Matheson, 1998). Each scolopidium consists of a bipolar sensory neuron, an attachment cell, a specialized scolopale cell, and a glial cell. Each organ comprises a cluster of scolopidia (the scoloparium), or sometimes a single scolopidium, embedded in connective tissue with additional accessory structures, which serve to anchor the organ and to amplify, attenuate, or channel the mechanical stimulus optimally (Figure 3.2). These organs are found at almost every joint in the exoskeleton (Figure 3.1) and also between joints in certain limb and body segments (Field and Matheson, 1998). They serve proprioceptive functions concerned with joint movement and position (Field and Burrows, 1982), as well as highly specialized mechanosensory functions, such as hearing and vibration detection (Suga, 1960; Schnorbus, 1971). Chordotonal organs can be classified into two main types:

1. Those at joints normally incorporate a connective tissue strand that spans the joint (Figure 3.1a,b,c); these are termed *connective chordotonal organs* (Moulins, 1976).
2. Those remote from joints are often associated with cuticular or tracheal specializations and are termed *nonconnective chordotonal organs* (Figure 3.1d).

3.1.2 The Scolopidium

The four scolopidial cells make up a lineage derived from a single epithelial sensory mother cell, which typically undergoes three divisions (Bodmer et al., 1989), but which may divide up to six times if the scolopidium contains two sensory neurons. The sensory neuron soma is enveloped by the glial cell, whereas the dendrite terminates in a modified cilium surrounded by the scolopale cell (Figure 3.2a). The latter secretes a set of stiff rods (scolopale rods) that form a longitudinal sleeve (scolopale) around the cilium, extending from the ciliary base to its tip. The scolopale is anchored to the dendrite basally by strong belt desmosomes. Distally, the scolopale inserts into a dense extracellular cap, which encloses the tip of the cilium, and to which the attachment cell forms a link to the cuticle. The scolopale is considered to form a stiff, barrellike enclosure holding the cilium taut between the cap and the base of the scolopale. Often the attachment cells contribute to one or more viscoelastic ligaments, which span a joint and are stretched by movement of the joint (Figure 3.1b). Variations occur in the number of neurons and the number and type of cilia in the scolopidium, form and presence or absence of a cap, extent of ciliary roots projecting into the neuron, and morphology of scolopale rods (Figure 3.2c,d).

3.1.3 Terminology

Terms frequently used to describe the different types of scolopidia are not always defined by contemporary authors. The definitions in Table 3.1 are assembled from reviews by Howse (1968) and Moulins (1976).

3.2 DISTRIBUTION, DIVERSITY, AND SENSORY MODALITIES SERVED

Although the variation in details of the various insect chordotonal organs is exhaustive, they tend to conform to general types characterized by anatomical location and function. These types are reviewed in the following sections and are elaborated where physiological knowledge is greatest.

3.2.1 Head and Associated Appendages

In mouthpart appendages, connective chordotonal organs occur in joints associated with the maxilla, labium, and mandibles (Wales, 1976). Extremely beautiful, detailed illustrations have been published for mouthparts of the cricket, *Gryllus domesticus* (Rosciszewska and Fudalewicz-Niemczyk, 1972). The lacinia of the maxilla contains a chordotonal organ with two scoloparia; in the maxillary palp, a single scoloparium occurs in the first segment, and a second occurs in the fourth segment. The labium contains three chordotonal organs with two scoloparia each: in the glossa, the paraglossa, and the second segment of the palp.

Presumably, mouthpart chordotonal organs code for joint position and movement. However, an unusual connective chordotonal organ that does not span a joint is the *apical sensory organ* (ASO) (Lee and Altner, 1986). This organ inserts into the tip of the labial and maxillary palps and is thought to code for the mechanical tapping or vibration made by the palps when deployed (Lee et al., 1988). In certain moths (Sphingidae), the palp pilifer contains a chordotonal organ modified to act as a hearing organ for the detection of bat sounds (Roeder et al., 1970).

The antennae contain chordotonal organs in the pedicel and scape, but not in the flagellum. Connective chordotonal organs span the joints distal to the pedicel and scape, with the ligament attached distally in both cases. Thus the pedicel chordotonal organ is attached to the flagellar base and lies within the *Johnston's organ* (described subsequently and in Chapter 2), as exemplified in the Neuroptera (Schmidt, 1969) and Orthoptera (Field et al., 1994). Little is known of the physiology of these organs; ultrastructurally, they contain only *mononematic* scolopidia (Table 3.1). These organs are morphologically different from the well-known Johnston's organ of the pedicel of most insect orders (Bode, 1986). The Johnston's organ is of the nonconnective type, is composed mainly of *mononematic heterodynal* scolopidia (Table 3.1), and is most highly elaborated in the Diptera, where up to 20,000 neurons may be found within the swollen pedicel (Field and Matheson, 1998). It consists of

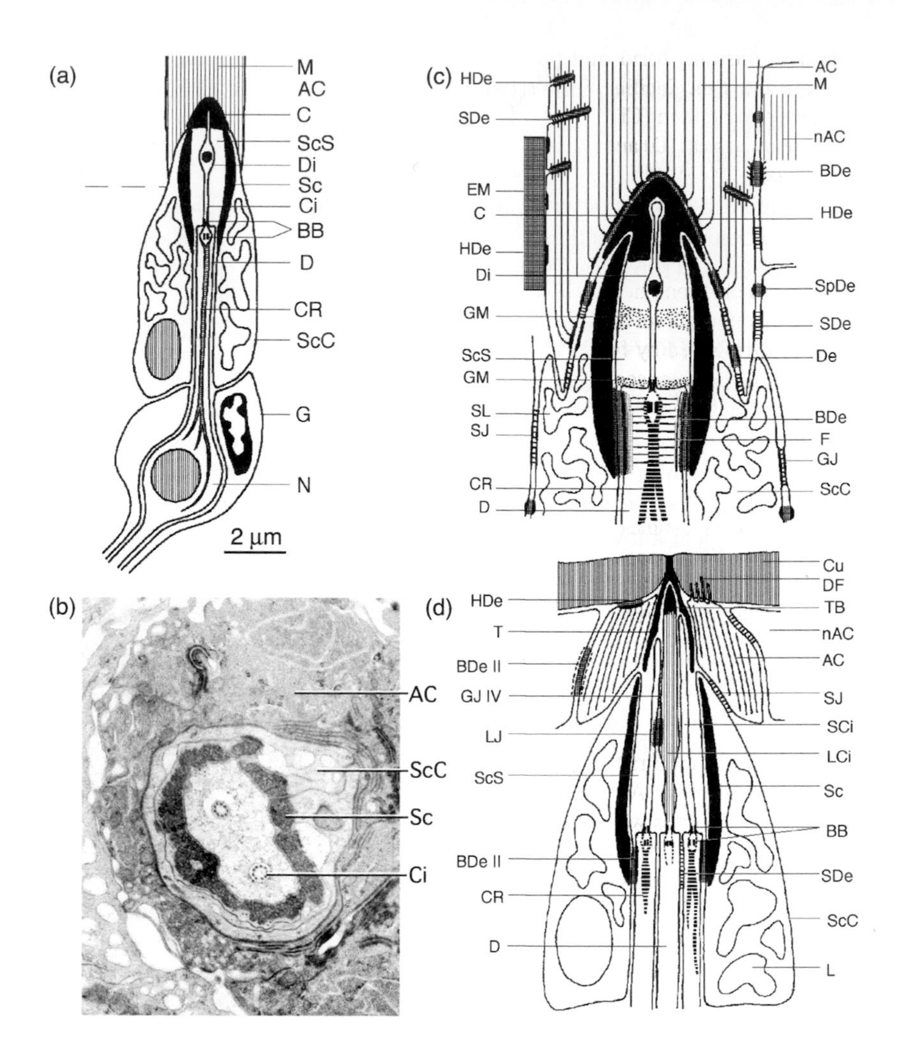

FIGURE 3.2 See caption on page 67.

a circular array of scolopidia inserted into a system of pleated prongs radiating from the base of the antennal flagellum (Belton, 1989). The Johnston's organ is used in male mosquitoes and midges to detect near-field air particle movement caused by the female's wingbeat sound. In honeybees, the 260 Hz sound generated during the waggle dance is detected by the Johnston's organ in foraging hive mates (Dreller and Kirchner, 1993).

3.2.2 Thorax

Three types of chordotonal organ occur in the thorax. Two are associated with joints and are of the connective type. These include (a) myochordotonal organs, in which scolopidial neurons are associated with small muscles, and (b) complex connective chordotonal organs at the leg bases. A third type includes a variety of sound-

FIGURE 3.2 Histology and ultrastructure of mononematic and amphinematic scolopidia, with Type 1 and Type 2 cilia (see Table 3.1 for definitions). **a**: Mononematic scolopidium with single Type 1 cilium (Ci). Four cells are shown: the bipolar sensory neuron (N) with an inner dendritic segment (D) and the outer dendritic segment (cilium). A glial cell (G) surrounds the neuron soma; the dendritic segments are surrounded by the highly vacuolated scolopale cell (ScC), which secretes the extracellular scolopale (Sc) around the cilium. Distal to the ciliary dilation (Di), the cilium inserts into an extracellular cap (C), which is produced by the microtubule (M) filled attachment cell (AC). Other abbreviations explained subsequently. **b**: Transmission electron micrograph of section through mononematic scolopidium with two cilia taken at about the level of dashed line in **a**. The cilia have a 9 + 0 axoneme configuration and are fully surrounded by fused scolopale rods forming the enclosing scolopale. The latter is surrounded by the paler scolopale cell. At this distal level, the attachment cell overlaps the scolopidium more than shown in **a.** The dark appearance of this cell is due to the high density of microtubules. **c–d**: Ultrastructure of mononematic (**c**) and amphinematic (**d**) scolopidia showing the variety of abundant intercellular junctions and details of dendritic segments. **c**: Neighboring attachment cells (nAC) are fused by two types of belt desmosomes (BDe and Bde II), septate desmosomes (SDe), and spot desmosomes (SpDe). Desmosomes (De) and hemidesmosomes (HDe) anchor the attachment cell to the cap and scolopale cell, and occasionally to the extracellular matrix (EM). Note the insertion of the cilium and scolopale rods into the cap without apparent attachment. The ciliary dilation (Di) often is associated with granular material (GM), seen in **b**. The scolopale space (ScS) is an extracellular region thought to have ionic content regulated by the scolopale cell, which itself is highly vacuolated with a scolopale labyrinth (SL). Neighboring scolopale cells are joined by septate junctions (SJ). The scolopale is strongly anchored to the inner dendritic segment (D) by belt desmosomes (BDe), which give rise to radial filaments (F) connecting to the ciliary root (CR) and basal body (BB in **a** and **d**). **d**: Amphinematic scolopidium typical of those in Johnston's organ, containing three dendrites with Type 1 (SCi) and Type 2 (LCi) cilia. The latter terminates in a dense tubular body (TB) enclosed by an extracellular tube (T) and may be anchored to adjacent cilia by Type IV gap junctions (GJIV) and laminate junctions (LJ). Note variations in ciliary roots. Additional abbreviations: Cu, cuticle; DF, dense filaments; L, labyrinth. (From Field, L. and Matheson, H.T., *Adv. Insect Physiol.* **27**:1–228, 1998. With permission.)

Table 3.1
Terminology Used for Insect Scolopidia

Monodynal	Scolopidia with a single neuron and cilium inserting into the cap
Heterodynal	Scolopidia with two or three cilia inserting into a single cap
Mononematic	Scolopidia with a clearly subepidermal cap into which a scolopale and one or more cilia insert
Amphinematic	Scolopidia that lack a cap, and in which the cilium is surrounded by a dense tubular sheath that is narrowed into a thread inserted into the cuticle or subepidermal tissue
Type 1	Scolopidia with uniform diameter cilium containing a 9 + 0 axoneme array of microtubules and inserted into the cap or tubular sheath
Type 2	Scolopidia with cilium containing 9 + 0 axoneme proximally, but which enlarges distally to become a microtubule-filled cylinder enclosed by the attachment cell

receiving, nonconnective chordotonal organs that are invariably associated with a cuticular tympanum and often with modified tracheal air sacs.

3.2.2.1 Neck

A single myochordotonal organ is found on either side of the neck, associated with a lateral longitudinal muscle 54 (Shepheard, 1973). Ablation of the elastic connectives of these organs demonstrates that they assist in postural motor control of the neck, as well as provide major proprioceptive feedback for controlling fast optokinetic movements of the head in the locust (Shepheard, 1973).

Two other connective chordotonal organs occur in the same ganglionic nerve root (Nerve II) of the locust: the anterior chordotonal organ (aCO) and the ventral chordotonal organ (vCO). In the prothoracic segment, the bilateral pair of aCOs connects to a unique medial cuticular plate (the cervicosternite) floating in the ventral neck membrane. These organs act both as neck movement detectors and as primitive ears (Pflüger and Field, 1999). To allow the latter function, the neck membrane is stretched taut by the probasisternite in the region containing the cervicosternite and thus acts as a tympanum that responds to sound energy (Figure 3.3). Novel chordotonal organs that also act as tympanal organs have been found recently in the dorsal

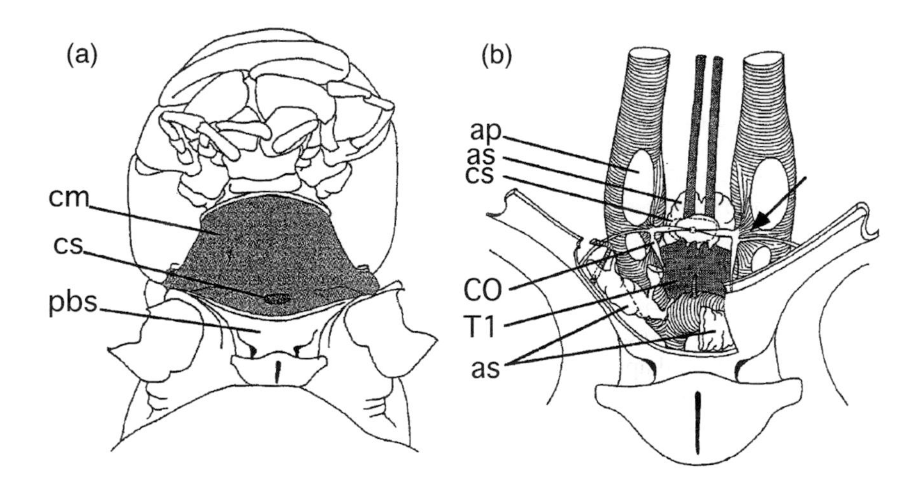

FIGURE 3.3 A prothoracic connective chordotonal organ in the locust neck acts as both a proprioceptor and a sound receptor (third ear). **a**: Ventral view of locust neck indicating location of cervicosternite (cs), a cuticular plate in taut region of cervical membrane (cm) bounded by probasisternite (pbs). The bilaterally paired anterior chordotonal organ (aCO) of the prothorax connects to the cervicosternite and monitors gross movement of the neck membrane as well as sound-induced vibration of the cervicosternite. **b**: Underlying anatomy reveals paired aCO and ventral chordotonal organ (vCO) forming an open V-shape (CO) with apex on longitudinal trachea and embedded among air sacs (as), which have been mostly removed, leaving openings (ap). Each aCO arises from Nerve II of the prothoracic ganglion (T1) and connects medially to the cervicosternite (cs). (From Pflüger, H.-J. and Field, L.H., *J. Comp. Physiol. A* **184**:169–183, 1999. With permission.)

neck region of tiger and scarab beetles (Spangler, 1988a,b; Yager and Spangler, 1995).

Two genera of parasitic flies (Diptera: Tachinidae) have a highly unusual modification of the prosternal neck membrane into an inflated tympanal organ. Within the inflated chamber is a nonconnective chordotonal organ that is particularly tuned to the calling-song spectrum (including ultrasonic frequencies) of the fly's orthopteran host. Female tachinids have much larger prosternal tympanal organs, which aid in location of the calling host, whereon she deposits maggots (Lakes-Harlan and Heller, 1992; Robert et al., 1994).

3.2.2.2 Wings

The radial vein in some pterygote insects, or radial + medial vein in crickets, contains one chordotonal organ (Fudalewicz-Niemczyk and Rosciszewska, 1972). When present, this occurs in the anterior wing as a small number of scolopidial neurons (10 in *Gryllus domesticus*), but in Orthoptera, the chordotonal organ is lacking in the posterior wing. Tettigonoidea and Grylloidea have the chordotonal organ, but Acridoidea appears to lack it.

In many insect orders, a small connective chordotonal organ, associated with a multiterminal stretch receptor cell, occurs at each wing base in thoracic segments 2 and 3. These are homologs of segmental chordotonal organs that occur serially along the abdomen (reviews: Field and Matheson, 1998; Prier and Boyan, 2000). While the multiterminal receptor detects wing movement and feeds back into the flight motor center, the wingbase chordotonal organs appear to be sensitive mostly to vibrations and low-frequency acoustic input (Möss, 1971; Pearson et al., 1989). In some moths (Lepidoptera), the wingbase chordotonal organ is associated with a tympanum and acts as a hearing organ rather than a wing proprioceptor (Yack and Fullard, 1993). The wingbase of acridid orthopterans also contains a cupola-shaped projection, the *tegula*, bearing a nonconnective chordotonal organ and external hair sensilla (see also Chapter 4). This organ responds to downward movement of the wing and is not involved in hearing; rather, it responds to phase and velocity of the wing downstroke and appears to control wing angular velocity (Kutsch et al., 1980; Fischer et al., 2002).

In the Diptera (flies and mosquitoes), the hind wings are reduced to slender stalks (*halteres*) invested with campaniform sensilla and two nonconnective chordotonal organs at the base (see also Chapter 4). The halteres move in synchrony with the wings during flight and act as a gyroscopic detector of angular velocity and acceleration through responses from the campaniform sensilla. No function is reported for the chordotonal organs, but because they have the same kind of scolopidia as found in Johnston's organ (Type 2; Table 3.1), they may well be vibration detectors (reviewed by Nalbach, 1993).

An unusual tympanal organ is found in the radius vein of wings of Neuroptera (lacewings), in which the nonconnective chordotonal organ resides in a fluid-filled structure sensitive to ultrasound (Miller, 1970). This is likely to represent a modification of the radial vein chordotonal organ found in other insect groups. The anatomy with a fluid-filled chamber is an exception to the general scheme of having an air-filled cavity behind the tympanum of insect hearing organs (Stumpner and von Helversen, 2001).

3.2.2.3 Tympanal Organs

Many chordotonal organs are modified into sound-receiving *tympanal organs*. These occur in the thorax and in many other sites in the insect body (Figure 3.4). Thoracic tympanal organs are especially well developed in Orthoptera, Lepidoptera, Homoptera, and Hemiptera (reviews: Michelsen and Larsen, 1985; Yager, 1999; Stumpner and von Helversen, 2001; Yack, 2004). Recent discoveries have expanded the list to Coleoptera, Diptera, and Dichtyoptera (Spangler, 1988a,b; Yager, 1989; Lakes-Harlan and Heller, 1992). In almost all of these organs, the anatomy consists of a nonconnective chordotonal organ associated with a tympanum and an air sac or chamber. Evidence from developmental, physiological, and comparative studies suggests that these tympanal organs (and others in the abdomen) are evolutionarily derived from connective chordotonal organ precursors. The transition with intermediate stages is represented in some moth species, in the primitive grasshopper *Bullacris,* and in the bifunctional locust aCO (von Staaden and Römer, 1998; review: Pflüger and Field, 1999; Yager, 1999).

Medial thoracic (cyclopean) tympanal organs are found in praying mantids (Dictyopteridae). Comparisons across a large number of species have shown that considerable variation occurs in sexual dimorphism, structural and histological complexity, and reduction in vestigial organs (Yager, 1989, 1990) The medial ear lies between the metathoracic legs and is broadly sensitive to ultrasound, thus allowing mantids to avoid bat predation. The anatomy is sufficiently different from other insect thoracic tympanal organs to suggest a separate evolutionary origin (Yager, 1999).

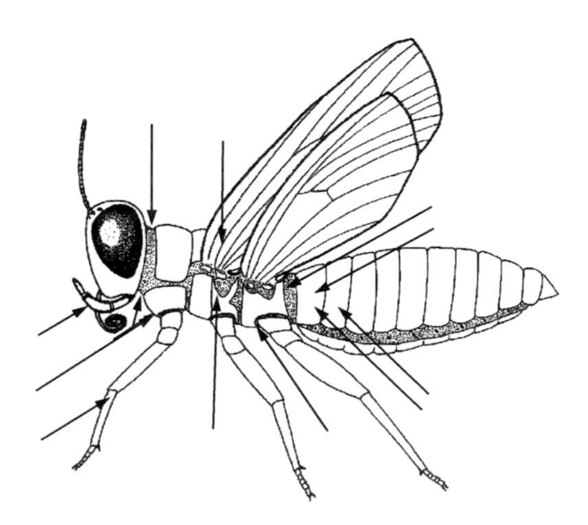

FIGURE 3.4 Generalized diagram of an insect indicating 12 anatomical sites where tympanal organs (functional ears) have been described for various species. (Modified from Yack, J.E. and Fullard, J.H., *Ann. Ent. Soc. Amer.* **86**:677–682, 1993. With permission.)

3.2.2.4 Thorax–Coxa Region

A complex array of up to seven connective chordotonal organs is associated with the ventral region of the thorax and the coxa of the leg (Hustert, 1982). Within these, the sternite chordotonal organs, aCO, vCO, and an apodeme chordotonal organ (apCO), are not connected to the leg. The aCO in the locust is modified into a hearing organ, whereas the vCO and apCO attach to the sternite and sternal apophysis. No functions have been ascribed to the latter two organs (Bräunig et al., 1981). The coxa of the leg is monitored by three chordotonal organs: the anterior joint chordotonal organ (ajCO), the posterior joint chordotonal organ (pjCO), and the coxal chordotonal organ (cCO) (reviewed in detail in Field and Matheson, 1998). Their distribution, presence or absence, and attachment sites differ among the three thoracic segments. The organs respond to different aspects of the three-dimensional motion of the coxa, and they mediate reflex activation of coxal muscles in motor control of the leg (Hustert, 1983). Finally, a myochordotonal organ occurs in the locust pro- and metathorax, where it monitors a functionally different muscle in the two respective segments. Nothing is known of its sensitivity or role in motor control (Bräunig, 1982; Field and Matheson, 1998).

3.2.3 Abdomen

The chordotonal organs in the abdomen comprise the pleural (lateral) connective chordotonal organs, and ventrolateral (stick insects) and midventral (cockroaches) chordotonal organs (e.g., Pill and Mill, 1981). These monitor ventilatory movements and are phasic detectors (Orchard, 1975). Specialized nonconnective chordotonal organs are usually modified into tympanal organs in certain orders (Orthoptera, Homoptera, Lepidoptera, Blattaria, Isoptera, Coleoptera, and Diptera) (Field and Matheson, 1998). A newly discovered chordotonal organ associated with the heart provides the first instance of an insect cardiac chordotonal organ (Dulcis and Levine, 2004). Specific reviews of abdominal chordotonal organs include Finlayson (1976), Michelsen and Larsen (1985), and Yack and Fullard (1993).

The cardiac chordotonal organ has two pairs of sensory neurones that project dendrites into a ligament spanning the caudal cardiac chamber and the dorsal body wall of the terminal abdominal segment. Thus, the chordotonal organ is of the connective type. The cardiac chordotonal organ was described in the moth *Manduca sexta*, where it appears to provide strong inhibitory input to one of the cardiac motorneurons and thereby induces reversal of the anterograde heartbeat (Dulcis and Levine, 2004). The chordotonal organ is unusual since the insect heart has been thought to lack intrinsic mechanoreceptors.

3.2.4 Legs

The legs of insects contain connective chordotonal organs in all major joints except the coxa–trochanter joint (which is monitored by a multiterminal Type II receptor) and the trochanter–femur joint, which is fused in most insects (Bräunig et al., 1981; Field and Matheson, 1998). In addition, the legs contain the nonconnective subgenual organ for vibration reception and the *crista acustica,* which is elaborated for sound reception in prothoracic tympanal organs (exemplified in Orthoptera).

3.2.4.1 Joint Chordotonal Organs

Three organs are usually found in all legs: the femoro-tibial organ (FeCO), the tibio-tarsal organ (TiCO), and the tarso-pretarsal organ. The best studied is the FeCO in Orthoptera (Figure 3.2a,b). It is known to comprise three scoloparia (Nishino and Sakai, 1997), which may or may not be fused in different taxa and different legs (Figure 3.5). At least one scoloparium mediates postural resistance reflexes that

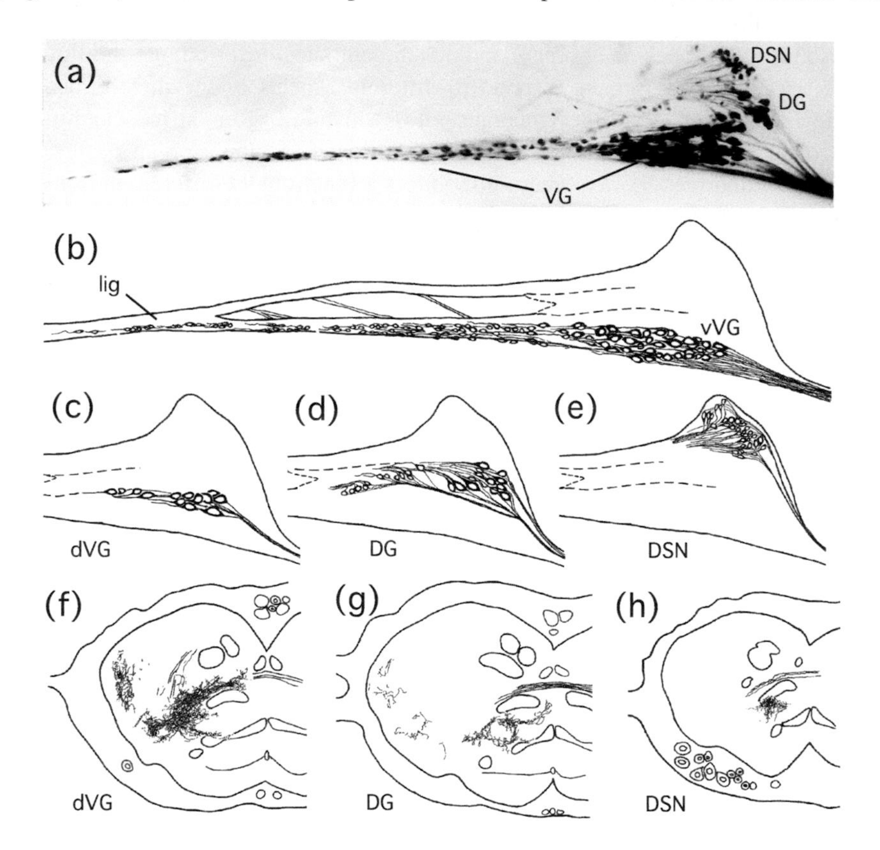

FIGURE 3.5 Anatomically differentiated subgroups in FeCO scoloparia also imply functional differentiation. **a**: $NiCl_2$–$CoCl_2$ backfilling of the metathoracic FeCO of the New Zealand tree weta reveals the dorsal scoloparium neurons (DSN), the dorsal group (DG), and the ventral group (VG). The ventral group is subdivided into the ventral subgroup (vVG, **b**) and the dorsal subgroup (dVG, **c**). **c–e**: Each subgroup arises from a distinct axon bundle. **f–h**: Different afferent projection patterns to the same level of the metathoracic ganglion support the peripheral distinction into subgroups. **f**: dVG axons project to the dorsolateral margin of the medial ventral association center; **g**: DG axons project deep into the lateral, medial, and ventral aspects of the mVAC; **h**: DSN axons project to a confined central region of the medial ventral association center (mVAC). (Modified from Nishino, H. and Field, L.H., *J. Comp. Neurol.* **464**:327–342, 2003. With permission.)

stabilize the joint during phasic perturbances (Burns, 1974); another appears to detect air vibration associated with the tracheal system (Field and Pflüger, 1989; Nishino et al., 1999). The FeCO also participates in reflex control of most other leg muscles, although the detailed circuitry is not understood (Field and Rind, 1981). The TiCO monitors tarsal movement and position and also controls resistance reflexes in the tarsal levator and depressor muscles (Young, 1970; Laurent, 1987). Little is known about the function of the pretarsal chordotonal organ. It has been described for species in four orders, but the physiology has not been studied (Field and Matheson, 1998). It holds interest for investigators because in Lepidoptera, at least, it has two scoloparia separately innervated by *amphinematic* and *mononematic* scolopidia (Table 3.1) (Faucheux, 1985). This may offer the possibility of studying the functional difference between these scolopidial types.

3.2.4.2 The Nonconnective Chordotonal Organs

The nonconnective chordotonal organs occur as a group of two or three, located in the proximal tibiae of all legs (with the possible exception of the Diptera and Coleoptera) (Howse, 1968). They are collectively known as the *complex tibial organ* (CTO), which includes the subgenual organ (SGO), the *crista acustica* (CA), and the *intermediate organ* (IO) (detailed descriptions and variety reviewed in Field and Matheson, 1998). A fourth collection of scolopidia (AO) is closely associated with the crista acustica (e.g., Blattodea, Gryllacridoidea [Rhaphidophoridae], Caelifera), and sometimes the intermediate organ is missing (Cökl et al., 1995).

The SGO consists of a transverse septum, usually membranous, that partially occludes the hemocoel in the proximal tibia. It is innervated by a semicircular array of mononematic scolopidia. The function is to detect substrate vibrations conducted through the legs to the tibia (Schnorbus, 1971, McVean and Field, 1996). Although many reports indicate that the septum is thin, in the primitive New Zealand tree weta (Orthoptera: Anostostomatidae) it is a thick, pillowlike structure (Nishino and Field, 2003). This would yield a different vibratory response mode from that which must be produced in a thin septum, which in turn, implies interesting differences in mechanical operation of the SGO among the insect families.

Nothing is known about the details of the physical induction of vibration of the septum, nor about the mechanisms of transduction in this system. Shaw (1994a) presented evidence that the SGO is closely associated with tracheation in the tibia and is sensitive to airborne vibratory energy. It is not clear how the energy pathways are apportioned between tracheae and tibial hemolymph. Recent studies (Mason et al., 1999; Pollack and Imaizumi, 1999) confirm Shaw's work (Shaw, 1994b) and show that the SGO includes not only vibration detectors but also (in Ensifera) low-frequency, high-threshold acoustic receptors that project to the acoustic neuropil. Other groups of SGO vibration detectors project to different neuropilar sites that also receive vibration-sensitive afferents from the dorsal scoloparium neurons of the FeCO (joint chordotonal organ). The latter implicates the SGO in motor control functions (Nishino and Field, 2003).

The CA has been intensively studied, as it is the acoustic detector in tibial tympanal organs found in the forelegs of many Orthoptera. A rich literature on insect hearing and communication, including the physiology of the CA, is reviewed elsewhere (e.g., Yager, 1999; see also Chapter 5). The CA mononematic scolopidia occur in a linear array with decreasing size along the top of the anterior trachea in the tibia. The trachea, in turn, is usually intimately joined to one or two tympana of thin cuticle (reviewed in Field and Matheson, 1998; Yager, 1999). The decrease in scolopidial size is associated with an increase in high frequency sensitivity (tonotopic organization: Oldfield, 1982; Lin et al., 1993), as in vertebrate cochlea. It is not clear to what extent the tuning of the receptors is based on local resonances compared with properties of individual scolopidia, and conflicting hypotheses have been proposed (Oldfield, 1985a,b; Kalmring et al., 1993).

Less elaborate CA organs are found in the meso- and metathoracic legs of Ensifera, where tympanal membranes or cuticular specializations are lacking. These reduced organs have been called tracheal organs (Ball and Young, 1974), inasmuch as the legs lack cuticular tympana. The meso- and metathoracic CA (and IO) are apparently sensitive to vibrations and low-frequency sound (as judged by indirect inference of origin from axonal recordings at nerve roots) (Rössler, 1992; Kalmring et al., 1994; Cökl et al., 1995). They are associated with tracheae in the tibia, which may provide a vibratory pathway initiated through the cuticle or hemolymph. Little information is available on details of excitation mechanisms in these chordotonal organs.

Scolopidial neurons in the IO are sensitive to low-frequency sound at high intensity (Kalmring et al., 1993) and are known to project partially to the same auditory neuropilar area as the axons of the CA (the posterior mVAC). IO also projects to a different area of the mVAC, which may indicate a different functional role for this component of the CTO (Nishino and Field, 2003).

The AO neurons have been little studied, but their projection to the dorso-anterior mVAC overlaps with that of some SGO neurons, whereas other AO projections overlap with projections of joint chordotonal organ vibration-sensitive neurons in the motor association area (LAC). These interesting results suggest a motor control role for vibratory information deriving from all three chordotonal organs (Nishino and Field, 2003). More research is required to understand fully the apparently diverse functions of the CTO.

3.3 VISUALIZING CHORDOTONAL ORGANS

Although often neglected in reviews, visualization of chordotonal organs is the first step in studying them in dissections and whole-mount and sectioned material. Specific histological stains can preferentially reveal whole chordotonal organs or scolopidia in live insects, and variously colored components in sectioned and whole-mount fixed material. Recent dye uptake and immunochemical techniques have increasingly proliferated to give ever more specific ways of visualizing whole chordotonal organs, scolopidia, and components of scolopidia. These have been reviewed (Field and Matheson, 1998), and the present review updates this knowledge.

3.3.1 Histochemical Staining of Fixed Tissue

The scolopales of chordotonal organs appear to take up preferentially stains that include acid fuchsin and appear in various shades of red or pink. This was first noticed by Young (1970), who used Baker's modification of Masson's Trichrome stain on cockroach material fixed in alcoholic Bouin's fixative. Scolopales in fixed material (e.g., Bouin's fixative) stain red in 1% acid fuchsin (Shelton et al., 1992) and pink in a 0.5% eosin solution in 95% ethanol (H. Nishino, personal communication). Lin et al., (1993) utilized a combination stain of acid fuchsin, analine blue, and orange G to stain crista acustica scolopidia in tettigoniids.

3.3.2 Intravital Staining of Chordotonal Organs

Methylene blue in reduced form gives deep blue staining of insect sensory neurons, as well as varying degrees of axonal staining. It is superb for anatomical and whole-mount studies and for rapid surveys of sensory neuron distribution. The dye is either injected into the hemocoel as a concentrated solution (0.5%) in saline, or the exposed tissue is submerged in a sky blue solution of the 0.5% stock solution, which is exposed to O_2 by leaving tissue near the surface or by bubbling air in the solution. A full schedule is given by Plotnikova and Nevmyvaka (1980).

The sheath and peripheral nerves of insect chordotonal organs are reported to stain with 0.02% Janus Green B while leaving surrounding tissue transparent (Yack and Fullard, 1993). I find that it also stains scolopale caps, connective chordotonal organ ligaments and strands, and attachment cells in the CA of tympanal organs.

3.3.3 Axonal Backfilling and Intracellular Dye Injection

The standard techniques of *in vivo* axonal diffusion (Pitman et al., 1972; Mücke, 1991) of cobalt and nickel salts (acetate, chloride, hexammine, lysine) work well with chordotonal organs (e.g., Figure 3.5a). It is also advantageous to perfuse different salts ($CoCl_2$ and $NiCl_2$) into separate nerves to achieve bicolored stains for differentiation of axonal pathways or peripheral chordotonal cell targets (e.g., Sakai and Yamaguchi, 1983). Heavy metal enhancement of the precipitated metal ions in backfills or intracellular fills greatly improves fine detail of dendrites or axonal projections into the CNS (Tyrer et al., 1980; Mesce et al., 1993). Other dyes used for injection or perfusion include horseradish peroxidase (HRP) (reviewed by Nässl, 1987), the fluorescent dye Lucifer yellow (reviewed by Strausfeld et al., 1983), and a range of carbocyanine fluorescent dyes that move along axon membranes due to their lipid solubility (Haugland, 1996). Multiply-colored fluorescent stains may be achieved with use of fluorescent dyes having different excitation and emission wavelengths (see, e.g., Pflüger and Field, 1999). In particular, lysine-fixable 3000 MW fluorophore-labeled dextrans[14] are attractive because they can be aldehyde-fixed and embedded in plastic for high-resolution histology, as well as photoconverted in the presence of diaminobenzidene (DAB) to yield electron-dense deposits for electron microscopy (Haugland, 1996).

The backfilling technique is not only useful to visualize whole chordotonal organs; a more powerful application has been the recent differentiation of scoloparial organization into recognizable subgroups of neurons that have different, well-ordered projections into the sensory neuropil of the CNS (Figure 3.5b,c) (Nishino, 2000; Nishino and Field, 2003). This approach takes advantage of the anatomical arrangement of separate axon bundles projecting to the scoloparia of chordotonal organs in certain insect species; the individual bundles can be filled to reveal both distal and central organization of subgroups. The resulting topographical maps indicate unsuspected complex relationships of sensory neurons to anatomical sources of mechanical stimulation in, for example, the SGO (Nishino and Field, 2003), and functional complexity as implied by the separate projection patterns within ganglionic sensory neuropil (e.g., the three subdivisions of scolopidial neurons in the FeCO of the weta leg (Figure 3.5f–h) (Nishino and Field, 2003).

3.3.4 Immunocytochemical Labeling

3.3.4.1 Whole Cell Staining

Immunofluorescent labeling of chordotonal organs received impetus from the discovery that antibodies to HRP specifically stain insect neurons (Jan and Jan, 1982). By incubating aldehyde-fixed tissue in fluorescein or rhodamine-coupled antibodies to HRP, cryostat sections can be made of embryos, or appendages of later stages, and viewed with fluorescence microscopy. By subsequent incubation in a dilute solution of HRP, and then exposure of the tissue to 3,3' diaminobenzidine, the tissue can be post-fixed in osmium tetroxide and prepared for electron microscopy. Transmission micrographs show that neuronal and chordotonal cell membranes are specifically labeled (Figure 3.6) (Jan and Jan, 1982).

Another technique for labeling entire sensory neurons utilizes neurobiotin (Horikawa and Armstrong, 1988), a low-molecular-weight molecule with a high affinity for avidin, which can be conjugated to fluorescein, rhodamine, Texas Red, Cy3, or HRP for subsequent visualization. Neurobiotin can be applied to cut chordotonal organ nerves or injected through intracellular micropipettes into sensory cells, and it has a rapid dispersion time. In some cases, it appears better than cobalt for revealing fine branches of projections. The technique takes advantage of the amplification of labeling intensity due to the multiple binding to biotin of avidin molecules, which are conjugated to a variety of fluorophores (fluorescein, rhodamine, Cy3 and 5, etc.). Avidin can also be conjugated to HRP and stained with 3,3' diaminobenzedine (as mentioned previously) for electron microscopy.

Enhancer trap labeling of chordotonal organ projections into central ganglionic neuropil has also been effective in studying afferent pathways (Smith and Shepherd, 1996; Tyrer et al., 2000; Tyrer, personal communication).

3.3.4.2 Subcellular Components and Associated Molecules

Specific labeling has been achieved using antibodies to cells and to subcellular components. Actin was discovered in scolopales by labeling with phalloidin conjugated with fluorescein isothiocyanate (FITC) or rhodamine (Wolfrum, 1990). The same technique was used to trace chordotonal organ projections into the embryonic *Drosophila* nervous system (Merritt et al., 1993). Later, actin and α-actinin were localized in scolopales using the antiactin MabC4 and two polyclonal antibodies raised against chicken gizzard actin and α-actinin (Wolfrum 1991a,b). Tropomyosin also was localized in scolopales using an antiserum against locust

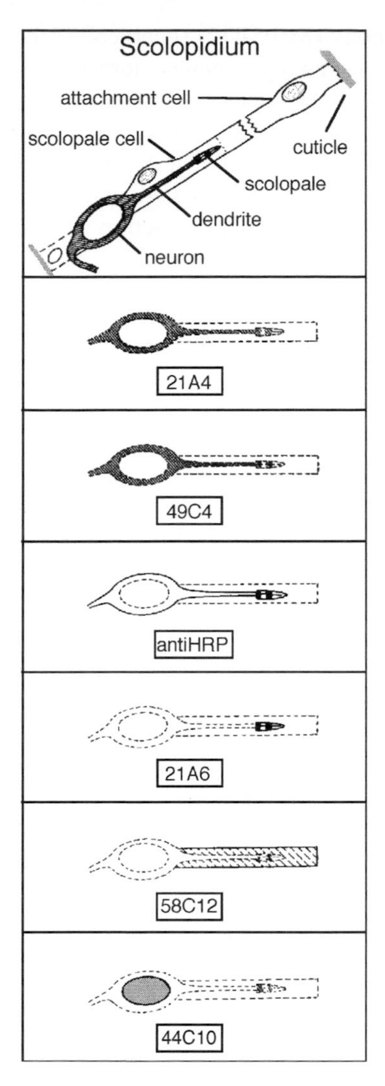

FIGURE 3.6 Summary of immunochemical labeling of chordotonal scolopidia in *Drosophila*. 21A4 antibody labels sensory neuron cytoplasm. 49C4 antibody labels only a subset of chordotonal organs. antiHRP (antihorseradish peroxidase) antibody labels the plasma membrane of all sensory neurons. 21A6 antibody labels scolopales only. 58C12 antibody labels the cap and sheath cells in older embryos. 44C10 antibody labels the nuclei of sensory neurons to allow cell counts. (Modified from Bodmer, R. et al., *Cell* **51**:293–307, 1987. With permission.)

tropomyosin (Wolfrum, 1991b). An even more exciting result came from Wolfrum's demonstration (1991a) of a centrinlike protein, as well as α-actinin, in the ciliary rootlets of scolopidia. Centrin is thought to play a role in the motility of flagella in green algae, which suggested a possible contractile role of ciliary roots of chordotonal organ neurons (Wolfrum, 1991a, 1992).

A monoclonal antibody (Mab5B12) specifically labels the protein glionexin in the extracellular matrix around mechanoreceptor (including chordotonal organ) glial cells (Field et al., 1994). The antibody was prepared from cricket ventral nerve cord (Meyer et al., 1987) and has provided insight into a possible ionic buffering role for glial cells associated with chordotonal/hair sensilla neurons.

Genetic studies of *Drosophila* increasingly have provided antibodies for immunohistological techniques to reveal gene action sites in scolopidia (Figure 3.6). The diversity of available labels in *Drosophila* chordotonal organs achieved by the specificity of antibodies (Figure 3.6) implies that differential labeling of scolopidial components may be achieved in a single preparation (Bodmer et al., 1987). Reviews covering molecular roles of gene products in insect sense organs include Jan and Jan (1993) and, more recently (dealing only with Johnston's organ), Eberl (1999) and Caldwell and Eberl (2002) (see also Chapter 9 and Chapter 10).

3.4 UNRAVELING ULTRASTRUCTURAL FUNCTION: ELECTRON MICROSCOPY, GENETICS, AND IMMUNOCHEMISTRY

Most unsolved mysteries of chordotonal organs concern the roles of ultrastructural components of the scolopidium. Here the micromechanical transduction of displacement occurs. It is likely to involve amplification mechanisms, and if the newly revealed moleculo-mechanical amplifiers in stereocilia of vertebrate ears are exemplary (Hudspeth, 1997), then the cilia of scolopidia are primary targets for study. Furthermore, there is indirect evidence that the scolopale and ciliary roots may alter tension in the scolopale/cilium apparatus and, thus, affect the neuron's sensitivity to mechanical stimulation (Wolfrum, 1991a,b, 1997), but direct proof is lacking. Many of these topics are amenable to investigation by electron microscopy, immunocytochemistry, genetic techniques, or a combination of these (see also Chapter 9 and Chapter 10 for multidisciplinary studies in the chemical senses). The following discussion will focus on ultrastructural aspects of scolopidia and cilia, as shown in the generalized diagrams of Figure 3.1c and 3.1d.

3.4.1 Electron Microscopy and Immunogold Labeling

Initial ideas about scolopidial function were derived from electron microscopical evidence of structural relationships among scolopidial components. Reviews that provide details of cellular and ultrastructural components of insect mechanoreceptors and chordotonal organs in particular, as well as common themes of functionality, include Moulins (1976), Keil (1997), and Field and Matheson (1998). A general scheme for scolopidial function posits that the scolopale rods are strongly attached by belt desmosomes to the proximal dendritic segment of the sensory neuron and, peripherally, to the scolopale cap and attachment cell (Gray, 1960; Yack and Root, 1992). The rods are thought to be stiff structures (Slifer and Sekhon, 1975) that encase the cilium and hold it taut. The cilium appears to be "snugly" inserted into the cap (Moran and Rowley, 1975). The resulting mechanical model (based on the mononematic scolopidium only) postulates a rigid scolopale cage in which the cilium either is able to actively bend or becomes passively bent by oblique or axial stretch of the organ, which exerts a pulling force on the cap (detailed review by Field and Matheson, 1998).

3.4.1.1 Evidence for Ciliary Bending

Because fixation affects ultrastructure, it is important to note that standard aldehyde fixation, staining, and embedding techniques have been used for most TEM studies of scolopidia. The ciliary bending hypothesis is based on a study of chordotonal organs stimulated by stretch and then plunged into liquid nitrogen before fixation and subsequent processing (Moran and Rowley, 1975). Cilia in stretched chordotonal organs were usually bent, while those in control organs were not. Additional support is the frequently observed bending at the base of the cilium in longitudinal TEM sections of scolopidia prepared by chemical fixation (e.g., Schmidt, 1969; Toh and Yokohari, 1985).

3.4.1.2 Internal Ciliary Mechanics

With standard chemical fixation, and more clearly with rapid cryofixation and high-pressure cryofixation (see next section), it is apparent that the axoneme microtubules are linked to the cilium plasma membrane by protein bridges (Moran and Rowley, 1975; Crouau, 1983; Keil, 1997). Similar protein bridges were originally found to make the same linkage between microtubule and cell membrane in the microtubule dense tubular body of hair and campaniform sensilla; the structures have been termed *membrane integrated cones* (MIC) by Thurm et al., (1983). It is now thought that stretch of the cilium in scolopidia may cause tilting of the MICs, which could in turn mechanically activate ion channels (Crouau, 1983; Keil, 1997). Various scenarios of mechanical distortion of the cilium by axonemal movement or displacement have been proposed, based on the MIC linkages (reviewed in Field and Matheson, 1998).

3.4.1.3 Fixation Artifacts and Cryotechniques

The preceding studies usually utilized chemical fixation with 2 to 4.5% glutaraldehyde and sometimes 2% formaldehyde, followed by 4% osmium tetroxide postfixation. Chemical fixation is known to cause fixation artifacts due to bulk water movement into or out of cells if the tissue is surrounded by extracellular fluid spaces (Steinbrecht, 1992). Two consequences have been observed in scolopidial electron micrographs: unpredictable distortion of membrane-bound structures and inclusion of membrane bits in the extracellular scolopale space (surrounding the cilium). In other chordotonal organs (especially those forming tympanal organs), tracheal air spaces may restrict the penetration of the fixative and allow poor fixation. This has been overcome by the addition of a small drop of wetting agent (Photoflo) to the fixative and fixing under a vacuum (Young, 1973).

Rapid cryofixation, rather than chemical fixation, may be a way to overcome the distortion, which hinders progress in solving current questions of ciliary and scolopidial function (Wolfrum, 1990; Steinbrecht, 1992). Techniques of rapid cryofixation, followed by cryoembedding and cryomicrotomy, have been reviewed by Steinbrecht and Zierold (1987) and Wolfrum (1997). Usually, the cryofixation technique is effective only in a shallow surface zone of the material (about 10 μm), and appropriate consideration should be given to selecting chordotonal organs that are amenable to the technique. A costly, but successful, alternative is to utilize high-pressure cryofixation, which allows analysis of tissue to depths of at least 40 μm. Steinbrecht (1993) reviewed this technique for use in insects.

3.4.1.4 Immunogold Localization

Although immunostaining with light microscopy gives strong hints about the locations of actin, tropomyosin, MAP2, actinin, and centrin, the definitive evidence to support the resultant hypotheses about the mechanical roles of these molecules came from localization with immunogold labeling by binding 1 nm or 10 nm particles to the appropriate antibodies and enhancing the label with silver intensification (Wolfrum, 1997).

3.4.1.5 Mechanical Model of Scolopale and Ciliary Root Action

It is now known that scolopales are composed of 10-nm filamentous actin with interposed microtubules and tropomyosin. These elements are crosslinked by MAP2 to form a mechanically integrated structure. Because any mechanical contraction would require the additional presence of myosin, and myosin is almost certainly lacking, Wolfrum (1997) concluded that scolopales serve as a stabilizing cylindrical complex enclosing the cilium, with sufficient elasticity to restore the resting position after undergoing slight flexion during any oblique distortion by stretch stimuli. An added feature is that the restoring force could be controlled by Ca^{2+} and binding by tropomyosin.

The presence of the Ca^{2+} binding phosphoprotein, centrin, has led to the hypothesis that the ciliary rootlets also serve a mechanical function in the scolopidium. Because centrin is the main protein involved in flagellar rootlet contraction in unicellular green algae, it could well undergo Ca^{2+}-mediated contraction in the scolopidium. This could affect the displacement of the ciliary basal body at the base of the cilium, which in turn would transmit force to the ciliary necklace attached to the outer dendritic segment (Figure 3.1c). Because mechanically activated channels are thought to reside at this site, the rootlet contractions could affect responsiveness of the sensory neuron.

3.4.2 Genetics

Rapid progress is being made in identifying genes that encode developmental, structural, and sensory transduction proteins in scolopidia of *Drosophila.* Caldwell and Eberl (2002) reviewed this research insofar as it relates to the hearing function of the Johnston's organ in the *Drosophila* antenna. Additional studies have also included new gene products of pleural chordotonal organs (e.g., lch5) in the abdomen, and the femoral chordotonal organ (FeCO) of the legs, all of which are easily seen in embryonic flies (see, e.g., Chung et al., 2001). Much of the genetic technique involved in this research is beyond the scope of the present chapter, and the reader is referred to original papers for details (e.g., Eberl et al., 2000; Dubruille et al., 2002; Sharma et al., 2002; Caldwell et al., 2003; Kim et al., 2003). General approaches and chordotonal organ-specific techniques are discussed in the following sections.

3.4.2.1 Mutagenic Screening

The first step in identifying genes has been to use a mutagenic screen to isolate hearing, proprioceptive, and mechanosensory mutants that may have dysfunctional chordotonal organs. Exposure to a diet of sucrose containing the mutagen ethyl methane sulfonate (EMS) induces mutants from which behavioral deficits are recognized (Eberl et al., 1997; Dubruille et al., 2002). Larval behavior may be uncoordinated and locomotion characterized by decreased linear paths, increased turning, reduced velocity, and so on, or mechanoreceptor responses may be reduced or absent (*unc, tilB* and *nomp* are examples of identified genes involved in such functions) (Caldwell et al., 2003). The same gene mutants yielded adults that showed uncoordinated movements in locomotion and responses to alarm stimuli and especially behavioral disruption in courtship and copulation latency. These phenotypes have been shown to have altered gene expression (see Section 3.4.2.3) in Johnston's organ (antennae), pleural chordotonal organs (abdomen), and the femoral chordotonal organ (legs). Most adults do not survive, and results must be taken from freshly emerged individuals (Eberl et al., 2000).

3.4.2.2 Localization of Gene Products

A major technical requirement for this work is to visualize and localize the gene products, or ultrastructural consequences of altered gene expression, in mutants. A most successful and frequently used method to localize gene products in chordotonal organs of *Drosopohila* is the GAL4 enhancer trap technique (Brand and Perrimon, 1993; Brand, 1999; van Rössel and Brand, 2002). This technique allows selective activation of any cloned gene in insect tissues, and it reveals the cell-specific patterns that can be visualized in light or fluorescent microscopy. In particular, the method generates fly lines that express a transcriptional activator, GAL4, in specific cell or tissue types. When a minimal promotor is combined with the GAL4 gene and inserted into a P-element vector (pCaSpeR), the resulting construct can be injected into embryos. The GAL4 gene is randomly integrated into the genome. If the construct lands downstream of a cell-specific enhancer, it will be expressed in that cell line. By inserting a reporter construct, such as a marker gene for green fluorescent protein (GFP) or LacZ, into the target cells downstream from GAL4 binding sites (UAS, yeast-derived upstream activating sites), the GAL4 expressing cell line will also express the marker (Brand, 1999).

By driving the GAL4 system with a specific enhancer fragment of interest cloned into the UAS construct, such transgenes may be localized to highly specific cell types, where their expression may be studied. A P-element vector for testing enhancer fragments has been developed for this purpose (Sharma et al., 2002). By incorporating a GFP construct, the technique can allow detailed study of gene action controlling development and function in *Drosophila* chordotonal organs (Johnston's organ and leg chordotonal organs) (see Chapter 9 and Chapter 10 for discussions of how this technology is used to study chemosensory systems). Thus, dendritic structure and morphology, axonal projections into the CNS (Tyrer et al., 2000), and synaptic ultrastructure of wild-type and gene mutants can be studied (Sharma et al., 2002).

A valuable variation of the GAL4 system incorporates reduced silver staining on serial semithin sections of *Drosophila* ganglia to study central chordotonal organ projections (Tyrer et al., 2000). The method combines the power of a molecular labeling technique with accurate staining of the ganglionic architecture to allow the best detailed localization of projection afferents described to date for *Drosophila*. It is possible to view individual GFP-labeled cells or central projections in sections using confocal microscopy, and then study the exact projection environment in relation to other cell processes and neuropil structures.

3.4.2.3 Molecular Components of Chordotonal Organs

This field is expanding rapidly, owing primarily to (a) the ability to visualize easily several chordotonal organ types in *Drosophila,* using techniques described previously, and (b) the ability to make electrophysiological recordings from a single chordotonal organ: Johnston's organ in the antenna. A description of genes controlling development of *Drosophila* chordotonal organs has been given previously (Field and Matheson, 1998), but since then additional genes and greater detail in knowledge have accrued. This has been well reviewed by Caldwell and Eberl (2002) and will not be covered here.

Of more immediate interest are genes that are involved in mechanotransduction: those controlling ciliary growth, composition, and attachment, as well as others controlling ion channel distribution and function. The following brief review deals mostly with results obtained for Johnston's organ, but it should be emphasized that the amphinematic scolopidial type often found in Johnston's organ is uncommon in the insects. Although genetic studies tend to generalize results to all chordotonal organs, they have given little or no attention to the most common type in insects: the mononematic scolopidium. In fact, Johnston's organ contains both types of scolopidium. The ultrastructure and organization of Type 1 and Type 2 cilia differ greatly (Table 3.1), yet this distinction has not been made in genetic studies to date. Thus, it is unclear whether the findings reviewed below extend to other chordotonal organs in *Drosophila* or other insects, with the exception of a few histological results that have been repeated in the abdominal pleural chordotonal organs of *Drosophila* embryos (e.g., Chung et al., 2001).

Figure 3.7 gives names and locations of gene action or gene products found in scolopidia of Johnston's organ. Three kinds of gene effects related to physiology have been isolated in the scolopidia:

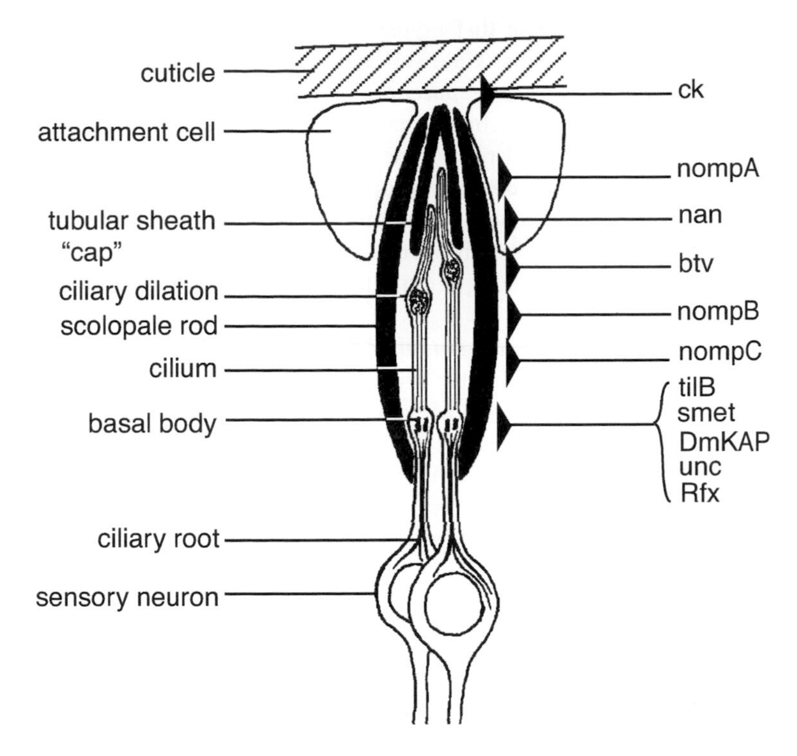

FIGURE 3.7 Specific labeling of gene products or sites of gene action in Johnston's organ scolopidia allow unprecedented insight into ultrastructural function of scolopidial components. *Ck* codes for myosin VIIa and is involved with attachment to the cuticle. *nompA* is involved with attachment of the cilia to the cap (tube). *Nan* codes for a Ca^{2+} channel thought to be in the ciliary membrane. *Btv* gene regulates structure of the ciliary dilation. *nompB* is concerned with control of ciliary assembly and axonemal traffic. *nompC* gene product may link ciliary components or be part of a transduction channel. *tilB* and *smet* are thought to provide ciliary integrity. *DmKAP* is apparently involved in ciliary assembly and growth. *unc* may regulate ciliary assembly through conversion of centrioles into basal bodies. *Rfx* is also involved in ciliary assembly.

- Control of attachment of cilia or scolopidia to the source of mechanical stimulation
- Control of ciliary or axonemal assembly and architecture
- Production or relationship to ion channels in the cilium

The following genes are relevant:

crinkled (ck) — In *Drosophila* this gene codes for myosin VIIa, which has an unknown function in chordotonal organs. Mutants are deaf, and the Johnston's organ is detached from the A2A3 joint. Hence, the mechanical stimulus arising from vibration of the antennal arista does not reach the scolopidia. It is possible that myosin is also involved in some unknown molecular motor function (Caldwell and Eberl, 2002).

nompA (no mechanoreceptor potential A) — This gene encodes a single transmembrane protein, which appears to be produced in the cap of the scolopidium. The mutant phenotype is deaf to sound stimulation, and the cilia tips are detached from the scolopidial cap. The function of the protein appears to relate to binding of the cilia to the cap and of the cap to the attachment cell. This gene product is also found in one chordotonal organ with a mononematic scolopidium: the pleural lch5 chordotonal organ (Chung et al., 2001).

nompB (no mechanoreceptor potential B) — A two-domain protein is produced, which is homologous to mouse, *Caenorhabditis elegans,* and *Chlamydomonas* proteins, all of which are involved in intraciliar transport of proteins along the axoneme. Mutants have lost the ability to hear, and cilia are missing or heavily malformed in scolopidia. The gene product is thought to control axoneme traffic, which may be crucial for cilium assembly (Caldwell and Eberl, 2002).

nompC (no mechanoreceptor potential C) — A six-domain transmembrane protein is encoded in *Drosophila* (Walker et al., 2000). This product distantly resembles TRP (transient receptor potential) ion channel proteins but also has many ankyrin repeats. The mutant phenotype shows a small reduction in the receptor potential response to sound stimuli, and motor control is generally uncoordinated. The function in scolopidia is unclear; the protein could serve to link ciliary or axonemal components or be part of a transduction channel.

btv (beethoven) — Candidate proteins for this gene include a cadherinlike sequence and a heavy chain dynein. The mutant flies have a hearing loss, are slightly uncoordinated, and are sedentary. The mechanoreceptor potential is reduced or absent. Defects have been found in the ciliary dilation within the scolopidium (Eberl et al., 2000). The gene product may function in an intercellular adhesion role (e.g., ciliary attachment to cap) or in transport within the cilium (Caldwell and Eberl, 2002).

tilB; smet (touch insensitive larva B; smetana) — The products of these two genes are unmapped. Both have mutant phenotypes characterized by lack of touch sensitivity in larvae, but no effect on the receptor potential response to sound stimulation. Motor coordination is only slightly affected, and no apparent defects were seen in scolopidial ultrastructure (Eberl et al., 2000). Because the mutant flies have gross defects in axonemes of sperm flagella, these genes are thought to be involved in providing ciliary integrity in chordotonal organs (Caldwell and Eberl, 2002).

unc (uncoordinated) — The product of *unc* is tentatively characterized (Baker et al., 2001) as a coiled-coil protein expressed in mechanoreceptors with Type 1 cilia. The mutants are deaf and have uncoordinated behavior. In scolopidia of Johnston's organ, the cilia are not connected to the cap.

The role of the protein appears to be in the conversion of centrioles into basal bodies required to assemble cilia (Baker et al., 2001).

nan (Nanchung) — This gene encodes a TRPV (transthyretin-related protein, vanilloid sensitive) ion channel for Ca^{2+} conductance in hypoosmotic conditions. Mutant flies are deaf but show no ultrastructural defects in the scolopidia of Johnston's organ. The protein occurs only in the outer dendritic segment of the neuron of both Johnston's organ and the lateral pleural chordotonal organ, lch5. This is thought to be a Ca^{2+} ion channel in the ciliary membrane that responds to osmotic stretch-related stress or perhaps to some unknown stimulus mechanism.

DmKAP (Drosophila melanogaster Kinesin associated protein) — The gene product is a non-motor associated protein subunit of kinesin II. The mutants are deaf, uncoordinated, and sluggish. The cilia are absent in the scolopidia of both Johnston's organ and the abdominal pleural chordotonal organs. The function is thought to be in assembling the axoneme in the process of cilium growth (Sarpal et al., 2003).

Rfx (Regulatory Factor X) — In *Drosophila,* RFX is a transcription factor characterized by a DNA-binding domain of wing-helix structure. Mutants for RFX cannot fly and are uncoordinated. In scolopidia, Johnston's organ, and lateral pleural chordotonal organs, cilia are short and swollen abnormally (or missing), and dendrites are disorganized. The axoneme is either disrupted or missing. The role of RFX is to assist in the intraflagellar transport (IFT) mechanism by which cilia are assembled (Dubruille et al., 2002).

3.4.2.4 Electrophysiological Recording in Genetic Studies

Genetic studies of chordotonal organs have relied on extracellular recording from the antennal nerve in *Drosophila*, using a sharpened tungsten electrode introduced through the arthrodial membrane between the first (scape) and second (pedicel) antennal segments. Although the records contain impulse traffic from Johnston's organ (Eberl et al., 2000), none of the published accounts acknowledges that two other (connective) chordotonal organs exist in the antenna (reviewed in Field and Matheson, 1998) or that these sense organs must contribute to the records obtained from the antennal nerve. Furthermore, because the two chordotonal organs contain mononematic scolopidia instead of the amphinematic scolopidia of Johnston's organ, genetic conclusions about molecular functions in the papers reviewed previously do not distinguish between these scolopidium types.

Thus far, it has not been possible to record from the pleural chordotonal organs of the abdomen in *Drosophila*; a technique for this could help to solve the problem of addressing functional differences of molecular components of mononematic and amphinemataic scolopidia.

3.5 TRANSDUCTION AND SPIKE GENERATION: PATCH CLAMPING AND INTRACELLULAR RECORDING

3.5.1 MECHANICALLY ACTIVATED CHANNELS

It would be highly exceptional if mechanically activated channels (MACs) did not occur in the transducer region(s) of chordotonal organ neurons. Two histological impediments interfere with physiologically accessing such ion channels:

- First, the whole chordotonal organ typically is surrounded by fibrous connective tissue that presumably ensures an optimal tissue pathway, allowing mechanical distortion to reach the stretch-sensitive scolopidia.
- Second, the scolopidial structure itself blocks access to the cilium and distal region of the inner dendritic segment of the chordotonal neuron.

One way to circumvent the first problem is to culture isolated chordotonal neurons and make patch clamp recordings *in vitro*. This was done for Type I neurons from antennal chordotonal organs (probably a mixture of neurons from Johnston's organ and the pedicel and scape connective organs) from the cockroach and the hawkmoth *Manduca*. Stockbridge and French (1989) made patch clamp recordings from cockroach antennal Type I neuronal somata with cation MACs permeable to potassium and sodium (approximately 100 pS conductance). The dissociated and cultured neurons lacked dendrites and scolopales. Because these channels were not located in the suspected mechanotransducer regions of the scolopidial neuron, they could represent channels with a different sensory function (French, 1992) or they could reflect nonlocalized spread of channels due to injury. A better culture preparation with healthy *Manduca* pupal neurons from Johnston's organ and hair sensilla was developed by Torkkeli and French (1999). The chordotonal neurons were distinguished from neurons of external cuticular sensilla by size of somata and immunolabeling. However, the latter seemed to lack scolopidia, and because the culturing process required dissociation of the neurons from other scolopidial support cells, fully developed scolopidia may have been precluded *in vitro*. Whole cell recordings gave rapid inward currents followed by a slower outward current. Single-channel patch clamp recordings from the somata and from neurites, including dendrites, indicated that MACs responded equally to mechanical stimuli using both positive and negative pressure. Interestingly, the channel currents have not been related to the mechanoreceptor bumps described subsequently, and details of both remain to be elucidated. No MAC recordings have been possible from identified inner and outer dendritic segments of chordotonal neurons.

MAC recordings have been made from Type II sensory neurons, including calcium-permeable MACs, and the results may be generalized to Type I neurons even though the receptor potentials of Type II neurons are graded and lack the mechanoreceptor bumps found in Type I neurons (reviewed by French et al., 2002). If it proves impossible to record from the outer or inner dendritic segments of chordotonal neurons, it may be possible to extract the MACs and study them in artificial membranes. Excellent instructions for recording from MACs are given by Erxleben et al., (1991).

3.5.2 Intracellular Studies and Transduction Potentials

Penetration of chordotonal neurons with intracellular pipettes has been successful in tympanal organs (crista acustica in tettigoniid and anostostomatid orthopterans, Müller's organ in locusts) and in femoral chordotonal organs (FeCO) of insect legs (Zill, 1985). In the former, the crista (anostomatid wetas, tettigoniid bush crickets) or Müller's organ (locusts) naturally rested on its large tracheal support and a window was opened in overlying cuticle to introduce the electrode into the stabilized leg (Hill, 1980, 1983a,b; Oldfield and Hill, 1986). The advantage of this system was that the sound stimulus did not physically dislodge the electrode, because the displacement amplitude was so small. Individual scolopidial neurons and attachment cells could be visualized and penetrated with the help of transillumination (fiber optic). Double recordings (one of each cell type) were possible, and recording sites could be marked by injection of Lucifer Yellow fluorescent dye.

The first electrical sign of transduction in the preceding tympanal preparations was a series of discrete subthreshold potential depolarizations similar to the quantum bumps described for visual receptors (Hill, 1983a). This is unlike the usual mechanoreceptor graded depolarization found in Type II neuron mechanoreceptor potentials (French et al., 2002) although the discrete potentials summate toward an equilibrium potential that is more positive than the neuron's resting potential. The summated potential acts as a receptor potential to depolarize the neuron to spiking threshold. The relationship of the mechanoreceptor bump potentials to MAC activity remains to be elucidated.

A similar subthreshold unitary event (Figure 3.8c,d) occurs in FeCO neurons subjected to restricted mechanical stretches (Field and Matheson, unpublished data). The chordotonal organ preparation rests on a supporting platform *in vivo,* and stretch is delivered by open-loop displacement of the FeCO apodeme.

3.5.3 Dendritic Spike Potentials and Soma Spike

Three types of spike potentials have been identified by intracellular recording in chordotonal neurons of tympanal organs (Hill, 1983b; Oldfield and Hill, 1986). The mechanoreceptor bumps summate to produce an apical spike that neither overshoots 0 mV nor undershoots the neuron's resting potential (Figure 3.8a,b). This spike appears to trigger electrotonically a basal dendritic spike that shows normal Na^+/K^+ conductances (e.g., overshooting 0 mV), which in turn triggers a conventional orthodromically conducted axon spike at the trigger zone.

3.5.4 Use of Blocking Agents

Dimethyl sulfoxide (DMSO) has been used to show that the transduction region of scolopidial neurons can be selectively blocked relative to the axonal spiking region (Theophilidis and Kravari, 1994). The analgesic effect of DMSO apparently blocks potassium transduction channels in the dendrite at a lower concentration (0.85%) than that which blocks axonal potassium channels (4.8%). A similar study utilized insecticides to demonstrate differential effects on spike generation and axonal conduction (Theophilidis et al., 1993). Bath applications of fluvalinate ($< 3.33 \times 10^{-7}$ M) and

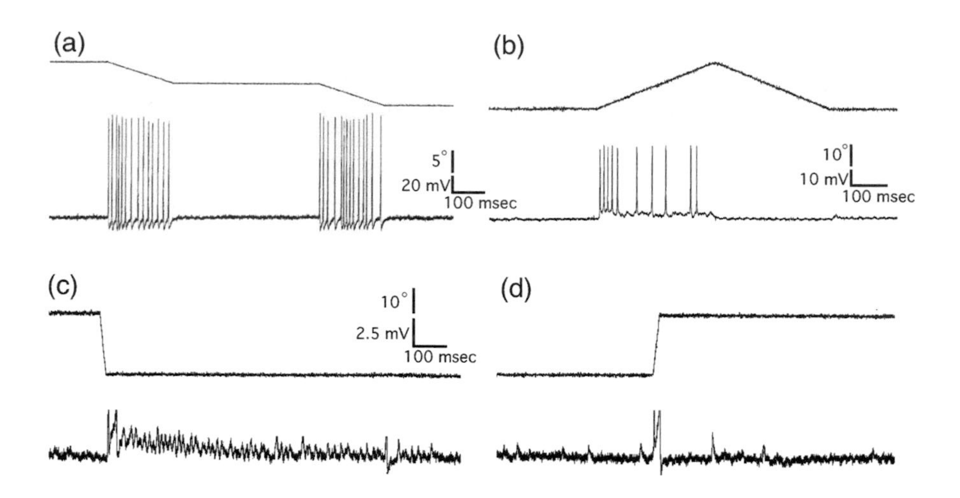

FIGURE 3.8 Unusual action potentials and transduction events occur in chordotonal neurons. **a**: A velocity-sensitive neuron in the locust FeCO shows normal action potentials with overshoot above zero and hyperpolarizing afterpotential. **b**: Another neuron displays nonovershooting, nonundershooting spike potentials believed to arise from the apical dendritic inner segment. It is depolarized by a plateau of summated small potentials. **c–d**: Discrete unitary potentials represent transduction events activated by a small displacement of the FeCO, and which show adaptation (**c**). Two truncated spike potentials fired initially. **d**: After reaching a low tonic frequency of unitary potentials, return to the initial position evokes a brief discharge and two spike potentials (truncated).

deltamethrin ($> 4.06 \times 10^{-8}$ M) affected spike-generating mechanisms (possibly sodium channels), but higher concentrations affected axonal conduction properties. The advantage of using bath application of these drugs is that the FeCO is not protected from hemolymph by the sheath that surrounds the CNS in insects, and thus the effects are seen directly on the sensory neurons.

3.6 SENSORY PHYSIOLOGY: MECHANICAL STIMULATION AND ANALYSIS

3.6.1 Mechanical Nature of the Stimulus

Regardless of which type of chordotonal organ is tested, the stimulus ultimately must be some form of mechanical displacement. The range and velocity of movement can vary enormously, from nanometers for hearing organs to about a millimeter for joint chordotonal organs. For connective chordotonal organs, it is important to determine the physiological range of displacement in order to scale the stimulus accordingly and not to overstimulate — and possibly destroy — the sensory tissue.

This applies to magnitude of stretch as well as velocity of movement. The latter is especially important for connective chordotonal organs because their connective tissue components have finite stress relaxation times and step displacement could cause severe disruption of this tissue or attachments to cuticle. In fact, mechanical step displacement occurs very rarely in biological organisms (Cruse, 1996); where it does, there are often special histological adaptations to cope with the extremely high acceleration involved (such as resilin in the joints of legs undergoing ballistic extension). Tympanal organs and vibration receptors detect very small displacements, and their accessory tissues allow damping of extreme sound or vibratory stimuli, and so the problem of tissue damage is not so important.

Although displacement is the initial stimulus, chordotonal neurons can detect different mechanical parameters of the movement and may be specifically sensitive to only a limited form of the stimulus energy. The specificity arises from the internal anatomy and associated tissues acting to filter the stimulus mechanically. Hence, ligaments of joint chordotonal organs may differentially filter stretch by having viscoelasticity differences among the attachment cells (Field, unpublished observations), air sacs may enhance specific frequency ranges through cavity resonance, and arrays of attachment ligaments may restrict responses through range fractionation. As a result, certain chordotonal neurons (e.g., FeCO) uniquely report position, velocity, or acceleration of the stimulus. Other neurons in vibration detectors (SGO) and tympanal organs detect only acceleration (review: Field and Matheson, 1998).

3.6.2 Stimulating Connective Chordotonal Organs

3.6.2.1 Open-Loop vs. Closed-Loop Stimulation

Most connective chordotonal organs occur in joints. In early research, it was thought that simple movement of the joint sufficed to stimulate the organ in order to elucidate its sensory responses. This method was called *closed-loop stimulation* because the chordotonal organ remained intact *in vivo,* its responses were recorded proximally in the appropriate nerve, and all natural feedback channels were intact in the nervous system.

The closed-loop approach is risky, and generally it is not recommended unless the following conditions are satisfied:

First, the method of attachment to a segment of the joint should prevent backlash, or sloppy movement, which can arise from poor tolerance in connections to the driving device or to the appendage. Sometimes a bent minuten pin attached to the driver is pushed through the appendage to allow a tight-fitting connection to the cuticle of the movable appendage segment (Burns, 1974).

Second, unintentional stimulation of other mechanoreceptors must be avoided. Researchers often overlooked the presence of other movement-sensitive organs at the same joint, and chordotonal organ sensory responses were contaminated by cuticular hair sensilla and Type II receptors. Furthermore, in studies of the proprioceptive feedback pathways and interneuron responses to chordotonal organ stimulation, spurious activation of additional sense organs could cast doubt on the results. Although the problem can be overcome by surgically ablating known receptors at the joint under study and possible neuromuscular feedback pathways, even this proves inadequate when new sense organs are subsequently discovered at the joint. An example is the femur–tibia joint in the locust leg. Movement of the tibia served to stimulate the FeCO (Usherwood et al., 1968; Burns, 1974); however, a nerve containing cuticular hair sensilla axons passes along the FeCO and merges into the FeCO nerve (N5B1). Inadvertant touching of the leg to any mounting or manipulation apparatus could stimulate these afferents. It was also known that tibial extension activated multipolar joint receptors (Coillot and Boistel, 1969) and a muscle receptor organ (Williamson and Burns, 1978). Subsequently, Bräunig (1985) discovered that part of the FeCO contains a strand receptor organ that is also activated by extension of the flexor–tibia (FT) joint. Then Matheson and Field (1995) showed that a multipolar receptor associated with the accessory flexor muscle (adjacent to the FeCO) also sends an axon to the FeCO nerve and is sensitive to FT joint movement. Ablation of all these organs in order to isolate stimulation of the FeCO alone would be very challenging.

A better approach is to expose and disconnect the chordotonal organ under study and apply the movement stimulus only to its attachment ligament or apodeme. This is an *open-loop stimulation* of the sense organ because, in systems analysis terms, feedback has been eliminated from affecting the chordotonal organ output. Although it has some pitfalls, this offers a solution to the preceding problems. The main pitfall is to move the sense organ in an extreme or an unnatural way. Use of a suitably small probe and careful inspection of the anatomy helps to avoid this problem.

3.6.2.2 Probes

To provide open-loop stimulation, a probe or miniature forceps mounted on the mechanical driver (see next section) must be attached to the distal part of the chordotonal organ (see, e.g., Field and Burrows, 1982). Forceps can be filed from a narrow tungsten rod, adapted from the tips of watchmaker's forceps, or constructed by grinding two insect pins flat on one side and soldering/gluing the flat sides together (Figure 3.9a–e). If removing the tips from watchmaker's forceps, be aware that these are usually case-hardened and may resist hole-drilling with normal drill bits. Tungsten is difficult to machine, but its very light mass makes it ideal. In all cases, it is useful to apply a small bend or to file a notch halfway along the shaft to make an opening between the inner surfaces to allow a miniature lever to be inserted to open the forceps apart when attaching to the chordotonal organ (Figure 3.9c–e). Probes may be made of stiff metal (tungsten wire) and can be etched and attached to a chordotonal organ with various cyanoacrylate adhesives.

The mass of the probe must be minimized in order to prevent the stimulus waveform from being distorted by inertial delays and momentum. As the velocity

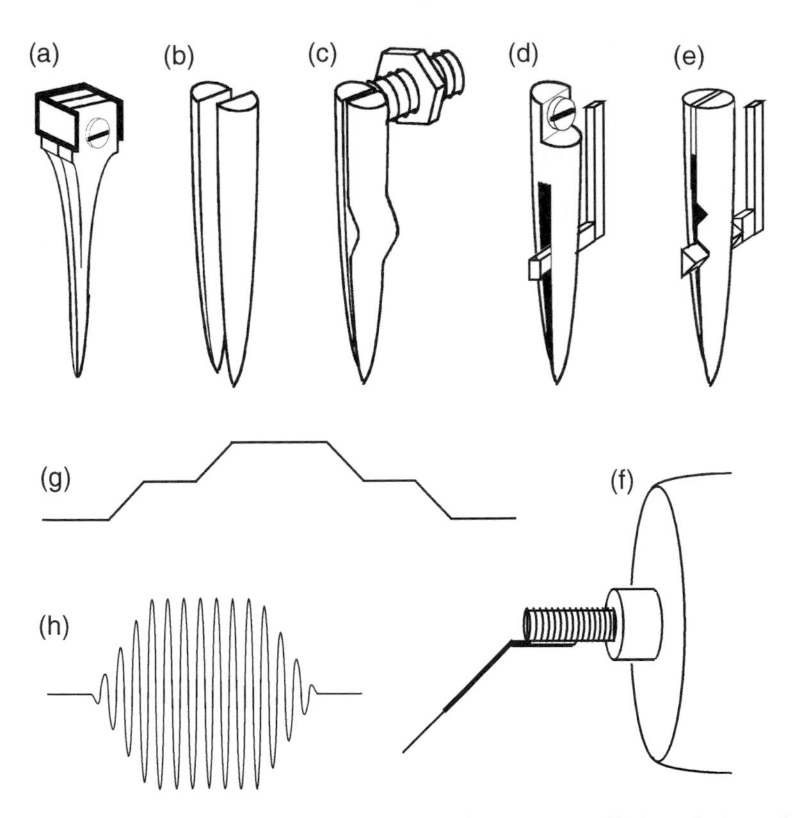

FIGURE 3.9 Techniques for displacing chordotonal organs. **a–e**: Various designs of mini-forceps made or adapted from commercial designs. **a**: Watchmaker's forceps (No. 5) with mounting hole drilled or ground while intact, then cut near tip, mounted in brass clamp, and bolted to holding rod. **b–c**: Two insect pins are sanded flat, cut to length, bound with fine copper wire at top (not shown), and soldered to a holding rod. One is bent to prevent a saline meniscus from rising into the solder joint. Note tightening nut on threaded holding rod, to allow fixing of the forceps at desired angle of rotation of rod. **d**: A tungsten rod is slotted and ground to a point. Note different method of attachment to rod. Sprung tips may be parted by inserting flat lever (right) and rotating slightly. **e**: Another variation assembled from two ground halves. Notch prevents saline from corroding solder joint at top. Lever inserted at right is held in place permanently in another notch. **f**: Movement probe soldered to threaded rod and attached to minivibrator for chordotonal organ stimulation (see text). A tightening nut should hold rod in place. **g**: Ramp and hold displacement sequence for stimulating a joint connective chordotonal organ. **h**: Waveform with damped onset and offset for acoustic stimulation of tympanal organs. Duration is usually short (10 to 200 msec).

increases, the mass of the probe at the end of the driver can cause erratic lateral movement and affect the linearity of the imposed waveform. These problems can be minimized by (a) using tungsten probes instead of steel, (b) keeping the driver/probe length as short as possible, and (c) avoiding gross off-axis imbalance of the probe at the end of the driving shaft.

3.6.2.3 Mechanical Drivers

Desired waveform voltages are usually fed to a mechanical driver with an extension to which the probe is attached. Drivers include servo pen-motors removed from oscillographic pen recorders, piezoelectric crystals (e.g., piezoceramic, lead zirconate–titanate, and barium titanate), minishakers or electromagnetic vibrators, and audio loudspeakers in which the cone has been reduced to four supporting legs for the central disk to which the probe is attached. Commercial lever systems offer a sophisticated and expensive way to produce highly controlled length changes with very short response times(2). Servo pen-motors offer a large displacement and have accurate position control, but the upper frequency is limited to a few hundred Hz (unsuitable for studying vibration receptors). Piezoelectric bending transducers(22) are narrow beams that have a high upper frequency limit (3 to 4 kHz) but limited displacement (~300 μm). Minishakers are typically made for vibration testing, but they are very accurate devices that produce high force and have upper frequency limits of around 3 kHz(11).

Regardless of which driver system is chosen, the following properties must be addressed carefully before the probe is used to test chordotonal organs.

Linearity — The movement of the probe must be monitored, or at least an input–output calibration curve must be constructed, to determine the linearity of the output of the system with the probe attached, for all signal functions and frequencies used. Probe movement may be monitored by a light beam–photocell system in which the probe, or a flag attached to it, interrupts the beam. Infrared photocells and LEDs are useful for this purpose. Usually, driver systems become nonlinear as frequency increases, and it is best to use frequencies in the linear part of the calibration curve.

Resonance — All movement devices must be checked for resonance within the frequency range to be used. As probe mass increases, resonances will become more apparent in the output. The probe is almost certainly going to cause some resonance in the system, and the user should determine amplitude and frequency limits within which resonance can be avoided. Resonance is determined by comparing the probe output movement amplitude (monitored with an infrared beam occluded by the probe) to the input waveform amplitude, as frequency is varied with a sine wave generator. If problems persist, the probe should be shortened or made lighter.

Drift — Any studies of position sensitivity of chordotonal organs require that the driver be checked for drift while being set at different static positions. Drift may arise in the driver or the signal generator. For example, electromagnetic solenoid shakers (vibrators) are usually meant for sinusoidal driving, although they work well when delivering other waveforms (e.g., ramps). However, they do not hold positions well when a steady DC voltage is applied to them. Drivers controlled by servo feedback electronics are more reliable.

Hysteresis — A final check should be made for hysteresis in the system. It should be set at a zero position and moved through a series of offsets, such as ramp and hold steps, which return to the zero setting. The ramp sequence should be given for both directions away from zero. If the probe does not return to the zero position, there is hysteresis in the system, which must be eliminated.

3.6.2.4 Stimulus Parameters and Waveforms

In insects, chordotonal organs normally are strain detectors (length change/initial length) that report Newtonian parameters of movement and position. A few examples of myochordotonal organs have been reported (Shepherd, 1973; Bräunig et al., 1981), in which the chordotonal scoloparia are located on a tiny receptor muscle in parallel with a working muscle. Strictly speaking, these are still strain detectors and not stress detectors (force/unit area exerted onto the muscle apodeme) of muscle tension, because they apparently monitor length changes in the receptor muscle which, in turn, would bias chordotonal output. True isometric stress detectors that are sensitive to muscle tension have been described only in Crustacea (apodeme sensory nerves, Macmillan, 1976). In general, documented cases in which chordotonal organs function as isometric stress detectors are lacking in insects. This topic has been discussed in detail by Macmillan (1976).

The distinction here between strain and stress receptors bears upon the form of stimulus used in experimentation with chordotonal organs. Because these organs measure position and length changes, stimuli should explore sensitivity to the physiological range of positions and movements imposed by the insect. Movement analysis includes studying direction, velocity, and acceleration sensitivity. In nonconnective chordotonal organs (hearing and vibration receptors), the movement component is normally acceleration, and the variables are amplitude and frequency of sine wave stimuli.

Two kinds of experimental question need to be addressed. First, the operational parameters of the chordotonal organ's sensitivity should be determined to elucidate the information supplied by the organ to the rest of the nervous system. Second, quantitative analyses of the physical processes (transduction, stimulus coding, and spike generation) underlying the organ's response are carried out in attempt to model the system's behavior.

The sensory response of a chordotonal organ is best studied by a set of ramp and hold steps that successively increase and decrease (Figure 3.9g). The linear ramp slope (velocity) may be varied, and the hold plateaus should be long enough to allow receptor adaptation to a steady plateau. Ramp and hold stimuli test sensitivity to position and adaptation to position, movement direction, and velocity. The test also reveals the presence of hysteresis and range fractionation among individual neurons (Matheson, 1992). An indication of acceleration sensitivity also is gained, although accurate response quantification requires a method of delivering constant acceleration (Hofmann and Koch, 1985). Variations in phasic and tonic responses should be studied to determine adaptation properties of the chordotonal organ.

Another way to study the operational contribution of chordotonal organs to behavior is to manipulate an organ mechanically or chemically. Bässler (1967, 1979) surgically detached the FeCO apodeme and reattached it to the opposite side of the femur–tibia joint axis in intact stick insects and locusts, thus demonstrating a resistance reflex reversal and illustrating the role of the FeCO in mediating the reflex *in vivo*.

Octopamine has been administered to FeCO preparations during mechanical stimulation to determine peripheral effects of such biogenic amines (Matheson, 1997). Injection of octopamine into intact insects demonstrated additional central effects on the reflex system mediated by the FeCO (but with possible unknown additional effects from other sense organs) (Buschges et al., 1993).

Modeling studies of chordotonal organ neurons involve delivering forcing functions to the mechanical input pathway of the chordotonal organ, or electrical waveforms to the neuron itself. Responses are analyzed with systems analysis procedures to model the output by calculating transfer functions in the time and frequency domains. Practical mechanical input functions include the sine function, the statistical (noise) function, and the ramp function. Outputs from steady-state sine stimuli are summarized as Bode or Nyquist plots of amplitude (gain) and phase. The statistical function is usually band-limited Gaussian white noise with a flat frequency content and a Gaussian amplitude distribution of sine wave amplitudes. The spike output is converted to a continuous function and analyzed by crosscorrelation with the input. This may be plotted as Wiener kernels of first and higher orders to reveal linear and nonlinear dynamic properties of the neuron. Details of the preceding method are provided by Kondoh et al., (1995), based on theoretical treatments by Marmarelis and Marmarelis (1978) and Naka et al., (1985). A useful suite of programs to convert spike data and perform Wiener kernel analysis is provided by Professor A.S. French (http://asf-pht.medicine.dal.ca/Downloads/Index.html). General treatments of systems analysis of mechanoreceptors are given by Mill and Price (1976) and Cruse (1996).

The Wiener kernel method has provided valuable insight into dynamic response properties of FeCO neurons. Tonic neurons were modeled as dynamic low-pass filters followed by static nonlinear filters with constant gain, whereas phasic neurons were modeled as dynamic band-pass filters followed by static nonlinear filters acting as rectifiers (Kondoh et al., 1995). An unexpected finding was a frequency-determined shift in response sensitivity from position to velocity, and from velocity to acceleration sensitivity, in some of the neurons.

3.6.3 Stimulating Nonconnective Chordotonal Organs

Nonconnective chordotonal organs include tympanal, subgenual, and Johnston's organs, as well as some unusual scolopidial organs embedded in soft abdominal integument (Sugawara, 1996). Except for the last, all others are stimulated by sine stimuli delivered either as sound or as substrate vibration.

3.6.3.1 Sound Stimulation

Acoustic stimulation includes near-field (particle displacement) and far-field sound directed to the sense organ. Near-field stimulation is used for hair sensilla and antennal (and, indirectly, Johnston's organ) stimulation in which the acoustic source resides within about 1 wavelength (1 λ) of the sound stimulus, and the considerable air particle movement in the sound field promotes little radiation of sound from the source (Peterson and Gross, 1967). Far-field stimulation utilizes the sound source at least 1 λ (lowest frequency used) from the sense organ, where the sound field dissipates by 6 dB/doubling of distance from source. This is the typical acoustic stimulus mode for tympanal organs.

3.6.3.2 Speakers, Amplifiers, and Microphones

Speakers and amplifiers used for sound delivery should have a flat frequency response over the bandwidth used for stimulation. Frequency response information of all acoustic stimulating and recording equipment is normally required when publishing. This can be tested with a calibrated microphone and a spectrum analyzer. Usually the microphone is provided with a calibration curve of its frequency response; if lacking, this information should be obtained from the manufacturer.

In near-field stimulation, it is important to position the opening of the sound delivery system < 1 λ from the preparation and to reduce or eliminate echoes from the source. Sound can be delivered via a tube placed near the center of the speaker, with the preparation within one diameter of the tube opening at the opposite end (Figure 3.10a) (Eberl et al., 2000), or with speakers placed close to the preparation (technical details are discussed by Gopfert et al., 1999; Gopfert and Robert, 2002).

3.6.3.3 Anechoic Conditions

Acoustic experiments should be carried out in an environment that absorbs the emitted sound from the speaker. In addition, external sound should not enter the experimental position. Two principles should be kept in mind when designing the experimental enclosure. First, sound reflects maximally from flat, solid surfaces. Second, sound is absorbed (transmitted less) by dense material. Thus, an enclosure should be lined with surface material that scatters and absorbs sound: a rough, broken surface (often pyramidal or conical foam material, even egg cartons), a porous or composite material (synthetic rubber or plastic foam, fiberglass, or chopped fiber battings), or hanging drapes and sheets of absorbing material. Walls of the enclosure should be solid (preferably without air gaps inside) and dense to prevent external sound from entering. One of the most cost-efficient linings on walls is a double thickness of plasterboard (gypsum board, Gib board) under the nonreflective surface. Doors to rooms should not be hollow and should be faced with the same double plasterboard lining. Air gaps around access (door, cable ports) should be acoustically blocked.

It is important to minimize the possibility of sound reflection returning to the preparation from a wall opposite the speaker. In ideal cases, this is done by having the preparation in a tunnel with good sound absorption at the end. The floor or bottom of the enclosure should be equally sound absorbing (e.g., a deep pile carpet).

A different approach was used by Hill (1980) to study the tympanal organ of the tree weta (Figure 3.10c). This was a close-field system in which the foreleg was

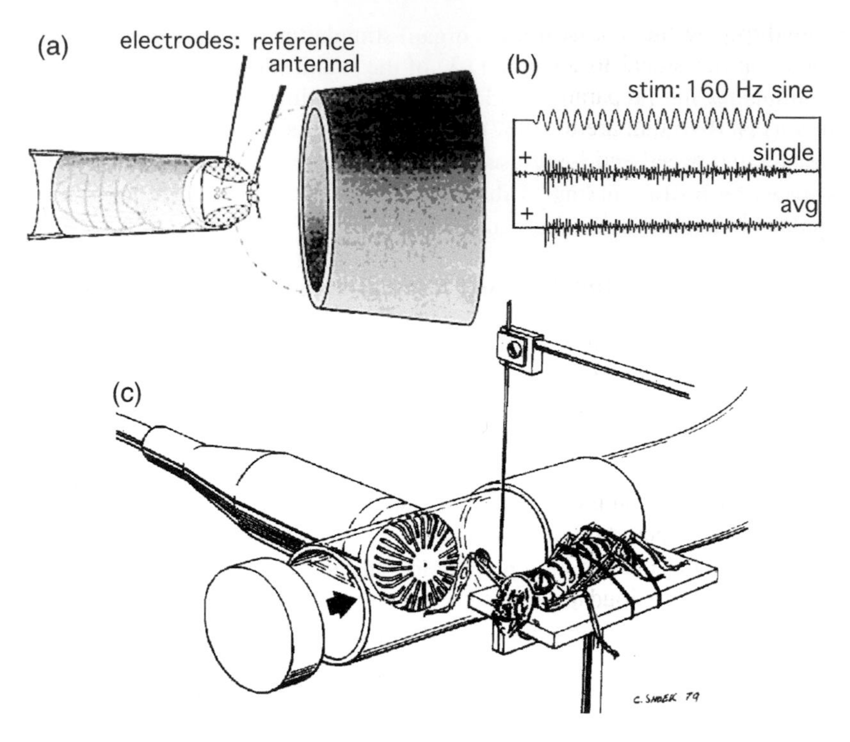

FIGURE 3.10 Recording methods for acoustic stimulation in near and far fields. **a**: Recording from Johnston's organ nerve with tapered tungsten electrodes while *Drosophila* is held in end of micropipette tip. The sound stimulus is delivered through a 5 mm tube placed a few millimeters from the fly to obtain near-field conditions (indicated by dashed line). **b**: Tapered sine stimulus to speaker produced damped extracellular spike discharge from organ for single and averaged recordings. **c**: Method of intracellular recording from afferent axons in the tympanal nerve in prothoracic leg of the New Zealand tree weta. Femur is immobilized after tibia (bearing the tympanal organ) is placed through a small hole in a sound pipe plugged at the end. A microphone monitors the stimulus through another port in the sound pipe. (**a, b** from Eberl, D.F. et al., *J. Neurosci.* **20**:5981–5988, 2000. With permission. **c** from Hill, K.G., *J. Comp. Physiol.* **152**:475–482, 1983a. With permission.)

inserted into a large flexible tube, blocked at one end and positioned next to the speaker at the other end. The sound field was entirely enclosed in the tube and monitored by a 1-in. microphone at the position of the leg.

3.6.3.4 Waveforms

The sine generator should deliver standard-length test pulses (~50 to 200 msec) over the full frequency range expected, which can extend from 0.5 to 20 kHz (sonic range) or even extend into the ultrasonic range (> 20 kHz). The onset and offset of the test pulse should be tapered to the baseline so that abrupt transitions are eliminated: the pulse shape is, therefore, a symmetrical double trapezoid (Figure 3.9h). Rise and fall times are usually on the order of 1 to 10 msec. This is easily achieved by multiplying the sine signal by the envelope shape using a computer, and then delivering the signal to an amplifier driving the speaker. The stimulus should be monitored at the preparation to determine whether it is distorted or contaminated by echoes. If stimuli are in the ultrasound range, special transducers are required to generate the sound and high-frequency microphones must be used (e.g., Bruel and Kjaer 4135 1/4 in. microphone[(5)]). Useful information and guidelines are found in Sales and Pye (1974).

3.6.3.5 Vibration Stimulation

Vibration-sensitive organs (e.g., subgenual organ) can be extraordinarily sensitive to surface vibrations delivered to the insect's appendages (Shaw, 1994a). Researchers have often used a peizoelectric element or a minishaker (electromagnetic vibrator, e.g., Ling V101, Bruel and Kjaer 4369) to deliver stimulus pulses to the tarsus of an intact insect's leg waxed (low–melting point colophonium/beeswax mixture) to a T-piece mounted on the driver (e.g., Kühne, 1982). The fidelity of the stimulus at the preparation should be checked with an accelerometer (e.g., Bruel and Kjaer Type 4338[(5)]; PCB Piezoelectronics M353B65[(22)]) mounted to the driven T-piece. Another method utilized a loudspeaker with the legs of a lacewing resting on the cone (Devetak and Amon, 1997), while the insect's body was immobilized on a static rod. Both methods also can generate airborne sound, and it is important to do control recording experiments with the leg(s) detached from the vibratory surface to determine the contribution of sound to the response. A third method involved releasing a bent spring steel platform to deliver a reproduceable damped vibratory stimulus, which was monitored by an infrared LED (Shaw, 1994a).

Frequencies for vibration studies tend to be restricted from around 20–30 Hz to 3–5 kHz. They are usually delivered in brief pulses (50 to 200 msec) with rise and fall times of ~10 msec, repeated 1/sec.

3.7 CONCLUSIONS AND OUTLOOK

The increase in knowledge of chordotonal organs since the last major review of this topic (1998) has been rapid and exciting; yet these complex mechanoreceptors continue to defy a complete understanding of their operation. Pressing questions remain. What is the role of the cilium and does it contain MACs? How is the scolopidium stimulated, and is it stimulated similarly in both types of scolopidium (mononematic and amphinematic) and in connective and nonconnective chordotonal organs? What are the roles of the proteins uncovered by recent genetic studies? Does Ca^{2+} have a role in the scolopidial processes leading to the output of action potentials? To overcome barriers to answering these questions, the analytical tools of the future must derive from multidisciplinary approaches using neurophysiology, genetics, immunocytochemistry, physics, and nanotechnology.

Another facet of chordotonal organ physiology is sure to remain open to continued investigation: the uncovering of evolutionary variety in form and function of these fascinating sense organs. The recent discoveries of cyclopean ears, deep tissue chordotonal organs in the genital chamber, tympanal organs in neck tissue, and even a cardiac chordotonal organ, all herald a wealth of new discoveries awaiting the curious and persistent scientist.

ACKNOWLEDGMENTS

I thank the following colleagues for helpful input and discussion of recent research findings: Ali Steinbrecht, Hiroshi Nishino, Andrew French, Päivi Torkkeli, Tom Matheson, Dan Eberl, Mark Tyrer, and Dave Yager. I also thank Tom Christensen for his solid support and extended efforts in editing this chapter.

REFERENCES

Baker JD, Esenwa V, Kernan M (2001) Uncoordinated is a novel protein required for the organization and function of ciliogenic centrioles. *Mol. Biol. Cell* 12 (Suppl.) 447a.

Ball EE, Young D (1974) Structure and development of the auditory system in the prothoracic leg of the cricket *Teleogryllus commodus* (Walker). II. Postembryonic development. *Z. Zellforsch.* 147:313–324.

Bässler U (1967) Zur Steuerung des Springens bei der Wanderheuschrecke Schistocerca gregaria. *Kybernetik* 4:112.

Bässler U (1979) Effects of crossing the receptor apodeme of the femoral chordotonal organ on walking, jumping and singing in locusts and grasshoppers. *J. Comp. Physiol.* **134**:173–176.

Belton P (1989) The structure and probable function of the internal cuticular parts of Johnston's organ in mosquitoes (*Aedes aegyptii*). *Can. J. Zool.* **67**:2625–2632.

Bode W (1986) Fine structure of the scolopophorous organs in the pedicel of three species of *Thysanoptera* (*Insecta*). *Int. J. Insect Morphol. Embryol.* **15**:139–154.

Bodmer R, Barbel S, Sheperd S, Jack JW, Jan LY, Jan YN (1987) Transformations of sensory organs by mutations of the cut locus of *D. melanogaster. Cell* **51**:293–307.

Bodmer R, Carretto R, Jan YN (1989) Neurogenesis of the peripheral nervous system in *Drosophila* embryos. *Neuron* **3**:21–32.

Brand AH (1999). GFP as a cell and developmental marker in the *Drosophila* nervous system. In: Sullivan KF, Kay S, Eds., *Green Fluorescent Proteins.* San Diego, Academic Press. *Methods in Cell Biology,* 58:165–181.

Brand AH, Perrimon N (1993) Targeted gene expression as a means of altering cell fates and generating dominant phenotypes. *Development* **118**:401–415.

Bräunig P, Hustert R, Pflüger H-J (1981) Distribution and specific central projections of mechanoreceptors in the thorax and proximal leg joints of locusts. I. Morphology, location and innervation of internal proprioceptors of pro- and metathorax and their central projections. *Cell Tissue Res.* **216**:57–77.

Bräunig P (1982) The peripheral and central nervous organization of the locust coxo-trochanteral joint. *J. Neurobiol.* **13**:413–433.

Bräunig P (1985) Strand receptors associated with the femoral chordotonal organs of locust legs. *J. Exp. Biol.* **116**:331–341.

Burns M D (1974) Structure and physiology of the locust femoral chordotonal organ. *J. Ins. Physiol.* **20**:1319–1339.

Buschges A, Kittman R, Ramirez J-M (1993) Octopamine effects mimick state-dependent changes in a proprioceptive feedback system. *J. Neurobiol.* **24**:598–610.

Caldwell JC, Eberl DF (2002) Towards a molecular understanding of *Drosophila* hearing. *J. Neurobiol.* **53**:172–189.

Caldwell, JC, Miller MM, Wing S, Soll DR, Eberl DF (2003) Dynamic analysis of larval locomotion in *Drosophila* chordotonal organ mutants. *Proc. Nat. Acad. Sci.* **100**:16053–16058.

Chung YD, Zhu J, Han Y-G, Kernan M J (2001) nompA encodes a PNS-specific, ZP domain protein required to connect mechanosensory dendrites to sensory structures. *Neuron* **29**:415–428.

Coillot JP, Boistel J. (1969) Localisation et déscription des récepteurs à l'étirement au niveau de l'articulation tibio-femorale de la patte sauteuse du criquet *Schistocerca gregaria. J. Insect. Physiol.* **14**:1661–1667.

Cökl A, Kalmring K, Roessler W (1995) Physiology of atympanate tibial organs in forelegs and midlegs of the cave-living ensifera, *Troglophilus neglectus* (Raphidophoridae, Gryllacroidea). *J. Exp. Zool.* **273**:376–388

Crouau Y (1983) Ultrastructure study of the 9 + 0 type cilia of insect and crustacean chordotonal sensilla. Generalisation of the hypotheses on the motility of 9 + 0 type cilia. *J. Submicrosc. Cytol.* **15**:295–299.

Cruse H (1996) *Neural Networks as Cybernetic Systems.* G. Thieme Verlag; Thieme Medical Publishers, Stuttgart.

Devetak D, Amon T (1997) Substrate vibration sensitivity of the leg scolopidial organs in the green lacewing *Chrysoperla carnea. J. Insect Physiol.* **43**:433–437.

Dreller C, Kirchner WH (1993) Hearing in honeybees: localization of the auditory sense organ. *J. Comp. Physiol. A.* **173**:275–279.

Dubruille R, Laurençon A, Vandaele C, Shishido E, Coulon-Bublex M, Swoboda P, Couble P, Kernan M, Durand B (2002) *Drosophila* regulatory factor X is necessary for ciliated sensory neuron differentiation. *Development* **129**:5487–5498.

Dulcis D, Levine RB (2004) Remodeling of a larval skeletal muscle motorneuron to drive the posterior cardiac pacemaker in the adult moth, *Manduca sexta. J. Comp. Neurol.* **478**:126–142.

Eberl DF (1999) Feeling the vibes: chordotonal mechanisms of insect hearing. *Curr. Opin. Neurobiol.* **9**:389–393.

Eberl DF, Duyk GM, Perrimon N (1997) A genetic screen for mutations that disrupt an auditory response in *Drosophila melanogaster. Proc. Nat. Acad. Sci. USA* **94**:14837–14842.

Eberl DF, Hardy RW, Kernan MJ (2000) Genetically similar transduction mechanisms for touch and hearing in *Drosophila. J. Neurosci.* **20**:5981–5988.

Erxleben C, Ubl J, Kolb H-A (1991) Identifying and characterizing stretch-activated channels. In: *Molecular Neurobiology: A Practical Approach.* Wheal H, Chad, J, Eds., IRL Press, Oxford, pp. 75–91.

Faucheux M J (1985) Structure of the tarso-pretarsal chordotonal organ in the imago of *Tineola bisselliella* Humm. (Lepidoptera: Tineidae). *Int. J. Insect Morphol. Embryol.* **14**:147–154.

Field LH, Burrows M (1982) Reflex effects of the femoral chordotonal organ upon leg motor neurones of the locust. *J. Exp. Biol.* **101**:265–285.

Field LH, Matheson T (1998) Chordotonal organs of insects. *Adv. Insect Physiol.* **27**:1–228.

Field LH, Pflüger H-J (1989) The femoral chordotonal organ: a bifunctional orthopteran (*Locusta migratoria*) sense organ. *Comp. Biochem. Physiol.* **93A**:729–743.

Field LH, Rind FC (1981) A single insect chordotonal organ mediates inter- and intra-segmental leg reflexes. *Comp. Biochem. Physiol.* **68A**:99–102.

Field LH, Meyer MR, Edwards JS (1994) Selective expression of glionexin, a glial glycoprotein, in insect mechanoreceptors. *J. Neurobiol.* **25**:1017–1028.

Finlayson LH (1968) Proprioceptors in the invertebrates. *Symp. Zool. Soc. Lond.* **23**:217–249.

Finlayson LH (1976) Abdominal and thoracic receptors in insects, centipedes and scorpions. In: *Structure and Function of Proprioceptors in the Invertebrates.* Mill PJ, Ed., Chapman & Hall, London pp. 153–212.

Fischer H, Wolf H, Büschges A (2002) The locust tegula: kinematic parameters and activity pattern during the wing stroke. *J. Exp. Biol.* **205**:1531–1545.

French AS (1992) Mechanotransduction. *Ann. Rev. Physiol.* **54**:135–152.

French, A.S., Torkkeli, P.H. and Seyfarth, E-A. (2002) From stress and strain to spikes: mechanotransduction in spider slit sensilla. *J. Comp. Phys. A* **188**:739–752.

Fudalewicz-Niemczyk W, Rosciszewska M (1972) The innervation and sense organs of the wings of *Gryllus domesticus L.* (Orthoptera). *Acta Biologoca Cracoviensia* **15**:35–51.

Gopfert MC, Robert D (2002) The mechanical basis of *Drosophila* audition. *J. Exp. Biol.* **205**:1199–1208.

Gopfert MC, Briegel H, Robert D (1999) Mosquito hearing: sound-induced antennal vibrations in male and female *Aedes aegypti. J. Exp. Biol.* **202**:2727–2738.

Gray EG (1960) The fine structure of the insect ear. *Phil. Trans. Roy. Soc. B* **243**:75–94.

Haugland RP (1996) *Handbook of Fluorescent Probes and Research Chemicals.* Sixth Ed. Molecular Probes Inc., Eugene, Oregon.

Hill KG (1980) Physiological characteristics of auditory receptors in *Hemideina crassidens* (Blanchard) (Ensifera: Stenopelmatidae). *J. Comp. Physiol.* **141**:39–46.

Hill KG (1983a) The physiology of locust auditory receptors. I. Discrete depolarizations of receptor cells. *J. Comp. Physiol.* **152**:475–482.

Hill KG (1983b) The physiology of locust auditory receptors. II. Membrane potentials associated with the response of the receptor cell. *J. Comp. Physiol.* **152**:483–493.

Hofmann T, Koch UT (1985) Acceleration receptors in the femoral chordotonal organ in the stick insect *Cuniculina impigra. J. Exp. Biol.* **114**:225–237.

Horikawa K, Armstrong WE (1988) A versatile means of intracellular labeling: injection of biocytin and its detections with avidin conjugates. *J. Neurosci. Methods* **25**:1–11.

Howse PE (1968). The fine structure and functional organization of chordotonal organs. *Symp. Zool. Soc. Lond.* **23**:167–198.

Hudspeth AJ (1997) How hearing happens. *Neuron* **19**:947–950.

Hustert R (1982) The propriceptive function of a complex chordotonal organ associated with the mesothoracic coxa in locusts. *J. Comp. Physiol.* **147**:389–399.

Hustert R (1983) Proprioceptor responses and convergence of proprioceptive influence on motoneurones in the mesothoracic thoraco-coxal joint of locusts. *J. Comp. Physiol.* **150**:77–86.

Jan LY, Jan, YN (1982) Antibodies to horseradish peroxidase as specific neuronal markers in *Drosophila* and grasshopper embryos. *Proc. Nat. Acad. Sci.* **79**:2700–2704.

Jan YN, Jan LY (1993) The peripheral nervous system. In: *The Development of* Drosophila melanogaster. Bate M, Arias AM, Eds., New York, Cold Spring Harbor Laboratory Press, pp. 1207–1244.

Kalmring K, Rössler W, Ebendt R, Ahi J, Lakes R (1993) The auditory receptor organs in the forelegs of bushcrickets: physiology, receptor cell arrangement, and morphology of the tympanal and intermediate organs of three closely related species. *Zool. J. Physiol.* **97**:75–94.

Kalmring K, Rössler W, Unrast C (1994) Complex tibial organs in the forelegs, midlegs and hindlegs of the bushcricket *Gampsocleis gratiosa* (Tettigoniidae): comparison of the physiology of the organs. *J. Exp. Zool.* **270**:155–161.

Keil TA (1997) Functional morphology of insect mechanoreceptors. *Microscopy Res. Tech.* **39**:506–531.

Kim J, Chung YD, Park D, Chol S, Shin DW, Soh H, Lee HW, Son W, Yi J, Park C-S, Kernan MJ, Kim C (2003). A TRPV family ion channel required for hearing in *Drosophila. Nature* **424**:81–84.

Kondoh Y, Okuma J, Newland PL (1995) Dynamics of neurones controlling movements of a locust hind leg: Wiener kernal analysis of the responses of proprioceptive afferents. *J. Neurophysiol.* **73**:1829–1842.

Kuhne R (1982) Neurophysiology of the vibration sense in locusts and bushcrickets: response characteristics of single receptor units. *J. Ins. Physiol.* **28**:155–163.

Kutsch W, Hanloser H, Reinecke M (1980) Light- and electron-microscopic analysis of a complex sensory organ: the tegula of *Locusta migratoria. Cell Tissue Res.* **210**:461–478.

Lakes-Harlan R, Hell, KG (1992) Ultrasound sensitive ears in a parasitoid fly. *Naturwissenschaften.* **79**:224–226.

Laurent G (1987) The role of spiking local interneurones in shaping the receptive fields of intersegmental interneurones in the locust. *J. Neurosci.* **7**:2977–2989.

Lee JK, Altner H (1986) Structure, development and death of sensory cells and neurons in the pupal labial palp of the butterflies *Pieris rapae* (L.) and *Pieris brassicae* (L.) (Insecta, Lepidoptera). *Cell Tiss. Res.* **244**:371–383.

Lee J-K, Kim C-W, Altner H (1988) Differences in degeneration of the apical scolopidial organ in the labial palp of Lepidoptera during pupal development. *Zoomorph.* **108**:77–83.

Lin Y, Kalmring K, Jatho M, Sickmann T, Rössler, W (1993) Auditory receptor organs in the forelegs of *Gampsocleis gratiosa* (Tettigoniidae): morphology and function of the organs in comparison to the frequency parameters of the conspecific song. *J. Exp. Zool.* **267**:377–388.

Macmillan DL (1976) Arthropod apodeme tensin receptors. In: Mill PJ, Ed. *Structure and Function of Proprioceptors in the Invertebrates.* Chapman & Hall, London, pp. 427–439.

Marmarelis PZ, Marmarelis VZ (1978) *Analysis of Physiological Systems. The White Noise Approach.* Plenum Press, New York.

Mason AC, Morris GK, Hoy RR (1999) Peripheral frequency mis-match in the primitive ensiferan *Cyphoderris monstra* (Orthoptera: Haglidae). *J. Comp. Physiol. A* **184**:543–551.

Matheson T (1992) Range fractionation in the locust metathoracic femoral chordotonal organ. *J. Comp. Physiol.* **170**:509–520.

Matheson T (1997) Octopamine modulates the responses and presynaptic inhibition of proprioceptive sensory neurones in the locust *Schistocerca gregaria. J. Exp. Biol.* **200**:1317–1325.

Matheson T, Field LH (1995) An elaborate tension receptor system highlights sensory complexity in the hind leg of the locust. *J. Exp. Biol.* **198**:1673–1689.

McIver, SB (1985). Mechanoreception. In: *Comprehensive Insect Physiology, Biochemistry and Pharmacology.* Kerkut GA and Gilbert LI, Eds., Pergamon, Oxford, **6**:71–132.

McVean A, Field LH (1996) Communication by substrate vibration in the New Zealand tree weta *Hemideina femorata* (Stenopelmatidae: Orthoptera). *J. Zool. London* **239**:101–122.

Merritt DJ, Hawken A, Whitington PM (1993) The role of the cut gene in the specification of central projections by sensory axons in *Drosophila. Neuron* **10**:741–752.

Mesce KA, Amos T, Clough SM (1993) A light insensitive method for contrast enhancement of insect neurons filled with a cobalt-lysine complex. *Biotechnic Histochem.* **68**:222–228.

Meyer MR, Reddy RG, Edwards JS (1987) Immunological probes reveal spatial and developmental diversity in insect neuroglia. *J. Neurosci.* **7**:512–521.

Michelsen A, Larsen ON (1985) Hearing and sound. In: *Comparative Insect Biochemistry and Physiology*, vol 6. Kerkut GA, Gilbert LI, Eds., Pergamon Press Oxford, UK.

Mill, PJ (1976). *Structure and Function of Proprioceptors in the Invertebrates.* London: Chapman & Hall.

Mill PJ, Price RN (1976) Analysis of proprioceptive information. In *Structure and Function of Proprioceptors in the Invertebrates.* Mill, PJ, Ed., London, Chapman & Hall, pp. 605–639.

Miller LA (1970) The structure of the green lacewing tympanal organ. *J. Morphol.* **131**:359.

Moran DT, Rowley III JC (1975) The fine structure of the cockroach subgenual organ. *Tissue Cell* **7**:91–105.

Möss D (1971) Sinnesorgane im Bereich des Flügels der Feldgrille (*Gryllus campestris* L.) und ihre Bedeutung für die Kontrolle der Singbewegung und die Einstellung der Flugellage. *Z. Verlag. Physiol.* **73**:53–83.

Moulins M (1976). Ultrastructure of chordotonal organs. In *Structure and Function of Proprioceptors in the Invertebrates.* Mill, PJ, Ed., Chapman & Hall, London, pp. 387–426.

Mücke A (1991) Innervation pattern and sensory supply of the midleg of *Schistocerca gregaria* (Insecta, Orthoptera). *Zoomorph.* **110**:175–187.

Naka K-I, Sakuranaga M, Ando Y-I (1985) White noise as a tool in vision physiology. In: *Progress in Clinical and Biological Research*, vol. 176, Contemporary Sensory Neurobiology. Correia MJ, Perachio AA, Eds., Alan R. Liss, New York, pp. 307–322.

Nalbach G (1993) The halteres of the blowfly *Calliphora.* I. Kinematics and dynamics. *J. Comp. Physiol.* **173**:293–300.

Nässel DR (1987) Strategies for neuronal marking in arthropod brains. In *Arthropod Brain: Its Evolution, Development, Structure and Functions.* Gupta, AP, Ed., John Wiley & Sons Inc., NY, pp. 549–570.

Nishino H, Field LH (2003) Somatotopic mapping of chordotonal organs in a primitive ensiferan, the New Zealand tree weta *Hemideina femorata*: II. Complex tibial organ. *J. Comp. Neurol.* **464**:327–342.

Nishino H, Sakai M (1997) Three neural groups in the femoral chordotonal organ of the cricket *Gryllus bimaculatus*: central projections and soma arrangement and displacement during joint flexion. *J. Exp. Biol.* **200**:2583–2595.

Nishino H, Sakai M and Field LH (1999) Two antagonistic functions of neural groups of the femoral chordotonal organ underlie thanatiosis in the cricket *Gryllus bimaculatus* DeGeer. *J. Comp. Physiol. (A)* **185**:143–155.

Oldfield BP (1982) Tonotopic organisation of auditory receptors in Tettigoniidae (Orthoptera: Ensifera). *J. Comp. Physiol. A* **147**:461–469.

Oldfield BP (1985a) The role of the tympanal membranes and the receptor array in the tuning of auditory receptors in bushcrickets. In: *Acoustic and Vibrational Communication in Insects.* Kalmring K, Elsner N, Eds., Paul Parey, Berlin, pp. 17–22.

Oldfield BP (1985b) The tuning of auditory receptors in bushcrickets. *Hearing Res.* **17**:27–35.

Oldfield BP, Hill KG (1986) Functional organisation of insect auditory sensilla. *J. Comp. Physiol. A* **158**:27–34.

Orchard, I. (1975). The structure and properties of an abdominal chordotonal organ in *Carausius morosus* and *Blaberus discoidalis. J. Insect Physiol.* **21**:1491–1499.

Pearson KG, Hedwig B, Wolf H (1989) Are the hindwing chordotonal organs elements of the locust flight pattern generator? *J. Exp. Biol.* **144**:235–255.

Peterson APG, Gross EE Jr (1967) *Handbook of Noise Measurement* 6th ed. General Radio Co., West Concord, MA.

Pflüger H-J, Field LH (1999) A locust chordotonal organ coding for proprioceptive and acoustic stimuli. *J. Comp. Physiol. A* **184**:169–183.

Pill CEJ, Mill PJ (1981) The structure and physiology of abdominal proprioceptors in larval dragonflies (*Anisoptera*). *Odonotologica* **10**:117–130.

Pitman RM, Tweedle CD, Cohen MJ (1972) Branching of central neurons: intracellular cobalt injection for light and electron microscopy. *Science* **176**:412–414.

Plotnikova SI, Nevmyvaka GA (1980) The methylene blue technique: classic and recent applications to the insect nervous system. In: *Neuroanatomical Techniques – Insect Nervous System.* Strausfeld NJ, Miller TA, Eds., Springer-Verlag, Berlin, pp. 1–15.

Pollack GS, Imaizumi K (1999) Neural analysis of sound frequency in insects. *BioEssays* **21**:295–303.

Prier KR, Boyan GS (2000) Synaptic input from serial chordotonal organs onto segmentally homologous interneurons in the grasshopper *Schistocerca gregaria. J. Insect Physiol.* **46**:297–312.

Rice, MJ (1975) Insect mechanoreceptor mechanisms. In *Sensory Physiology and Behavior.* Galun R, Hillman P, Parnas I, and Werman R, Eds., Plenum, New York. pp. 135–165.

Robert D, Read MP, Hoy, RR (1994) The tympanal hearing organ of the parasitoid fly *Ormia ochracea* (Diptera, Tachinidae, Ormiini). *Cell Tissue Res.* **275**:63–78.

Roeder KD, Treat AE, Van de Berg JS (1970) Distal lobe of the pilifer: an ultrasonic receptor in choerocampine hawkmoths. *Science* **170**:1098–1099.

Rosciszewsk M, Fudalewicz-Neimczyk W (1972) The peripheral nervous system of the larva of *Gryllus domesticus* L. (Orthoptera. Part II. Mouthparts. *Acta Biol. Cracov.* **17**:19–39.

Rössler W (1992) Functional morphology and development of tibial tympanal organs in the legs I, II and III of the bushcricket *Ephippiger ephippiger* (Insecta, Ensifera). *Zoo-morph.* **112**:181–188.

Sakai M, Yamaguchi T (1983) Differential staining of insect neurones with nickel and cobalt. *J. Ins. Physiol.* **29**:393–397.

Sales GD, Pye JD (1974) *Ultrasonic Communication by Animals.* Chapman & Hall, London.

Sarpal R, Todi SV, Sivan-Loukianova E, Shirolikar S, Subramanian N, Raff EC, Erickson JW, Ray K, Eberl DF (2003) *Drosophila* KAP interacts with Kinesin II motor subunit KLP64D to assemble chordotonal sensory cilia, but not sperm tails. *Curr. Biol.* **13**:1687–1696.

Schmidt K (1969) Der Feinbau der stiftfürhenden Sinnesorgane im Pedicellus der Florfliege Chrysopa Leach (Chrysopidae, Plannipennia). *Zeit. Zellforsch. Mikrosk. Anat.* **99**:357–388.

Schnorbus H (1971) Die Subgenualen Sinnesorgane von *Periplaneta americana*: Histologie und Vibrationsschwellen. *Z. Ver. Physiol.* **71**:14–48.

Sharma Y, Cheun U, Larsen EW, Eberl DF (2002) pPTGAL, a convenient Gal4 p-element vector for testing expression of enhancer fragments in *Drosophila. Genesis* **34**:115–118.

Shaw SR (1994a) Re-evaluation of the absolute threshold and response mode of the most sensitive known vibration detector, the cockroach's subgenual organ: a cochlea-like displacement threshold and a direct response to sound. *J. Neurobiol.* **25**:1167–1185.

Shaw SR (1994b) Detection of airborne sound by a cockroach 'vibration detector': a possible missing link in insect auditory evolution. *J. Exp. Biol.* **193**:13–47.

Shelton PMJ, Stephen RO, Scott JJA, Tindall AR (1992) The apodeme complex of the femoral chordotonal organ in the metathoracic leg of the locust *Schistocerca gregaria. J. Exp. Biol.* **163**:345–358.

Shepherd P (1973) Musculature and innervation of the desert locust *Schistocerca gregaris* (Forskal). *J. Morph.* **139**:439–464.

Slifer EH, Sekhon SS (1975) The femoral chordotonal organs of a grasshopper, Orthoptera, Acrididae. *J. Neurocytol.* **4**:419–438.

Smith SA, Shepherd D (1996) The central organisation of proprioceptive sensory neurons in *Drosophila* revealed with the enhancer-trap technique. *J. Comp. Neurol.* **364**:311–323.

Spangler HG (1988a) Hearing in tiger beetles (Cicindelidae). *Physiol. Entomol.* **12**:447–452.

Spangler HG (1988b) Moth hearing, defense and communication. *Ann. Rev. Entomol.* **33**:59–81.

Steinbrecht RA, Zierold K, Eds. (1987). *Cryotechniques in Biological Electron Microscopy.* Springer-Verlag, Berlin.

Steinbrecht RA (1992) Cryotechniques with sensory organs. *Microsc. Anal.* (September), 21–23.

Steinbrecht RA (1993) Freeze-substitution for morphological and immunocytochemical studies in insects. *Micr. Res. Tech.* **24**:488–508.

Stockbridge LL, French AS (1989) Ion channels in isolated mechanosensory cells from the connective chordotonal organ in the pedicel of the American cockroach. *Neurosci. Abs.* **15**:1287.

Strausfeld NJ, Seyan HS, Wohlers DJ, Bacon JP (1983) Lucifer yellow histology for insects. In: *Functional Neuroanatomy.* Strausfeld NJ, Ed., Springer, Heidelberg, pp. 132–155.

Stumpner A, von Helversen D (2001) Evolution and function of auditory systems in insects. *Naturwissenschaften* **88**:159–170.

Suga N (1960) Peripheral mechanisms of hearing in the locust. *Jap. J. Physiol.* **10**:533–546.

Sugawara T (1996) Chordotonal sensilla embedded in the epidermis of the soft integument of the cricket, *Teleogryllus commodus. Cell Tissue Res.* **284**:125–142.

Theophilidis G, Kravari K (1994) Dimethylsulfoxide (DMSO) eliminates the response of the sensory neurones of an insect mechanoreceptor, the femoral chordotonal organ of *Locusta migratoria*, but blocks conduction of their sensory axons at much higher concentrations: a possible mechanism of analgesia. *Neurosci. Letters* **181**:91–94.

Theophilidis G, Pappa A, Papadopoulou-Mourkidou E (1993) The neurophysiological effects of deltamethrin and fluvalinate on an insect mechanoreceptor, the metathoracic femoral chordotonal organ of *Locusta migratorta. Pestic. Biochem. Physiol.* **45**:198–209.

Thurm U, Erler G, Gödde J, Kastrup H, Keil Th, Völker W, Vohwinkel B (1983) Cilia specialized for mechanoreception. *J. Submicrosc. Cytol.* **15**:151–155.

Toh Y, Yokohari F (1985) Structure of the antennal chordotonal sensilla of the American cockroach. *J. Ultrastruct. Res.* **90**:124–134.

Torkelli PH, French AS (1999) Primary culture of antennal mechanoreceptor neurons of *Manduca sexta. Cell Tiss. Res.* **297**:301–309.

Tyrer NM, Shaw MK, Altman JS (1980) Intensification of cobalt-filled neurons in sections (light and electron microscopy). In: *Neuroanatomical Techniques – Insect Nervous System.* Straulfeld NJ, Miller TA, Eds., Springer, Berlin.

Tyrer NM, Shepherd D, Williams DW (2000) Methods for imaging labeled neurons together with neuropil features in *Drosophila. J. Histochem. Cytochem.* **48**:1575–1581.

Usherwood PNR, Runion HI, Campbell I (1968) Structure and physiology of a chordotonal organ in the locust leg. *J. Exp. Biol.* **48**:305–323.

van Rössel P, Brand AH (2002) Imaging into the future: visualizing gene expression and protein interactions with fluorescent proteins. *Nature Cell Biol.* **4**:E15–E20.

von Staaden MJ, Römer H (1998) Evolutionary transition from stretch to hearing organs in ancient grasshoppers. *Nature* **394**:773–776.

Wales W (1976) The receptors of the mouthparts and gut of arthropods. In: *Structure and Function of Proprioceptors in the Invertebrates.* Mill PJ, Ed., Chapman & Hall, London, pp. 213–242.

Walker RG, Willingham AT, Zucke CS (2000) A *Drosophila* mechanosensory transduction channel. *Science* **287**:2229–2234.

Williamson R, Burns MD (1978) Multiterminal receptors in the locust leg. *J. Insect. Physiol.* **24**:661–666.

Wolfrum U (1990). Actin filaments: the main components of the scolopale in insect sensilla. *Cell Tissue Res.* **261**:85–96.

Wolfrum U (1991a) Centrin- and α-actinin-like immunoreactivity in the ciliary rootlets of insect sensilla. *Cell Tissue Res.* **266**:231–238.

Wolfrum U (1991b) Tropomyosin is co-localised with the actin filaments of the scolopale in insect sensilla. *Cell Tissue Res.* **265**:11–17.

Wolfrum U (1992) Cytoskeletal elements in arthropod sensilla and mammalian photoreceptors. *Biol. Cell.* **76**:373–381.

Wolfrum U (1997) Cytoskeletal elements in insect sensilla. *J. Ins. Morphol. Embryol.* **26**:191–203.

Wright BR (1976) Limb and wing receptors in insects, chelicerates and myriapods. In *Structure and Function of Proprioceptors in the Invertebrates.* Mill PJ, Ed., Chapman & Hall, London, pp. 323–386.

Yack JE (2004) The structure and function of auditory chordotonal organs in insects. *Microsc. Res. Tech.* **63**:315–337.

Yack JE, Fullard JH (1993) What is an insect ear? *Ann. Ent. Soc. Amer.* **86**:677–682.

Yack JE, Root BI (1992) The metathoracic wing-hinge chordotonal organ of an atympanate moth, *Actias luna* (Lepidoptera, Saturnidae): a light and electron microscopic study. *Cell Tissue Res.* **267**:455–471.

Yager DD (1989) A diversity of mantis ears: evolutionary implications. *Proc. 5th Int.. Mtg. Orthopterists Soc.*, Segovia, Spain. Orthopterists Society, Ann Arbor, MI.

Yager DD (1990) Sexual dimorphism of auditory function and structures in praying mantises (Mantodea: Dictyoptera). *J. Zool.* (Lond.) **221**:517–537.

Yager DD (1999) Structure, development and evolution of insect auditory systems. *Microsc. Res. Tech.* **47**:380–400.

Yager DD, Spangler HG (1995) Characterization of auditory afferents in the tiger beetle, *Cicindela marutha* Dow. *J. Comp. Physiol. A* **176**:587–599.

Young D (1970) The structure and function of a connective chordotonal organ in the cockroach leg. *Phil. Trans. Roy. Soc. Lond. B* **256**:401–426.

Young D (1973) Fine structure of the sensory cilium of an insect auditory receptor. *J. Neurocytol.* **2**:47–58.

Zill SN (1985) Plasticity and proprioception in the insects. I. Responses and cellular properties of individual receptors of the locust metathoracic femoral chordotonal organ. *J. Exp. Biol.* **116**:435–461.

4 Mechanosensory Integration for Flight Control in Insects

Mark A. Frye and John R. Gray

CONTENTS

4.1 INTRODUCTION

Flying insects show a degree of behavioral robustness and flexibility that even the most sophisticated human-engineered robots cannot match. By taking an integrative "reverse engineering" approach, we may be able to extract and formalize the functional algorithms that drive flight and thereby uncover general principles by which nervous systems coordinate complex behavior. The neural architecture of insects is characterized by massive amounts of feedback from a suite of sensory modalities converging onto sparse efferent motor systems. For example, for flies navigating an odor plume toward a food source, input from thousands of sensors encoding olfactory and visual signals is rapidly integrated with hundreds of thoracic mechanosensory

0-8493-2024-0/05/$0.00+$1.50

channels to shape the activity pattern of only about a dozen wing-steering motoneurons (for a review of flight muscles, see Dickinson and Tu, 1997). The motor pattern must serve to maintain an upright and stable posture while simultaneously propelling the animal toward attractive features of its sensory landscape. Maintaining stability and maneuverability requires continuous feedback about the kinematics and dynamics of moving appendages. Therefore, understanding the neural mechanisms of locomotor behavior in insects, or in any animal, requires an understanding of mechanosensory integration at the cellular, systems, and behavioral levels of organization.

Recent advances in experimental aerodynamics have provided a clear picture of the physical principles by which insects generate the forces necessary both to stay aloft and to execute steering maneuvers. Thus, it is now possible to examine the neural mechanisms of aerodynamic control. Specifically, we seek to integrate free-flight behavior, constituent wing kinematics, and mechanosensory feedback with neural processes of sensorimotor integration. In this overview of some of the most recent experimental advances, we discuss general principles of the aerodynamics of free-flight maneuvers, the neuromuscular control of wing kinematics, and the cellular processes of mechanosensory integration. We conclude with several specific experimental analyses that examine the link between visual and mechanosensory signals for the control of equilibrium reflexes. We emphasize how an integrative research program illuminates the mechanosensory control of flight. This is not meant to be a comprehensive review of the literature; rather, we highlight select model systems that exemplify progress within each topic of discussion.

4.2 THE CONTROL OF FLIGHT TRAJECTORY

In flies, tracking single animals in free flight has revealed the phenomenological characteristics of flight behavior, and reconstructing the patterns of optic flow on the eye has revealed the visual cues that trigger collision avoidance stability maneuvers (see also chapters in Section III). High-speed video analyses have subsequently revealed the detailed wing kinematics and body dynamics of free-flight turns. Finally, replaying natural wing kinematics through dynamically scaled mechanical flapping devices, equipped with force transducers and flow imagers, has revealed the fundamental aerodynamic mechanisms of hovering and free-flight maneuvers.

The next challenge is to examine how the nervous system coordinates subtle changes in wing kinematics and aerodynamics for stability and control of hovering, forward flight, and steering maneuvers. By incorporating elements of free-flight tracking, experimental aerodynamics, and classical electrophysiology, we are beginning to understand the neural mechanisms by which insects perform their deft aerial acrobatics.

4.2.1 THE INITIATION OF FREE-FLIGHT MANEUVERS

A commercially available infrared video tracking system (Fry et al., 2000) enabled experiments that directly examine the influence of visual background texture on flight control in the fruit fly *Drosophila melanogaster* (Figure 4.1a). Tammero and Dickinson (2002b) quantified a pattern of straight trajectories punctuated with rapid

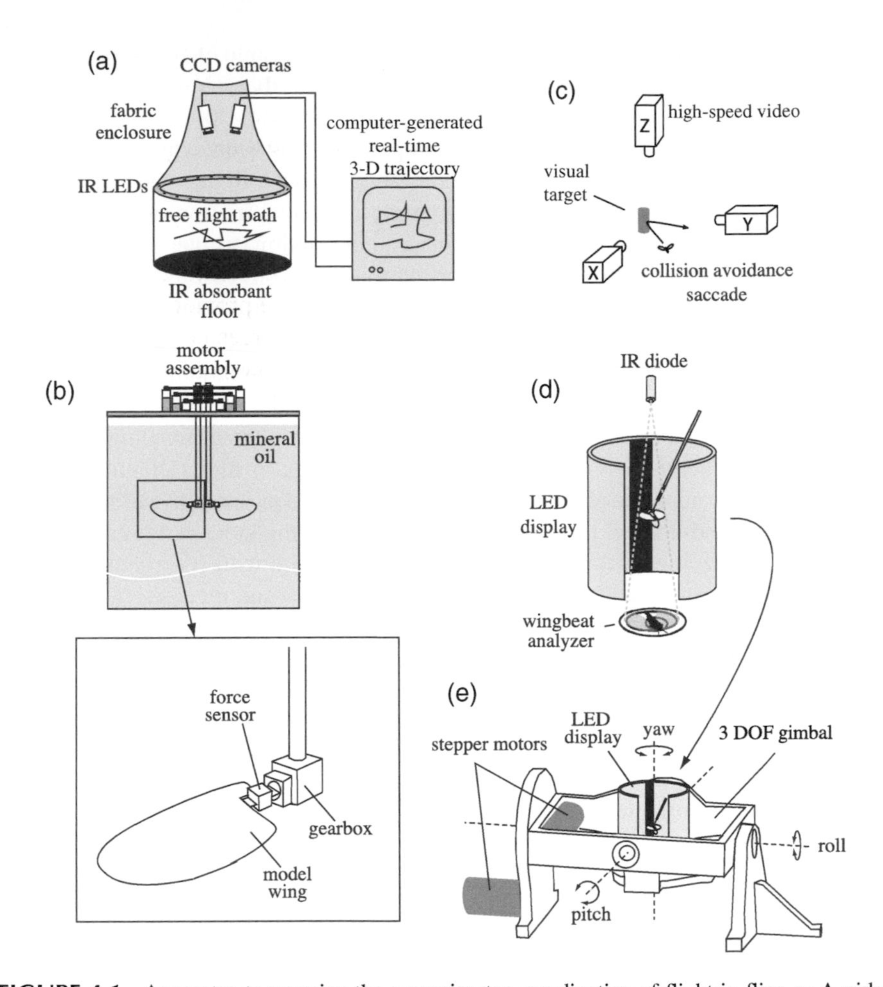

FIGURE 4.1 Apparatus to examine the sensorimotor coordination of flight in flies. **a**: A video tracking system suspended above a 1 m diameter arena records the three-dimensional position of an individual fruit fly as it explores varying sensory landscapes. The infrared tracking system illumination is invisible to the fly, enabling experimenters to manipulate independently the animal's visual surroundings. (Adapted from Tammero, L.F. and Dickinson, MH., *J. Exp. Biol.* **205**:327–343, 2002b. With permission.) **b**: A dynamically scaled flapping wing, suspended in one ton of mineral oil, flaps slowly to match the forces acting on rapidly flapping wings of real insects. The robotic wing is equipped with force transducers to quantify the time course of lift and drag during flight. (Adapted from Dickinson, M.H. et al., *Science* **284:**1954–1960, 1999. With permission.) **c**: High-speed video cameras capture the body and wing kinematics of a collision-avoidance saccade (see text). The wing kinematics of the saccade are replayed through the flapping robot in **b** to determine how aerodynamic forces are coordinated during free-flight maneuvers. **d**: A "virtual reality" flight arena tracks instantaneous changes in wing kinematics with an optical sensor. The wingbeat analyzer output is linked to the motion of the pattern projected on the LED array to close the feedback loop between the fly's attempts to turn and resultant visual motion. **e**: The visual flight arena is mounted within a rotational gimbal. This arrangement simulates both mechanical and visual rotation stimuli of free flight. The optical wingbeat analyzer quantifies the fly's attempts to maintain visual and mechanosensory equilibrium. (Adapted from Sherman, A. and Dickinson, M.H., *J. Exp. Biol.* **207**:133–142, 2004. With permission.)

90-degree turns in freely flying flies (Figure 4.2a). These rapid changes in the angular velocity of the animal's heading are termed *saccades* for their similarity to vertebrate gaze-stabilizing eye movements (Tammero and Dickinson, 2002b). In *Drosophila*, saccades are not initiated by the classic optomotor rotation equilibrium system. Reconstructing the "fly's-eye view" of the arena in free flight, and replaying the image dynamics through a physiological model of visual motion processing in the brain, shows that patterns of visual expansion, not rotation, trigger saccades (Figure 4.2b). As the animal approaches the wall of the arena on a straight trajectory, the image on the retina continues to expand until a threshold value is reached — triggering a collision-avoidance saccade directed away from the near wall (Tammero and Dickinson, 2002a). But what are the cues used (1) to maintain a straight flight course between saccades and (2) to terminate a saccade once it is initiated? The answer lies in mechanosensory circuits crucial for the control of wing kinematics and associated aerodynamics of equilibrium responses in flies (Dickinson, 1999). Insight into the neuromuscular saccade control first requires that we diverge from neurobiology and discuss the aerodynamics of insect flight.

4.2.2 Aerodynamics

Several laboratories have developed large, robotic flapping devices to measure both aerodynamic forces and the structure of fluid flow produced by flapping insect wings. In both cases, large mechanical wings flap such that the magnitude of flight forces are dynamically scaled to match those experienced by the small, quickly beating wings of real insects. The slowed-down time course and scaled size of the robot models make it possible to image directly the flow of fluid around the wing with either smoke (Ellington et al., 1996) or bubbles (Dickinson et al., 1999). Furthermore, force transducers mounted directly on the wing record the time course of lift and drag production within the stroke cycle (Figure 4.1b) (Dickinson et al., 1999; Sane and Dickinson, 2001; Usherwood and Ellington, 2002). Natural wing kinematics, as well as systematic variations of natural trajectories, can be "replayed" through the robot to quantify the influences on force production. Such analyses have revealed several basic aerodynamic mechanisms by which insects stay aloft: *delayed stall*, *rotational circulation*, and *wing–wake interactions* (Sane, 2003). Each mechanism can potentially augment lift by increasing the pressure and velocity gradients produced by fluid moving faster above the wing than below (i.e., by increasing the net circulation of fluid around the wing). Insects orient the wing at a steep angle relative to the direction of ambient fluid flow during each half-stroke. The large angle of attack causes flow to shear, separate from, and reattach to the wing, forming a leading-edge vortex (LEV), which radiates toward the wing tip (Ellington, 1995). The high angle of attack and LEV generate increased downward fluid momentum and greater circulatory force, resulting in increased lift. By visualizing the fluid flow around the robot wing, recent analyses have shown that the LEV remains stable throughout the stroke cycle (Figure 4.2c, part I) (Dickinson et al., 1999; Ellington, 1995). This results from the kinematics particular to flapping insect wings. By contrast, a wing translating linearly in one direction (i.e., not revolving about a hinge) generates an LEV that grows in size until it is shed from the wing, causing

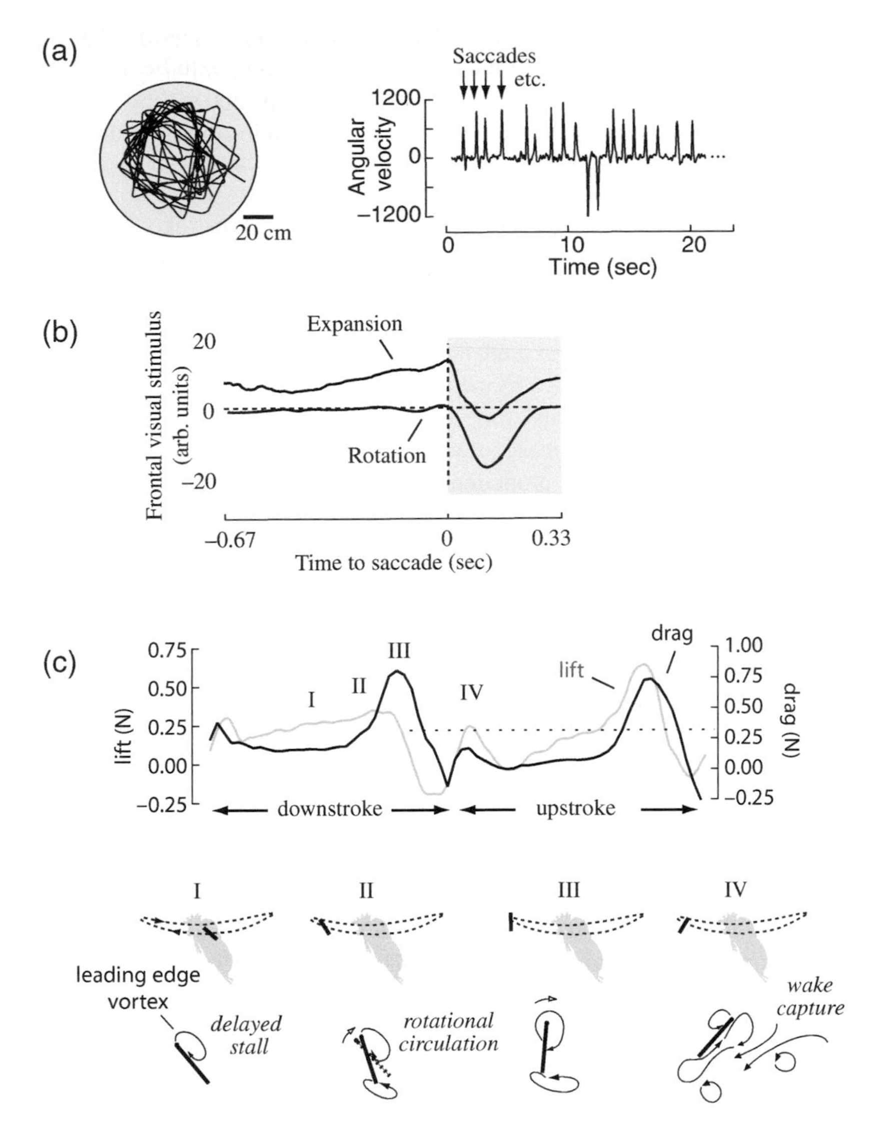

FIGURE 4.2 Visual and aerodynamic mechanisms for flight control in fruit flies. **a**: *Drosophila's* free-flight trajectory (left panel, viewed from above) is characterized by straight segments interspersed with fluctuations in the angular velocity of the animal's heading (right panel). The rapid turns are called *saccades* for the functional similarity to vertebrate ballistic eye movements. **b**: Peristimulus histograms synchronized to the execution of saccades indicate that saccades are triggered by a threshold value in image expansion, not rotation. Plots are means gathered from 9 individual flies and 760 saccades. (Adapted from Tammero, L.F. and Dickinson, M.H., *J. Exp. Biol.* **205**:327–343, 2002b. With permission.) **c**: Time course of lift (left ordinate, gray) and drag (right ordinate, black) production indicated for a complete wingstroke cycle in *Drosophila*. Diagrams illustrating changes in the structure of vortex fields are viewed along the span of the wing from tip to base.

a stall in circulatory force production (Dickinson and Götz, 1993). The physical basis for the stability of the LEV on a revolving wing remains to be fully resolved; however, the magnitude of enhanced circulatory force depends on the angle of attack and velocity of the moving wing. Thus, by varying the angle of attack, stroke trajectory, or wingbeat frequency, the neuromuscular system can modulate the size and strength of the LEV.

Insects generate lift during both the upstroke and the downstroke. This is accomplished as the wing rotates or flips at the end of each half-stroke to maintain a positive angle of attack. Like the LEV formed during the translation phase of the upstroke or downstroke, a wing rotating during pronation and supination can also enhance circulation, due to the tendency of the fluid to resist shear and maintain smooth flow at the trailing edge (Sane and Dickinson, 2001; Figure 4.2c, parts II, III). In flies, the magnitude and time course of rotational circulation is strongly determined by the timing, or phase, of wing rotation (Dickinson et al., 1999). Thus, by controlling the timing of pronation and supination, insects such as flies and butterflies actively modulate the time course and magnitude of rotational lift on a cycle-by-cycle basis (Dickinson et al., 1993; Srygley and Thomas, 2002).

Insects, especially small ones, tend to beat their wings back and forth very rapidly relative to their forward flight velocity (*advance ratio*; see Dudley, 2000). Thus, as the wing completes a half-stroke, it collides with the wake from the previous stroke. The vortices shed at the end of a stroke induce fluid momentum that is imparted to the returning wing, thus wing–wake interactions generate higher forces at stroke reversal (Figure 4.2c, part IV).

Given the basic aerodynamics of staying aloft, how are these forces modulated during visually triggered free-flight saccades? Fry et al. (2003) used a synchronized three-camera system (Figure 4.1c), recording at 5000 frames per second, to capture wing and body kinematics of free-flight saccades in *Drosophila*. Using custom graphical user interface routines developed in MATLAB®(13), the authors fit an animated wireframe model to each wing, in each of three fields of view, for the 20 or so wing strokes comprising a saccade (Figure 4.3a). They were therefore able to reconstruct precisely the temporal variation in wing kinematics, such as the angle of attack and stroke amplitude, within the fly-centered frame of reference during changes in body yaw, pitch, and roll. During a saccade, the body rotates an average of 90 degrees in the yaw plane and also shows some simultaneous roll producing a banked turn. However, the two wings show remarkably subtle differences from presaccade kinematics (Figure 4.3b). Nevertheless, by replaying the pattern of saccade kinematics through the robotic fly, the authors found that very small changes in stroke amplitude and the inclination of the stroke plane are sufficient to produce the yaw torque necessary for a saccade. Furthermore, the trajectory of torque generated by a fly is closely approximated by a theoretical prediction based on body morphology and inertial forces rather than frictional forces acting on the body (Figure 4.3c). The dramatic changes in forces, moments, and the animal's body orientation, resulting from very subtle cycle-by-cycle modulations of wing kinematics, illustrate that the sensorimotor system must exert very tight temporal control of musculoskeletal mechanics. How are the wing kinematics of flight maneuvers so precisely executed?

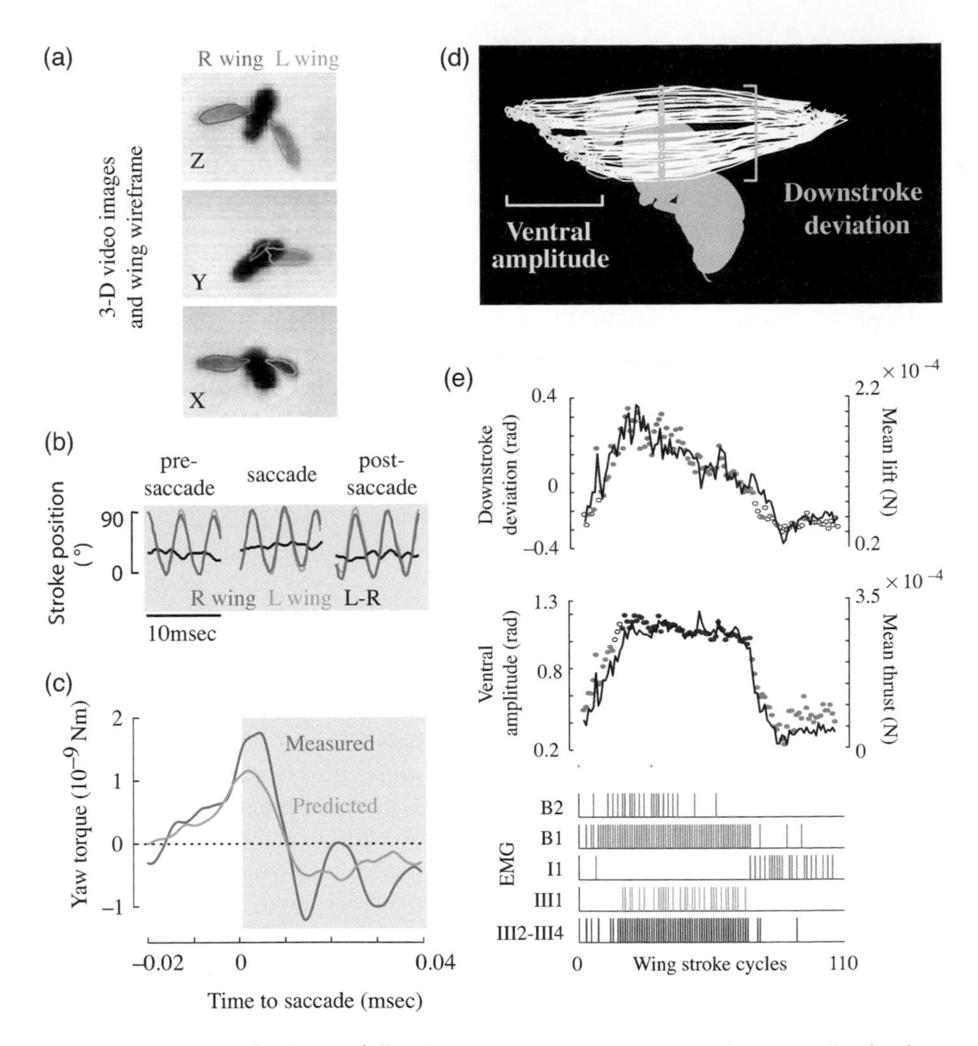

FIGURE 4.3 (See color insert following page 202.) Neuromuscular control of wing kinematics and flight forces. **a**: Video screen shots of a fruit fly executing a free-flight saccade. Wire frame models are fit to the right and left wings to track temporal changes in wing position. **b**: During a saccade, flies show very subtle asymmetrical changes in wing position compared with pre- and postsaccade trajectories. Note the slight increase in the difference between wing position, L–R. **c**: Replaying the changes in wing kinematics through the robotic fly produces a strong deviation in torque. (**a–c** adapted from Fry, S. et al., *Science* **300**:495–498, 2003. With permission.) **d**: The wing path traced by a tethered blowfly (facing left) shows changes in both the ventral amplitude and deviation of the downstroke. **e**: Patterns of steering muscle activation are mapped onto wing kinematics and flight forces. Downstroke deviation (left ordinate, dots) is tightly correlated with mean lift (right ordinate, solid line). Changes in ventral amplitude (left ordinate, dots) correspond to changes in mean thrust (right ordinate, solid line). The dots correspond to the activity of select steering muscles shown in spike rasters below. (Adapted from Balint, C.N., Ph.D. thesis, 2003. With permission.)

4.2.3 Motor Control of Wing Kinematics

In insects, the muscles that control steering kinematics of the wings are each innervated by a single motor neuron. For the larger flying insects, the control parameters of the dozen or so members of the steering motor pool include muscle recruitment, activation timing within the stroke cycle (i.e., phase), and activation frequency. However, smaller insects such as flies beat their wings hundreds of times per second such that steering muscles only have enough time to fire a single action potential per wingstroke cycle. Thus, a given motor unit is either active or not; when active, the phase of activation within the stroke cycle controls the mechanical power to steer the wing. For example, the first basalare muscle (B1) fires once each wingbeat, and its activation phase within the stroke cycle determines the path the wing takes through the air (Tu and Dickinson, 1996).

The activity of B1, as well as other steering muscles, coordinates wing motions with a combinatorial code such that any single muscle has varying effects on wing path, depending on both the background temporal pattern of other steering muscles (Balint and Dickinson, 2001) and the current location of the wing within the stroke cycle at the time of activation (Tu and Dickinson, 1994). How are aerodynamic forces controlled by the ensemble of motor activities in steering muscles? Mapping this relationship is accomplished by coupling multiple electromyograhic (EMG) recordings with high-speed videography during tethered flight maneuvers. By programming a flapping robot with EMG-coordinated patterns of wing kinematics, Balint has determined how changes in an ensemble of motor activities modulate the aerodynamic forces and moments associated with a turn (Balint, 2003). Mapping individual muscle spikes onto both wing kinematics and resultant force trajectory shows that steering moments are produced by the differential motor control of the rectilinear forces exerted by each wing. For example, activation of the basalare muscles B1 and B2 results in increased downstroke deviation — a measure of how much the wing plunges downward relative to the body as it moves anteriorly during the downstroke (Figure 4.3d). Increased downstroke deviation is strongly coupled with changes in lift (Figure 4.3e). By contrast, activation of muscles III-2,3,4 increases the ventral extent of each downstroke — a parameter called *ventral amplitude* — which results in increased thrust (Figure 4.3d). During a turn, increased downstroke deviation and lift on the outside wing is coupled with decreased magnitude of both parameters on the inside wing to evoke changes in yaw and roll that would result in a banked turn during free flight. Indeed, blowflies show banked turns during free-flight saccades (Schilstra and van Hateren, 1999).

Coupling EMG recordings with kinematics and aerodynamic analyses has, for the first time, identified the relationship between neural control parameters and the orientation and magnitude of the force vector produced by each wing. However, there appears to be no correspondence between the flight motor pattern and any single kinematic mechanism of aerodynamic force production, such as wingtip velocity, angle of attack, or wing rotation. Finally, temporal precision of the motor pattern is crucial because small variations in wing kinematics have large influences on the forces acting on the body during free flight (Figure 4.3b, c) (Fry et al., 2003).

The motor code is synchronized with the kinematics and forces produced by the wings through a cascade of mechanosensory feedback that directly accesses the membrane properties of thoracic interneurons and wing muscle motoneurons. These synaptic interactions are powerful and very fast. For example, mechanosensory stimulation of the blowfly haltere organ results in a mere 3 msec delay to a compensatory motor response (Sandeman and Markl, 1980). By contrast, a visually elicited turn during flight in hoverflies (animals with an extraordinarily fast and acute visual system) takes an order of magnitude longer (Land and Collett, 1974). To fully understand the neuromuscular control mechanisms of flight in insects, it is crucial to integrate the physiological mechanisms by which tonic multimodal sensory signals are fused and phase-locked with the motor control of wing kinematics.

4.3 ENCODING AND PROCESSING MECHANOSENSORY FEEDBACK

4.3.1 Mechanosensory Organs

Each mechanoreceptor associated with flight control is activated phasically within the wingstroke cycle. This temporal structure is very likely responsible for gating or somehow synchronizing relatively slow and tonic exteroceptive feedback signals descending from the brain onto the flight motor rhythm. We review several key mechanosensory organs and their roles in shaping the cycle-by-cycle changes in wing kinematics.

4.3.1.1 Orthopteran Head Hairs

Fine hairs on the heads of most insects deform when exposed to fluctuations in ambient airflow. Feedback from the sensory neurons at the base of each head hair initiates flight and provides phasic feedback to motor circuits by encoding periodic changes in air flow induced by the beating wings (Figure 4.4a). Signals from wind hair sensory neurons project to central flight motoneurons via the *tritocerebral commissural giant* (TCG) interneuron (Bacon and Tyrer, 1979). In locusts, the TCGs are maximally sensitive to air oscillations at about 20 to 25 Hz — the natural wingbeat frequency — and fire in rhythmical bursts phase-locked to wing motion (Figure 4.4a) (Bacon and Möhl, 1983).

Indirect evidence that wind hairs phasically structure motor output comes from immobilizing the hairs, which results in decreased wingbeat frequency. More direct evidence comes from electrically stimulating the sensory receptors, which results in reliable depolarizations in the TCG (Figure 4.4b, top panel). Activation of the TCG, in turn, results in depolarizations in a wing depressor motoneuron (Figure 4.4b, lower panel). Thus, the wind hair–TCG circuit phasically structures the rhythm of flight interneurons and motoneurons via cycle-by-cycle feedback from wing kinematics (Bacon and Möhl, 1979, 1983; Bacon and Tyrer, 1979).

Several wind-sensitive head-hair afferents project to the prothoracic ganglia, where they make excitatory synapses to interneuron A4I1. The cell body of A4I1 is located in the fourth abdominal ganglion, and its axon projects anteriorly to the

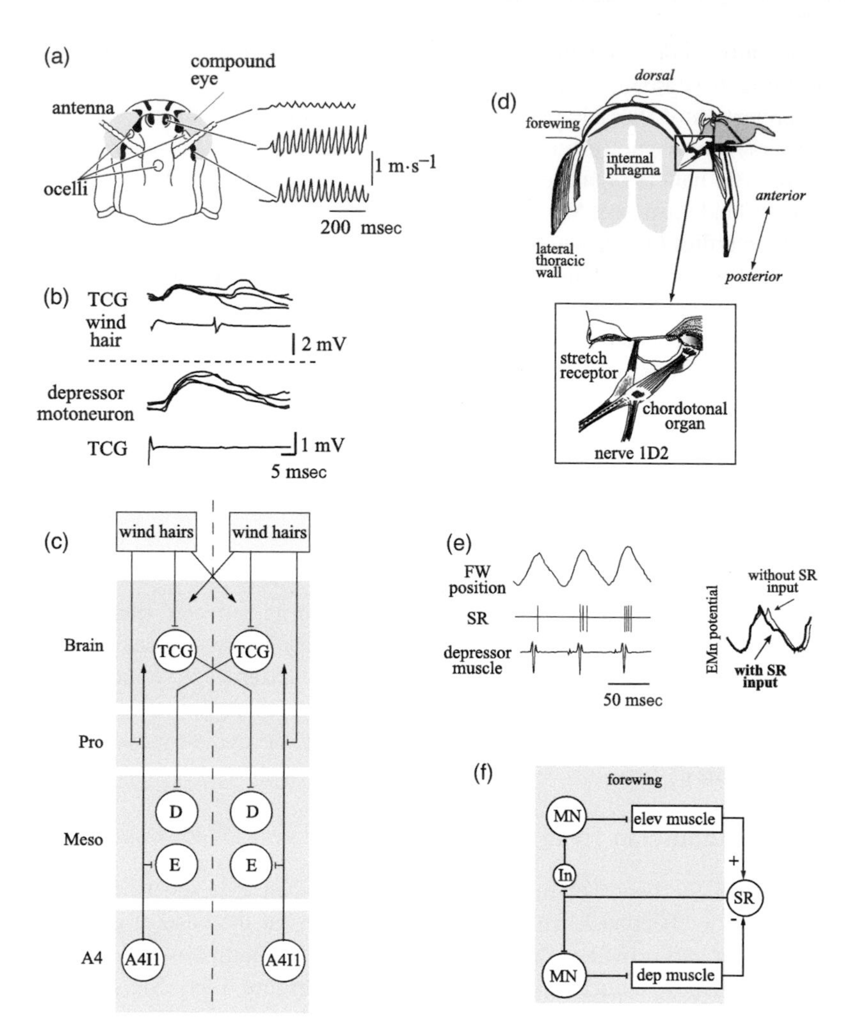

FIGURE 4.4 Anatomy, physiology, and model circuits for wind hairs and wing-hinge stretch receptors in locusts. **a**: Location of wind-sensitive hair fields on the head (regions shaded black), and receptor responses to fluctuations in wind generated by the beating wings. (Adapted from Reichert, H. et al., *Nature* **315**:142–144, 1985. With permission.) **b**: Electrical stimulation of a wind hair sensory axon elicits excitatory postsynaptic potentials (EPSPs) in the TCG (tritocerebral commissural giant axon). In turn, excitation of the TCG initiates depolarizations in depressor motoneurons. (Adapted from Bacon, J. and Möhl, B., *J. Comp. Physiol. A* **150**:439–452, 1983. With permission.) **c**: Diagram of synaptic connectivity of wind hair feedback within the brain, prothoracic (Pro), mesothoracic (Meso), and fourth abdominal (A4) ganglia. Depressor (D) and elevator (E) motoneurons, and interneuron A4I1 indicated. **d**: Morphology of the wing-hinge stretch receptor (SR) in locusts. (From Pfau, H.K., *Biona-report 1: Insektenflug I.* Nachtigall W, Gustav-Fischer W, Eds., 1983. With permission.) **e**: During flight, forewing elevation triggers spikes in the SR (left panel), as well as early depolarizations in elevator muscle motoneurons (EMn, right panel). (Adapted from Möhl, B., *J. Comp. Physiol. A* **156**:103–116, 1985. With permission.) **f**: Diagram of synaptic connectivity between the SR, local interneurons (In), and motoneurons (MN).

brain (Burrows and Pflüger, 1992). This anatomical finding is intriguing because it suggests that reverberating mechanosensory pathways may serve to prestructure or synchronize tonic descending signals with phasic control of wing kinematics. In the prothoracic ganglion, the A4I1 axon also receives excitatory input from head-hair afferents, and its mesothoracic collateral drives a pleuroaxillary motoneuron that modulates wing rotation at the end of the upstroke (Figure 4.4c) (Burrows and Pflüger, 1992).

The head hair system illustrates a classic example of how wingbeat synchronous mechanosensory feedback is integrated with other slower sensory modalities for flight control. For example, tonic input descending from the eyes and ocelli is gated by wingbeat synchronous phasic input from wind hairs to correct course deviations during flight (Reichert et al., 1985; Reichert and Rowell, 1986). Future experimental work could complete the picture of how synaptic interactions among the wind hairs, TCG, and thoracic motoneurons (Figure 4.4c) modulate stable forward-flight and turning maneuvers.

4.3.1.2 Orthopteran and Lepidopteran Wing-Hinge Stretch Receptors

Perhaps the most thoroughly examined mechanoreceptor in insects is the *wing-hinge stretch receptor* (SR). In locusts, the SR sensory organ consists of the cell body and dendritic arbor of a single neuron embedded within a strand of connective tissue that spans the base of the wing to an internal phragma (Figure 4.4d). Elevation of the wing stretches the strand and elicits action potentials in the sensory neuron. Both artificial rhythmic elevation of the wing and changes in wing elevation during tethered flight result in brief spike bursts centered near the top of the elevation cycle (Figure 4.4e) (Wilson and Gettrup, 1963; Pabst, 1965; Möhl, 1985). During tethered flight in locusts, information encoded within the burst includes the timing of wing elevation and the total extent of the upstroke (Möhl, 1985). *Manduca sexta* SRs also respond to imposed wing elevation with bursts phase-locked with wing elevation (Yack and Fullard, 1993), suggesting that SRs may be a common feature of flight control systems. Recent experiments with tethered flying animals have confirmed that the encoding properties of the hawkmoth SR are similar to those of locusts and phasically signal the timing of wing elevation, the interval between strokes, and elevation amplitude (Frye, 2001a,b).

The anatomical projection patterns and central synaptic physiology of SRs are unknown in moths. In locusts, each SR forms mono- and polysynaptic connections with flight interneurons and motor neurons (Burrows, 1975; Peters et al., 1985; Reye and Pearson, 1987). The forewing stretch receptor extends massive dendritic arborizations into all three thoracic ganglia, whereas hindwing SR inputs are restricted to the mesothoracic and metathoracic ganglia (Burrows, 1975; Altman and Tyrer, 1977). In locusts, the SRs form excitatory synapses with flight depressor muscle motoneurons, as well as polysynaptic inhibitory synapses with elevator motoneurons (Figure 4.4f) (Burrows, 1975; Reye and Pearson, 1987).

Now-classic studies on the influence of SR activity on the flight system in locusts showed that the SR can reset and entrain the centrally patterned flight rhythm when activated in phase with depressor motoneurons and interneurons (Pearson et al., 1983). That the SR fires near the top of the upstroke and forms monosynaptic excitatory connections with depressor motoneurons had been interpreted as evidence that the sensory input advances the timing of depressor muscle activation — thus maintaining a stable wingstroke cycle in the presence of a perturbation, such as a gust of wind. More recently, however, by simultaneously recording extracellular SR activity and intracellular motoneuron potentials, Pearson and Ramirez (1990) found that the SR burst occurs *after* the EPSPs in wing depressor motoneurons. This finding suggests that the main role of the SR is to reduce the extent of hyperpolarization in depressor motoneurons and advance the phase of the following wave of depolarization. In addition, they found that SR input enhances repolarization in elevator motoneurons (Figure 4.4e, right panel). Taken together, these two results help explain how the SR entrains wingbeat frequency and maintains the temporal precision of elevator and depressor activation patterns in the presence of external aerodynamic perturbations or internal morphological or neural asymmetries.

4.3.1.3 Orthopteran Tegulae

The locust *tegula* is a polyneuronal sensory organ associated with the base of each wing (Figure 4.5a). It consists of a cluster of external mechanosensory hairs and an internal chordotonal organ (Altman et al., 1978) that deforms along a three-dimensional trajectory during the wingstroke (Dawson et al., 2004). Pearson and Wolf (1988) proposed that the overall function of the tegula is to phase-lock the timing of initial depolarizations in elevator motoneurons to wing depression (for more information about chordotonal organs, see Chapter 3). The compound response of the tegula organ is phase-locked to the bottom of the downstroke (Figure 4.5b) (Wolf, 1993), specifically encoding the timing of stroke reversal and the angular velocity of the wing during the downstroke; both parameters directly modulate aerodynamic force production. The functional significance of tegula feedback for flight control is evident in experiments in which surgical ablation results in delayed upstroke timing and concomitant reduced lift production (Wolf, 1993). Two untested suggestions by Fischer and Ebert (1999) posit the functional role of these flight mechanosensors. First, the timing of tegula burst termination coincides with the posterior deviation of the wingstroke, indicating that the organ is more sensitive to wing motion perpendicular to the stroke plane than to motion within the average plane. If this is true, the tegula may encode the magnitude of downstroke deviation (see Figure 4.3d). Second, although the ensemble of tegula afference does not encode stroke amplitude specifically, the individual chordotonal organ component may signal a position-crossing event within the downstroke. The pressure-sensitive hairs and the deformation-sensitive chordotonal organ likely encode different components of wing motion or yield different frequency responses to a single component. In any event, tegula stimulation results in a phase advance in the depolarization of elevator motoneurons during fictive flight (Figure 4.5b, right).

The synaptic architecture of tegulae has been examined in several species. In *Schistocerca gregaria,* the hindwing tegula has two distinct inputs to flight motoneurons (Kien and Altman, 1979). A monosynaptic input excites an elevator motoneuron

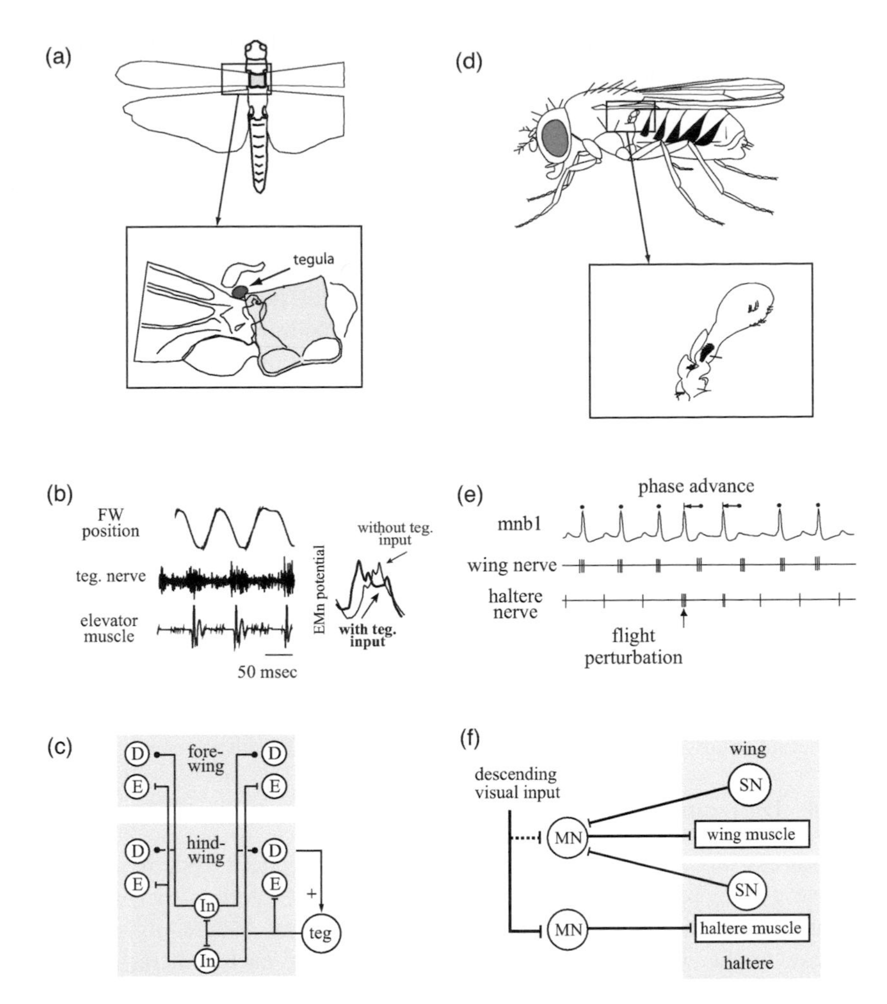

FIGURE 4.5 **a**: Location of the wing-hinge tegula organ in locusts. (Adapted from Fischer et al., *J. Exp. Biol.* **205**:1531–1545, 2002. With permission.) **b**: Depression of the forewing (FW) during tethered flight triggers a compound field potential from the tegula sensory nerve (left panel) and an advance in the depolarization of elevator muscle motoneuron (right panel). (Adapted from Fischer et al., *J. Exp. Biol.* **205**:1531–1545, 2002 and Wolf, H., *J. Exp. Biol.* **182**:229–253, 1993b. With permission.) **c**: Diagram of synaptic connectivity between the tegula organ (teg), depressor (D) and elevator (E) motoneurons, and local interneurons (In). **d**: The haltere is a gyroscopic sense organ in flies that beats antiphase to the wings during flight. (Adapted from Dickinson, M.H. et al., *Science* **284**:1954–1960, 1999. With permission.) **e**: Intracellular recording from the motoneuron of muscle B1, while stimulating the wing and haltere nerves. Haltere nerve stimulation elicits subthreshold EPSPs in mnb1, whereas wing nerve input provides superthreshold input. Increasing the instantaneous activation frequency of the haltere nerve — simulating a flight perturbation — results in a transient phase advance in mnb1. (Adapted from Fayyazuddin, A. and Dickinson, M.H., *J. Neurophysiol.* **82**:1916–1926, 1999. With permission.) **f**: Diagram of synaptic connectivity between the sensory nerves (SN) of the haltere, wing, and steering muscle motoneurons (MN), including direct visual feedback projecting from the brain (see text for details).

and inhibits a depressor motoneuron, whereas a second polysynaptic input inhibits the elevator and excites the depressor in the anterior thoracic segment (Figure 4.5c). The hindwing and forewing inputs to flight motoneurons are synergistic, which likely serves to amplify the individual sensory inputs. Conversely, in *Locusta migratoria*, Pearson and Wolf (1988) failed to find EPSPs in depressor motoneurons or inhibitory postsynaptic potentials (IPSPs) in elevator motoneurons in response to electrical tegula stimulation. Instead, their results suggest that the primary role of the tegulae is to generate the initial depolarization in elevator motoneurons. Subsequently, Wolf (1993) showed that removal of the tegulae caused a decrease in lift production by delaying the upstroke. Consistent with findings in tethered flight experiments, recent radiotelemetric flight muscle recordings from free-flying locusts demonstrate that although the tegulae are not required for generating the flight motor pattern, their removal results in decreased lift generation during flight (Dawson et al., 2004). In general, the pattern of tegula input to flight motoneurons reflects excitation of elevators and inhibition of depressors (Figure 4.5c).

Tegula ablation studies have provided evidence of robust neuroanatomical and functional plasticity within the locust flight control system. Following microsurgical ablation of the hindwing tegula organs, wingbeat frequency decreases, the interval between activation of depressor and elevator muscles increases, and the phase of elevator activation is delayed (Büschges and Pearson, 1991). Within two weeks of surgical ablation, the flight motor pattern functionally recovers, driven by rapid invasion of remaining forewing tegula afferents into the motor neuropil of the sensory depauperate hindwings (Büschges and Pearson, 1991; Büschges et al., 1992).

4.3.1.4 Dipteran Halteres

Flies possess an extreme form of mechanosensory feedback system consisting of sensitive gyroscopic organs called *halteres*. Through evolution, the hindwings of flies have been radically modified into the club-shaped haltere organs that beat antiphase to the wings (Figure 4.5d). At the base of each haltere, an array of biological strain gauges called *campaniform sensilla* are excited by cuticular deformation resulting from changes in the beating plane of the haltere. Thus, the morphology of the haltere system suggests that it encodes Coriolis forces acting to vary the haltere beating plane as the animal rotates in space (Pringle, 1948). The importance of haltere feedback is apparent upon their excision, whereby animals crash to the ground, unable to remain airborne.

Recent analyses using tethered fruit flies directly examined the role of haltere feedback in flight control. By optically tracking wing kinematics in response to mechanical rotation during flight, Dickinson (1999) showed that halteres mediate equilibrium responses to the rotation of the body along roll, pitch, and yaw axes. Equilibrium motor responses reflect simple trigonometric functions of the orientation of the rotation stimulus and are abolished by surgical ablation of the halteres.

4.3.2 Mechanisms of Sensorimotor Integration

One of the dominant contributors to the control of steering during equilibrium responses is the first basalare muscle, B1. Both haltere and wing campaniform axons synapse on the B1 motoneuron (mnB1). Recently, intracellular recordings from mnB1 have shown that calcium-mediated synaptic input from wing afferents sets the phase of mnB1 during the wing stroke; by contrast, electrotonic input from the haltere nerve transiently advances the firing phase mnB1 (Figure 4.5e). Thus, during flight, the activity of B1 is synchronized to wing motion via wing mechanosensory input, whereas haltere input rapidly phase-shifts the activation of B1 to counteract a perturbation and restore equilibrium.

The synaptic convergence of wing and haltere afferents onto steering muscle motoneurons helps to explain how a fly maintains a steady flight trajectory in the face of internal morphological or motor pattern asymmetries and external aerodynamic perturbations. Are these reflexive interactions somehow overridden or suppressed by visual cues that trigger the initiation of free-flight saccades? How the control of turns is integrated with steady-state equilibrium reflexes is a fascinating, yet vexing, avenue of current research. However, one peculiar result pushed forward our understanding of the functional interactions between multiple sensory inputs to the flight motor system in flies. As with the wings, the halteres are equipped with a set of very small muscles that fine-tune the trajectory taken by the haltere as it beats through the air. In quiescent animals, extracellular recordings from the haltere steering muscles exhibit suprathreshold responses to visual motion patterns projected on the retina (Chan et al., 1998). This visual input to the fly's gyroscope is remarkable because physiological evidence of visual input to the wing steering muscles has never been reported. The structure of mechanosensory convergence from wing and haltere inputs onto wing steering muscle motoneurons, combined with descending visual information targeted to the haltere motor system, suggests that visually elicited saccades are coordinated by an indirect reflex arc. In short, information signaling an impending collision is relayed through the descending pathway and stimulates the haltere steering muscles, causing haltere afferents to advance the phase of the B1mn, evoking a turn.

4.4 BEHAVIORAL PSYCHOPHYSICS OF VISUO-MECHANOSENSORY INTEGRATION

By modulating the membrane potential of interneurons or motoneurons, synaptic input from mechanoreceptors helps to maintain the temporal precision of the flight motor pattern to compensate for internal morphological asymmetries and external aerodynamic perturbations to the wings. The next challenge is to understand how mechanosensory feedback is integrated into a functional flight-control algorithm. Unfortunately for neurobiologists using reduced recording preparations to investigate synaptic interactions, much of the flight control system (e.g., visual afference to wing steering muscles in flies) may be inoperative unless the animal is actually flying. Therefore, tethering intact animals within electronic arenas that simulate the sensory conditions of free flight (Figure 4.1d) has proved an immensely valuable tool for examining how combinations of mechanosensory and visual cues influence the motor program, wing kinematics, and forces of flight maneuvers. Recent tethered-flight analyses have shown that mechanosensory feedback is integrated for the coordination of visually elicited compensatory visuomotor reflexes in hawkmoths and expands the dynamic range of equilibrium reflexes in flies.

When tethered to a lift sensor behind a visual grating pattern that moves up and down, hawkmoths modulate wing kinematics and flight forces in register with the moving frontal image (Frye, 2001a) (Figure 4.6). During upward pattern motion, ventral amplitude and lift increase; when the pattern moves down, ventral amplitude and lift decrease. During free flight, or in a closed-loop feedback system, these systematic changes in kinematics and forces reduce image slip on the retina (a classic optomotor response). Remarkably, the absence of feedback from the hindwing stretch receptors results in reduced-amplitude optomotor responses. Compared with a surgical sham control in which the sensory nerves were exposed but not cut, animals with transected stretch receptor axons show smaller changes in ventral amplitude during upward image motion (Figure 4.6a). Reduced modulation of ventral amplitude results in 35% lower lift responses to image motion (Frye, 2001a) (Figure 4.6c). Stretch receptor ablation does not alter the temporal dynamics of visual responses. Rather, the gain of lift modulation is attenuated equally across all velocities tested (Figure 4.6c). We know little about the central synaptic physiology of adult hawkmoths, but mechanosensory ablation does not alter patterns of abdominal ruddering or major power muscle activation. Thus, stretch receptor feedback likely projects to wing steering muscle motoneurons. Taken together, these results are intriguing because they suggest that wingbeat synchronous mechanosensory feedback not only shapes the motor rhythm generated in the thoracic pattern generator; it also gates visual input descending from visual centers in the brain.

Visuo-mechanosensory fusion is likely a general design theme for flying insects, especially those that display a high degree of aerial agility. Flies have the fastest visual kinetics known in insects (Autrum, 1958), as well as arguably the most complex and specialized mechanosensory organs devoted to flight control: the halteres. The functional value of vision is self-evident, but animals deprived of their halteres cannot fly freely (Derham, 1714) and show greatly attenuated compensatory wing kinematics in response to body rotation during tethered flight (Dickinson, 1999). The halteres and the visual system both encode the velocity of the rotating body (i.e., the direction and magnitude of motion in each of three rotational degrees of freedom). Therefore, mechanically rotating a fly in space or rotating a visual pattern around a stationary fly elicits compensatory equilibrium reflexes. Are the two sensory modalities redundant?

By imbedding an electronic visual display and optical wingbeat analyzer (Figure 4.1d) within a mechanical gimbal (Figure 4.1e), Sherman and Dickinson (2003) were able to control visual and mechanical stimuli independently along three degrees of freedom — all while the animal was flying under visual closed-loop conditions. As an aside, it is worth mentioning that the most important device for this experiment is the wingbeat analyzer. A custom-built system, the wingbeat analyzer tracks the motion of the wings optically. Unlike a conventional force beam, it can be rotated within the fly's frame of reference and its signals are not corrupted by the inertial forces generated by the rotation. Furthermore, the output of the analyzer is directly proportional to yaw torque (Tammero et al., 2004). The wingbeat analyzer is com-

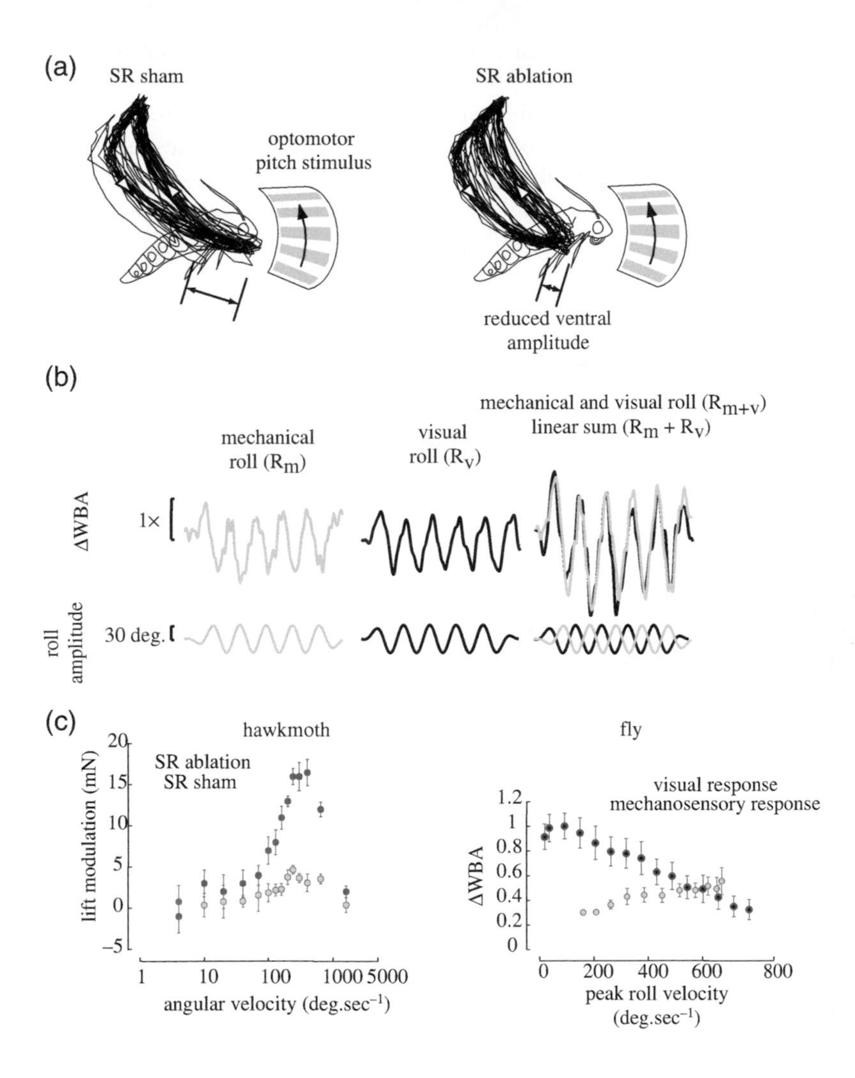

FIGURE 4.6 Psychophysical analysis of the control of equilibrium responses in hawkmoths and flies. **a**: Surgical ablation of the wing-hinge stretch receptors (SR), results in decreased ventral amplitude of the wingstroke in response to an upward optomotor pitch stimulus. **b**: Flies show strong equilibrium responses to mechanical roll, visual roll, and both stimuli presented concurrently. Equilibrium responses are indicated by differences in the wingbeat amplitude (ΔWBA) of the left and right wings. Motor responses to concurrent visual and mechanical roll (right column, black lines) reflect the linear superposition of the individual components (right panel, gray lines). (Adapted from Sherman, A. and Dickinson, M.H., *J. Exp. Biol.* **207**:133–142, 2004. With permission.) **c**: During tethered flight, moths increase ventral amplitude and lift in response to upward image rotation. Lift modulation (change in lift evoked by periodic changes in image direction) peaks at approximately 300 deg sec^{-1}. Lift modulation is uniformly attenuated by 30% in the absence of mechanosensory feedback from the SRs (left panel). In response to visual roll, flies show attenuated responses at high stimulus frequencies, whereas mechanosensory responses increase with increasing roll velocity (right panel).

posed of a pair of position sensors.[32] For details, see Lehmann and Dickinson (1997).

In response to visual roll, flies show a compensatory optomotor response that tracks the trajectory of image motion (Figure 4.6b). Similarly, in response to mechanical rotation, flies show equilibrium torque responses in the form of changes in the difference between right and left wingbeat amplitude. During a free-flight roll maneuver, the phase offset between visual and mechanical roll is exactly 180 degrees. Presenting the two stimuli concurrently at this natural phase offset results in greater amplitude wingbeat responses than for either stimulus presented in isolation, suggesting a summation of the two sensorimotor pathways (Figure 4.6b, right panel). The mathematical sum of the two individual responses is indistinguishable from the response to both stimuli presented concurrently. Therefore, during free flight, the visual and mechanosensory equilibrium reflexes are integrated to produce a larger equilibrium response than is mediated by either modality individually. An enhanced operational response gain is not the only advantage of the two velocity-encoding modalities. The two systems effectively expand the bandwidth of rotational velocity over which a fly can counteract a perturbation. Visual responses are maximal at low rotational velocities, whereas mechanosensory responses push the frequency sensitivity much higher than is detectable by the visual system (Figure 4.6c, right panel). The high frequency sensitivity and rapid transduction processes of the halteres likely enable flies to execute the straight paths, punctuated by rapid saccades, characteristic of many dipteran flight patterns.

A very intriguing result emerges from rotating the mechanical and visual stimuli along orthogonal stimulus axes (e.g., mechanical pitch and visual roll). Flies respond more strongly to the mechanical component than to the visual component (Sherman and Dickinson, 2004). Responses along the visual axis are 38% smaller than with the same visual stimulus applied in the absence of competing mechanosensory input. This suggests that visual responses are somehow attenuated or inhibited by mechanosensory feedback, and this phenomenon occurs at all frequencies and all stimulus axes tested. There are several reasons that the flight control system may be designed to weigh mechanosensory input more heavily than visual input. Owing to the kinetics of visual transduction, the visual system is slow compared with the mechanosensory system. Also, velocity estimation in the visual system depends on the spatial structure of the visual world, whereas the halteres encode velocity unambiguously. Regardless of the functional reason for the interaction, the physiological basis for the mechanosensory inhibition to the visual pathways is completely unknown.

4.5 CONCLUSIONS AND OUTLOOK

This review illustrates a general design feature of robust flight control in insects: subtle cycle-by-cycle variations in wing kinematics are precisely encoded by phasic mechanosensory feedback and result in large changes in aerodynamic forces and moments. Several exciting and challenging issues impede our understanding of the neurobiology of flight control in insects. First, what are the cellular and synaptic mechanisms by which tonic exteroceptive signals from the olfactory and visual centers of the brain are gated onto wingbeat synchronous motor rhythms? The answer may be hinted by the structure of anatomical projections of mechanosensory afferents (see also Chapter 1). For example, in flies the haltere sensory nerve sends a major projection to the brain (Chan and Dickinson, 1996). The neuroanatomical evidence supports the enticing hypothesis that mechanosensory feedback might prestructure descending signals to be phase-locked to the motor rhythm. Second, we have only indirect evidence showing how aerodynamic control parameters, such as delayed stall, rotational circulation, and wake capture, are encoded by mechanosensory organs. Discovering and formalizing these relationships between sensorimotor fusion and aerodynamics might best be examined within robotic test beds. Robotics engineers have sought designs that emulate the robustness and flexibility of insects — perhaps the pendulum swings both ways. Neurobiologists can use robots as hypothesis generators and test beds (see also Chapter 8). In any event, technological limitations are rapidly receding, and future work will undoubtedly answer these challenges and enhance our understanding of how nervous systems coordinate such remarkably complex behaviors as flight in insects.

ACKNOWLEDGMENTS

The authors would like to thank Dr. Claire Balint and Dr. Sanjay Sane for reviewing the manuscript. Dr. Frye would like to thank Dr. Michael Dickinson and members of his lab for informal contributions to the conceptual framework of this work. This work was supported in part by Office of Naval Research grant N00014-03-1-0604 to Michael H. Dickinson.

REFERENCES

Altman JS, Tyrer NM (1977) The locust wing hinge stretch receptors. I. Primary sensory neurones with enormous central arborizations. *J. Comp. Neurol.* **172**:409–430.

Altman JS, Anselment E, Kutsch W (1978) Postembryonic development of an insect sensory system: ingrowth of axons from hindwing sense organs in *Locusta migratoria. Proc. Roy. Soc. Lond. B* **202**:497–509.

Autrum H (1958) Electrophysiological analysis of the visual systems in insects. *Expl. Cell Res. Suppl.* **5**:426–439.

Bacon J, Möhl B (1979) Activity of an identified wind interneurone in a flying locust. *Nature* **278**:638–640.

Bacon J, Möhl B (1983) The tritocerebral commissure giant (TCG) wind-sensitive interneurone in the locust. I. Its activity in straight flight. *J. Comp. Physiol. A* **150**:439–452.

Bacon J, Tyrer NM (1979) Wind interneurone input to flight motor neurones in the locust *Schistocerca gregaria. Naturwissenschaften* **66**:116.

Balint CN (2003) The conversion of steering muscle activity to steering forces in *Calliphora vicina*. Ph.D. thesis, Univ. California, Berkeley.

Balint CN, Dickinson MH (2001) The correlation between wing kinematics and steering muscle activity in the blowfly *Calliphora vicina. J. Exp. Biol.* **204**:4213–4226.

Burrows M (1975) Monosynaptic connexions between wing stretch receptors and flight motoneurones of the locust. *J. Exp. Biol* **62**:189–219.

Burrows M, Pflüger H-J (1992) Output connections of a wind sensitive interneurone with motor neurones innervating flight steering muscles in the locust. *J. Comp. Physiol. A* **171**:437–446.

Büschges A, Pearson KG (1991) Adaptive modifications in the flight system of the locust after the removal of wing proprioceptors. *J. Exp. Biol.* **157**:313–333.

Büschges A, Ramirez J-M, Pearson KG (1992) Reorganization of sensory regulation of locust flight after partial deafferentation. *J. Neurobiol.* **23**:31–43.

Chan WP, Dickinson MH (1996) Position-specific central projections of mechanosensory neurons on the haltere of the blowfly, *Calliphora vicina. J. Comp. Neurol.* **369**:405–418.

Chan WP, Prete F, Dickinson MH (1998) Visual input to the efferent control system of a fly's "gyroscope." *Science* **280**:289–292.

Dawson JW, Kutsch W, Robertson RM (2004) Auditory-evoked evasive manoeuvres in free-flying locusts and moths. *J. Comp. Physiol. A* **190**:69–84.

Derham W (1714) *Physico-theology.* W & J Innys, London.

Dickinson MH (1999) Haltere-mediated equilibrium reflexes of the fruit fly, *Drosophila melanogaster. Phil. Trans. Roy. Soc. Lond. B* **354**:903–916.

Dickinson MH, Götz KG (1993) Unsteady aerodynamic performance of model wings at low Reynolds numbers. *J. Exp. Biol.* **174**:45–64.

Dickinson MH, Tu MS (1997) The function of Dipteran flight muscle. *Comp. Biochem. Physiol. A* **116**:223–238.

Dickinson MH, Lehmann FO, Götz KG (1993) The active control of wing rotation by *Drosophila. J. Exp. Biol.* **182**:173–189.

Dickinson MH, Lehmann FO, Sane SP (1999) Wing rotation and the aerodynamic basis of insect flight. *Science* **284**:1954–1960.

Dudley R (2000) *The Biomechanics of Insect Flight.* Princeton University Press, Princeton, NJ.

Ellington CP (1995) Unsteady aerodynamics of insect flight. *Symp. Soc. Exp. Biol.* **49**:109–129.

Ellington CP, vandenBerg C, Willmott AP, Thomas ALR (1996) Leading-edge vortices in insect flight. *Nature* **384**:626–630.

Fayyazuddin A, Dickinson MH (1999) Convergent mechanosensory input structures the firing phase of a steering motor neuron in the blowfly, *Calliphora. J. Neurophysiol.* **82**:1916–1926.

Fischer H, Ebert E (1999) Tegula function during free locust flight in relation to motor pattern, flight speed and aerodynamic output. *J. Exp. Biol.* **202**:711–721.

Fischer, H., Wolf, H., and Büschges, A. (2002) *J. Exp. Biol.*, **205**:1531–1545.

Fry S, Bichsel M, Muller P, Robert D (2000) Tracking of flying insects using pan-tilt cameras. *J. Neurosci. Methods* **101**:59–67.

Fry S, Sayaman R, Dickinson M (2003) The aerodynamics of free-flight maneuvers in *Drosophila. Science* **300**:495–498.

Frye MA (2001a) Effects of stretch receptor ablation on the optomotor control of lift in the hawkmoth *Manduca sexta. J. Exp. Biol.* **204**:3683–3691.

Frye MA (2001b) Encoding properties of the wing hinge stretch receptor in the hawkmoth *Manduca sexta. J. Exp. Biol.* **204**:3693–3702.

Kien J, Altman JS (1979) Connections of the locust tegulae with metathoracic flight motoneurons. *J. Comp. Physiol. A* **133**:299–310.

Land MF, Collett TS (1974) Chasing behavior of houseflies (*Fannia canicularis*): description and analysis. *J. Comp. Physiol.* **89**:331–357.

Lehmann FO, Dickinson MH (1997) The changes in power requirements and muscle efficiency during elevated force production in the fruit fly *Drosophila melanogaster. J. Exp. Biol.* **200**:1133–1143.

Möhl B (1985) The role of proprioception in locust flight control. II. Information signalled by forewing stretch receptors during flight. *J. Comp. Physiol. A* **156**:103–116.

Pabst H (1965) Elektrophysiologiske Untersuchung des Streck-Rezeptors am Flugelgelenk der wnderheuschrecke *Locusta migratoria. Z. Vgl. Physiol.* **50**:498–541.

Pearson KG, Ramirez JM (1990) Influence of input from the forewing stretch receptors on motoneurones in flying locusts. *J. Exp. Biol.* **151**:317–340.

Pearson KG, Wolf H (1988) Connections of hindwing tegulae with flight neurones in the locust, *Locusta migratoria. J. Exp. Biol.* **135**:381–409.

Pearson KG, Reye DN, Robertson RM (1983) Phase dependent influences of wing stretch receptors on flight rhythm in the locust. *J. Neurophysiol.* **49**:1168–1181.

Peters BH, Altman JS, Tyrer NM (1985) Synaptic connections between the hindwing stretch receptor and flight motor neurones in the locust revealed by double cobalt labelling for electron microscopy. *J. Comp. Neurol.* **233**:269–284.

Pfau HK (1983) Mechanik und sensorische Kontrolle der Flugel-Pronation und Supination. In: *Biona-report 1: Insektenflug I.* Nachtigall W, Gustav-Fischer W, Eds., pp. 61–77.

Pringle JWS (1948) The gyroscopic mechanism of the halteres of *Diptera. Phil. Trans. Roy. Soc. Lond. B* **233**:347–384.

Reichert H, Rowell CHF (1986) Neuronal circuits controlling flight in the locust: how sensory information is processed for motor control. *Trends Neurosci.* **9**:281–283.

Reichert H, Rowell CHF, Griss C (1985) Course correction circuitry translates feature detection into behavioural action in locusts. *Nature* **315**:142–144.

Reye DN, Pearson KG (1987) Projections of the wing stretch receptors to central flight neurons in the locust. *J. Neurosci.* **7**:2476–2487.

Sandeman DC, Markl H (1980) Head movements in flies (*Calliphora*) produced by deflexion of the halters. *J. Exp. Biol.* **85**:43–60.

Sane S (2003) The aerodynamics of insect flight. *J. Exp. Biol.* **206**:4191–4208.

Sane SP, Dickinson MH (2001) The control of flight force by a flapping wing I: Lift and drag production. *J. Exp. Biol.* **204**:2607–2626.

Schilstra C, van Hateren JH (1999) Blowfly flight and optic flow: I. Thorax kinematics and flight dynamics. *J. Exp. Biol.* **202**:1481–1490.

Sherman A, Dickinson MH (2003) A comparison of visual and haltere-mediated equilibrium reflexes in the fruit fly *Drosophila melanogaster. J. Exp. Biol.* **206**:295–302.

Sherman A, Dickinson M (2004) Summation of visual and mechanosensory feedback in *Drosophila* flight control. *J. Exp. Biol.* **207**:133–142.

Srygley RB, Thomas ALR (2002) Unconventional lift-generating mechanisms in free-flying butterflies. *Nature* **420**:660–664.

Tammero LF, Dickinson MH (2002a) Collision-avoidance and landing responses are mediated by separate pathways in the fruit fly, *Drosophila melanogaster. J. Exp. Biol.* **205**:2785–2798.

Tammero LF, Dickinson MH (2002b) The influence of visual landscape on the free flight behavior of the fruit fly *Drosophila melanogaster. J. Exp. Biol.* **205**:327–343.

Tammero LF, Frye MA, Dickinson M (2004) Spatial organization of visuomotor reflexes in *Drosophila. J. Exp. Biol.* **207**:113–122.

Tu MS, Dickinson MH (1994) Modulation of negative work output from a steering muscle of the blowfly *Calliphora vicina. J. Exp. Biol.* **192**:207–224.

Tu MS, Dickinson MH (1996) The control of wing kinematics by two steering muscles of the blowfly, *Calliphora vicina. J. Comp. Physiol. A* **178**:813–830.

Usherwood JR, Ellington CP (2002) The aerodynamics of revolving wings II. Propeller force coefficients from mayfly to quail. *J. Exp. Biol.* **205**:1565–1576.

Wilson DM, Gettrup E (1963) A stretch reflex controlling wingbeat frequency in grasshoppers. *J. Exp. Biol.* **40**:171–185.

Wolf H (1993) The locust tegula: significance for flight rhythm generation, wing movement control, and aerodynamic force production. *J. Exp. Biol.* **182**:229–253.

Yack JE, Fullard JH (1993) Proprioceptive activity of the wing-hinge stretch receptor in *Manduca sexta* and other atympanate moths: a study of the noctuoid moth ear B cell homologue. *J. Comp. Physiol. A* **173**:301–307.

5 Auditory Processing of Acoustic Communication Signals: Sensory Biophysics, Neural Coding, and Discrimination of Conspecific Songs

Andreas V.M. Herz, Jan Benda, Tim Gollisch, Christian K. Machens, Roland Schaette, Hartmut Schütze, and Martin B. Stemmler

CONTENTS

0-8493-2024-0/05/$0.00+$1.50

5.1 INTRODUCTION

Evolution has led to acoustic communication behaviors of fascinating complexity (see, e.g., Hauser, 1996; Bradbury and Vehrenkamp, 1998), which are made possible by sophisticated neural systems in both sender and receiver. Remarkably, even small insect auditory systems are capable of astounding computations. Some grasshoppers, for example, reliably detect gaps in conspecific songs as short as 1 to 2 msec (von Helversen, 1972), a performance level similar to that reached by birds and mammals.

These observations raise the question of how a minute insect auditory system processes auditory signals reliably and with high temporal precision. Important insight will come from understanding the auditory periphery. It serves as a strategic bottleneck between the external world and further neural processing stages; every computation and behavioral decision must be based on the primary stimulus representation at the level of auditory receptors.

Understanding the interplay between the dynamics and function of these neurons requires answers to a broad range of questions, such as the following:

- Which physical sound attribute (e.g., sound pressure or energy) actually drives the receptor?
- What are the essential processing steps of auditory transduction and encoding?
- How do neural noise sources limit the system's performance?
- Is it possible to "read" the sensory input from the output of a single receptor?
- Are receptor neurons specifically tuned to behaviorally relevant features?

Grasshoppers of the species *Chorthippus biguttulus* provide a suitable model system to study these questions. Their calling and courtship songs are based on broadband carrier signals with amplitudes that are strongly modulated in time. Although lacking tonal elements, the songs possess an elaborate temporal structure, rhythmically arranged into distinct syllables that are separated by short pauses (Figure 5.1a). Several differences between individual songs from the same species can be noted: the frequency content of the broadband carrier, the syllable length, and the precise temporal pattern of amplitude modulations within a syllable (Figure

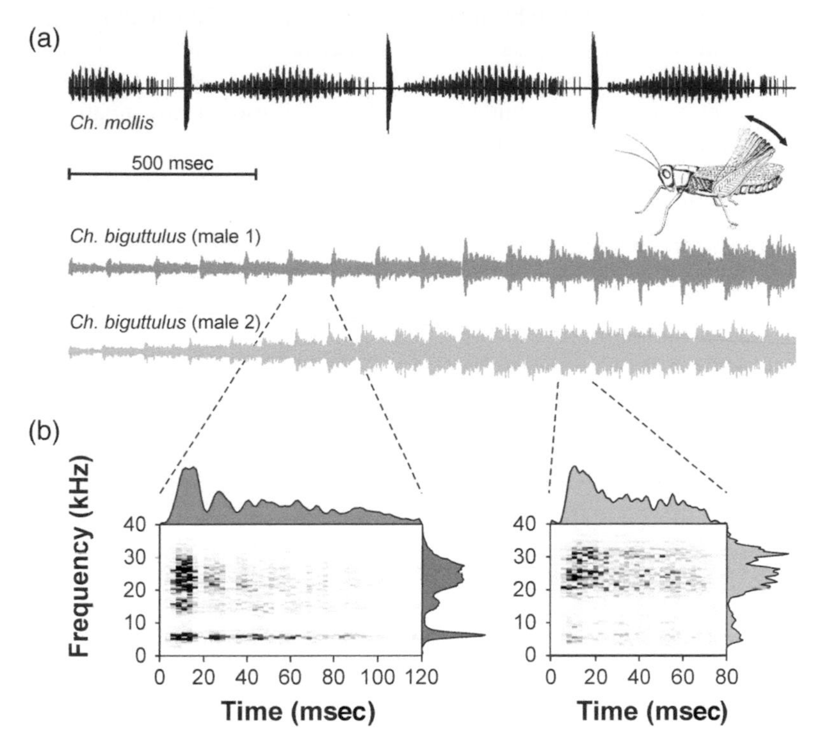

FIGURE 5.1 Acoustic communication signals. **a:** Grasshoppers produce species-specific sound patterns by rasping their hind legs across their forewings. Songs generated by one *Chorthippus mollis* and two *Chorthippus biguttulus* males are shown. Each song consists of many repetitions of a basic pattern, termed a *syllable*. **b:** Within each syllable, the amplitude of the high-frequency broadband carrier is strongly modulated in time, as illustrated by the spectrograms.

5.1b). This song variability could allow females to choose among different males and thus provide a basis for mate preference and sexual selection (Kriegbaum and von Helversen, 1992).

On the receiver side, the songs are encoded by a total of roughly 100 auditory receptors (in *Ch. biguttulus*, R. Lakes-Harlan, personal communication) into trains of action potentials. The cells are located at the two *tympana* on both sides of the animal; their axons extend through the tympanic nerves to the metathoracic ganglion, where auditory information is processed by local interneurons and then sent to the brain via ascending neurons (see also Chapter 1). The tympanic nerve is easily accessible for intracellular recordings from single axons. This allows one to study the output of individual neurons without damaging the animal's ear. Detailed investigations are best carried out in locusts (*Locusta migratoria*), whose larger but homologous structures (Stumpner and Ronacher, 1991) facilitate long experiments.

In this chapter, we will explore various aspects of the dynamics and signal-processing capabilities of a single type of neuron, the auditory receptor, using a variety of modern and partly novel experimental and theoretical techniques. The methods cover a wide range of disciplines — from biophysics to information theory

— and demonstrate that a tight interplay of experiment, data analysis, and theory can yield valuable new insights. All approaches have one feature in common: no neural parameter or variable, apart from the acoustic input and the final spike output, needs to be measured. The techniques may thus be of use for investigations of other neural systems that only allow axonal or extracellular recordings.

At this stage, our goal is *not* to produce a single, unified computational model. Rather, we will present a collection of independent studies that aim at elucidating key processes of auditory receptor dynamics — signal transduction (Section 5.2), spike-frequency adaptation (Section 5.3), and spike-train variability (Section 5.4) — and their consequences for neural coding (Section 5.5) and song discrimination (Section 5.6). Furthermore, we concentrate on the underlying concepts and main results. Detailed descriptions of the experiments and data-analysis methods can be found in the original articles cited in this chapter.

5.2 BIOPHYSICS OF AUDITORY SIGNAL TRANSDUCTION

Auditory receptors transform an incident sound wave into a train of action potentials. Several sequential steps are required for the process of *signal transduction*:

1. **Mechanical coupling.** The acoustic stimulus induces vibrations of the eardrum.
2. **Mechanosensory transduction.** These vibrations cause the opening of mechanosensory ion channels in the membrane of the receptor neuron, which in grasshoppers and locusts is directly connected to the eardrum via short dendrites.
3. **Electrical integration.** The electrical charge of the ions accumulates at the cell membrane.
4. **Spike generation.** Action potentials are triggered by voltage-dependent currents and travel down the auditory nerve.

Each of these four steps transforms the signal in a specific way, ranging from nearly linear (the eardrum response) to strongly nonlinear (the generation of action potentials). Spike generation itself is influenced by at least two further processes: *spike-frequency adaptation* (i.e., the gradual reduction of the output activity despite constant stimulation) and *stochastic fluctuations* of the membrane potential due to internal noise sources, such as ion-channel noise. These fluctuations cause jittered spike times and "missing" or "additional" spikes from trial to trial.

The entire sequence of signal transduction is completed after about one millisecond, not counting spike-frequency adaptation, which occurs on a longer time scale. As the eardrum is delicate and the sensory periphery highly vulnerable, how can one dissect the individual steps in the auditory transduction chain? We are often left with no choice but to make do with the final output, the spike. Nonetheless, much more information can be gathered about the details of transduction than one might surmise at first.

5.2.1 Conceptual Framework: Analysis and Comparison of Iso-Response Stimuli

To dissociate and identify the individual steps of auditory signal transmission, we revisit and extend an experimental strategy for measuring threshold curves in neurobiology (Evans, 1975) and for applying equivalence criteria in psychophysics (Jameson and Hurvich, 1972). Instead of estimating the full input–output function F describing all four stages of the transduction chain sketched previously, we focus on F's level surfaces. That is, we vary stimulus parameters such that the investigated neuron stays at a *constant level of final output activity* (Gollisch et al. 2002). For each point on such an "iso-response manifold in stimulus space," the nonlinear process of generating the specified output is identical. Thus, by examining the invariances in the input–output relations, we may uncover information about the earlier stages of the transduction chain.

In addition, even low-dimensional stimulus subspaces will contain a clear signature of the invariant stimulus manifolds. This is shown by a simple example: if the iso-response manifolds of some system form concentric ellipsoids in a high-dimensional input space, any measurement within a two-dimensional planar subspace will reveal ellipses as invariant regions (Figure 5.2a). These ellipses can readily

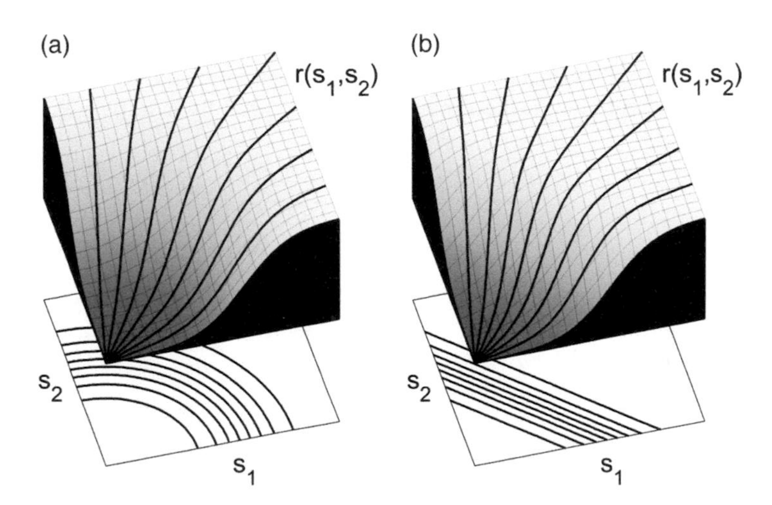

FIGURE 5.2 Characterization of input–output relations for two different response scenarios. A hypothetical two-dimensional stimulus space is parameterized by the variables s_1 and s_2. **a,b:** The surfaces drawn represent the resulting response $r(s_1,s_2)$ for two alternative models, which take the quadratic **(a)** and linear sum **(b)** as the argument of a sigmoid nonlinearity. Although the two scenarios are fundamentally different, both produce exactly the same one-dimensional response functions $r(s_1)$ and $r(s_2)$, respectively, as seen by the black areas at the sides of the surface block. Furthermore, any measurement along a radial direction, as is common in experimental practice, will produce similar sigmoid response curves in both cases, as seen by the thick black lines running along the surfaces. The iso-response manifolds $r = const$ (here: one-dimensional curves) below the surface plots, however, give a clear signature of the different underlying processes.

be distinguished from the straight lines that result from iso-response manifolds consisting of hyperplanes (Figure 5.2b).

Compared with other approaches that involve presenting a whole range of stimuli meant to cover all possible relevant inputs, the approach of comparing only stimuli producing the *same* output is fairly parsimonious. Experimental time is often limited; not needing to take the final nonlinear mapping into account can be a highly valuable benefit.

5.2.2 Mathematical Framework for Sensory Processing: LNLN Signal Cascades

Sensory neurons convert an incoming stimulus into a neural response, such as a firing rate of the occurrence probability of a single action potential. From a mathematical point of view, the operation of a single sensory neuron can be written as

$$r(t) = F[\vec{s}(\cdot)]\ , \tag{5.1}$$

where the dot $(\cdot)$ in the argument of the stimulus $\vec{s}$ emphasizes that the response r at time t depends not only on the current stimulus but also on its history.

A realization of this general mapping could consist of a linear filter operation L acting on the stimulus in the time domain, followed by a static nonlinearity f:

$$r(t) = f\left[\int_0^\infty L(\tau)\cdot s(t-\tau)d\tau\right]. \tag{5.2}$$

Such a sequence of a linear filter and a nonlinear map is also known as an *LN cascade* (Hunter and Korenberg, 1986). If the elementary processes are arranged in the opposite order, an *NL cascade* results

$$r(t) = \int_0^\infty L(\tau)\cdot f\left[s(t-\tau)\right]d\tau. \tag{5.3}$$

Transduction processes with more complicated dynamics can be modeled as combinations of elementary cascades. For example, an LNLN cascade corresponds to two nested LN cascades:

$$r(t) = f_2\left\{\int_0^\infty L_2(\tau_2)\cdot f_1\left[\int_0^\infty L_1(\tau_1)\cdot s(t-\tau_1-\tau_2)d\tau_1\right]d\tau_2\right\}. \tag{5.4}$$

5.2.3 Firing Rates Reveal the Functional Form of the Intermediate Static Nonlinearity

To illustrate how the iso-response method can be used to find models for the operation of auditory receptors, we first consider sound pressure waves $s(t)$ that consist of superimposed pure tones:

$$s(t) = \sum_{n=1}^{N} A_n \sin\left(2\pi\nu_n t + \varphi_n\right), \tag{5.5}$$

where v_n denotes the frequencies, φ_n the phase offsets, and A_n the respective amplitudes. Due to mechanics of the eardrum, $s(t)$ is linearly filtered and thereby turned into

$$s(t) = \sum_{n=1}^{N} \frac{A_n}{C_n} \sin\left(2\pi\nu_n t + \varphi_n\right), \tag{5.6}$$

This means that every tone receives a frequency-dependent gain factor, $1/C_n$. In addition, the phase may change from φ_n to $\tilde{\varphi}_n$, but this shift is not relevant for stimulus encoding because the investigated receptors do not phase-lock to the sound's carrier (Suga, 1960), but rather generate irregular spike trains. Because the resulting discharge rate $r(t)$ stays approximately constant after a brief transient of heightened activity following stimulus onset, the input–output relation (Equation 5.1) can be approximated by a static nonlinearity

$$r = f(J) \tag{5.7}$$

where the effective stimulus intensity J describes the cell's spectral integration process.

Turning to an ongoing debate about which stimulus attribute of a sound actually triggers neural output activity (see, e.g., Garner, 1947; Tougaard, 1996; or Heil and Neubauer, 2001), the iso-response framework allows us to test three rival hypotheses about the physical nature of J:

Amplitude hypothesis: J corresponds to the maximum amplitude of $\tilde{s}(t)$. This is the common view of a threshold; a response occurs once the signal reaches a certain value. In the case of few frequency components, J is given by the sum of the scaled amplitudes:

$$J_{AH} = \sum_{n=1}^{N} \frac{A_n}{C_n}. \tag{5.8}$$

Energy hypothesis: J corresponds to the temporal mean of the squared signal:

$$J_{EH} = \langle \tilde{s}(t)^2 \rangle = \frac{1}{2}\sum_{n=1}^{N} \frac{A_n^2}{C_n^2}. \tag{5.9}$$

Because the square of the amplitude of a sinusoidal oscillation is proportional to the energy contained in the oscillation, this hypothesis reflects an energy-integration mechanism.

Pressure hypothesis: J corresponds to the temporal mean of the absolute value of $\tilde{s}(t)$:

$$J_{PH} = \langle |\tilde{s}(t)| \rangle. \tag{5.10}$$

This describes a pressure-integration mechanism after full-wave rectification.

Based on the iso-response paradigm, we measured firing-rate responses to superpositions of two sine wave stimuli (Equation 5.5) with different relative contributions, A_1 and A_2, of the two tones. The stimulus intensities were adjusted during the experiment such that the same firing rate was obtained in response to any mixture of the two tones. This led to different combinations of A_1 and A_2, which can be compared with the predictions of the three hypotheses. As suggested by Figure 5.3a, the amplitude and pressure hypotheses do not adequately predict the data (Gollisch et al. 2002). The energy hypothesis, on the other hand, provides a good description of the data in the two-tone case.

As an additional test of the energy hypothesis, we also investigated how iso-firing-rate curves (which were obtained separately for different firing rates) are related to one another. Figure 5.3b shows pairs (A_1, A_2) corresponding to several different firing rates. Pairs corresponding to the same firing rate are accurately fitted by ellipses. To a good approximation, all ellipses are scaled versions of one another. This result is consistent with the energy hypothesis, as the ratio of the ellipses' half-axes should always equal the ratio of the filter constants C_1 and C_2. In addition, the energy model holds for superpositions of multiple pure tones, and it even accurately predicts receptor responses to stationary noise stimuli (data not shown).

5.2.4 Firing Probabilities Reveal the Time Scales of Mechanical and Electrical Integration

As the experiments agree with the energy hypothesis (Equation 5.9), we can model the auditory transduction dynamics of the investigated receptors as a sequence of four elementary processes. The first two stages are given by the linear filter operation (Equation 5.6) and a static square nonlinearity. A second linear filter with a flat kernel accounts for the time average in Equation 5.9, and a final nonlinearity

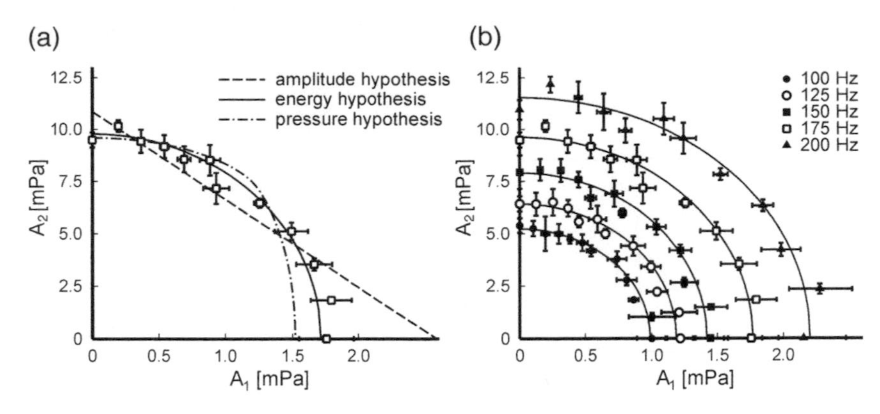

FIGURE 5.3 Identification of iso-response manifolds within a firing-rate description. For superpositions of two pure tones (amplitudes A_1 and A_2), measured pairs of amplitudes corresponding to a discharge rate of 175 Hz (open squares) are shown in **(a)** together with the iso-firing-rate curves for three rival hypotheses about the governing stimulus attribute. The dashed line denotes the fit of the amplitude hypothesis, the solid line the fit of the energy hypothesis, and the dash-dotted line the fit of the pressure hypothesis. Whereas the curves for the amplitude and the pressure hypothesis deviate systematically, the ellipse obtained from the energy hypothesis corresponds well with the data. The different scales on the axes are due to the strong dependence of the sensitivity on the sound frequency and thus reflect the neuron's tuning curve. The points in **(b)** display measured pairs of amplitudes, and the solid lines are corresponding ellipses fitted to the data in accordance with the energy hypothesis. The feedback-adjusted firing rates rise from 100 to 200 Hz in steps of 25 Hz. Note that the fits agree with the data regardless of the firing rate and that ellipses for different firing rates are scaled versions of each other, as predicted by the energy hypothesis. (From Gollisch, T. et al., *J. Neurosci.*, 22, 10442, 2002. With permission.)

describes the firing-rate encoding of the effective sound intensity J_{EH}. Within the general signal-transduction framework, this corresponds to an LNLN cascade (Equation 5.4) with a quadratic f_1 and a yet undetermined f_2. Because the amplitudes of the sound stimuli used so far did not vary in time, temporal details of the transduction process were beyond the experiment's reach.

To uncover the time-resolved dynamics, we extended the iso-response paradigm to the occurrence probability of single spikes by using short time-dependent stimuli that trigger at most one spike (Gollisch and Herz, 2004a). A series of two clicks with amplitudes A_1 and A_2 (click duration: 20 μsec; interclick interval < 1 msec) was presented repeatedly while adjusting the stimulus intensity online so that the probability of eliciting the single spike remained constant. Spike probabilities were measured using temporal windows extending to 10 msec after the first stimulus — long enough to make the approach insensitive to the variability in spike timing.

If the interclick interval is sufficiently large (Figure 5.4a), the iso-response manifolds are nearly circular (Figure 5.4c, filled circles). This is in accordance with the energy hypothesis, which predicts a dependence of the response on the

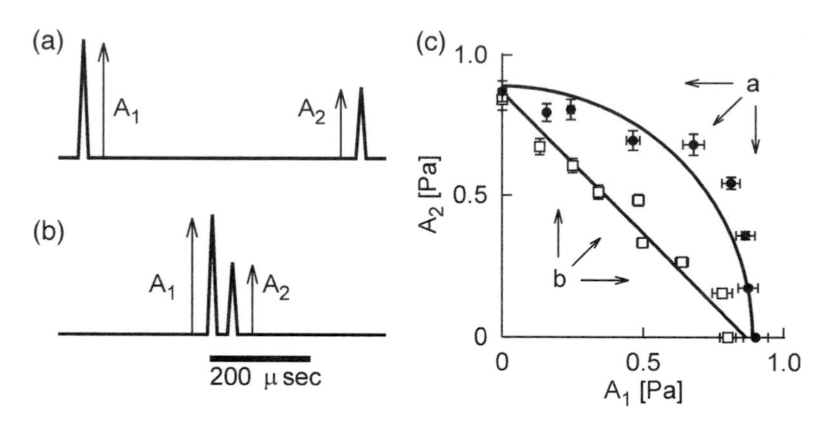

FIGURE 5.4 Identification of iso-response manifolds on short time scales. **a,b:** The acoustic stimuli consisted of two short clicks with amplitudes A_1 and A_2 separated by a peak-to-peak interval of **(a)** 550 μsec or **(b)** 40 μsec. **c**: By adjusting the overall intensity for fixed ratios of A_1 and A_2, stimulus combinations yielding spike probabilities of 70% were obtained (filled circles for stimulus a, open squares for stimulus b). All error measures display 95% confidence intervals. For the long interval, the data are well fitted by a circle, whereas for the short interval, a straight line yields a good fit (solid lines in both cases). This example demonstrates that on different time scales, different stimulus variables can be relevant for the transduction dynamics, here given by the energy A^2 of the sound stimulus (curved line) or simply its amplitude A (straight line). (From Gollisch, T. et al., *J. Neurosci.* 22, 10442, 2002. With permission.)

sound energy, $A_1^2 + A_2^2$. For very small interclick intervals (Figure 5.4b), however, the iso-response manifolds are nearly a straight line (Figure 5.4c, open squares). This indicates that the sum of both click amplitudes, $A_1 + A_2$, governs the transduction on very short time scales and corresponds to the expectation that the eardrum operates approximately as a linear filter on the sound-pressure wave (Schiolten et al. 1981).

Together, these findings imply that the amplitude of the sound pressure and its square (i.e., its energy) are the relevant stimulus variables for two subprocesses within the auditory transduction chain. Within an LNLN cascade (Equation 5.4), the minimal model compatible with the data, the detailed temporal features of the two processes are reflected by the time courses of L_1 and L_2, respectively. Based on the model structure and the squaring nonlinearity connecting L_1 and L_2, exact descriptions of these two functions can be derived from appropriately designed iso-response experiments (Gollisch and Herz, 2004a). Measurements show that L_1 behaves like a damped oscillator with a natural oscillation frequency close to the receptor's best frequency, which lies between 5 and 40 kHz, and a damping time constant in the range of 50 to 500 μsec. Thus, L_1 most likely reflects the mechanical properties of the eardrum at the receptor's attachment site. Quite differently, L_2 exhibits the exponential characteristics of a leaky integrator and captures the electrical properties of the cell membrane. With values as low as 200 μsec, the membrane time constants are very short. These very fast mechanical and electrical processes suggest a tuning of the auditory receptors to the rapid amplitude modulations of behaviorally relevant stimuli, such as conspecific communication calls (see also Section 5.5 and Section 5.6).

5.3 ADAPTATION

Spike-frequency adaptation (Figure 5.5a) is a common phenomenon observed in many spiking neurons. It operates on time scales that range from tens of milliseconds to several seconds (i.e., roughly 1000 times longer than the signal-transduction processes considered in the previous section) and serves many functions, including gain control, high-pass filtering (Nelson et al. 1997; French et al. 2001; Benda et al. 2004), forward masking (Sobel and Tank, 1994; Wang, 1998), and synchronizing network activity (Crook et al. 1998; Fuhrmann et al. 2002). Prominent sources of adaptation are the activation of calcium-dependent or slow voltage-dependent potassium currents and the inactivation of fast sodium currents (for an overview, see Benda and Herz, 2003). What all these sources have in common is that they are driven by the output of the neuron, such that spikes provide a "negative feedback" by opening ionic currents to downregulate the frequency of spiking. Adaptation may, however, also contain components that are directly driven by the sensory or synaptic input in a feedforward way. In receptors, for example, adaptation mechanisms that are independent of the neuron's output act on all steps of the signal-processing chain: the coupling, transduction, and encoding of the primary sensory signal. For higher-order neurons, synaptic mechanisms and inhibitory inputs contribute to an input-driven adaptation effect. Different sources of spike-frequency adaptation may have different effects on the coding properties of a sensory neuron. A complete understanding of the functional aspects of adaptation, therefore, requires the localization of its biophysical sources and the identification of the causal relations between sensory input, neural activity, and level of adaptation.

5.3.1 Disentangling Input-Driven and Output-Driven Adaptation *in Vivo*

To discriminate between input-driven and output-driven sources for adaptation in an auditory receptor, one should use an experimental technique that permits the measurement of one component independently of the other. This can be done by adjusting the intensities for different sound frequencies in such a way that the neuron's steady-state firing rate is the same (see Figure 5.3). In this situation, the level of output-driven adaptation must also be the same. Sudden switches between those sound stimuli then reveal input-driven components of adaptation because these will approach a new equilibrium value after such a switch. This results in transient deflections of the firing rate, which can be observed in electrophysiological recordings of the spiking activity. The deflections allow the measurement of features of the input-driven adaptation component, such as its strength, its time constants, and its correlation with different stimulus and activity parameters (Gollisch and Herz, 2004b). Similar insight is to be expected for other neurons exhibiting a mixture of input-driven and output-driven adaptation.

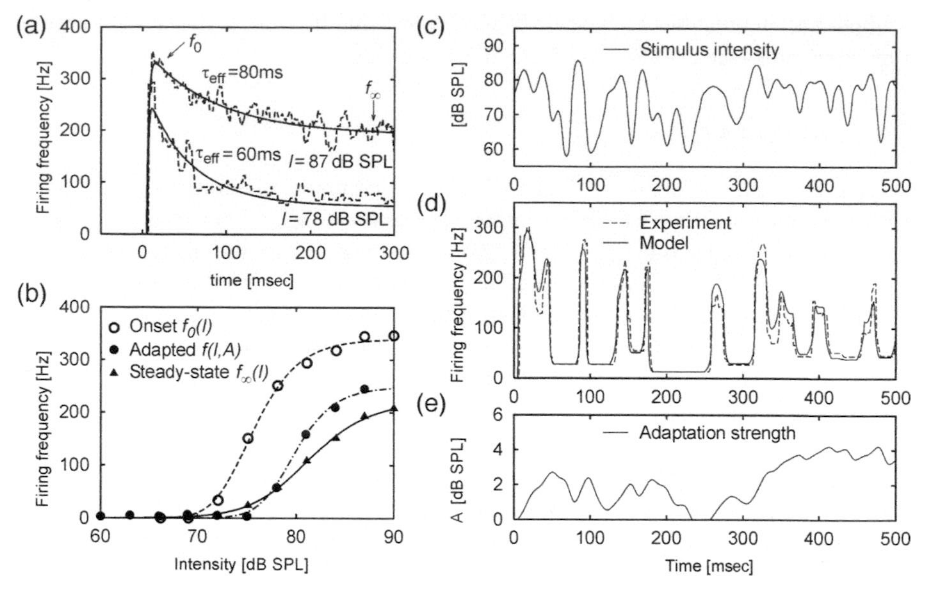

FIGURE 5.5 Spike-frequency adaptation. **a:** Step responses (dashed lines) of a locust auditory receptor to two constant sound stimuli switched on at time zero; sound intensities (I) are indicated by the labels. For simplicity, we assume that the intensity measured in dB SPL is proportional to the input current driving the spike generator. **b:** Onset f-I curve (f_0, open circles) and steady-state f-I curve ($f_\infty(I)$, triangles) determined from the measurements shown in (**a**). The filled circles display the f-I curve for a fixed level of adaptation (A = *const*) corresponding to I = 78 dB. **c–e:** Performance of the model. Neural response (**d**, dashed line) to the amplitude-modulated stimulus in **c** (carrier frequency: 5 kHz, modulation: broadband white noise with cut-off frequency 50 Hz). The f-I curves from (**b**) and an adaptation constant τ = 150 msec resulting from Equation 5.12 were used to calculate the firing frequency from the adaptation model in Equation 5.11 (**d**, solid line). Notice how the prediction closely follows the experimental data. The firing frequency was computed as the averaged inverse interspike interval at any time bin. **e:** Adaptation strength A vs. time. Note that a 5 dB shift of the f_0 (I) curve may alter the firing rate by more than 100 Hz (see **b**).

5.3.2 Phenomenological Model for Output-Driven Adaptation

Within most cellular types of output-driven adaptation, spikes directly or indirectly activate ionic currents. Such adaptation currents act subtractively on the time-dependent input current $I(t)$ because ionic currents are flowing in parallel through the cell membrane. Because the dynamics of adaptation are usually slow with respect to the time scale of interspike intervals, we approximate adaptation currents by their temporal average over an interspike interval. To model the firing frequency output of an adapting neuron, we describe its spiking dynamics by the onset f-I curve, $f_0(I)$, which we obtain by measuring the onset response f_0 of the unadapted neuron (Figure 5.5a,b). This approach allows one to derive a universal model for the firing frequency (Benda and Herz, 2003) that is independent of the specific biophysical mechanism of the spike-driven adaptation process. In its simplest form, the model reads

$$\begin{aligned} f(t) &= f_0(I(t) - A(t)) \\ \tau \frac{d}{dt} A(t) &= A_{\max}(f(t)) - A(t). \end{aligned} \tag{5.11}$$

The adaptation strength A denotes the averaged adaptation current. Depending on the size of A, the onset *f-I* curve is shifted to higher input currents (Figure 5.5b). The maximum adaptation strength $A_{\max}$ depends on the output firing frequency f and is approached on a time scale governed by the adaptation time constant τ. $A_{\max}$ is given by the difference between the onset *f-I* curve $f_0(I)$ and the steady-state *f-I* curve $f_\infty(I)$. The latter is obtained by measuring the steady-state firing frequencies f_∞ for different constant input currents I.

Often, the time course of adaptation is estimated from an exponential fit to the firing-frequency response to current steps (Figure 5.5a). The "effective" adaptation time constant τ_{eff} obtained from such a measurement differs in general from the adaptation time constant τ. Both parameters are approximately related via the slopes $f'_0(I)$ and $f'_\infty(I)$ of the *f-I* curves, or

$$\tau \approx \tau_{\text{eff}} \frac{f'_0(I)}{f'_\infty(I)}. \tag{5.12}$$

It follows that the observed effective time constant (10 to 75 msec in the auditory receptors of *L. migratoria*) is smaller than the true time constant of the adaptation mechanism.

The adaptation model (Equation 5.11) is completely specified by the onset *f-I* curve, the steady-state *f-I* curve, and the adaptation time constant. For a specific neuron, all three quantities can easily be obtained from the firing-frequency responses to step-current inputs (Figure 5.5a,b). Once they are determined, the response of that neuron to different stimuli, such as band-limited white noise stimuli or conspecific grasshopper songs, can be predicted (Figure 5.5c). As the figure demonstrates, the model is in good agreement with the measurements showing that, on the level of firing frequencies, a single adaptation mechanism is sufficient to describe the data. Without the adaptation process, the data cannot be explained. In addition, the time course of the adaptation strength, and thus the position of the current *f-I* curve, is obtained. This provides a tool for analyzing how adaptation shifts the neuron's *f-I* curve according to the stimulus.

The approaches presented so far either treat neural responses on the level of mean firing rates or do not describe spike generation at all. The following section presents a stochastic model for the generation of action potential that also captures the observed response variability.

5.4 SPIKE GENERATION AND TRIAL-TO-TRIAL RESPONSE VARIABILITY

Even under well-controlled laboratory conditions, repeated presentations of the same stimulus lead to different neural responses. This trial-to-trial variability is caused by stochastic components of the single-cell dynamics, such as ion-channel noise, and results in spike-time jitter and additional or missing spikes. Trial-to-trial variability limits the amount of information that is transmitted by a sensory neuron about the external stimulus. Reliable accounts of this phenomenon are a prerequisite for the proper interpretation of neural dynamics and coding principles. Models that accurately describe neural variability over a wide range of stimulation and response patterns are thus highly desirable, especially if they can explain the variability in terms of basic neural observables and parameters, such as firing rate and refractory period.

5.4.1 MATHEMATICAL FRAMEWORK FOR STOCHASTIC SPIKE GENERATION: RENEWAL PROCESSES

The simplest mathematical description of spike-time variability treats a neuron as a random point process, that is, a process that stochastically generates a series of events (spikes) in time. In general, the probability of generating a spike at time t could depend on a variety of factors. Modeling is greatly facilitated if one can assume that only an "effective" stimulus strength $q(t)$ at that very moment and refractoriness caused by the last spike at t_{last} determine the generation of an action potential. For time-independent stimuli, that is, $q(t) = q$, spike generation becomes a "renewal process" (Cox, 1962), so that interspike intervals (ISIs) are independent. For time-dependent stimuli, the strict independence of ISIs is no longer true and one rather speaks of a "modulated renewal process" (Reich et al. 1998).

The influence of refractoriness on the generation of an action potential depends on the interval Δ that has passed since the last spike was elicited at time $t_{last} = t - \Delta$. Let us use $w(\Delta)$ to denote this memory term, or "recovery function" (Berry and Meister, 1998; for a comparison with the "hazard function," see Johnson, 1996, or Gerstner and Kistler, 2002). Note that the memory is independent of the strength of the effective stimulus $q(t)$. Mathematically, spike generation is, thus, described by a probability per unit time (the "hazard") $\rho(t|t_{last})$, which is conditional on the last spike occurring at time t_{last}:

$$\rho(t|t_{last}) = q(t) \cdot w(t - t_{last}) \tag{5.13}$$

Values of the recovery function w range between zero and unity. Within the absolute refractory period τ_a, even arbitrarily large stimuli cannot elicit a spike. To capture this property, $w(\Delta)$ vanishes for $0 \leq \Delta \leq \tau_a$. It then rises monotonically to account for the relative refractory period during which the neuron relaxes back to its normal level of excitability (Figure 5.6a, lower panel).

For a renewal process, it is possible to compute the recovery function directly from the ISI distribution $P_{ISI}(\Delta)$ obtained under constant stimulation ($q(t) = q$), as has been discussed in the literature (Johnson, 1996; Berry and Meister, 1998; Gerstner and Kistler, 2002):

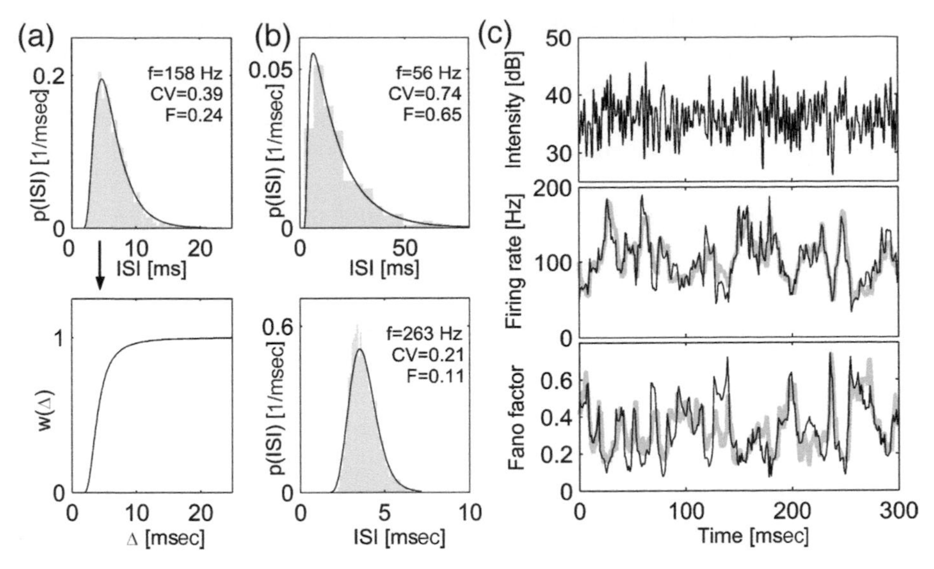

FIGURE 5.6 Response variability for constant and time-varying stimuli. **a:** Upper panel: interspike interval (ISI) distribution (gray histogram) of a locust auditory receptor resulting from a constant stimulus with intermediate sound intensity (evoked firing rate: 158 Hz). This distribution was used to calculate the recovery function (see Equation 5.19) depicted in the lower panel. **b:** Two examples of ISI distributions from a low (56 Hz, upper panel) and a very high firing rate (253 Hz, lower panel), recorded from the same cell as in (**a**) but at different intensity levels. The black lines are predictions from the recovery function depicted in (**a**) that were obtained by adjusting the stimulus strength q such that experimental and model firing rates matched. Also shown are CV values (the ratio between the standard deviation of the ISI distribution and the mean ISI) and Fano factors (the ratio between the variance and the average of the spike count, both calculated for a certain time window, here 10 msec). With decreasing firing rates, the CV value and the Fano factor increase because refractoriness plays a smaller role at low firing rates. **c:** Upper panel: amplitude modulation of a Gaussian white-noise sound stimulus (carrier frequency: 4 kHz). Middle panel: time course of the firing rate determined with a 10-msec sliding window. As in the lower panel of this figure, thick gray lines depict experimental data and thin black lines denote the model result. Lower panel: spike-count variability as measured by the Fano factor. As for constant stimuli, spike-train variability is anticorrelated with the firing rate on short time scales; high-activity episodes come with low variability (high precision), and vice versa.

$$w(\Delta) = \frac{1}{q} \cdot \frac{P_{ISI}(\Delta)}{1 - \int_0^{\Delta} d\tau \, P_{ISI}(\tau)} \tag{5.14}$$

Equation 5.14 can be inverted to calculate the ISI distribution from the recovery function for given stimulus strength:

$$P_{ISI}(\Delta) = q \cdot w(\Delta) \cdot \exp\left[-q \cdot \int_0^{\Delta} d\tau \, w(\tau)\right] \tag{5.15}$$

To use this model framework, one must first investigate whether the spike trains of a specific neuron are indeed consistent with a renewal process (independent ISIs) for constant stimulation. Second, the effective input q must be determined as a function of the stimulus intensity — such that the observed and the predicted mean firing rates match (for details, see Schaette et al., 2004).

5.4.2 Summary of Results from *Locusta migratoria*

The variability of receptor spike trains depends strongly on the evoked firing rate. Low firing rates (below 50 Hz) are accompanied by high variability, as reflected in a broad ISI distribution and a coefficient of variation (CV) near unity. The coefficient of variation is defined as the ratio between the standard deviation of the ISI distribution and the mean ISI. It is unity for Poisson spike trains. For higher firing rates, the ISI distribution becomes sharper (Figure 5.6a,b) and the CV decreases to values of 0.2 (very regular spike trains) for maximum firing rates of around 300 Hz. A similar dependence is observed for dynamic stimuli, but the situation is more complicated because there is also an influence of the temporal structure and modulation depth of the stimulus amplitude modulation.

Extensive data from locust auditory receptors suggest that the renewal assumption is well justified for this system once adaptation effects are taken care of. Each neuron can then be characterized by one unique recovery function $w(\Delta)$. This minimal description provides accurate predictions for ISI distributions caused by input intensities over the entire range of firing-rate responses (see Figure 5.6a,b). To cover responses to dynamic stimuli, the renewal process is driven by a time-dependent effective stimulus strength $q(t)$. The model captures the spike-count variability and salient features of the fine temporal structure in response to dynamic stimuli although the recovery functions were always calculated from ISI distributions obtained under constant stimulation (Figure 5.6c).

These results demonstrate that key ingredients of stochastic responses are faithfully captured by the renewal model and that a clear mathematical separation is possible between external stimulus and stimulus-independent cell dynamics. Furthermore, spike variability can be modeled using the same stochastic process for constant and strongly time-varying stimuli. This indicates that there is no principal difference between the dynamics underlying responses to both stimulus classes. Differences in the encoding quality (see Section 5.5) may simply arise from the specific usage of the neuron's dynamic range by the particular stimulus.

Including a detailed adaptation model, such as the one presented in Section 5.3, would enhance the model's predictive power for stimuli varying on multiple time scales, but at the cost of substantially longer recordings needed to calibrate the additional model parameters. The degree of realism could be further enhanced by incorporating the signal-transduction model of Section 5.2. Together, these elements might yield a biophysically motivated and simple, yet highly accurate, description of stimulus encoding.

5.5 DECODING SPIKE TRAINS

As shown in Section 5.2 through Section 5.4, mathematical models of single-cell dynamics provide valuable insight about the response dynamics of auditory receptors. We now go one step further and investigate their signal-processing capabilities using methods from systems analysis and information theory (see also Bialek et al. 1991; Rieke et al. 1997; Borst and Theunissen, 1999). These methods provide a quantitative means to estimate the information contained in a spike train that was evoked by a given stimulus. In particular, one can ask whether auditory receptors encode a large range of acoustic stimuli or whether they are specifically tuned to behaviorally relevant features, such as the temporal structure of a grasshopper calling song (Machens et al. 2001).

5.5.1 STIMULUS RECONSTRUCTION

In what follows, we focus on linear stimulus reconstruction, that is, the attempt to recover the original stimulus *s(t)* from a spike train *y(t)*, as illustrated in Figure 5.7. To do so, each spike is replaced by a filter function $h(\tau)$, resulting in the signal estimate $s_{est}(t)$:

$$s_{\text{est}}(t) = h_0 + \int_0^T d\tau\, h_1(\tau) y(t-\tau)\ , \tag{5.16}$$

where h_0 is the mean signal level in the absence of spiking. The parameters h_0 and $h(\tau)$ are determined by minimizing the mean-square error $\langle n_{\text{mse}}(t)^2 \rangle$, where the angular brack-

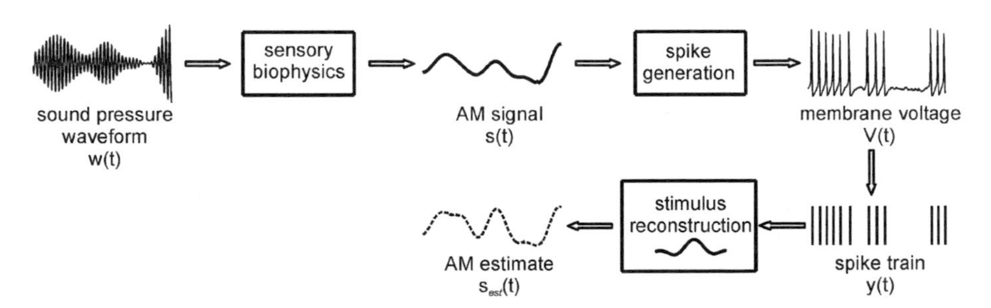

FIGURE 5.7 Stimulus preprocessing and reconstruction. The mechanics of the receiver's ear extract the slow amplitude modulation *s(t)* of a rapidly oscillating sound-pressure wave *w(t)*. Auditory receptor neurons then encode *s(t)* into the membrane voltage *V(t)*. As a first step in the stimulus reconstruction, the spike train *y(t)* is extracted from the voltage trace. Within linear reconstruction, each spike is replaced by an optimal filter function to yield $s_{est}(t)$, the estimate of *s(t)*. As shown by this example, stimulus reconstruction does *not* aim at recovering the original, complete physical stimulus *w(t)* but instead requires the identification of a representation of the stimulus that is relevant for the animal, in the present case the AM signal *s(t)*. (From Machens, C.K., *J. Neurosci.*, 21, 3218, 2001. With permission.)

ets denote a time average over the section of the experiment used for parameter estimation and $n_{mse}(t)$ is the time-dependent reconstruction error $n_{mse}(t) = s(t) - s_{est}(t)$.

The use of stimulus reconstruction methods does not imply that we think the auditory system is trying to reconstruct acoustic stimuli from spike trains. Rather, by comparing the reconstruction quality for different stimulus ensembles, we seek to find out which characteristics of acoustic signals are encoded faithfully and which features are discarded.

Auditory receptor neurons of grasshoppers are sensitive to the amplitude modulation (AM) of broadband sound-pressure waves that exceed a certain threshold. Below this threshold, the cells remain silent. Therefore, the appropriate stimulus $s(t)$ for applying reconstruction techniques (Figure 5.8) is *not* the original sound-pressure wave $w(t)$ but rather that part of the AM signal that lies in the sound intensity range covered by the particular receptor. Within the stimulus reconstruction algorithm, therefore, the AM signal should be half-wave rectified at the threshold of each cell and then used for the stimulus reconstruction algorithm. From now on, the thresholded AM signal will simply be referred to as the *signal*.

A nonlinear relationship between the AM signal and the firing rate could require higher-order reconstruction filters for adequate signal reconstruction. Such filters seem to suggest relational codes (i.e., coding schemes that involve higher orders of the spike-train statistics, as in ISI-based codes). Because the firing-rate responses of auditory receptors of grasshoppers are approximately threshold-linear if amplitude modulations are measured on a logarithmic scale (Römer, 1976; Stumpner and Ronacher, 1991; Ronacher and Krahe, 2000), this potentially misleading interpretation of higher-order kernels can be obviated by transforming $s(t)$ and $s_{est}(t)$ into the

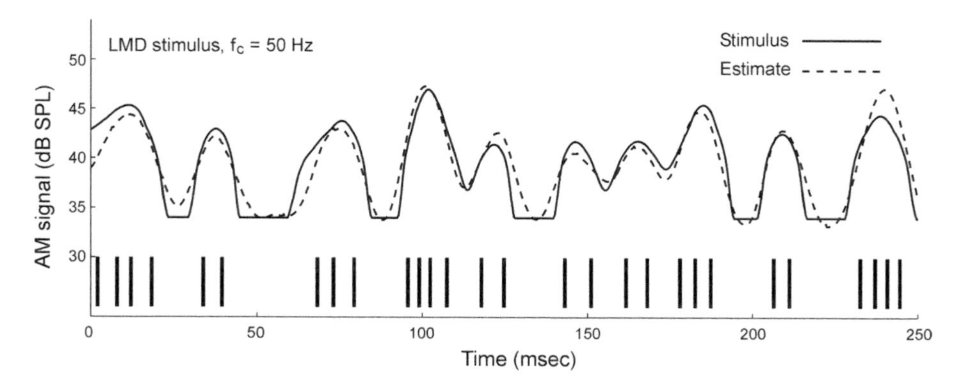

FIGURE 5.8 "Reading" a neural spike train. Shown is the linear reconstruction of an LMD signal with 50 Hz cut-off frequency from the responses of a single locust auditory receptor. The stimulus was thresholded at 34 dB, the threshold of this particular neuron. The close correspondence between stimulus and linear stimulus estimate demonstrates that the sound signal can be faithfully inferred from the time-resolved firing rate; higher-order correlations of the spike train are not needed to reconstruct the stimulus. However, both the timing of individual spikes and their local rate contribute to the high reconstruction quality. (From Machens, C.K. et al., *J. Neurosci.*, 21, 3220, 2001. With permission.)

decibel scale. In a more rigorous approach, one could also first measure the *f-I* curve of the neuron under study and use this information to rescale the input signal such that the firing rate depends linearly on the transformed input.

Spike-frequency adaptation (Section 5.3) can be described, as a rule, by higher-order or time-dependent reconstruction filters, but their estimation requires enormous amounts of data. In most studies, such complications are circumvented by studying the responses of fully adapted neurons, discarding their responses during the initial phase.

The reconstruction error $n_{mse}(t)$ can be separated into random and systematic components. Systematic errors occur if one attempts to reconstruct a signal $s(t)$ that is incompatible with the signal the neuron actually encodes. For instance, if only a low-pass-filtered version of the signal is encoded, any attempts to reconstruct higher frequencies must fail. Systematic errors can be corrected by introducing a frequency-dependent gain $g(f)$ such that $s_{est}(f) = g(f)[s(f) + n_{eff}(f)]$, where $n_{eff}(f)$ denotes the random errors, or "effective noise," as referred to the input (Theunissen et al. 1996; Rieke et al. 1997).

Given the effective noise $n_{eff}(f)$, the reconstruction success in each frequency band can be measured by the frequency-resolved *signal-to-noise ratio* (SNR):

$$SNR(f) = \frac{S(f)}{N_{eff}(f)} = \frac{s(f)s^*(f)}{n_{eff}(f)n^*_{eff}(f)}, \tag{5.17}$$

where $S(f)$ and $N_{eff}(f)$ are the power spectral densities of the signal and the effective noise, respectively. A high $SNR(f)$ indicates an accurate reconstruction of the particular frequency component, while an $SNR(f)$ of zero implies chance level. The frequency-resolved SNR, therefore, allows one to assess which stimulus components are best decoded by signal reconstruction. Reconstruction of signals with high bandwidth serves to estimate the cut-off frequency of the system; this natural cut-off will be unveiled as the frequency where the signal-to-noise ratio approaches zero. To measure the overall reconstruction success, the ratio of the total power of signal and noise will be used, $SNR = \int df\, S(f)/\int df\, N_{eff}(f)$.

5.5.2 Information Theory

The *mutual information rate* R_{info} quantifies how many bits of information about the signal $s(t)$ are carried by a spike train per second. For example, a value of $R_{info} = 1$ bit / sec means that the uncertainty about the stimulus can be halved every second by reading the corresponding spike train. Note that the mutual information rate can be large even if the signal is only poorly reconstructed, as might occur for stimuli with high bandwidth. If $s(t)$ is a Gaussian random signal, a lower bound on R_{info} (Rieke et al. 1997) is given by

$$R_{info} \geq \int_0^\infty df \log_2[1 + \mathrm{SNR}(f)]. \tag{5.18}$$

Given a time resolution Δt, the efficiency of a neuron at transmitting information can be measured by comparing the estimated mutual information rate $R_{info}(\Delta t)$ with the information-theoretic limit $R_{max}(\Delta t)$, which is reached if the spike train is maximally disordered (i.e., Poisson; see Rieke et al. 1995, for details). The *coding efficiency* $\varepsilon(\Delta t)$ is then defined as

$$\varepsilon(\Delta t) = \frac{R_{info}(\Delta t)}{R_{max}(\Delta t)}, \tag{5.19}$$

which takes on values between zero and one. Although $R_{max}(\Delta t)$ tends to infinity for $\Delta t \to 0$, this is not the case for $R_{info}(\Delta t)$, which will instead achieve the value R_{info} given in Equation 5.18. To yield nontrivial results, the coding efficiency, therefore, must be evaluated at a finite time resolution that reflects spike-timing variability due to intrinsic noise sources. This time resolution can be estimated by crosscorrelation analysis and is about $\Delta t \approx 1$ msec for grasshopper auditory receptor cells.

5.5.3 Summary of Results

To identify essential features of grasshopper songs and their neural representations, artificial stimuli were designed to vary the most salient statistical properties of grasshopper sounds (see Figure 5.1). The communication signals alternate between noise bursts and pauses, resulting in a characteristic double-peak distribution of sound amplitudes (see, e.g., Machens et al. 2001). To study the importance of this structural aspect, two different classes of stimuli were generated.

The first class consisted of random stimuli that have the same amplitude distribution as a typical grasshopper song and thus imitate the gap-infiltrated structure of these songs. Due to their modulation depth of ~24 dB, these stimuli are called *large modulation depth* (LMD) stimuli. Within the second class, stimuli have a Gaussian amplitude distribution with a modulation depth of 10 dB and are called *small modulation depth* (SMD) stimuli. These random stimuli simulated the combined sound pattern of a group of five to ten grasshoppers singing simultaneously, such that the song pauses of individual songs are filled by the other songs.

Because the shortest behaviorally relevant time scales of the AM signals are around 1 to 2 msec (von Helversen, 1972), frequency components of at least 250 to 500 Hz are required in the random stimuli. To analyze the neural representation at these short time scales, LMD and SMD stimuli were designed with piecewise flat spectral characteristics and cut-off frequencies of up to 800 Hz. Additionally, to test whether the specific mix of frequency components found in natural songs might be of importance, one of the LMD stimuli exhibited a songlike spectrum (SLS). In all experiments, the amplitude distribution for each stimulus was kept constant by fixing the integrated AM signal power. A larger bandwidth, therefore, corresponds to a lower power spectral density.

Reconstructions from the recorded spike trains of single locust receptor neurons demonstrate that even single cells are capable of encoding amplitude modulations with signal-to-noise ratios of up to 10:1. In this regime, the original stimulus can be faithfully "read" (Bialek et al. 1991) from the spike train (Figure 5.8). Most importantly, the success of the linear reconstruction method suggests that the studied auditory receptors encode acoustic stimuli as a strongly time-varying firing rate; we did not find any indications of coding schemes that involve higher orders of the spike-train statistics. The data also show that sounds with large modulation depth are encoded with higher signal-to-noise ratios, information rates (up to 180 bits/sec) and coding efficiencies (up to 0.4) than stimuli with small modulation depth. While LMD stimuli exhibit greater raw amplitude variations, the encoding of SMD stimuli is still poorer, even when the AM signal power above threshold is identical in the two classes of stimuli. Matching the spectral properties of the songs as in the SLS stimulus, on the other hand, does not increase signal-to-noise ratios.

Spikes are triggered with high reliability and temporal precision when the sound intensity rapidly passes the firing threshold — as occurs at the beginning of a syllable of the grasshopper calling song (see Figure 5.1). This phenomenon emphasizes the paramount importance of gaps and pauses for the recognition of acoustic stimuli, as the precision in spike timing leads to a faithful representation of the suprathreshold sound pattern. Grasshoppers seem to utilize this effect in the design of their songs, which consist of repeated patterns of sound and (relative) quiet.

The highest information transfer rates of single cells are observed for stimuli with large modulation depth and a cut-off frequency of 200 Hz. This finding should be compared with behavioral studies in which various artificial auditory stimuli were presented that were generated by filtering the Fourier components of model songs with regular or irregular syllable composition (von Helversen and von Helversen, 1998). These studies demonstrate that, depending on the original syllable structure, Fourier components between 150 and 300 Hz are required by *Ch. biguttulus* females to detect gap signals reliably. Together, these two results suggest that the response properties of single receptor neurons are optimized for features of the acoustic environment that are of prime importance for behavioral decisions.

5.6 SONG DISCRIMINATION

The experimental results summarized in the previous section clearly indicate that single auditory receptors are well suited to transmit information about conspecific communication signals. This raises the question of whether they can also discriminate between individual songs from the same species, which would provide a basis for mate preference and sexual selection at the single-cell level.

5.6.1 Spike-Train Metrics and Discrimination Matrices

Successful discrimination requires that spike trains elicited by repeated presentations of the same song be more similar to each other than spike trains elicited by different songs. To compute the similarity between two spike trains, each spike was first replaced by an alpha function that mimics the time course of an *excitatory postsynaptic potential* (EPSP) in a hypothetical downstream neuron of the grasshopper's auditory system. The distance between the two spike trains was then defined as the mean square distance between their EPSP convolved traces (van Rossum, 2001). By varying the width τ of the EPSP function, effects of the temporal resolution can be studied: if τ is large, only differences in the average spike rate contribute to the distance measure; if τ is small, even small differences in spike timing matter.

To discriminate the songs based on their evoked responses, one spike train was arbitrarily chosen as a template for each of the eight presented songs. The remaining spike trains were classified by assigning each spike train to the closest of the eight templates. Averages were then computed by permuting all possible template choices, yielding classification matrices (Figure 5.9a). The diagonal elements of the matrices correspond to correctly classified spike trains, the off-diagonal elements to misclassified spike trains. Replacing the alpha function with exponential or Gaussian kernels of the same width led to similar results, as did the use of cost-based metrics (Victor and Purpura, 1997). Using more sophisticated supervised and unsupervised cluster-

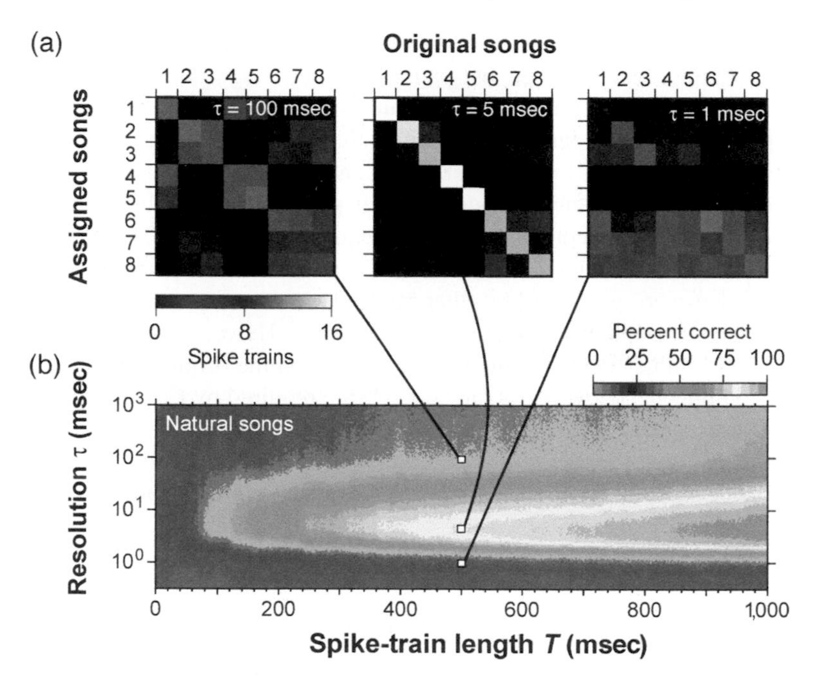

FIGURE 5.9 (See color insert following page 202.) Song discrimination. **a:** Discrimination matrices computed at $T = 500$ msec for $\tau = 100$ msec, $\tau = 5$ msec, and $\tau = 1$ msec. T denotes the spike-train length used for the data analysis, and τ represents the time scale on which time-varying firing rates are resolved. Rows and columns of the matrices denote the consecutively numbered songs. The matrix entries N(a,b) give the number of spike trains that were originally elicited by song # b (horizontal axis) and in turn classified as belonging to song # a (vertical axis). A spike train is correctly classified if a = b. Perfect discrimination is achieved if all spike trains are assigned onto the diagonal elements. **b:** Discrimination performance as a function of the response duration T and resolution τ. Within the red area, at least 95% of the spike trains are classified correctly; the half-width of the EPSP-like functions used to filter the spike trains before discriminating the songs are then within the range of $\tau = 3$–10 msec.

ing algorithms to assign spike trains to the songs that elicited them provided, at best, a marginal improvement in overall discrimination performance.

5.6.2 Summary of Results

Too broad or too narrow an EPSP filter will impair the discrimination of songs, leading to many off-diagonal entries in the corresponding matrices. However, for an EPSP width of $\tau \approx 5$ msec, most spike trains are correctly assigned. This time scale also corresponds to the shortest unique song features that auditory receptors accurately encode. Signal variations on a scale of a few milliseconds are thus important for song evaluation; downstream neurons with much longer EPSPs will not be suitable to process the information relevant for discrimination.

To analyze how the discrimination capability unfolds in time, we varied the time window T in which spikes were used to discriminate, extending T from 0 to 1000 msec. Calculating the percentage of correctly classified spike trains for every (τ,T) pair yields the contour plot shown in Figure 5.9b. The eight tested songs can be discriminated with high reliability (> 80%) after a mere four or five syllables ($T = 400$ to 500 msec).

The question arises as to which song cues are relevant for evaluating potential mates. For instance, the songs differ in their syllable length (60 to 120 msec) and broadband carrier frequency content (Figure 5.1b). However, such variations are unlikely to play a role in the females' assessment of males: syllable duration is greatly affected by body temperature and can vary more than twofold when the male moves from sun to shadow. Similarly, the frequency content depends strongly on the distance of the listening female grasshopper to the singer, with higher frequencies decaying more quickly with distance. Differences in spectral shape result in different firing rates, thereby facilitating discrimination. In fact, close inspection of Figure 5.9b reveals that using a moving average of the neuron's firing rate alone (filter width $\tau = 1000$ msec), almost 40% of the spike trains can be assigned to the correct song.

To eliminate any spurious discrimination cues that simply reflect differences in syllable length or broadband frequency content, we complicated the discrimination task and generated a set of rescaled songs that each had a common syllable length of 100 msec and the same carrier spectrum. The overall stimulus power was adjusted such that the last second of each song had the same mean sound intensity. The detailed structure of amplitude modulations within a syllable, which differed from song to song, remained as a discrimination cue.

Surprisingly, the ability to differentiate these artificial songs does not decrease substantially (data not shown, but see Machens et al. 2003). A few hundred milliseconds, or 40 to 50 spikes, suffice to distinguish more than 80% of the stimuli. Indeed, two (instead of eight) songs can often be discriminated perfectly after a few tens of milliseconds, or four to five spikes. This demonstrates that the remarkable temporal precision of the receptor neurons allows one to recover even slight differences between the rescaled songs from the neural response.

With a population of 40 to 50 receptors per ear, even more information is available about the fine details of the amplitude modulation pattern. To discriminate conspecific from heterospecific signals, however, such a high precision is superfluous. Instead, the acquired information could be used to evaluate male singers. In this context, it should be noted that a female grasshopper does not need to recognize an individual male; it is sufficient to respond to those males whose signals indicate a good genetic constitution.

Grasshoppers produce their songs by a rhythmic movement of the hind legs against the forewings. During a syllable, each hind leg moves up and down three to four times. Each of these strokes lasts for about 7 to 12 msec and is succeeded by a short gap (~ 2 msec) at the movement reversal point. In a healthy male, these brief gaps are camouflaged by systematic phase shifts between the movement of both legs, leading to a fairly smooth total amplitude modulation pattern within a syllable (von Helversen and von Helversen, 1997). However, should the sound produced by one of the hind legs be weaker, the modulation pattern of the other side will dominate. Males with developmental faults often damage a hind leg or cripple a forewing when molting. Their songs exhibit 2-msec gaps within each syllable. Females refuse to mate with such impaired singers (Kriegbaum, 1989). Poor genetic quality may also result in inadequate coordination between the two hemisegmental central pattern generators that are responsible for the stridulation movement, creating a deviant amplitude modulation pattern (Ronacher, 1989). It may thus be speculated that it would be advantageous for a female grasshopper to carefully "inspect" the amplitude modulations of the song of a potential mate. Our data demonstrate that the information required to do so is faithfully encoded in the auditory periphery.

5.7 CONCLUSIONS AND OUTLOOK

As we have tried to demonstrate in this chapter, several fundamental questions about the dynamics and signal processing capabilities of auditory receptors can be quantitatively answered by closely combining experiment, modeling, and theory. For example, we have been able to show that sound energy, but not sound pressure, governs the auditory transduction process. This finding sets tight constraints for any future biophysical transduction model. The time scales of the mechanical and electrical integration have been revealed with high accuracy from recording neural output activity more than 1 mm away from the relevant processes, thus leaving the vulnerable mechanical structures of the ear unimpaired. The measured electrical time constants are well below 1 msec and reflect the high demand for temporal resolution in the auditory periphery. Traditional correlation techniques are limited by spike-time variability and thus cannot capture the dynamics on time scales as short as in our example.

Together with the low internal noise level, short integration times also explain the large information rates of the receptor neurons and their astounding discrimination capability. It thus appears that various microscopic biophysical parameters are tuned to encode conspecific communication signals optimally. Similarly, stimulus reconstructions from spike trains evoked by grasshopper songs suggest that even the time scale of adaptation may be tuned such that it balances the rising overall song intensity (see Figure 5.1) to provide a fairly invariant sound representation for downstream neurons (Machens et al. 2001).

These findings indicate that the dynamics and coding strategies of grasshopper auditory receptors are well adjusted to important behavioral tasks. Whether the auditory receptors have evolved to process grasshopper songs, or selection pressure has forced the song patterns to match the properties of receptors, or both, is a currently unanswered question. One may therefore ask whether superoptimal stimuli exist that receptor spike trains can encode even better than conspecific calling songs. To investigate this question, we have started to search systematically for such stimuli. Instead of exploring the huge space of natural and naturalistic stimuli, we use information from the responses of a given receptor neuron to identify those regions in stimulus space that are encoded best. An information-theoretic foundation (Machens, 2002), reliable online analysis, and automatic feedback to the stimulus generation are central aspects of this approach.

From a conceptual point of view, the underlying strategy is similar to the iso-response method (Section 5.2) whose key ingredient is a systematic exploration of stimuli causing the *same* output. Investigating such regions implies a radical change of the traditional perspective regarding neural input–output relations. Instead of asking what output is produced by a given input, one seeks to identify input ensembles that give rise to a fixed response. Fast online techniques and feedback-driven stimulation are again crucial ingredients of this approach. With growing computer power already integrated into the modern lab bench, both types of investigation will soon become suitable for various other sensory systems.

These methodological observations suggest that future advances in sensory neurobiology will benefit strongly from close interactions between theoreticians and experimentalists. In fact, the requirements of an approach grounded in modern data-analysis techniques need to be considered long before the first experiment is carried out and necessitate a tight integration of theoreticians into the experimental design process. It is our strong belief that many of the currently unsolved questions regarding the dynamics and coding strategies of auditory and other sensory systems will be solved through such interactions.

Computer-aided searches for invariant stimulus regions may also help us to better understand the astounding computations that follow the first steps of transducing and encoding the external stimulus discussed in the present contribution. Female grasshoppers, for example, seem to use the ratio between syllable length and pause length for recognizing male songs as "conspecific" (von Helversen and von Helversen, 1997). They even tolerate a global time warp of more than 100% if the syllable-to-pause ratio remains the same. This implies that small insect neural systems are capable of performing nontrivial division operations in the temporal domain. They also solve similarly complex computations when localizing a sound source. Discovering the physical principles and biological mechanisms underlying these behaviors remains a fascinating challenge.

ACKNOWLEDGMENTS

Astrid Franz, Olga Kolesnikova, Rüdiger Krahe, Petra Prinz, and in particular Bernd Ronacher were strongly involved in various projects summarized in this review. We would like to thank them for their specific contributions and for many fruitful discussions.

REFERENCES

Benda J, Herz AVM (2003) A universal model for spike-frequency adaptation. *Neural Comp.* **15:**2523–2564.

Benda J, Longtin A, Maler L (2004) Rapid spike-frequency adaptation enhances the detection of transient communication signals. Preprint.

Berry MJ, Meister M (1998) Refractoriness and neural precision. *J. Neurosci.* **18:**2200–2211.

Bialek W, Rieke F, de Ruyter van Steveninck RR, Warland D (1991) Reading a neural code. *Science* **252**:1854–1857.

Borst A, Theunissen FE (1999) Information theory and neural coding. *Nat. Neurosci.* **2**:947–957.

Bradbury JW, Vehrenkamp SL (1998) *Principles of Animal Communication.* Sinauer, Sunderland, MA.

Cox DR (1962) *Renewal Theory.* Methuen, London.

Crook SM, Ermentrout GB, Bower JM (1998) Spike frequency adaptation affects the synchronization properties of networks of cortical oscillators. *Neural Comput.* **10**:837–854.

Evans EF (1975) Cochlear nerve and cochlear nucleus. In: *Handbook of Sensory Physiology, Vol. 5.2, Auditory Systems.* Keidel WD, Neff WD, Eds. Springer, Berlin, pp. 1–108.

French AS, Höger U, Sekizawa S-I, Torkkeli P H (2001) Frequency response functions and information capacities of paired spider mechanoreceptor neurons. *Biol. Cybern.* **85**:293–300.

Fuhrmann G, Markram H, Tsodyks M. (2002) Spike frequency adaptation and neocortical rhythms. *J. Neurophysiol.* **88**:761–770.

Garner WR (1947) The effect of frequency spectrum on temporal integration of energy in the ear. *J. Acoust. Soc. Am.* **19**:808–814.

Gerstner W, Kistler WM (2002) *Spiking Neuron Models.* Cambridge University Press, Cambridge, U.K.

Gollisch T, Herz AVM (2004a) Disentangling sub-millisecond processes within an auditory transduction chain. Preprint.

Gollisch T, Herz AVM (2004b) Input-driven components of spike-frequency adaptation can be unmasked *in vivo. J. Neurosci.* **24**:7435–7444.

Gollisch T, Schütze H, Benda J, Herz AVM (2002) Energy integration describes sound-intensity coding in an insect auditory system. *J. Neurosci.* **22**:10434–10448.

Hauser MD (1996) *The Evolution of Communication.* MIT Press, Cambridge, MA.

Heil P, Neubauer H (2001) Temporal integration of sound pressure determines thresholds of auditory-nerve fibers. *J. Neurosci.* **21**:7404–7415.

Hunter IW, Korenberg MJ. (1986) The identification of nonlinear biological systems: Wiener and Hammerstein cascade models. *Biol. Cybern.* **55**(2–3):135–144.

Jameson D, Hurvich LM (1972) Eds. *Handbook of Sensory Physiology*, Vol. 7.4, Visual Psychophysics, Springer, Berlin.

Johnson (1996) Point process models of single-neuron discharge. *J. Comp. Neurosci.* **3**:275–299.

Kriegbaum H (1989) Female choice in the grasshopper *Chorthippus biguttulus. Naturwissenschaften* **76**:81–82.

Kriegbaum H, von Helversen O (1992) Influence of male songs on female mating behaviour in the grasshopper *Chorthippus biguttulus. Ethology* **91**:248–254.

Machens CK (2002) Adaptive sampling by information maximization. *Phys. Rev. Lett.,* **88**:228104.

Machens CK, Stemmler MB, Prinz P, Krahe R, Ronacher B, Herz AVM (2001) Representation of acoustic communication signals by insect auditory receptor neurons. *J. Neurosci.* **21**:3215–3227.

Machens CK, Schütze H, Franz A, Kolesnikova O, Stemmler MB, Ronacher B, Herz AVM (2003) Single auditory neurons rapidly discriminate conspecific communication signals. *Nature Neurosci.* **6**(4):341–342.

Nelson ME, Xu Z, Payne JR (1997) Characterization and modeling of p-type electrosensory afferent responses to amplitude modulations in a wave-type electric fish. *J. Comp. Physiol. A* **181**:532–544.

Reich DS, Victor JC, Knight BW (1998) The power ratio and the interval map: spiking models and extracellular recordings. *J. Neurosci.* **18**:10090–10104.

Rieke F, Bodnar DA, Bialek W (1995) Naturalistic stimuli increase the rate and efficiency of information transmission by primary auditory afferents. *Proc. R. Soc. Lond. B. Biol. Sci.* **262**:59–265.

Rieke F, Warland D, de Ruyter van Steveninck RR, Bialek W (1997) *Spikes: Exploring the Neural Code.* MIT Press, Cambridge, MA.

Römer H (1976) Die Informationsverarbeitung tympanaler Rezeptorelemente von Locusta migratoria (Acrididae, Orthoptera) *J. Comp. Physiol. A* **109**:101–122.

Ronacher B (1989) Stridulation of acridid grasshoppers after hemisection of thoracic ganglia: evidence for hemiganglionic oscillators. *J. Comp. Physiol. A* **164**:723–736.

Ronacher B, Krahe R (2000) Temporal integration vs. parallel processing: coping with the variability of neuronal messages in directional hearing of insects. *Eur. J. Neurosci.* **12**:2147–2156.

Schaette R, Gollisch T, Herz AVM (2004) Spike-train variability of auditory neurons *in vivo*: dynamic responses follow predictions from constant stimuli. Preprint.

Schiolten P, Larsen ON, Michelsen A (1981) Mechanical time resolution in some insect ears. *J. Comp. Physiol.* **143**:289–295.

Sobel EC, Tank DW (1994) *In vivo* Ca^{2+} dynamics in a cricket auditory neuron: an example of chemical computation. *Science* **263**:823–826.

Stumpner A, Ronacher B (1991) Auditory interneurones in the metathoracic ganglion of the grasshopper *Chorthippus biguttulus.* I. Morphological and physiological characterization. *J. Exp. Biol.* **158**:391–410.

Suga N (1960). Peripheral mechanisms of hearing in locusts. *Jpn. J. Physiol.* **10**:533–546.

Theunissen FE, Roddey J, Stufflebeam S, Clague H, Miller J (1996) Information theoretic analysis of dynamical encoding by four identified primary sensory interneurons in the cricket cercal system. *J. Neurophysiol.* **75**:1345–1364.

Tougaard J (1996) Energy detection and temporal integration in the noctuid A1 auditory receptor. *J. Comp. Physiol. A* **178**:669–677.

van Rossum M. (2001) A novel spike distance. *Neural Comp.* **13**:751–763.

Victor JD, Purpura KP (1997) Metric-space analysis of spike trains: theory, algorithms, and application. *Netw. Comput. Neural Syst.* **8**:127–164.

von Helversen D (1972) Gesang des Männchens und Lautschema des Weibchens bei der Feldheuschrecke Chorthippus biguttulus. *J. Comp. Physiol.* **81**:381– 422.

von Helversen O, von Helversen D (1994) Forces driving coevolution of song and song recognition in grasshoppers. In: *Neural Basis of Behavioural Adaptations.* Schildberger K, Elsner N, Eds. Gustav Fischer, Stuttgart, pp. 253–284.

von Helversen O, von Helversen D (1997) Recognition of sex in the acoustic communication of the grasshopper *Chorthippus biguttulus. J. Comp. Physiol. A* **180**:373–386.

von Helversen D, von Helversen O (1998) Acoustic pattern recognition in a grasshopper: processing in the time or frequency domain? *Biol. Cybern.* **79**:467–476.

Wang XJ (1998) Calcium coding and adaptive temporal computation in cortical pyramidal neurons. *J. Neurophysiol.* **79**:1549–1566.

Section III

Vision

6 Modern Optical Tools for Studying Insect Eyes

Doekele G. Stavenga

CONTENTS

6.1 INTRODUCTION

Insects are highly visual animals, having evolved an elaborate visual system with which to sample a wide range of complex optical signals in their varied environments. The main input organs, a pair of large *compound eyes*, often cover a large fraction of the insect's head. In addition, three single-lens eyes (*ocelli*) are usually positioned on the upper part of the head, between the compound eyes. Research on ocelli has been relatively scanty, and therefore, most progress in understanding insect vision has been achieved through studies on the compound eyes, the main topic of the present chapter.

0-8493-2024-0/05/$0.00+$1.50

Compound eyes consist of hundreds or even many thousands of more or less identical building blocks, the *ommatidia* (Figure 6.1). The ommatidia are tapered cylinders, hexagonally packed in the outer shell of the eye, forming approximately

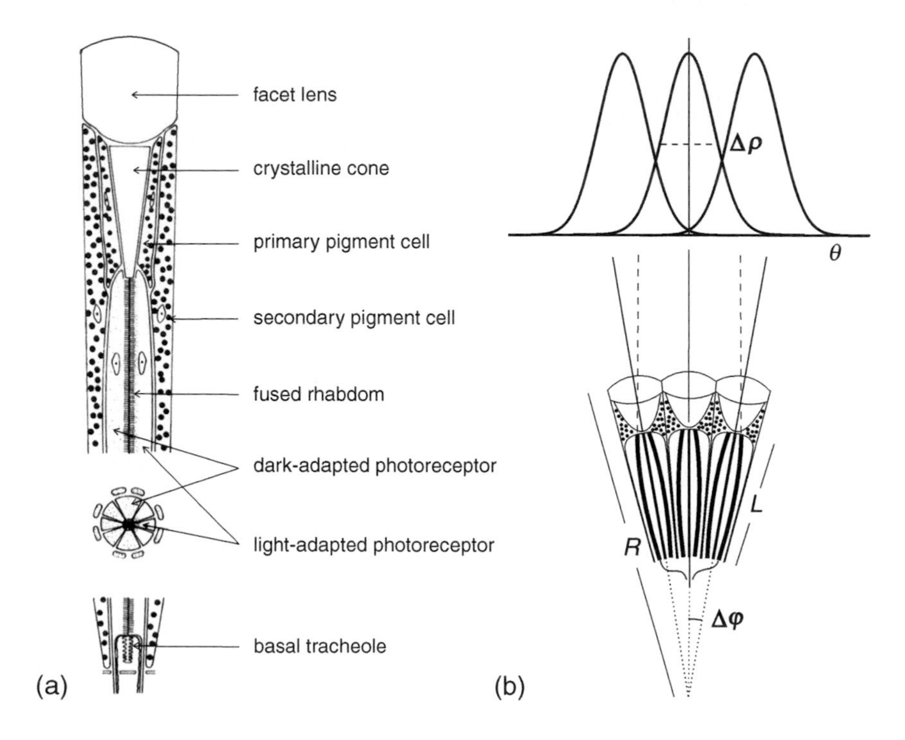

FIGURE 6.1 Different types of insect eyes. **a:** Diagram of an ommatidium in the apposition eye of the butterfly. The facet lens and the crystalline cone channel incident light into the fused rhabdom, consisting of the rhabdomeres of nine photoreceptors. Pigment granules in the photoreceptor cell bodies are remote from the rhabdom in the dark-adapted state, but they migrate near the rhabdom upon light adaptation. Primary and secondary pigment cells surround the photoreceptors. Proximally of the rhabdom, basal tracheoles form an interference reflection multilayer: the tapetum. **b:** Diagram of three ommatidia in the neural superposition eye of the fruit fly, *Drosophila*. The visual axes of the ommatidia are determined by the central rhabdomeres. The angle between the ommatidial axes is $\Delta\varphi$. When the diameter of a facet lens is D_l and the radius of curvature is R, then $\Delta\varphi = D_l/R$. Incident light first enters the biconvex facet lens, which focuses the light into the rhabdomeres, the organelles of the photoreceptors that contain the visual pigment. The peripheral photoreceptors, R1–6, have fatter rhabdomeres than the central photoreceptors, R7, R8. The rhabdomeres of photoreceptors R1–6 each have a length L; the rhabdomeres of R7 and R8 share this length. A specific set of photoreceptors of neighboring ommatidia have the same visual direction (dashed lines), and their axons project to the same cartridge of second order neurons in the lamina, the first optical ganglion processing the optical information sampled by the retina. Pigment cells, filled with strongly light-absorbing pigment granules, optically shield the ommatidia from each other. The sensitivity of a photoreceptor depends on the angle θ between the direction of the light source and the visual axis of the photoreceptor. $\Delta\rho$ is the acceptance angle (i.e., the half-width of the angular sensitivity curve).

a hemisphere. In each ommatidium, the dioptric apparatus, consisting of a facet lens and a cone-shaped structure (the crystalline cone), focuses incident light from a limited visual field onto the underlying *photoreceptor cells*. The surrounding pigment cells form an optical shield for stray light that would otherwise deteriorate the directional sensitivity of the photoreceptors. The distal pigments are often black (i.e., are highly absorbent), but many compound eyes have a very distal layer of pigment cells with little absorbing or strongly colored pigment. The pigment granules then predominantly scatter light, thus creating patterned or colored eyes. Well over a century ago, Exner (1891, 1989) realized that the *pseudopupils*, intriguing optical phenomena observable in the compound eyes of insects and other arthropods, are due to the specific anatomical arrangements of the dioptric system and the pigment cells. Leydig (1855) coined the term *pseudopupils*, because lightly pigmented compound eyes feature a dark spot resembling the dark pupil of human eyes. The pupil functions as a light-controlling diaphragm, but light-flux control in compound eyes is executed in a fundamentally different way (Stavenga, 1979).

There are two main types of compound eyes. In the *apposition eyes* of diurnal insects such as bees and butterflies (Figure 6.1a), each facet lens focuses light from a limited visual space onto a set of nine photoreceptor cells. Because the photoreceptors of that set share the same visual field, the sampling lattice equals the ommatidial lattice. In the *optical superposition eye* of night-flying insects, photoreceptors are also grouped in ommatidia, but because of special optics and a clear zone beneath the lenses, the photoreceptors in one ommatidium receive light via the facet lenses of several, neighboring ommatidia. In the fruit fly's *neural superposition eye* (Figure 6.1b), each ommatidium contains eight photoreceptors that have slightly deviating visual fields. Nevertheless, the number of spatial sampling points equals the number of ommatidia, due to precise anatomical arrangement and axonal wiring (Nilsson, 1989).

The photoreceptors detect light with their visual pigment molecules (see Section 6.2), which are concentrated in a special organelle, the *rhabdomere*. The assembly of the rhabdomeres of one ommatidium is called the *rhabdom*. In a bee or butterfly ommatidium, the rhabdomeres are closely juxtaposed and together function as a united optical waveguide, thus forming a so-called *fused rhabdom* (Figure 6.1a). The rhabdomeres of the photoreceptors in a fly ommatidium are spatially separate and function as individual optical waveguides; flies thus have an *open rhabdom* (Figure 6.1b).

Light channeled by the dioptric system into the rhabdom(eres) propagates there until it is absorbed by visual pigment. Photon absorption induces transformation of the energized visual pigment molecule, which then triggers the *phototransduction* process. The light-induced electrical signal, which is subsequently relayed to higher-order neurons, is proportional to the number of absorbed photons per unit time. That number, integrated over all photoreceptors, is a measure for the absolute light sensitivity of the eye (Land, 1981). Eyes, being sensors for light, may be expected to absorb as many photons as possible. Several constraints and conditions play a modulatory role in the realization of a well-designed eye, however. First, light absorption by a photoreceptor should be maximized only for light from a narrow spatial field because spatial acuity is important for discriminating vital objects such as food, predators, prey, or mates. Similarly, spectral discrimination will also be important, and this is enhanced by proper tuning of the visual pigments' sensitivity spectra to the spectral characteristics of the environment and objects of interest. The angular and spectral sensitivities of the photoreceptors of an eye are, therefore, central to understanding eye performance (Snyder, 1979; Land, 1981; Stavenga, 2004).

The visual and ecological functions of compound eyes have been investigated with various methods: anatomical, molecular biological, electrophysiological, and behavioral, as will be highlighted in the following sections. Optical tools fulfill a central role in insect vision research. This chapter presents a number of the modern measurement and analysis techniques and illustrates their value as part of a review of insect eyes and recently gained results. The treatment focuses mainly on two insect families, flies (Diptera) and butterflies (Lepidoptera), because their compound eyes are the most thoroughly investigated, and thus, the best understood. See Chapter 2 for further discussion on the ecology of insect vision.

6.2 VISUAL PIGMENTS

6.2.1 Molecular Biology and Spectral Sensitivity

Insect visual pigments form, together with the visual pigments of all animals, a subclass of the *G-protein coupled receptors*, which are integral membrane proteins (Minke and Hardie, 2000). The chromophores of insect visual pigments known so far are retinal(dehyde), 3,4-didehydroretinal, and 3-hydroxyretinal, derivatives of vitamin A1, vitamin A2, and vitamin A3, respectively (Schwemer, 1989; Vogt, 1989; Gärtner, 2000). The best-characterized insect visual pigments (*rhodopsins*) are those of the fruit fly, *Drosophila*, as a result of extensive molecular biological research. Table 6.1 shows that *Drosophila*, in addition to possessing the main blue-green rhodopsin (Rh1), has a violet rhodopsin (Rh2), two UV-rhodopsins (Rh3 and Rh4), a blue rhodopsin (Rh5), and a green rhodopsin (Rh6).

All rhodopsins of the fruit fly are known in complete molecular detail, and much progress has recently been made in unraveling the relation between their molecular

TABLE 6.1
Visual Pigments of the Fruit Fly *Drosophila*

Rhodopsin	Rh1	Rh2	Rh3	Rh4	Rh5	Rh6
Location	R1–6, R7r	ocelli	R7p, R7,8marg	R7y	R8p	R8y
R	486	418	331	355	442	515
M	566	506	468	470	494	468

Note: The third and fourth rows show λ_{max}, the wavelength (in nm) where the rhodopsin (R) and metarhodopsin (M) absorb maximally (Salcedo et al., *J. Neurosci.*, 19, 10716–10726, 1999).

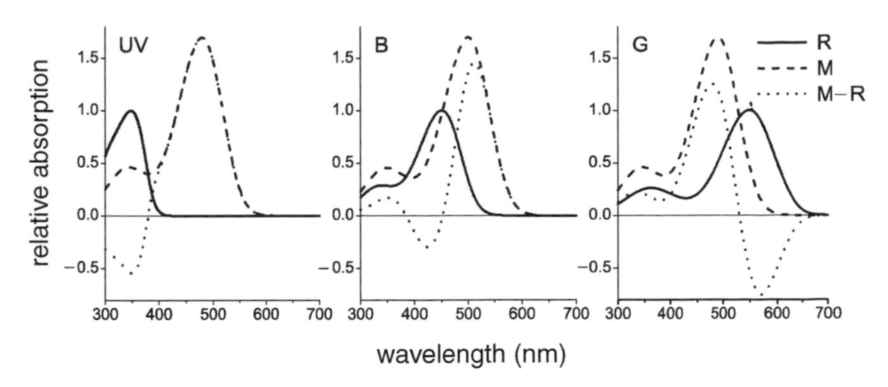

FIGURE 6.2 The set of three visual pigments found in many insect eyes, namely an ultraviolet- (UV), blue- (B), and green- (G) absorbing rhodopsin, absorbing maximally at 350, 450, and 550 nm, respectively. The metarhodopsins (thermostable forms of the rhodopsins) are assumed to absorb maximally at 480, 500, and 490 nm, respectively. Spectrophotometry of insect visual pigments has invariably shown that the peak absorption of the metarhodopsins relative to the peak absorption of their rhodopsin is distinctly larger than 1, with an average ratio of about 1.7, the value used in calculating the metarhodopsin spectra. The shapes of both the rhodopsin and the metarhodopsin spectra were derived from the template formulas (Equation 6.1 and Equation 6.2). The difference spectrum (M–R) is the absorption spectrum of metarhodopsin (M) minus that of rhodopsin (R).

composition and absorption spectrum (Salcedo et al., 1999). The absorption spectrum is a crucial piece of knowledge in vision research because it largely determines the photoreceptor's spectral sensitivity. For many experimental purposes, mathematical templates that formally describe the shape of these spectral curves represent important diagnostic tools (Figure 6.2). Several different template formulas have been offered in recent decades (see Stavenga et al., 2000 for review). Notably, Mansfield (1985) and MacNichol (1986) showed that two unique invariant shapes, for vitamin A1- and A2-based visual pigments, emerge when experimental spectra are plotted at a frequency scale relative to the peak frequency (i.e., f/f_{max}). This basic notion has been extended by assuming that the absorption spectrum of a visual pigment is an algebraic sum of its absorption bands (Stavenga et al., 1993). The bands are called α, β, γ, etc., going from long to short wavelengths. The α-band of the absorption spectrum of vitamin A1-based visual pigments is described by Govardovskii et al., (2000):

$$S_a = \{\exp[A(a-x)] + \exp[B(b-x)] + \exp[C(c-x)] + D\}^{-1} \tag{6.1}$$

where $A = 69.7$, $a = 0.8795 + 0.0459\exp[-(\lambda_{max} - 300)^2/11940]$, $B = 28$, $b = 0.922$, $C = -14.9$, $c = 1.104$, $D = 0.674$, and $x = f/f_{max} = [\lambda/\lambda_{max}]^{-1}$, with the wavelength values (λ) expressed in nm.

The β-band is assumed to be Gaussian-shaped (Govardovskii et al., 2000):

$$S_\beta = A_\beta \exp\left\{-\left[\left(\lambda - \lambda_{m\beta}\right)/d\right]^2\right\} \tag{6.2}$$

where A_β is the amplitude of the β-band relative to that of the α-band, $\lambda_{m\beta}$ is the wavelength of the β-peak, and d is a bandwith parameter. Experimental spectra yielded as optimal parameter values: $A = 0.26$, $\lambda_{m\beta} = 189 + 0.315\lambda_{max}$, and $d = -40.5 + 0.195\lambda_{max}$, again with the peak wavelength of the α-band, λ_{max}, in nm. Using λ_{max} as the only free parameter, the template formulas then can be fitted to measured absorption spectra.

Frequently, sensitivity spectra obtained by electrophysiological or behavioral experiments closely approximate the visual pigment absorption spectra, and thus, visual pigment templates are widely used to interpret experimental spectra heuristically (e.g., Salcedo et al., 1999, 2003; Kelber et al., 2002). Considering the limited accuracy attainable for most experimental spectra, it appears to be not very critical that the latter authors used the template of Stavenga et al. (1993), rather than those given by Equation 6.1 and Equation 6.2. A physical–chemical basis for the templates (i.e., a theory predicting the spectral characteristics of visual pigments from the known primary structure and the interaction of the protein and the chromophore) is eagerly awaited.

6.2.2 Photochemistry of Visual Pigments

When an insect rhodopsin (R) molecule absorbs a photon, the absorbed energy sends the molecule into an excited state. The chromophore then isomerizes from the 11-*cis* to the all-*trans* state, which is accompanied by the transition of the visual pigment molecules through a few thermolabile intermediate states, transforming finally to a thermostable *metarhodopsin* (M) state. Insect visual pigments hence are photochromic substances. Because the intermediate states are short-lived, for all practical purposes the photochemistry is that of a two-state molecular process:

$$R \underset{k_M}{\overset{k_R}{\rightleftarrows}} M \tag{6.3}$$

where the rate constants are given by $k_P = \int I(\lambda)\, \beta_P(\lambda)\, d\lambda$ (P = R, M), with I the light intensity and $\beta_P = \alpha_P \lambda_P$ the photosensitivity; α_P is the molecular absorption coefficient of the two states, P = R or P = M, and λ_P is the quantum efficiency (i.e., the probability that the absorbed photon results in photoconversion). Both λ_R and λ_M can be assumed to be wavelength independent and about equal (Hamdorf, 1979). Prolonged illumination of a photoreceptor cell and its visual pigment yields a photosteady state. The number of rhodopsin-to-metarhodopsin conversions per unit time then equals the number of opposite conversions. The fraction of the molecules existing in the rhodopsin conformation in the photosteady state equals $f_R = k_M/(k_R + k_M)$. The rhodopsin–metarhodopsin ratio hence depends on the spectral composition of the illumination, $I(\lambda)$, and the absorption spectra of the two thermostable visual pigment states, $\alpha_R(\lambda)$ and $\alpha_M(\lambda)$ (Schwemer, 1989; Stavenga et al., 2000).

The interaction of the protein and the chromophore determines the absorption spectrum of the visual pigment states, and therefore, the two absorption spectra of the rhodopsin–metarhodopsin pair are intimately linked. Figure 6.2 shows the spectra for a set of three visual pigments as an exemplary case for many insect eyes, namely, an ultraviolet- (UV), blue- (B), and green- (G) absorbing insect rhodopsin, absorbing maximally at 350, 450, and 550 nm, with metarhodopsins absorbing maximally at 480, 500, and 490 nm, respectively. The peak absorption of the metarhodopsins relative to the peak absorption of their rhodopsin is always distinctly larger than 1; experimental data yield a ratio of about 1.7, the value used in Figure 6.2. These spectral curves were derived from the template formulas (Equation 6.1 and Equation 6.2) for both the rhodopsins and the metarhodopsins. A striking property of insect visual pigments is the relative position of the peak wavelengths, λ_{max}, of the two thermostable states, which can be formally described by the following rule (Stavenga, 1992): when $\lambda_{max}(R) < 500$ nm, then $\lambda_{max}(M) > \lambda_{max}(R)$. That is, for UV and blue rhodopsins, the metarhodopsin is bathochromic (long-wavelength) shifted; and if $\lambda_{max}(R) > 500$ nm, then $\lambda_{max}(M) < \lambda_{max}(R)$. For green rhodopsins, the metarhodopsin is hypsochromic (short-wavelength) shifted. (See, e.g., Table 6.1 for the visual pigments of the fruit fly *Drosophila.*)

6.3 THE FLY NEURAL SUPERPOSITION EYE

6.3.1 Anatomy

The best-characterized insect eye is that of the fruit fly *Drosophila*. About 700 ommatidia are arranged in an approximately half-sphere with radius R = 180 μm (Franceschini and Kirschfeld, 1971). A *Drosophila* ommatidium is capped by a biconvex facet lens, with diameter D_l = 16 μm (Franceschini and Kirschfeld, 1971) and thickness t = 8 μm (Figure 6.1b). The value of the interommatidial angle, which determines the spatial sampling lattice, is = D_l/R (Figure 6.1b), yielding for *Drosophila* $\Delta\varphi = 5.1°$. The eight photoreceptors of an ommatidium form two classes: the peripheral photoreceptors R1–6 and the central photoreceptors R7 and R8. The distal ends of the rhabdomeres of R1–7 are at a distance of about 20 μm from the facet lens. The rhabdomeres of R1–6 have distally a diameter of about 2.0 μm and taper to a diameter of about 1.0 μm at the proximal end of the retinal layer, with a total length L = 80 μm (Hardie, 1985). The rhabdomeres of R7 and R8 together make up this distance; their diameter is more or less constant at about 1.0 μm. The refractive index of the rhabdomeres is higher than that of the surrounding medium, so the rhabdomeres act as optical waveguides.

Each fly photoreceptor expresses one of several rhodopsins (Table 6.1), which are embedded in the rhabdomeric membrane. The membrane is folded into cylindrical microvilli, with a cross section of about 0.06 μm and a length of 1 to 2 μm, depending on the width of the rhabdomere; the long axis of a microvillus is perpendicular to that of the rhabdomere. The large, peripheral R1–6 rhabdomeres always express the same blue–green absorbing rhodopsin (Rh1), whereas the rhodopsins of the slender R7 and R8 rhabdomeres come in pairs, depending on the ommatidium.

There are two main classes of ommatidium that are distinguished by their rhabdomere–rhodopsin combinations: R7p/Rh3 resides with R8p/Rh5, or R7y/Rh4 with R8y/Rh6 (Table 6.1). A third ommatidial class exists in the so-called dorsal margin of the eye. This class detects polarized light by using UV rhodopsins in the R7 and R8 rhabdomeres. A fourth class shows the R7/Rh1 combination and is localized to the dorsal area of the eye in males. It has been dubbed the "love spot" for its special role in tracking females in flight (Hardie, 1986). The rhodopsin Rh2 is expressed in the ocelli.

Figure 6.3 shows the absorption spectra of the main visual pigment of the blowfly, that is, the visual pigment of the R1-6 photoreceptors, when existing in the rhodopsin (R) or metarhodopsin (M) state (Hamdorf, 1979; Stavenga, 2002a). The high absorption band in the UV is due to a sensitizing pigment, 3-hydroxyretinol, which absorbs exclusively in the ultraviolet wavelength range. It is so tightly bound to the visual pigment molecule that it transfers absorbed energy to the chromophore, 3-hydroxyretinal, the excitation of which then leads to photoconversion. This holds for the photoconversion of rhodopsin into metarhodopsin and vice versa (Minke and Kirschfeld, 1980). Which of these photoconversions is favored depends on the spectral content of the illumination. In the natural situation, where red screening pigment covers the photoreceptors (Figure 6.3, S), stray red light will stream through the eye. The visual pigment molecules existing in the metarhodopsin state will then be preferentially reconverted into the rhodopsin state. In other words, the red screening pigment of fly eyes serves a photoregenerative role. The yellow pigment granules in the cell body of the photoreceptor perform a similar function (Figure 6.3, P) (Stavenga, 2002a).

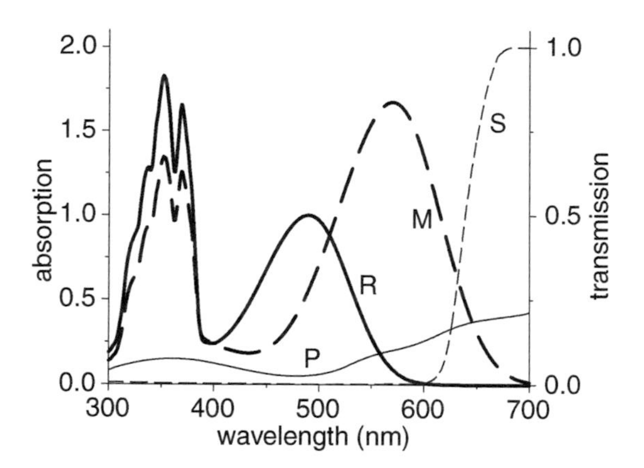

FIGURE 6.3 Absorption spectra of pigments in the eyes of the blowfly, *Calliphora vicina*. The photoreceptors R1–6 have a rhodopsin (R) absorbing maximally at about 490 nm. The metarhodopsin (M) absorbs maximally at about 570 nm. The fine-structured UV bands are due to a sensitizing pigment, 3-hydroxyretinol, which transfers the energy of absorbed UV light to the visual pigment's chromophore, 3-hydroxyretinal. Red-transparent screening pigment (S) lets red stray light roam through the eye to photoconvert existing metarhodopsin molecules back into rhodopsin. The yellow pigment (P) of the granules inside the soma of the photoreceptors plays a similar photoregenerative role.

6.3.2 Phototransduction and Measuring Calcium in Fly Photoreceptors

Photoconversion of a rhodopsin into its metarhodopsin state triggers phototransduction, the conversion of light into a cellular signal. Extensive research on fly photoreceptors, especially in *Drosophila* (Minke and Hardie, 2000; Huber, 2001; Hardie and Raghu, 2001), has shown that phototransduction starts with coupling of the metarhodopsin to a G-protein. This causes GDP–GTP transfer, activation of a phospholipase C (PLC), and finally opening of transient receptor potential (TRP) and TRP-like (TRPL) channels, which are located, like the visual pigment molecules, in the membrane of the rhabdomeric microvilli. Channel opening results in a massive influx of Na^+ and Ca^{2+} ions. The latter enact an important feedback control function on various molecular components of the phototransduction chain, adapting the photoreceptor to the illumination (Hardie and Raghu, 2001).

The light-induced influx of Ca^{2+} in a fly photoreceptor can be measured in a virtually intact, living animal using a newly developed method (Oberwinkler, 2000). The procedure is described briefly here (Figure 6.4): a fly is first immobilized by wax and mounted on the stage of a fluorescence microscope equipped with an electrode

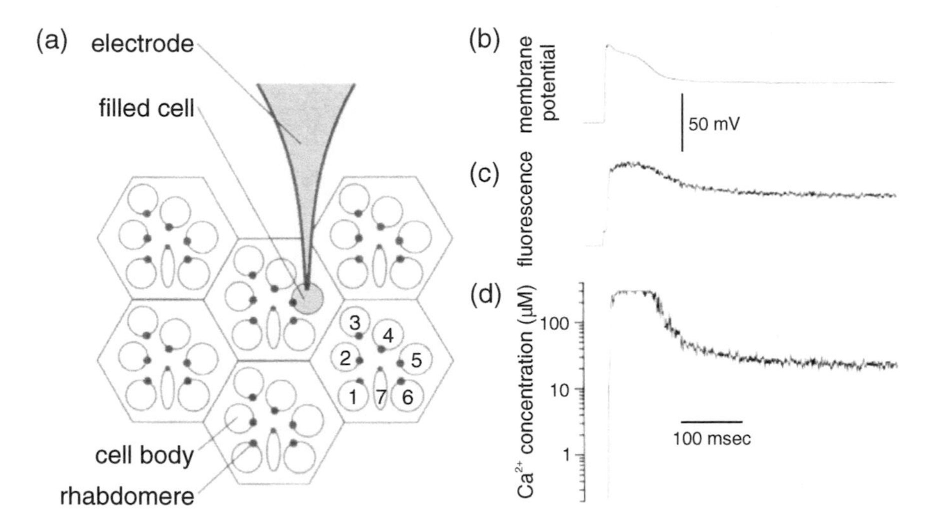

FIGURE 6.4 Measurement of the calcium concentration in single fly photoreceptors. **a**: Fly photoreceptors are arranged in a characteristic trapezoidal pattern, with six large, peripheral photoreceptors, R1–6, and two smaller, central photoreceptors, R7 and R8. The latter photoreceptors are arranged in tandem (see Figure 6.1b). A photoreceptor is impaled by a microelectrode, filled with the calcium indicator dye OG5N (Molecular Probes)[(14)], and then filled electrophoretically, with fly and microelectrode holder on the stage of a fluorescence microscope. **b**: Illumination of a dark-adapted photoreceptor with bright light depolarizes the cell by about 60 mV. **c**: Depolarization is accompanied by a massive influx of calcium, resulting in a sudden increase in the fluorescence of the photoreceptor. **d**: The light-induced change in calcium concentration is calculated by using the calcium concentrations that cause minimal and maximal fluorescence and the dye's K_d (see Oberwinkler and Stavenga, 2000a,b).

micromanipulator. The tip of the electrode, filled with a Ca^{2+}-sensitive dye, is then positioned in focus, and subsequently the fly is moved (and not the electrode!) until the electrode impales a photoreceptor. The cell can then be filled with the dye by electrophoretic injection (Figure 6.4a). With this technique, the receptor potential (i.e., the light-induced change in membrane potential; Figure 6.4b) can be recorded at the same time that an increase in the calcium signal in the photoreceptor is recorded as a change in fluorescence intensity (Figure 6.4c). The light-induced increase in calcium in fly photoreceptors appears to result fully from influx from extracellular calcium and not from calcium release from intracellular stores (Hardie and Raghu, 2001).

The method of optically monitoring calcium changes in a photoreceptor in an intact eye of a living animal requires bright light, however, and this drives closure of the pupil (see Section 6.3.4). This problem can be avoided by using white-eyed mutants that lack the pupillary pigment (Oberwinkler and Stavenga, 2000a). Another problem, the calibration of the light-induced changes in the intracellular calcium concentration, can be solved by applying dyes with different affinities for calcium. The fluorescence signal then saturates at different calcium concentrations, depending on the characteristics of the dye. Correlating the fluorescence data with the dependence of the receptor potential on the light intensity then yields the calcium concentration induced by a given illumination intensity (Oberwinkler and Stavenga, 1998).

The changes in intracellular calcium concentration that occur during phototransduction can be enormous, especially when a dark-adapted photoreceptor is suddenly illuminated with bright light, as in the experiment of Figure 6.4b,c. The phototransduction machinery, including the light-driven channels, is compacted in the extremely slender microvilli, so a single photon can cause a jump in the calcium concentration in one microvillus up to the millimolar range (Postma et al., 1999). This high increase is short-lived, because the calcium ions rapidly diffuse into the photoreceptor soma. From the fluorescence data of Figure 6.4c, one can calculate that the bright light causes the calcium concentration in the soma to increase from about 0.2 μM in the dark-adapted state to values exceeding 200 μM (Figure 6.4d). Upon light adaptation, the concentration levels fall off to 5 to 10 μM (Oberwinkler and Stavenga, 2000a; Hardie et al., 2001).

Withdrawal of the electrode leaves the cell filled with the dye, and therefore, long-term optical measurements can be performed on that cell, for instance, with a *confocal laser-scanning microscope* (Oberwinkler and Stavenga, 2000b). When the optical power of the facet lens is reduced by using a water immersion objective, the influx of calcium ions in the rhabdomere and the subsequent diffusion into the cell soma can be imaged and followed in real time (Oberwinkler and Stavenga, 2000b).

6.3.3 Fluorescence Microscopy and the Deep Pseudopupil of Mutant Fly Eyes

From his studies on the fruit fly *Drosophila*, Franceschini (1975) deduced that the image observed in the plane of the eye center, the so-called deep pseudopupil (DPP), represents a superposition of images originating from several ommatidia. Due to the spherical shape of the eye and the almost perfect crystalline anatomical organization of the fly eye, the superimposed images in the eye center faithfully represent the structures in the ommatidial focal planes. The DPP has proved to be a most useful tool in compound eye research, and observation of the DPP has become a standard approach in optically screening *Drosophila* mutants (Pichaud and Desplan, 2001).

Although the first optical studies on fly eyes have been performed with conventional transmission and reflection microscopy (applying epi-illumination), fluorescence spectroscopy has gained tremendous popularity in the recent decades. In epi-illumination microscopy, an object is illuminated via the objective, by using a semitransparent mirror in between the objective and the ocular. Virtually the same technique is, in fact, used in fluorescence microscopy, with three minor modifications:

- An excitation filter is inserted into the illumination beam.
- The wavelength-neutral semitransparent mirror is replaced by a dichroic mirror.
- A barrier filter is put in the observation path.

To unravel developmental processes and the molecular mechanisms of phototransduction, molecular biological studies on the fly retina have amply used fluorescent markers. Extensive studies on fruit fly mutants using the pseudopupil methods and electrophysiology have greatly contributed to the present detailed knowledge of the molecular details of phototransduction, although a few tantalizing details concerning the opening of the involved ion channels in the rhabdomeric membrane await final elucidation (Huber, 2001; Hardie and Raghu, 2001).

Figure 6.5a shows the eye of a white-eyed *Drosophila* mutant where the TRPL channels in the rhabdomeres of R1–6 photoreceptors are tagged with eGFP (Meyer et al., 2003). The fluorescence microscope was focused at the level of the eye center. There, the images of several ommatidia superimpose, causing the bright pattern of R1–6 rhabdomeres (see Figure 6.5a). The pattern is bright, because part of the fluorescence is guided distally by the rhabdomere, so that the tip of the rhabdomere acts as a light source (indicated in Figure 6.5b by solid lines), which is seen imaged at the eye center. The number of the superimposed images equals the number of ommatidia with visual fields within the aperture of the microscope objective. In addition to the bright central rhabdomere pattern, a crown of several somewhat fainter patches is seen. The patches are caused by fluorescence that escapes the rhabdomeric waveguide sideways (indicated in Figure 6.5b by dashed lines). The fluorescent rhabdomeres are not imaged by their overlying facet lens, but by a neighboring facet lens (Figure 6.5b,c). Using the nomenclature of Exner (1989), we can call the central pattern the principal (deep) pseudopupil, and the crown of patches are accessory pseudopupils (Stavenga, 1979).

6.3.4 Fly Photoreceptor Optics

The optical properties of the fly eye, arguably the compound eye with the simplest optics, determine the light sensitivity of the photoreceptors. The diffraction optics of the facet lens (Stavenga and van Hateren, 1991) and the waveguide optics of the rhabdomeres (Snyder, 1979; van Hateren, 1989) are well understood, and formalisms describing the integrated optics of the facet lens–rhabdomere system are available

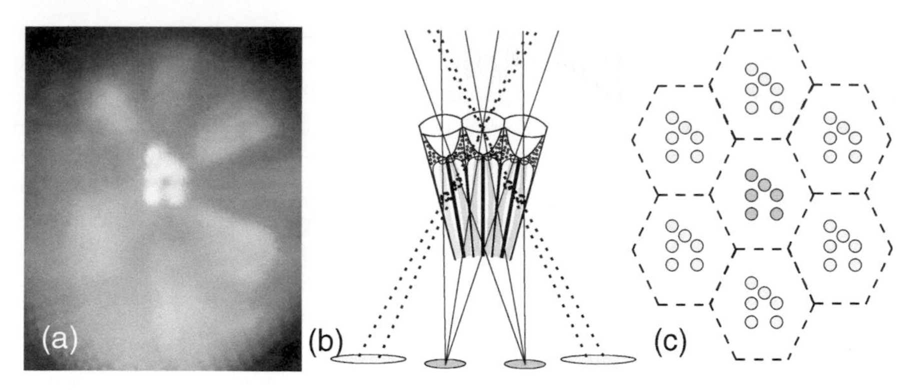

FIGURE 6.5 Pseudopupils in the eye of a white-eyed mutant fruit fly *Drosophila*, where the TRPL channels are tagged with eGFP. **a**: In the center, the fluorescent rhabdomeres of photoreceptors R1–6 are seen. The induced fluorescence leaves the rhabdomeres in several ommatidia. **b**: The fluorescence of corresponding photoreceptors (continuous lines) seems to emerge from the same spot in the deep pseudopupil, seen at the level of the center of the eye, because of the regular arrangement of the rhabdomeres and the spherical shape of the eye. When the fluorescent rhabdomeres are imaged by facet lenses of adjacent ommatidia (dotted lines), the superposition of images of corresponding rhabdomeres causes a crown of faint patches. **c**: The patches surrounding the principal (deep) pseudopupil (in the center) are called accessory pseudopupils (Exner, 1989; Stavenga, 1979).

(Barrell and Pask, 1979; van Hateren, 1984; Stavenga, 2003a). The facet lens–rhabdomere system appears to be quite robust to defocusing (Stavenga, 2003a), and therefore, the rhabdomere entrance can be assumed to coincide with the focal plane of the facet lens (van Hateren, 1985). The crucial parameter determining the facet lens optics is the *F-number*, that is, the ratio between the focal length and the lens diameter (Kirschfeld, 1974; Land, 1981). The F-number of the facet lenses of blowfly eyes is about 2.0 (Stavenga et al., 1990).

The light-receiving tips of the rhabdomeres are positioned at a distance of a few μm from each other in the focal plane of the facet lens, so the photoreceptors of one ommatidium have distinct, though partially overlapping, visual fields (Figure 6.1b). Incident light, after being focused by the facet lens into the rhabdomere, is guided through the rhabdomere until the visual pigment absorbs it. Due to the wave properties of light, the propagated light travels in distinct patterns, so-called modes. The modes exist partly outside the rhabdomere border as the so-called boundary waves, the extent of which is very limited. They decay exponentially with a space constant of a few tenths of a micrometer (Stavenga, 2004). The boundary wave can couple with another waveguide when it is sufficiently nearby. The distance between the rhabdomeres of an ommatidium, therefore, rapidly increases from distal to proximal (Figure 6.1b), thus avoiding the danger of optical crosstalk (Wijngaard and Stavenga, 1975).

Fly R1–6 photoreceptors contain distally in the soma small pigment granules with a cross section of about 0.15 μm. The granules are remote from the rhabdomere in the dark-adapted state, but they migrate toward the rhabdomere boundary upon illumination. When they approach the rhabdomere border within about 0.5 μm, the granules start to absorb light from the boundary wave. The assembly of photoreceptor pigment granules thus acts as a light-control system, called the *pupil mechanism* (Franceschini, 1975; Stavenga, 2004). The pupil in the fly eye has many functions. First, it reduces the light flux, so that the working range of the photoreceptor is extended. Second, it reduces the photoreceptor acceptance angle, because the pupil predominantly absorbs modes that are excited by increasingly off-axis light. The pupil of fly R1–6 photoreceptors predominantly absorbs short-wavelength light, and it, therefore, also plays a photoregenerative role in the photochemistry of the visual pigment (Section 6.2.2; Figure 6.3) (Stavenga, 1989).

6.3.5 Angular and Spectral Sensitivity

The light sensitivity of photoreceptors is experimentally characterized by measuring their angular and spectral sensitivity, via intracellular recording of the receptor potential. The spectral sensitivity function is measured by stimulating the photoreceptor with a spatially fixed, monochromatic light source. The number of photons necessary to elicit a criterion response in the photoreceptor is then determined as a function of wavelength. Subsequent normalization of this function at the peak wavelength, λ_{max}, yields the spectral sensitivity. The angular sensitivity function is experimentally determined with a spectrally fixed point source, the spatial direction of which is varied. The number of photons necessary for eliciting a criterion response in the photoreceptor is then determined as a function of the angle of illumination. The resulting function is usually a Gaussianlike, symmetric function. Subsequent normalization to the value at the photoreceptor's visual axis yields the angular sensitivity. The half width of the angular sensitivity curve is called the *acceptance angle*, $\Delta\varphi$ (Figure 6.1b).

Detailed angular and spectral sensitivity measurements by intracellular electrophysiological recordings have been performed in the housefly *Musca* and the blowfly *Calliphora*. The angular sensitivity appears to depend on the light wavelength, but also the spectral sensitivity depends on the spatial properties of the light source and the eye's optics (Pask and Barrell, 1980a,b; Smakman et al., 1984; van Hateren, 1984; Stavenga, 2003b). Figure 6.6a–c presents angular sensitivity curves of a dark-adapted (DA) *Calliphora* photoreceptor measured with monochromatic light of 355, 494, and 588 nm, yielding acceptance angles of 1.27°, 1.22°, and 1.25°, respectively (Smakman et al., 1984). In the light-adapted photoreceptor (LA), the acceptance angles appear to be 1.04°, 1.07°, and 1.25°, respectively (Figure 6.6d–f). The angular sensitivity curves are well described by an optical model incorporating all known optical components, that is, the diffracting facet lens, the waveguiding rhabdomere, and the light-absorbing rhodopsin and sensitizing pigment (Stavenga, 2004). The light-induced narrowing of the angular sensitivity in the UV and blue is now well understood: the second order mode, which substantially contributes to the angular sensitivity of the dark-adapted photoreceptor, is preferentially absorbed by the pupil (Smakman et al., 1984; Stavenga, 2004). The angular sensitivity of the quite small *Drosophila* photoreceptors has only been derived indirectly, from behavioral experiments, giving a value of $\Delta\rho = 3.5°$ (Götz, 1964).

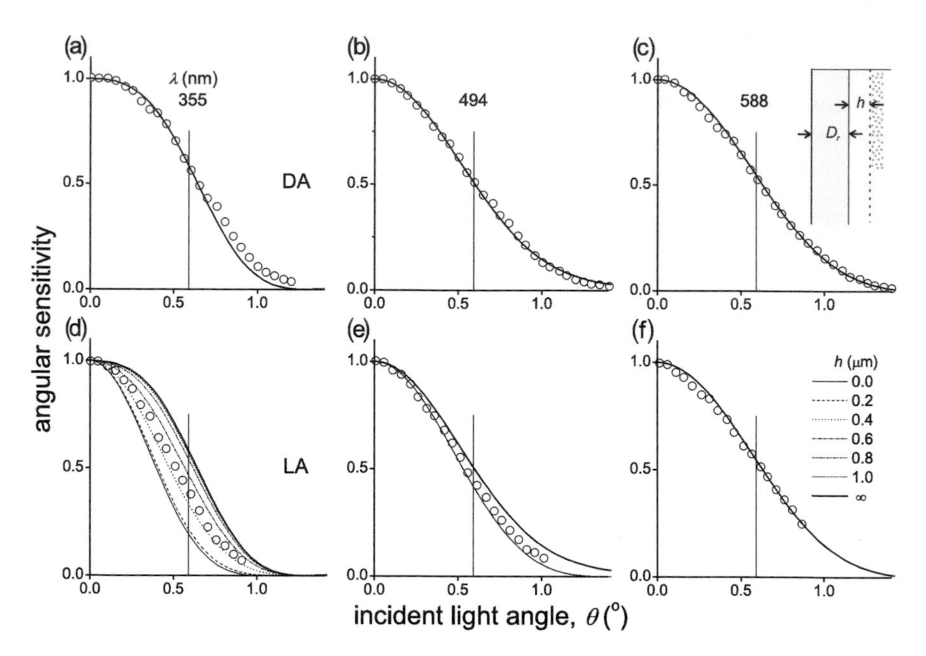

FIGURE 6.6 Experimental and theoretical angular sensitivity curves of a blowfly photoreceptor. The angular sensitivity was measured by intracellular recordings at wavelengths $\lambda =$ 355 (**a**,**d**), 494 (**b**,**e**), and 588 nm (**c**,**f**) in the dark-adapted (DA; **a**,**b**,**c**) and light-adapted state (LA; **d**,**e**,**f**) (cell C of Smakman et al., 1984) (open symbols). Inset: diagram of a photoreceptor with rhabdomere, diameter D_r, and distal cell soma, where a pupil, existing of pigment granules, has a distance h to the rhabdomere boundary. The angular sensitivity curves are described with a model that incorporates the diffraction at the facet lens, the waveguide optics of the rhabdomere, and the absorbing visual pigment. The pupil is open in the dark-adapted state; then $h = \infty$ (bold curves). Light-adaptation causes pupil closure (i.e., a decreasing pupil distance h). This results in a gradually increasing absorption of modes. Assuming that $D_r =$ 1.6 μm, two modes exist at 355 and 494 nm. The pupil then causes narrowing of the angular sensitivity at these wavelengths. The pupil leaves the angular sensitivity curves unchanged at 588 nm because only one mode is allowed there. The thin vertical lines indicate the angle of the rhabdomere border, which is 0.59° when the facet lens focal distance is taken to be $f =$ 77.5 μm (for details see Stavenga, 2004).

The optical modeling of the fly's integrated optical system of facet lens and rhabdomere has shown that the acceptance angle of the slender R7 and R8 photoreceptors increases monotonically with wavelength, because diffraction by the facet lens determines the angular sensitivity. In dark-adapted R1–6 photoreceptors, the wavelength dependence of diffraction is compensated by an opposite wavelength dependence of the waveguide properties (van Hateren, 1989; Stavenga, 2003a,b), resulting in acceptance angles that vary little with wavelength throughout the visible range, including the UV. The depression of the waveguide modes by the pupil in light-adapted photoreceptors reduces that compensating effect, resulting in a smaller acceptance angle at shorter light wavelengths (Stavenga, 2004).

The pupil affects not only the angular sensitivity but also the spectral sensitivity. A closing pupil causes a hypsochromic spectral shift of the spectral sensitivity, experimentally found in *Calliphora* (Hardie, 1979) and *Musca* (Vogt et al., 1982), which can also be well understood from the wave-optics photoreceptor model (Stavenga, 2004).

6.4 BUTTERFLY EYES

6.4.1 Anatomy

Butterflies have typical apposition eyes, where in each ommatidium, nine photoreceptors (R1–9) contribute their rhabdomeres to a fused rhabdom (Figure 6.1a). The rhabdomeres of photoreceptors R1–8 fill up most of the rhabdom space; the rhabdomere of R9 is always positioned proximally in the rhabdom. In most Nymphalidae, the R1–8 rhabdomeres stretch parallel to each other throughout the main, distal part of the rhabdom. In, for instance, the Papilionidae, however, R1–4 rhabdomeres together make up the distal two-thirds of the rhabdom, and the R5–8 rhabdomeres occupy most of the remaining one-third of the rhabdom space (Bandai et al., 1992). Papilionidae thus have a *tiered rhabdom*, and its organization is virtually identical to that found in Pieridae. Papilionidae lack the light-reflecting *tapetum*, a multilayer created by tracheoles proximally of the rhabdom, which is a characteristic element of the eyes of diurnal butterflies (Miller and Bernard, 1968).

Recent research into the Papilionidae and Pieridae has revealed that the eyes of these butterflies share several important structural features. The cell bodies of the R3–8 photoreceptors contain clusters of pigment granules positioned very close to the rhabdom border, indicating that these pigments affect the rhabdoms' boundary waves. Concerted research using several approaches (namely, anatomy, electrophysiology, molecular biology, and computational modeling) has demonstrated that the ommatidia contain three classes of ommatidia, arranged in a random pattern (Arikawa and Stavenga, 1997; Kitamoto et al., 2000; Qiu et al., 2002; Qiu and Arikawa, 2003). The following sections treat a number of exemplary data that illustrate the findings.

6.4.2 Retinal Heterogeneity

The heterogeneity of insect eyes, which first clearly emerged from studies on fruitfly eyes (see Section 6.3.1) (Franceschini, 1975; Hardie, 1985; Salcedo et al., 1999), was strikingly revealed by further work on the Japanese yellow swallowtail, *Papilio xuthus* (Bandai et al., 1992; Arikawa and Stavenga, 1997) and the small white *Pieris rapae crucivora* (Qiu et al., 2002; Qiu and Arikawa, 2003). A beautiful anatomical study also demonstrated the retinal heterogeneity of the moth *Manduca sexta* (White et al., 2003). The heterogeneity can be easily overlooked, because the ommatidia are usually arranged in a very regular hexagonal lattice, where the ommatidial fine structure is repeated in rigid detail. Painstaking electrophysiology on single photoreceptors, followed by light-microscopical identification of the recorded cell type, gave convincing proof that the photoreceptors in the butterfly ommatidia do not form a unique, repetitive set (Arikawa et al., 1999a,b). Freeze microtome sections of the *Papilio xuthus* retina unequivocally demonstrated that the pigment clusters in the R3–8 photoreceptors of one and the same ommatidium always have the same color (i.e., either red or yellow). The red and yellow pigmented ommatidia are randomly arranged. This can be directly demonstrated by observing a sliced-off eye cap using transmission microscopy. Furthermore, fluorescence microscopy revealed that part of the red-pigmented ommatidia contains a UV-absorbing, whitish emitting pigment (Arikawa and Stavenga, 1997).

6.4.3 Epi-Illumination Microscopy and Eye Shine

Autofluorescence of retinal tissue is virtually always extremely low; thus, intense illumination is required for attaining an acceptable image. In the naturally highly light-sensitive eyes, fluorescence cannot be used without extreme caution. For butterfly eyes with a light-reflecting tapetum (Figure 6.1a), epi-illumination is a most attractive alternative that can be applied with great benefit. Observing a dark-adapted butterfly eye with an epi-illumination microscope reveals that a number of ommatidia stand out by showing a bright reflection. It concerns those ommatidia that have their visual axis within the objective aperture (Bernard and Miller, 1970; Stavenga et al., 2001). The reflected light, called the *eye shine*, is the remaining part of incident light that has traveled down the rhabdom, reached the proximal tapetum, been bounced back at the tapetum, and left the rhabdom and the dioptric apparatus again after a second traverse without having been caught by the visual pigments in the rhabdom. The eye shine is short-lived, however (Stavenga, 1979). Within a few seconds of illumination, a "pupil," consisting of pigment granules in the photoreceptors, closes in at the rhabdom and absorbs light from the boundary wave (Figure 6.1a). As in fly eyes, the pupil in butterfly eyes causes narrowing of the acceptance angle of the photoreceptors (Land and Osorio, 1990).

Eye shine is a quite striking phenomenon, but in a normal epi-illumination microscope, it is only observable when the objective has a small aperture, for example, with a 5×, NA 0.1 objective, which has an aperture of 5.7°. It is important to realize that the eye shine from each individual ommatidium has an aperture of about 1°, corresponding to the aperture of the visual field of the photoreceptors. A conventional 10×, NA 0.3 objective has an aperture of 17.5°, and illumination with such a large angle creates substantial reflection from the surfaces of the corneal facet lenses and wide-angle scattering from cellular structures (e.g., the pigment granules in the distal pigment cells). These factors easily obscure the eye shine. Furthermore, large-aperture, high-power objectives consist of several lenses, each with reflecting surfaces, and even with the best antireflection coatings, substantial reflections remain, which makes conventional epi-illumination microscopy of butterfly eye shine quite cumbersome. These drawbacks are (largely) removed in a setup especially designed for spatially filtering the background reflections (Figure 6.7). The design is based on the observation that the light beams emerging from the individual ommatidia coalesce in the DPP; that is, they merge in the center of the eye, due to the spherical shape of the eye. Reflections from the facet lens surface and other structures (Figure 6.7, dotted line) are effectively removed by focusing a light source,

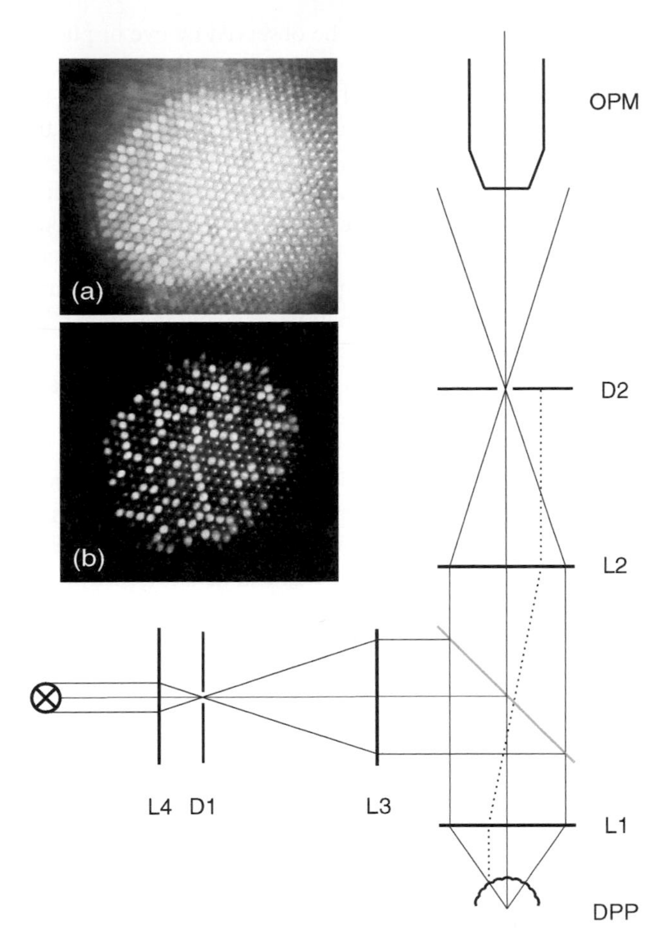

FIGURE 6.7 Setup for visualizing butterfly eye shine. The heart of the setup is a telescope, consisting of a microscope objective L1 and lens L2. A butterfly is adjusted so that the deep pseudopupil of the eye (DPP) coincides with the focal plane of L1. Lens L3 projects a small light source, created by lens L4 and diaphragm D1, via a half-mirror onto the DPP. The reflected light (i.e., the eye shine) then is collected by diaphragm D2. The eye shine can be observed at the corneal level by proper focusing of a photomicroscope, the objective of which is shown (OPM), at the image of the corneal facet lenses. **a,b**: Demonstration of the effect of spatial filtering on reflected light seen at the corneal level of a *Heliconius melpomene*. **a**: Diaphragms D1 and D2 are slightly open so that reflections at the facet lenses together with scattering at distal pigment obscure the eye shine. **b**: The diaphragms are carefully adjusted, thus selecting the eye shine (i.e., the light reflected by the ommatidial tapeta). Minor reflections at the front surface of the facet lenses remain. The individual ommatidia reflect either yellow or red light. The signals in the green channel of RGB photographs are shown.

created by a small diaphragm D1, with a telescope consisting of objective L1 and lens L3 onto the DPP, and then selecting the light emerging from the DPP via telescope L1 and lens L2 with diaphragm D2. Well-adjusted diaphragms leave the eye shine unobstructed, which then can be observed by eye or photographed with a photomicroscope.

Figure 6.7a,b illustrates how spatial filtering improves the eye shine image of a *Heliconius melpomene* (i.e., how one can block the reflections of the corneal facet lenses that normally veil the eye shine). Figure 6.7a was obtained with diaphragms D1 and D2 slightly open, but in Figure 6.7b, they were adjusted so that the reflections from cornea (and objective lens surfaces) were strongly suppressed. The reflection of an individual ommatidium of *Heliconius* is either yellow or red. Figure 6.7a,b shows the reflection of the eye in the green (i.e., in the green channel of an RGB photograph). The yellow-reflecting ommatidia stand out in Figure 6.7b, and the small dot in each facet is the remaining reflection at the outer facet lens surface that still passes diaphragm D2. The yellow- and red-reflecting ommatidia appear to be randomly arranged in the extremely regular ommatidial lattice.

The heterogeneous distribution of several hundred ommatidia, as seen in Figure 6.7b, can be studied with a 10× NA 0.3 objective. Of course, the number of reflecting ommatidia increases proportionally with the microscope objective's aperture. Unfortunately, with an increasing aperture, the working distance of microscope objectives decreases. The smallest working distance that can be allowed without damaging the eye equals, of course, the radius of a butterfly eye, which is usually on the order of 1 mm. Microscope objectives with high numerical aperture (NA) as well as acceptable working distance (WD) fortunately exist. With an increasing power of the objective, the field of view is rapidly limited by the photomicroscope, but this problem is solved with a field lens (Stavenga, 2002b). The eye shine of several butterfly species was observed with such an extended setup, using a Leitz LM32 objective(12) with NA = 0.6 and WD = 3 mm (Figure 6.8).

As shown in Figure 6.8, each butterfly species has its characteristic eye shine pattern (Bernard and Miller, 1970; Miller, 1979; Stavenga et al., 2001). Occasionally, the colors emerging from the individual facet lenses are very similar, yielding a homogeneous eye shine (Figure 6.8a,b). More frequently, a homogeneous color occurs only in a more or less limited area in the dorsal eye, while the remaining, ventral part of the eye is very heterogeneous (Figure 6.8e,f,i,j,k,m). The heterogeneity sometimes occurs throughout all eye regions, with the same color combinations everywhere (Figure 6.8c,d) or with different combinations (Figure 6.8g,k). In some cases, the dorsal and ventral areas are both more or less homogeneous, although quite differently colored (Figure 6.8m). The usual eye shine colors are in the long wavelength range (Figure 6.8a–g), but occasionally a dominant green is seen (Figure 6.8h,l). Blue or even violet eye shines can be seen dorsally (Figure 6.8k,n).

The physical origin of the observed colors is not yet fully established, although some likely hypotheses can be put forward. The homogeneous orange eye shine of several nymphalid butterflies is due to the dominant absorption by green rhodopsins. Intense and prolonged illumination bleaches the visual pigment (i.e., causes its breakdown and removal from the membrane) (Bernard, 1983), resulting in a yellow to pale coloring (Vanhoutte, 2003). The remaining broadband reflectance spectrum, which always cuts off in the red, is due to the tapetum. This conclusion is in line with the fine structure of the tapetal multilayer, where the layer thickness is nonuniform but has a gradient from distal to proximal (Miller and Bernard, 1968).

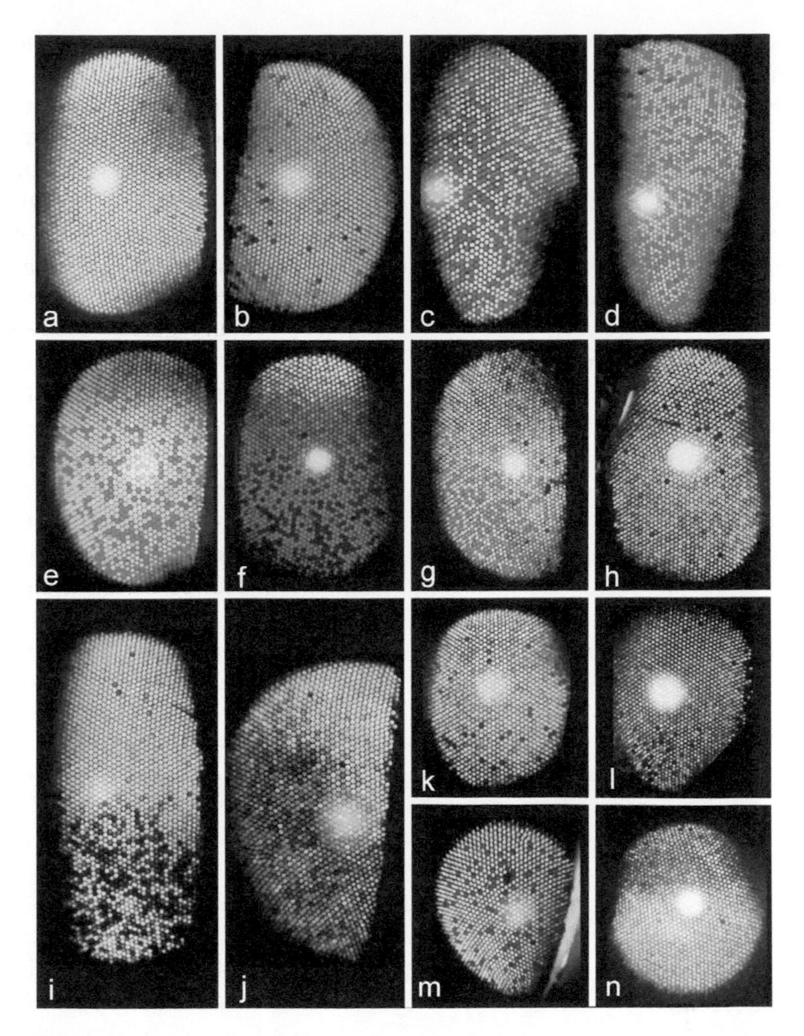

FIGURE 6.8 (See color insert following page 202.) Butterfly eye shine photographed with a large aperture epi-illumination setup. **a**: comma, *Polygonia c-album*; **b**: peacock, *Inachis io*; **c**: glider, *Cymothoe herminae*; **d**: forest pearl charaxes, *Charaxes fulvescens*; **e**: lycaenid, *Narathura japonica*; **f**: small white, *Pieris rapae*; **g**: speckled wood, *Pararge aegeria*; **h**: ringlet, *Aphantopus hyperantus*; **i**: variable eggfly, *Hypolimnas anthedon*; **j**: blue mother-of-pearl, *Salamis temora*; **k**: lycaenid, *Pseudozizeeria maha argia*; **l**: satyrine, *Ypthima argus*; **m**: lycaenid**,** *Everes argiades hellotia*; **n**: map butterfly, *Araschnia levana*. The eyeshine pattern can be virtually homogeneous yellow/orange (**a**, **b**), or a random mixture of yellow and orange (**c**, **d**) in all eye regions, but more often a distinct dorsal region exists, that is rather small (**e–g**, **k**, **n**), or large (**i**, **j**, **l**), having a more or less homogeneously colored eyeshine. The ventral area is usually marked by a rich mixture of different colored ommatidia, often with a distinct red component (**e**, **g**, **j–m**). The dark ommatidia in **f** are well reflecting in the deep red. The bright spots are reflections from the lens surfaces of the microscope objective. (From Stavenga, D.G., *J. Comp. Physiol. A* **188:**337–348, 2002. With permission.)

Eyes with a strongly heterogeneous eye shine often feature brightly red-reflecting ommatidia. Measurements of the reflectance spectra of individual ommatidia indicate that red screening pigments then border the rhabdom, as in *Papilio xuthus* and *Pieris rapae* (Ribi, 1979; Qiu et al., 2002), and that the pigments filter the light propagating in the rhabdom and reflected by the tapetum. Males of the Japanese small white, *Pieris rapae crucivora*, display a red eye shine in the major, frontal–ventral part of the eye, whereas the reflection in a restricted dorsal area is yellow (Figure 6.8f). A closer look reveals that the main eye part contains two classes of red and deep-red ommatidia, which reflect maximally at about 610 nm (Figure 6.9a) and 670 nm (Figure 6.9b), respectively. The latter class emits a distinct white-greenish fluorescence when irradiated with UV (370 nm) light (Figure 6.9c). Detailed anatomy has demonstrated that the 610 nm reflecting class is in fact composed of two different subclasses. The two types of red and deep-red reflecting ommatidia generally occur in *Pieris rapae*, males and females, but the fluorescence in the deep-red reflecting ommatidia has so far been encountered only in male *Pieris rapae crucivora* (Qiu et al., 2002).

The optical methods described here have been instrumental for the characterization of the photoreceptors studied by electrophysiology and for identifying their ommatidial class (Qiu et al., 2002; Qiu and Arikawa, 2003). Further analyses are required to clarify their function in processing the spatial and spectral information of the animal's environment.

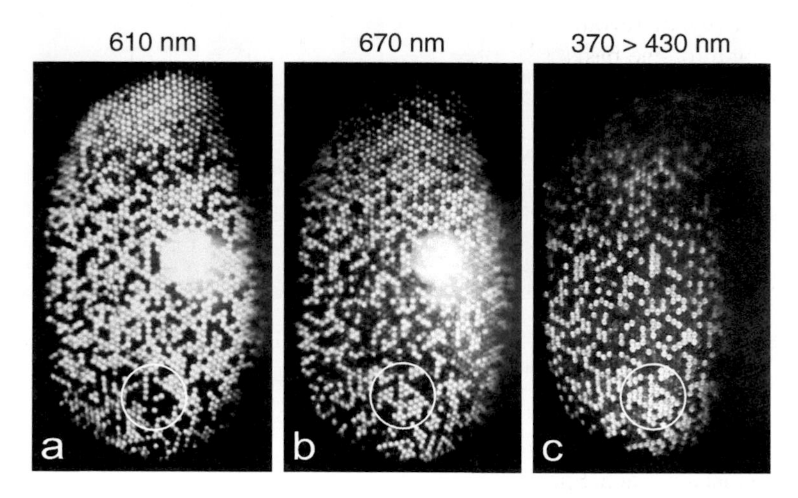

FIGURE 6.9 (See color insert following page 202.) Eye shine of a male small white, *Pieris rapae crucivora*, created by reflection (**a**, **b**) and fluorescence (**c**), using the extended set-up of Figure 6.7 (Stavenga, 2002b). Epi-illumination creates a reddish eye shine in the main (frontal–ventral) part of the eye. Using monochromatic light with wavelengths 610 (**a**) or 670 nm (**b**) shows that the ommatidia reflect either in red (**a**) or in deep red (**b**). The latter ommatidia fluoresce under UV excitation, emitting white-greenish light (**c**), emphasized by the circles (see also Qiu et al., 2002). The ommatidia in the dorsal area have a yellow eye shine and are nonfluorescent.

6.4.4 *In Vivo* Microspectrophotometry of Butterfly Visual Pigments

With an open pupil, the eye shine allows optical probing of basic properties of butterfly eyes. A principal piece of knowledge is, of course, the nature of the visual pigments. This can be studied by measuring reflectance spectra in different photochemical states of the visual pigment (Bernard, 1983). Photochemical conversions can be monitored with the setup of Figure 6.7, modified into a microspectrophotometer, which is realized by positioning the tip of a light guide in the plane of diaphragm D2 and connecting the other end to a photodiode-array spectrophotometer. Reflectance spectra (Figure 6.10a) measured during the photoconversion of the visual pigment in the comma butterfly, *Polygonia c-album*, yielded the absorbance difference spectra of Figure 6.10b. The spectra demonstrate that the comma has a dominant green rhodopsin (Figure 6.2c), absorbing maximally at λ_{max} = 532 nm, which is photointerconvertible with a metarhodopsin, absorbing maximally at λ_{max} = 492 nm (Vanhoutte, 2003).

6.4.5 Image Analysis of Butterfly Eye Shine

So far, butterfly eye shine has been studied mainly by manual analysis of photographs, made by conventional (Miller and Bernard, 1968; Stavenga et al., 2001) or digital (Stavenga, 2002b) cameras, which is generally a time-consuming process. The recent digital revolution has now opened new vistas for quantitative image analysis on a routine basis. For butterfly eye shine, a novel image analysis tool has recently been developed, which exploits the crystalline packing of the ommatidia in insect eyes and the correspondence of the facet lenses with Wigner–Seitz cells or the cells in a Voronoi diagram (Stavenga, 1979; Vanhoutte et al., 2003). The tool allows quantitative microspectrophotometry of visual pigment bleaching, as well as measurements of the pupil working range via digital imaging (Vanhoutte et al., 2003).

6.5 CONCLUSIONS AND OUTLOOK

Understanding how insect eyes process visual information requires knowledge of the spatial and spectral characteristics of the photoreceptors, their distribution over the eye, and their projections to the higher-order neurons. Classical anatomy is still indispensable, but the recent development of new optical techniques has distinctly expanded that approach. Specifically, fluorescent dyes, applied by intracellular injection or by molecular biological methods, are powerful additional tools (Section 6.3.3). Ion-sensitive dyes can be used to monitor intracellular ion concentrations quantitatively, even in virtually intact, living animals (Section 6.3.2).

Present research activities demonstrate that combinations of a variety of methods are essential for rapid progress. For instance, the important research achievements in the retina of *Drosophila* would probably not have been reached without the basic optical analyses of the last decades (e.g., Franceschini, 1975) expanded with modern molecular biology. Also, the recent discovery of widespread heterogeneity in insect eyes would not have been demonstrated so convincingly without the combination

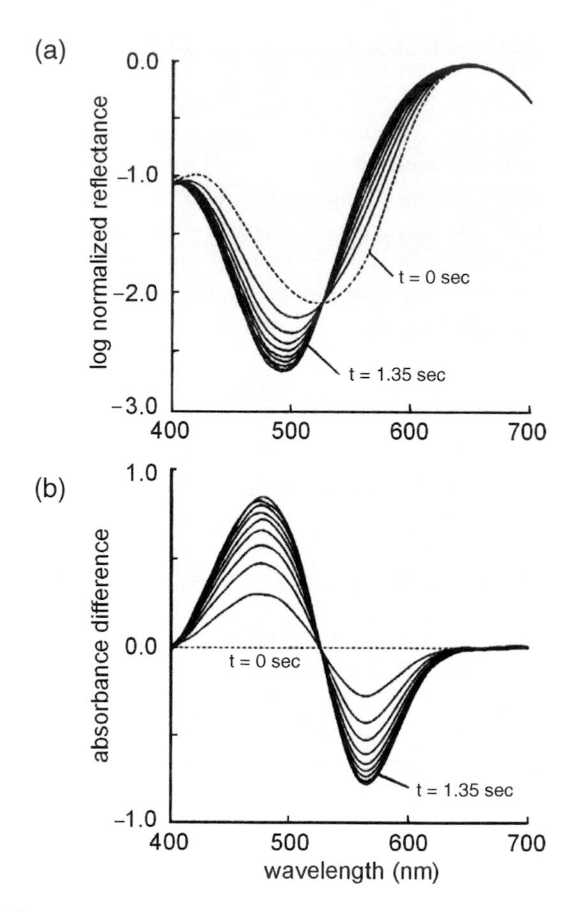

FIGURE 6.10 Microspectrophotometry of the eye shine of a comma, *Polygonia c-album*, using the setup of Figure 6.7, but with the tip of a light guide positioned in diaphragm D2. The other end of the light guide is connected to a photodiode-array spectrophotometer, allowing measurement of spectra with interval $t = 0.15$ sec. The animal is initially long-term dark-adapted, creating a maximal rhodopsin population of the visual pigments. **a**: The reflectance spectra measured at $t = 0$ sec show a deep trough at about 530 nm, indicating a main absorption by a green rhodopsin. The intense, broadband illumination necessary for obtaining a spectrum photoconverts the rhodopsin into metarhodopsin, resulting in a characteristic change in the reflectance spectrum, until a photosteady state is reached, in about 1.35 sec. **b**: Absorbance difference spectra obtained by taking the family of log reflectance spectra and subtracting each resulting spectrum from the initial spectrum. The calculated difference spectra are proportional to the difference in molecular absorption spectra of the metarhodopsin and rhodopsin state (see Figure 6.2c), except for some minor modulatory optical effects caused by the optical system of facet lens and rhabdom waveguide (Vanhoutte, 2003).

of classical anatomy, electrophysiology, and novel optical tools (Arikawa and Stavenga, 1997) (Section 6.4.2). The latter work will undoubtedly sprout new branches, specifically in the direction of color vision research. The common assumption that lepidopterans have excellent color vision has recently been substantiated by behavioral experiments in both diurnal butterflies and moths (Kelber and Pfaff, 1999; Kinoshita et al., 1999; Kinoshita and Arikawa, 2000; Kelber et al., 2002), but how they do it with a retina where the color receptors seem to be randomly organized forms a challenging question for future research. A possible answer might be expected from *Drosophila*, given its genetic potential. Color vision of flies is most likely mediated by the central photoreceptors (Troje, 1993), which have a random organization similar to that of butterflies (Salcedo et al., 1999). However, although the analysis of spatial vision in flies has seen great progress (see also Chapter 4), unfortunately, color-driven behavior of flies has thus far been difficult to study. The virtually untapped, large variety of butterfly species will at least offer a rich source for new insights in insect vision.

ACKNOWLEDGMENTS

I thank Dr. A. Huber for providing the *Drosophila* mutant shown in Figure 6.4, and Drs. J. Oberwinkler and K.J.A. Vanhoutte for their collaboration.

REFERENCES

Arikawa K, Stavenga DG (1997) Random array of colour filters in the eyes of butterflies. *J. Exp. Biol.* **200**:2501–2506.

Arikawa K, Mizuno S, Scholten DGW, Kinoshita M, Seki T, Kitamoto J, Stavenga DG (1999a) An ultraviolet absorbing pigment causes a narrow-band violet receptor and a single-peaked green receptor in the eye of the butterfly *Papilio. Vision Res.* **39**:1–8.

Arikawa K, Scholten DGW, Kinoshita M, Stavenga DG (1999b) Tuning of photoreceptor spectral sensitivities by red and yellow pigments in the butterfly *Papilio xuthus. Zool. Sci.* **16**:17–24.

Bandai K, Arikawa K, Eguchi E (1992) Localization of spectral receptors in the ommatidium of butterfly compound eye determined by polarization sensitivity. *J. Comp. Physiol. A* **171**:289–297.

Barrell KF, Pask C (1979) Optical fibre excitation by lenses. *Optica Acta* **26**:91–108.

Bernard GD (1983) Dark-processes following photoconversion of butterfly rhodopsins. *Biophys. Struct. Mech.* **9**:227–286.

Bernard GD, Miller WH (1970) What does antenna engineering have to do with insect eyes? *IEEE Student J.* **8**:2–8.

Exner S (1891) *Die Physiologie der facettirten Augen von Krebsen und Insecten.* Deuticke, Leipzig.

Exner S (1989) *The Physiology of the Compound eyes of Insects and Crustaceans* (trans. Hardie RC). Springer, Berlin.

Franceschini N (1975) Sampling of the visual environment by the compound eye of the fly: fundamentals and applications. In: *Photoreceptor Optics*, Snyder AW, Menzel R, Eds., Springer, Berlin, pp. 98–125.

Franceschini N, Kirschfeld K (1971) Étude optique *in vivo* des éléments photorécepteurs dans l'oeil composé de *Drosophila. Kybernetik* **8**:1–13.

Gärtner W (2000) Invertebrate visual pigments. In: *Molecular Mechanisms in Visual Transduction. Handbook of Biological Physics, Vol. 3*, Stavenga DG, DeGrip WJ, Pugh Jr. EN, Eds., Elsevier, Amsterdam, pp. 297–388.

Govardovskii VI, Fyhrquist N, Reuter T, Kuzmin DG, Donner K (2000) In search of the visual pigment template. *Vis. Neurosci.* **17**:509–528.

Götz KG (1964) Optomotorische Untersuchung des visuellen Systems einiger Augenmutanten der Fruchtfliege *Drosophila. Kybernetik* **2**:77–92.

Hamdorf K (1979) The physiology of invertebrate visual pigments. In: *Handbook of Sensory Physiology, Vol. VII/6A*, Autrum H, Ed., Springer, Berlin, pp. 145–224.

Hardie RC (1979) Electrophysiological analysis of the fly retina. I. Comparative properties of R1–6 and R7 and R8. *J. Comp. Physiol. A* **129**:19–33.

Hardie RC (1985) Functional organization of the fly retina. In: *Progress in Sensory Physiology, Vol. 5*, Ottoson D, Ed., Springer, Berlin, pp. 1–79.

Hardie RC (1986) The photoreceptor array of the dipteran retina. *Trends Neurosci.* **9**:419–423.

Hardie RC, Raghu P (2001) Visual transduction in *Drosophila. Nature* **413**:186–193.

Hardie RC, Raghu P, Moore S, Juusola M, Baines RA, Sweeney ST (2001) Calcium influx via TRP channels is required to maintain PIP2 levels in *Drosophila* photoreceptors. *Neuron* **30**:149–159.

Huber A (2001) Scaffolding proteins organize multimolecular protein complexes for sensory signal transduction. *Eur. J. Neurosci.* **14**:769–776.

Kelber A, Pfaff M (1999) True colour vision in the orchard butterfly, *Papilio aegeus. Naturwissenschaften* **86**:221–224.

Kelber A, Balkenius A, Warrant EJ (2002) Scotopic colour vision in nocturnal hawkmoths. *Nature* **419**:922–925.

Kinoshita M, Arikawa K (2000) Colour constancy in the swallowtail butterfly *Papilio xuthus. J. Exp. Biol.* **203**:3521–3530.

Kinoshita M, Shimada N, Arikawa K (1999) Colour vision of the foraging yellow swallowtail butterfly *Papilio xuthus. J. Exp. Biol.* **202**:95–102.

Kirschfeld K (1974) The absolute sensitivity of lens and compound eyes. *Z. Naturforsch. C* **29**:592–596.

Kitamoto J, Ozaki K, Arikawa K (2000) Ultraviolet receptors and violet receptors express identical mRNA encoding an ultraviolet-absorbing opsin: identification and histological localization of two mRNAs encoding short wavelength-absorbing opsins in the retina of the butterfly *Papilio xuthus. J. Exp. Biol.* **203**:2887–2894.

Land MF (1981) Optics and vision in invertebrates. In: *Handbook of Sensory Physiology, Vol. VII/6B*, Autrum H, Ed., Springer, Berlin, pp. 472–592.

Land MF, Osorio DC (1990) Waveguide modes and pupil action in the eyes of butterflies. *Proc. R. Soc. Lond. B* **241**:93–100.

Leydig F (1855) Zum feineren Bau der Arthropoden. *Müller's Arch. Anat. Physiol.* 406–444.

MacNichol Jr. EF (1986) A unifying presentation of photopigment spectra. *Vision Res.* **26**:1543–1556.

Mansfield RJW (1985) Primate photopigments and cone mechanisms. In: *The Visual System*, Fein A, Levine JS, Eds., Liss, New York, pp. 89–106.

Meyer N, Paulsen R, Huber A (2003) Analysis of light-regulated relocation of the TRPL ion channel in Drosophila photoreceptors using an eGFP reporter gene. *Eur. J. Cell. Biol. Suppl.* **82**:53–114.

Miller WH (1979) Ocular optical filtering. In: *Handbook of Sensory Physiology, Vol. VII/6A*, Autrum H, Ed., Springer, Berlin, pp. 69–143.

Miller WH, Bernard GD (1968) Butterfly glow. *J. Ultrastruct. Res.* **24**:286–294.

Minke B, Hardie RC (2000) Genetic dissection of *Drosophila* phototransduction. In: *Molecular Mechanisms in Visual Transduction. Handbook of Biological Physics, Vol. 3*, Stavenga DG, DeGrip WJ, Pugh Jr EN, Eds., Elsevier, Amsterdam, pp. 449–525.

Minke B, Kirschfeld K (1980) Fast electrical potentials arising from activation of metarhodopsin in the fly. *J. Gen. Physiol.* **75**:381–402.

Nilsson D-E (1989) Optics and evolution of the compound eye. In: *Facets of Vision*, Stavenga DG, Hardie RC, Eds., Springer, Berlin, pp. 30–73.

Oberwinkler J (2000) *Calcium influx, diffusion and extrusion in fly photoreceptor cells,* Ph.D. thesis, University of Groningen.

Oberwinkler J, Stavenga DG (1998) Light dependence of calcium and membrane potential measured in blowfly photoreceptors *in vivo. J. Gen. Physiol.* **112**:113–124.

Oberwinkler J, Stavenga DG (2000a) Calcium imaging demonstrates colocalization of calcium influx and extrusion in fly photoreceptors. *Proc. Natl. Acad. Sci. USA* **97**:8578–8583.

Oberwinkler J, Stavenga DG (2000b) Calcium transients in the rhabdomeres of dark- and light-adapted fly photoreceptor cells. *J. Neurosci.* **20**:1701–1709.

Pask C, Barrell KF (1980a) Photoreceptor optics I: introduction to formalism and excitation in a lens-photoreceptor system. *Biol. Cybern.* **36**:1–8.

Pask C, Barrell KF (1980b) Photoreceptor optics II: application to angular sensitivity and other properties of a lens-photoreceptor system. *Biol. Cybern.* **36**:9–18.

Pichaud F, Desplan C (2001) A new visualization approach for identifying mutations that affect differentiation and organization of the *Drosophila* ommatidia. *Development* **128**:815–826.

Postma M, Oberwinkler J, Stavenga DG (1999) Does Ca^{2+} reach millimolar concentrations after single photon absorption in *Drosophila* photoreceptor microvilli? *Biophys. J.* **77**:1811–1823.

Qiu X, Arikawa K (2003) The photoreceptor localization confirms the spectral heterogeneity of ommatidia in the male small white butterfly, *Pieris rapae crucivora. J. Comp. Physiol. A* **189**:81–88.

Qiu X, Vanhoutte KJA, Stavenga DG, Arikawa K (2002) Ommatidial heterogeneity in the compound eye of the male small white, *Pieris rapae crucivora. Cell Tissue Res.* **307**:371–379.

Ribi WA (1979) Coloured screening pigments cause red eye glow hue in pierid butterflies. *J. Comp. Physiol. A* **132**:1–9.

Salcedo E, Huber A, Henrich S, Chadwell LV, Chou WH, Paulsen R, Britt SG (1999) Blue- and green-absorbing visual pigments of Drosophila: ectopic expression and physiological characterization of the R8 photoreceptor cell-specific Rh5 and Rh6 rhodopsins. *J. Neurosci.* **19**:10716–10726.

Salcedo E, Zheng L, Phistry M, Bagg EE, Britt SG (2003) Molecular basis for ultraviolet vision in invertebrates. *J. Neurosci.* **26**:10873–10878.

Schwemer J (1989) Visual pigments of compound eyes: structure, photochemistry, and regeneration. In: *Facets of Vision*, Stavenga DG, Hardie RC, Eds., Springer, Berlin, pp. 112–133.

Smakman JG, van Hateren JH, Stavenga DG (1984) Angular sensitivity of blowfly photoreceptors: intracellular measurements and wave-optical predictions. *J. Comp. Physiol. A* **155**:239–247.

Snyder AW (1979) Physics of vision in compound eyes. In: *Handbook of Sensory Physiology, Vol. VII/6A*, Autrum H, Ed., Springer, Berlin, pp. 225–313.

Stavenga DG (1979) Pseudopupils of compound eyes. In: *Handbook of Sensory Physiology, Vol. VII/6A*, Autrum H, Ed., Springer, Berlin, pp. 357–439.

Stavenga DG (1989) Pigments in compound eyes. In: *Facets of Vision*, Stavenga DG, Hardie RC, Eds., Springer, Berlin, pp. 152–172.

Stavenga DG (1992) Eye regionalization and spectral tuning of retinal pigments in insects. *Trends Neurosci.* **15**:213–218.

Stavenga DG (2002a) Colour in the eyes of insects. *J. Comp. Physiol. A* **188**:337–348.

Stavenga DG (2002b) Reflections on colourful butterfly eyes. *J. Exp. Biol.* **205**:1077–1085.

Stavenga DG (2003a) Angular and spectral sensitivity of fly photoreceptors. I. Integrated facet lens and rhabdomere optics. *J. Comp. Physiol. A* **189**:1–17.

Stavenga DG (2003b) Angular and spectral sensitivity of fly photoreceptors. II. Dependence on facet lens F-number and rhabdomere type. *J. Comp. Physiol. A* **189**:189–202.

Stavenga DG (2004) Angular and spectral sensitivity of fly photoreceptors. III. Dependence on the pupil mechanism in the blowfly *Calliphora. J. Comp. Physiol. A* **190**:115–129.

Stavenga DG, van Hateren JH (1991) Focusing by a high power, low Fresnel number lens: the fly facet lens. *J. Opt. Soc. Am. A* **8**:14–19.

Stavenga DG, Kruizinga R, Leertouwer HL (1990) Dioptrics of the facet lenses of male blowflies *Calliphora* and *Chrysomia. J. Comp. Physiol. A* **166**:365–371.

Stavenga DG, Smits RP, Hoenders BJ (1993) Simple exponential functions describing the absorbance bands of visual pigment spectra. *Vision Res.* **33**:1011–1017.

Stavenga DG, Oberwinkler J, Postma M (2000) Modeling primary visual processes in insect photoreceptors. In: *Molecular Mechanisms in Visual Transduction. Handbook of Biological Physics, Vol. 3*, Stavenga DG, DeGrip WJ, Pugh Jr EN, Eds., Elsevier, Amsterdam, pp. 527–574.

Stavenga DG, Kinoshita M, Yang E-C, Arikawa K (2001) Retinal regionalization and heterogeneity of butterfly eyes. *Naturwissenschaften* **88**:477–481.

Troje N (1993) Spectral categories in the learning behaviour of blowflies. *Z. Naturforsch. C* **48**:96–104.

van Hateren JH (1984) Waveguide theory applied to optically measured angular sensitivities of fly photoreceptors. *J. Comp. Physiol. A* **154**:761–771.

van Hateren JH (1985) The Stiles–Crawford effect in the eye of the blowfly, *Calliphora erythrocephala. Vision Res.* **25**:1305–1315.

van Hateren JH (1989) Photoreceptor optics, theory and practice. In: *Facets of Vision*, Stavenga DG, Hardie RC, Eds., Springer, Berlin, pp. 74–89.

Vanhoutte KJA (2003) *Butterfly visual pigments: molecular cloning and optical reflections,* Ph.D. thesis, University of Groningen.

Vanhoutte KJA, Michielsen KFL, Stavenga DG (2003) Analyzing the reflections from single ommatidia in the butterfly compound eye with Voronoi diagrams. *J. Neurosci. Meth.* **131**:195–203.

Vogt K (1989) Distribution of insect visual chromophores: functional and phylogenetic aspects. In: *Facets of Vision*, Stavenga DG, Hardie RC, Eds., Springer, Berlin, pp. 134–151.

Vogt K, Kirschfeld K, Stavenga DG (1982) Spectral effects of the pupil in fly photoreceptors. *J. Comp. Physiol. A* **146**:145–152.

White RH, Xu H, Münch T, Bennett RR, Grable EA (2003) The retina of *Manduca sexta*: rhodopsin-expression, the mosaic of blue-green and UV-sensitive photoreceptors. *J. Exp. Biol.* **206**:3337–3348.

Wijngaard W, Stavenga DG (1975) On optical crosstalk between fly rhabdomeres. *Biol. Cybern.* **18**:61–67.

7 Novel Approaches to Visual Information Processing in Insects: Case Studies on Neuronal Computations in the Blowfly

Martin Egelhaaf, Jan Grewe, Katja Karmeier, Roland Kern, Rafael Kurtz, and Anne-Kathrin Warzecha

CONTENTS

7.1 INTRODUCTION

Vision guides behavior in virtually all animals, especially in numerous insects. The array of photoreceptors in the eye typically receives a wildly fluctuating pattern of image flow when the animal moves through its environment. It is the

0-8493-2024-0/05/$0.00+$1.50

task of the brain to interpret this complex spatio-temporal input and to make use of it in guiding behavior. Biological nervous systems outperform existing artificial vision systems with regard to their capabilities of processing retinal image flow. Insects make efficient use of visual information, for instance, during the aerobatic chasing maneuvers of male flies in the context of mating behavior (Land and Collett, 1974; Wagner, 1986; Boeddeker et al., 2003; Boeddeker and Egelhaaf, 2003a, b) and the ability of bees to assess, on the basis of visual motion cues, the distance traveled in an unknown environment (Esch and Burns, 1996; Srinivasan et al., 2000; Esch et al., 2001). These extraordinary capabilities are most remarkable given the small number of neurons in insect brains and the astonishing speed with which retinal images are processed.

Because the nervous systems of insects are well amenable to electrophysiological and imaging techniques under in vivo conditions, insects have served for many years as model systems for analyzing the processing of retinal image flow (review: Egelhaaf and Kern, 2002). Novel experimental approaches at both the behavioral and the neuronal levels, as well as new techniques for data analysis, are beginning to unveil the mechanisms underlying the amazing visual capabilities of insects (see also Chapters 1, 2, 6, and 8). In particular, blowflies have been used as a model system for understanding how visual motion information is processed (Borst and Haag, 2002; Egelhaaf et al., 2002; Kurtz and Egelhaaf, 2003). Results on blowflies will, therefore, form the basis of this review, but wherever possible, we also refer to the work done on other insect groups.

Visual motion has been shown to play an important role in behavioral control in blowflies. Examples are visual course control, landing behavior, and object–background discrimination. In mediating these behavioral maneuvers, the nervous system of blowflies relies on extracting behaviorally relevant information from the continually changing brightness patterns that are generated on the eyes during locomotion. This so-called optic flow contains information both about the direction and speed of the animal's self-motion and about the environmental layout (Koenderink, 1986; Lappe, 2000). The behavioral significance of motion vision in blowflies and other insects is reflected in an abundance of motion-sensitive neurons in their visual system (reviews: Wehner, 1981; Hausen, 1981; Hausen and Egelhaaf, 1989; Rind and Simmons, 1999; see also Chapter 8).

An ensemble of large visual interneurons, the so-called tangential cells (TCs), in the blowfly's third visual neuropil (the lobula complex) has been characterized in particular detail and is assumed to play a key role in processing visual motion information in the context of visually guided behavior. Most TCs receive their main input from two sets of retinotopically organized, local, motion-sensitive interneurons with opposite preferred directions. As a consequence, they respond in a directionally selective manner to motion in large parts of the visual field (reviews: Hausen and Egelhaaf, 1989; Egelhaaf and Warzecha, 1999; Egelhaaf et al., 2002; Borst and Haag, 2002). Neither the preferred directions of motion of TCs nor their motion sensitivities are homogeneous, but these change in a systematic way across the visual field. These intricate receptive field structures, which represent a phylogenetic adaptation rather than the outcome of sensory experience (Karmeier et al., 2001), suggest that TCs are tuned to certain patterns of optic flow, as seen in the blowfly during specific types of self-motion (Hengstenberg, 1982; Krapp and Hengstenberg, 1996; Krapp et al., 2001).

Some TCs receive, either exclusively or in addition to their retinotopic input, excitatory or inhibitory input from other TCs (Hausen and Egelhaaf, 1989; Egelhaaf et al., 1993; Warzecha et al., 1993; Horstmann et al., 2000; Haag and Borst, 2001, 2002). This input is assumed to enhance the selectivity of TCs for optic flow. As a consequence, some TCs respond best during coherent wide-field motion, as may occur while an animal turns around a particular body axis (reviews: Hausen, 1981; Hausen and Egelhaaf, 1989; Egelhaaf and Warzecha, 1999; Krapp, 2000; Egelhaaf et al., 2002). Others respond best to object motion, as may occur while the animal pursues a moving target or passes a stationary object in its environment (Collett, 1971; Olberg, 1981, 1986; Egelhaaf, 1985a, b; Gilbert and Strausfeld, 1991; Gauck and Borst, 1999; Kimmerle and Egelhaaf, 2000a, b; Olberg and Pinter, 1990).

By employing novel approaches, three different but related aspects of motion computation have been addressed in recent years:

1. Visual information processing in small neuronal networks
2. The reliability of visual motion encoding by neuronal populations
3. The neuronal representation of natural optic flow

The merits and limitations of these novel approaches will be summarized in the following pages, and we discuss how they have contributed to our understanding of how behaviorally relevant visual information is processed in the insect brain.

7.2 METHODS FOR ANALYZING VISUAL INFORMATION PROCESSING IN SMALL NEURAL NETWORKS

From a methodological point of view, the intricate network of TCs is a particularly advantageous model system to analyze the cellular basis of neural computation under *in vivo* conditions. Dendritic information processing and synaptic transmission can be analyzed by electrophysiological and optical recording techniques while the animal is stimulated by its natural visual input. Imaging of the intracellular activity distribution with calcium-sensitive dyes (see also Chapter 6 and Chapter 13) is feasible in the virtually intact brain, because TCs arborize in a plane almost parallel and close to the brain surface. In this way, it has been possible not only to unravel major parts of the neuronal circuitry at this level of the visual pathway, but also to elucidate how these neuronal circuits process information under their normal operating conditions.

7.2.1 Identifying Neural Networks

Dual electrophysiological recordings (see also Chapter 12 and Chapter 14) have been successfully applied to analyze synaptic connections within a neuronal circuit and are still the technique of choice because they provide a means of establishing synaptic connections between two cells and characterizing their functional properties. For instance, one of the cells of the so-called horizontal system, the HSE cell (horizontal system equatorial cell), was shown using this method to receive input from two TCs that reside in the contralateral half of the visual system. Both are excited by back-to-front motion. Owing to its ipsilateral retinotopic input, HSE is excited by front-to-back motion, so it can be expected to respond best to wide-field motion as is generated on the eyes when the animal turns about its vertical axis (Figure 7.1) (Horstmann et al., 2000; Krapp et al., 2001).

Recently, two other experimental approaches were employed for network analysis. These approaches allow us to establish synaptic connectivity without requiring simultaneous recording from two cells. The presumptive presynaptic neuron is recorded intracellularly after the presumed postsynaptic cell was intracellularly injected with an activity-dependent fluorescent dye (e.g., a calcium-sensitive dye such as fura 2, calcium-green, or Oregon-green)[(14)]. Synaptic connectivity is established by depolarizing the presynaptic cell by current injection via the recording electrode and monitoring the resulting fluorescence changes in the postsynaptic cell (Haag and Borst, 2001, 2002, 2003). Synaptic network interactions have been further established by the so-called fill-and-kill technique (Miller and Selverston, 1979; Warzecha et al., 1993; Farrow et al., 2003). The presynaptic cell is injected with a fluorescent dye, 6-carboxy-fluorescein[(27)], which becomes phototoxic after laser illumination. After photoablation of the presynaptic neuron, the functional consequences for the postsynaptic cell can be characterized. Using this method, it was established that a particular TC, a so-called *figure detection neuron* (FD1), is inhibited by one of the GABAergic *centrifugal horizontal cells* (CH cells) (Figure 7.2) (Warzecha et al., 1993). As a consequence of this inhibition, the FD1 neuron responds best to object motion but shows little activity during wide-field motion (Figure 7.2b). This circuit was further analyzed by photoablation of neurons that are presynaptic to the CH cells: CH cells receive their main ipsilateral motion input not from retinotopic small-field elements, as is the case for most other TCs, but via dendro–dendritic interactions from the HS cells. After laser ablation of HS cells, the CH cells could

FIGURE 7.1 **a**: Schematic of synaptic input organization of the HSE cell in the right half of the brain. It receives input from many retinotopically organized local motion-sensitive input elements (indicated by thin lines) in the ipsilateral visual field. Information about back-to-front motion in the contralateral visual field is mediated by the H1 and the H2 cell. The H2 cell contacts the HSE cell close to its output terminal; the H1 cell is likely to make a multitude of synaptic connections with its extended terminal region on the dendritic tree of the HSE cell. Gray arrows indicate the direction of signal flow in the cells. The light gray insets illustrate, seen from above, the fly looking at various motion stimuli and indicate the preferred directions of motion of the different cells. The graphs (below) display the EPSPs recorded in the HSE cell as evoked by the H1 and the H2 cells, respectively. Vertical arrows mark the occurrence of the H1 and H2 spikes, respectively. **b**: Time courses of responses of the HSE cell in the right half of the brain to rotational optic flow, as well as the corresponding monocular components. The arrows indicate front-to-back and back-to-front motion in the right and left visual field, respectively. Although the cell mainly shows graded depolarizations during ipsilateral front-to-back motion (first trace), there are many response transients during contralateral back-to-front motion (second trace) and clockwise rotational optic flow (third trace). The duration of motion is indicated by the black horizontal bar. (From Horstmann, W. et al., *Eur. J. Neurosci.* **12**:2157–2165, 2000. With permission.)

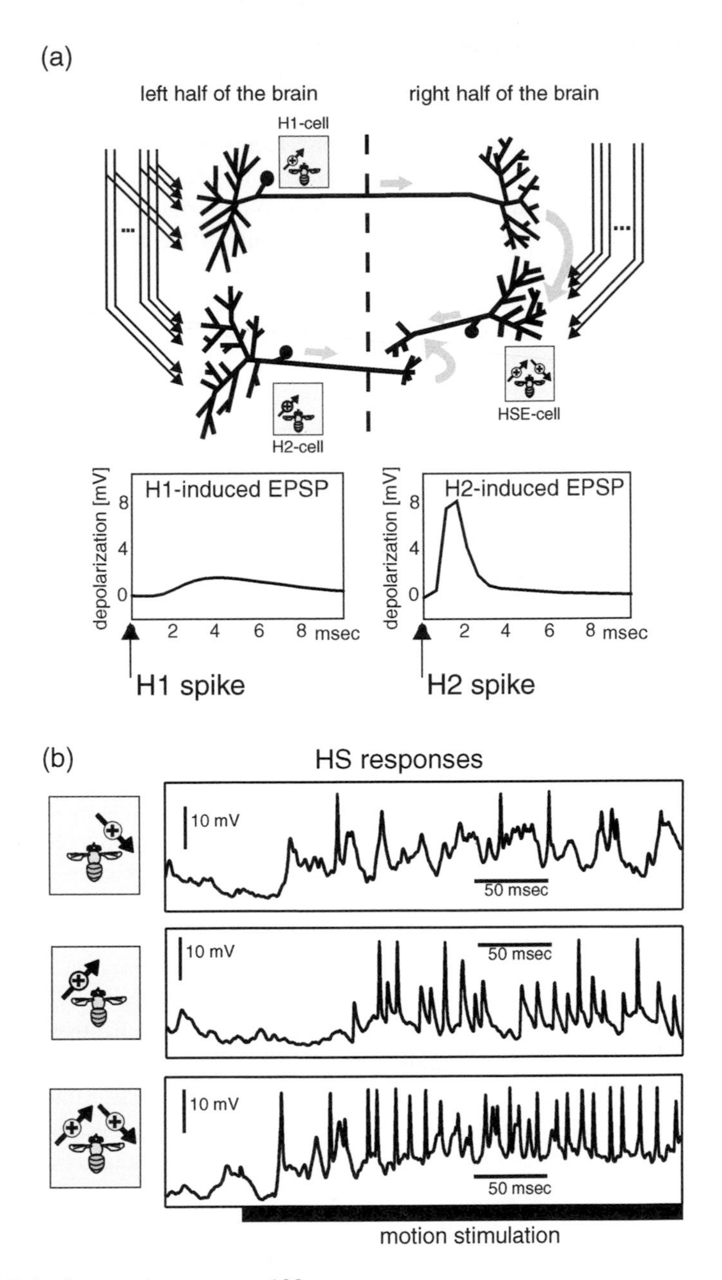

FIGURE 7.1 See caption on page 188.

no longer respond to motion in front of the ipsilateral eye, and only responses to motion in the contralateral visual field remained (Farrow et al., 2003).

By applying these approaches to neuronal network analysis, it has been possible to unravel major parts of the wiring diagram of the intricate network of TCs in the blowfly's third visual neuropil. The interactions that take place within the network are thought to play an important role in tuning TCs to particular types of optic flow, as are generated on the eyes when the animal turns around a particular body axis or passes a nearby object in front of a more distant background. The network interactions comprise heterolateral interactions, recurrent inhibitory interactions, and feed-forward inhibition. For instance, the HSE cells are major output elements of the visual system. In parallel they act, via dendro-dendritic synapses, as excitatory input elements of the CH cells, the inhibitory inputs of the object-motion sensitive FD1 cell (see previous), as well as of an excitatory heterolateral element, the H1 cell (see Chapter 8). The latter neuron in turn excites, among other cells, the contralateral HSE cell and the contralateral CH cells (Figure 7.3). Modeling is required to infer the functional significance of this intricate connection pattern.

7.2.2 Computational Properties of Synaptic Interactions

Understanding the computational properties of a neuronal network requires knowledge about details of the neuronal wiring scheme and also about the functional properties of the synaptic connections. In general, synapses are particularly important sites of cellular information processing because they may have peculiar nonlinear properties and may even change their transmission properties depending on their activation history (Juusola et al., 1996; Sabatini and Regehr, 1999; Paulsen and Sejnowski, 2000; Thomson, 2000; Fortune and Rose, 2001; Simmons, 2002).

Meaningful representations of optic flow are often achieved only by specific synaptic interactions between TCs (see previous section). To be beneficial, these synaptic interactions must be carefully adjusted to the natural operating range of the system. Otherwise, synaptic transmission may severely distort the information being transmitted. This hazard is particularly daunting because synaptic transmission is inherently noisy and the underlying biophysical processes have been found in many systems to be intrinsically nonlinear. Moreover, the transformation of the postsynaptic potential into spike activity may also be nonlinear. Combined electrophysiological and *in vivo* optical imaging experiments were performed in the blowfly to analyze the relationship between the activity of a given presynaptic TC and the spike rate of its postsynaptic target in the contralateral visual system. It was found that the entire range of presynaptic depolarization levels that can be elicited by motion in the "preferred direction" is transformed approximately linearly into the postsynaptic spike rate (Figure 7.4) (Kurtz et al., 2001). This is surprising, especially given the potential nonlinearities mentioned earlier. Linearity characterizes the transmission of membrane potential fluctuations up to frequencies of 10 Hz (see Section 7.3.1 (Warzecha et al., 2003). Thus, the linear synaptic regime covers most of the dynamic range within which visual motion information is transmitted with high gain (Haag and Borst, 1997; Warzecha et al., 1998). In addition to slow graded membrane potential changes, rapid presynaptic depolarizations, such as spikes, are also transmitted reliably at this synapse (Warzecha et al., 2003). As a consequence of the computational properties of the analyzed synapse, visual motion information is transmitted largely undistorted to the contralateral visual system. This ensures that the characteristic dependence of neural responses on stimulus parameters such as velocity or contrast is not affected by the intervening synapse.

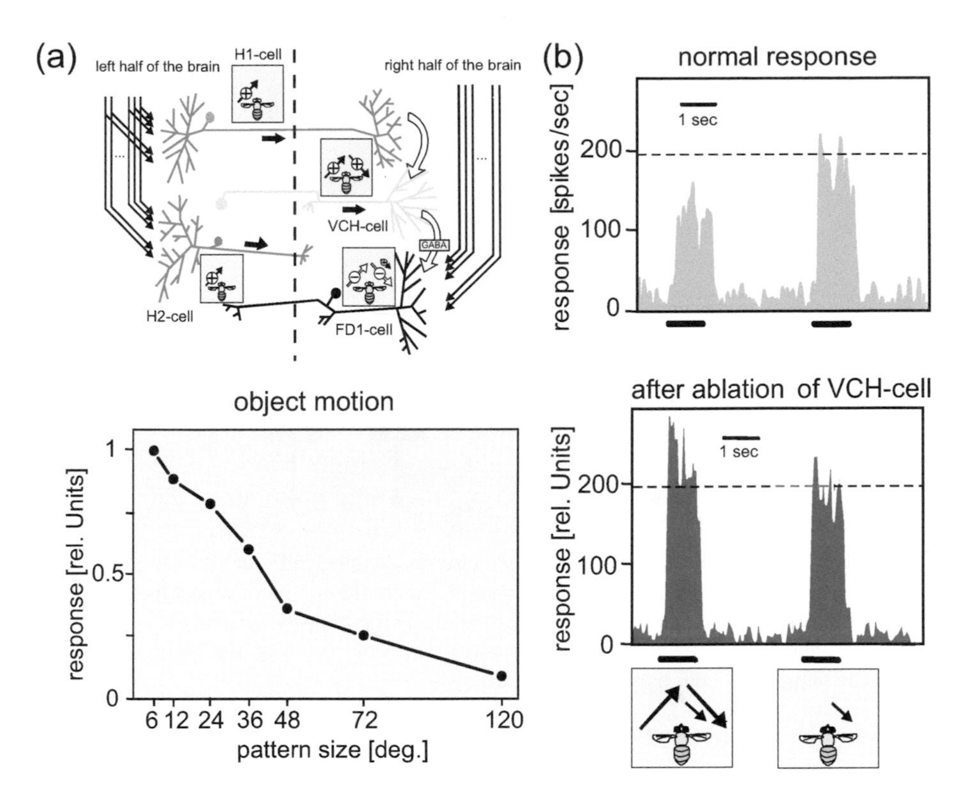

FIGURE 7.2 **a**: Top: Schematic of input organization of one of the figure detection cells, the FD1 cell. It receives input from many retinotopically organized local motion sensitive input elements (indicated by thin lines) in the ipsilateral visual field. It is inhibited via GABAergic synapses by one of the CH cells, the VCH cell. The VCH cell receives excitatory input from the contralateral eye via the H1 and the H2 cells. Black arrows indicate the direction of signal flow within the cells. Insets illustrate, seen from above, the fly looking at various motion stimuli and indicate the preferred directions of motion (black arrows) of the different cells or of the inhibitory input of the FD1 cell (open arrows). Bottom plot: Response of the FD1 cell as a function of pattern size, illustrating that it is most sensitive to the motion of a small pattern. (From Egelhaaf, M., *Biol. Cybern.* **52**:195–209,1985a. With permission.) **b**: Spike frequency histogram of responses of FD1 cell to wide-field and small-field motion (indicated in the insets below the diagrams) before and after photoablation of the ipsilateral VCH cell. Horizontal bars below the responses indicate the duration of motion. Arrows in insets symbolize the size and direction of the moving pattern. Dotted horizontal lines indicate mean response amplitudes during small-field motion. (From Warzecha, A.-K. et al., *J. Neurophysiol.* **69**:329–339, 1993. With permission.)

Presynaptic transmitter release is thought to be controlled by changes in the presynaptic calcium concentration. Therefore, presynaptic calcium concentration changes were monitored after the presynaptic cell was injected with a calcium-sensitive fluorescent dye. This allows us to monitor the time-dependent changes in calcium concentration in different parts of the axon terminal (Figure 7.4a). Interestingly, a linear relationship was found between presynaptic depolarization as is

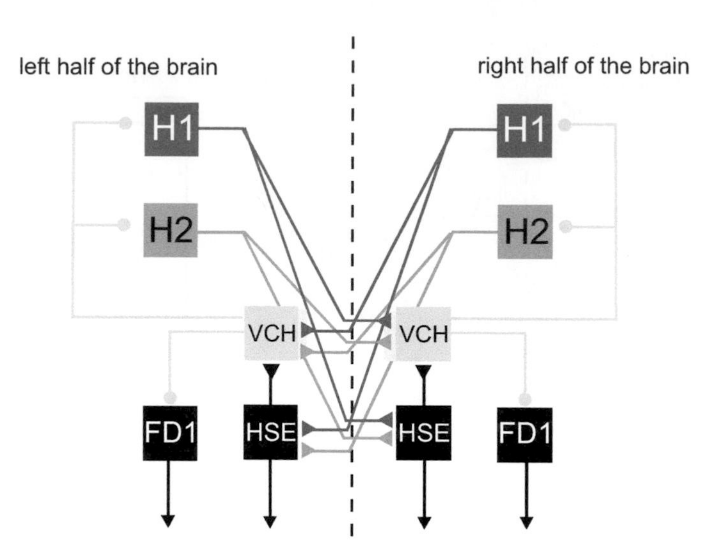

FIGURE 7.3 Relationship of the two neural circuits sketched in Figure 7.1a and Figure 7.2a, tuning blowfly tangential cells either to coherent wide-field motion or object motion, respectively. The cells are indicated by boxes. Excitatory and inhibitory synapses are indicated by triangles and circles, respectively. Note the reciprocal recurrent inhibitory connections between neurons in both halves of the visual system.

induced during preferred direction motion and the presynaptic calcium concentration. Moreover, the postsynaptic spike rate was also linearly related to presynaptic calcium concentration increases (Figure 7.4b) (Kurtz et al., 2001). Although these findings are in accordance with the overall characteristics of synaptic transmission, one should be careful in interpreting the measured presynaptic calcium signals to reflect precisely the calcium concentration that is relevant in the control of transmitter release: with conventional *in vivo* imaging techniques, only the bulk calcium concentration changes within presynaptic arborizations can be monitored (see also Chapter 13). In contrast, the calcium concentration changes that are relevant for transmitter release are likely to be regulated very close to the presynaptic membrane (review: Neher, 1998).

Because conclusions on the role of calcium in synaptic transmission are critically dependent on the exact determination of presynaptic calcium changes, two new technical approaches are currently being tested. First, to overcome spatial resolution limits of conventional intracellular imaging, two-photon laser scanning fluorescence microscopy has recently been adjusted for *in vivo* analysis of presynaptic calcium concentration changes at high spatial and temporal resolution (Kalb et al., 2004). Second, to relate presynaptic calcium concentration changes and postsynaptic activity in a more systematic way than is possible by visual stimulation, the flash-photolysis technique is used to release calcium in the presynaptic neuron. Here, the presynaptic neuron is injected with a photolabile calcium cage (BAPTA-1 hexapotassium salt or NP-EGTA tetrapotassium salt)(14). This compound is loaded with calcium that is rapidly released upon illumination

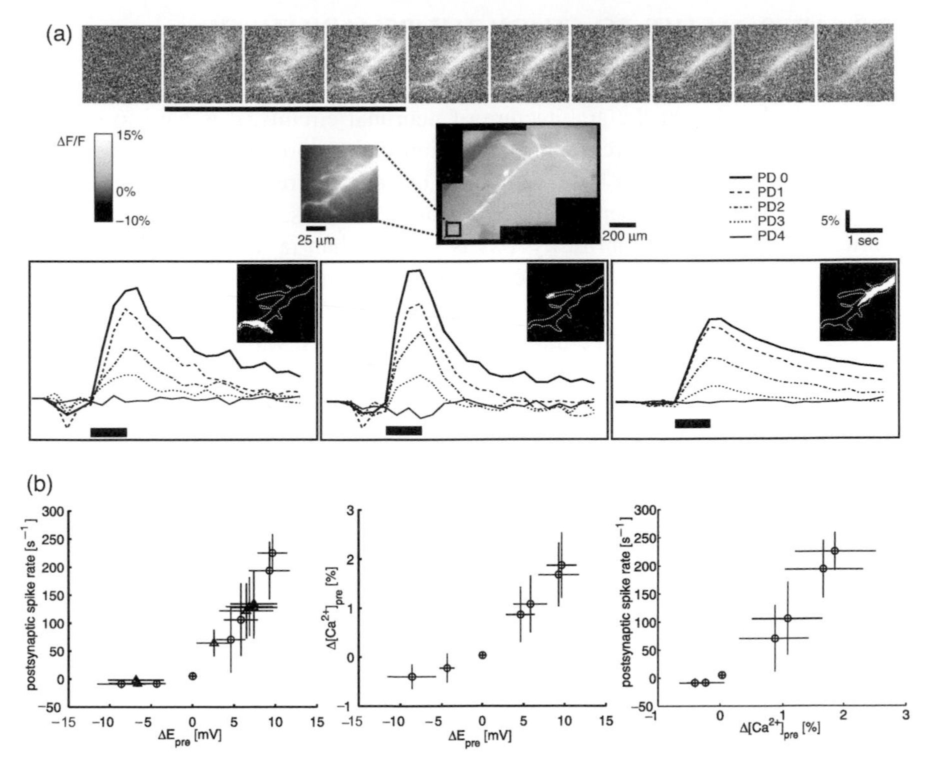

FIGURE 7.4 Transmission of optic flow information between a pair of TCs. **a**: Upper diagram: Presynaptic calcium accumulation in a VS cell filled with a calcium-sensitive fluorescent dye (raw fluorescence images of the entire cell and of the presynaptic region, left diagrams) during presentation of preferred direction motion (black horizontal bar). White and light gray values in the images correspond to increases in calcium concentration (measured as relative change in fluorescence: ΔF/F). Bottom diagrams: Time courses of presynaptic calcium concentration changes in three regions of the axon terminal (indicated in the insets) for variable stimulus strengths (different line types). The insets correspond to the terminal region as seen on the left raw fluorescence image (center). **b**: Linearity of the transfer of preferred direction motion. Left: Postsynaptic spike rate (relative to resting activity) is plotted vs. the presynaptic membrane potential change (ΔE_{pre}) for visual stimuli of variable strengths, moving either in the preferred direction (positive ΔE_{pre} values) or in the null direction (negative ΔE_{pre} values). The gain of signal transfer is about constant for the entire range of visually induced excitations, resulting in a linear relationship between presynaptic potential and postsynaptic spike rate upon motion in the preferred direction. A rectification is prominent for null direction motion. Linear dependencies for preferred direction motion are also present in the relationship between changes in presynaptic calcium and in presynaptic membrane potential (middle) and in that between postsynaptic spike rate and changes in presynaptic calcium (right). (From Kurtz, R. et al., *J. Neurosci.* **21**:6957–6966, 2001. With permission.)

of the preparation with a brief, high-intensity flash of the appropriate wavelength (Kurtz, 2004). These techniques are currently adjusted for their use in the intact fly brain to unravel the role of calcium in synaptic transmission — and thus in neuronal processing of optic flow information.

7.3 APPROACHES TO STUDY THE RELIABILITY OF ENCODING VISUAL MOTION INFORMATION

By simply looking at the architecture of neuronal circuits, it is hard to infer how reliably they extract the relevant information under normal behavioral conditions. One reason for this is the large variability of neuronal responses, even if these are evoked by repeated presentation of the same stimulus. Although the across-trial variance of the spike count in fly TCs is smaller than in motion-sensitive neurons in the primate cortex (see, e.g., de Ruyter van Steveninck et al., 1997; Warzecha and Egelhaaf, 1999; Barberini et al., 2000; Warzecha et al., 2000), it is hard to deduce from individual spike trains the extent to which variations in the interspike interval are caused by the stimulus or by sources that are not time-locked to the stimulus (*noise*). Neuronal response variability limits the timescale on which time-varying optic flow can be conveyed.

The performance of a neuron in real time can only be understood by scrutinizing individual response traces, not on the basis of ensemble averages. This requires assessing the variability in a quantitative way by appropriate statistical measures. The spike count variance across trials is the most straightforward measure of neuronal variability. However, this measure does not take into account the time course of neuronal activities that may be particularly important when analyzing responses to time-varying stimuli. In the following section, we briefly summarize some methods to quantify the variability of time-varying neuronal data and to analyze the timescale on which sensory information is encoded by the nervous system.

To understand the specificity of sensory coding, the analysis must be extended from individual neurons to populations of neurons that all respond to somewhat different aspects of the stimulus (see Chapter 14). The significance of population coding can be analyzed particularly well on blowfly TCs because the neurons makingup the population are largely known and are accessible to experimental analysis. Elaborate theoretical tools have been developed to assess (1) how well different optic flow patterns, as induced by the animal's self-motion, can be distinguished from each other and (2) how specifically these optic flow patterns are encoded on the basis of the population responses. Some of these theoretical tools will be summarized with regard to optic flow encoding by fly TCs.

7.3.1 Reliability of Neural Coding by Individual Nerve Cells

There are various measures to quantify the reliability of neuronal responses to time-varying stimuli. If repetitive stimulation with a given stimulus led always to identical neuronal responses, the timing of spikes would be determined exclusively by the stimulus. Obviously, sensory neurons do not respond with absolute accuracy, and the timing of spikes varies considerably from trial to trial. Thus, it is most likely that spike timing depends on noise sources, such as the stochastic absorption of photons by the photoreceptors or the stochastic nature of transmitter release at synapses and the subsequent opening and closing of ionic channels.

Several approaches to quantify neuronal variability are based on information theory (reviews: Rieke et al., 1997; Borst and Theunissen, 1999). Applying these measures allows us to analyze how much information about the stimulus is encoded by a neuron. Information theory implies that a stimulus that assumes many states contains more information than a stimulus assuming, for instance, only two states. Moreover, the information content of a stimulus depends also on the probability of the different stimulus states. Accordingly, the information conveyed by a neuron depends on the probability distribution of the neuronal response levels. The mutual information between stimulus and response specifies how much information about the different states of the stimulus is represented by the neural response (Figure 7.5). The mutual information between stimulus and response is constrained by the variability of neuronal responses upon repetitive presentation of a given stimulus. The mutual information can be determined from the relationship of the probability distribution of response levels when a particular stimulus is presented (conditional probability) and the overall probability distribution of the different responses levels obtained for all stimulus conditions (Figure 7.5).

Although, in principle, this procedure for determining the mutual information is simple, it can be applied in practice only with some difficulty (Borst and Theunissen, 1999). A major reason for this difficulty is that the different stimulus conditions and the different response states need to be defined explicitly. This may be feasible for simple stimuli (e.g., a pattern that moves only to the left or to the right, or a moving pattern whose orientation is varied in discrete steps), but it is hardly possible for complex time-dependent stimuli. Therefore, more practical approaches to quantify stimulus–response relationships have been developed that simplify matters but, as a trade-off, are based on certain assumptions about the properties of the stimuli or of the neuronal responses. One type of approach, *linear reconstruction techniques*, will be summarized briefly without going into formal details (for details, see Marmarelis and Marmarelis, 1978; Rieke et al., 1997; Borst and Theunissen, 1999; Bendat and Piersol, 2000).

Here, the neuronal encoding of a signal is not quantified on the basis of probability distributions of stimulus conditions and response levels, but by relating the measured responses to the output of an encoding model, that is, a model that accounts for sensory information processing in formal terms. From a formal point of view, the simplest encoding model is a linear one. The optimal linear model that allows one to estimate a time-dependent signal from another signal can be determined by the so-called *reverse reconstruction approach* (review: Rieke et al., 1997) (Figure 7.6a). This approach has frequently been applied to sensory systems, including blowfly TCs, to investigate how well stimulus velocity can be estimated on the basis of neuronal responses (Bialek et al., 1991; Haag and Borst, 1997, 1998). It was shown that time-dependent stimulus velocity is linearly encoded by blowfly TCs as long as pattern velocity and the velocity changes are relatively small. Otherwise, acceleration and higher-order temporal derivatives play an increasingly large role in shaping the time course of the motion response (Egelhaaf and Reichardt, 1987; Haag and Borst, 1997). Recently, the reverse reconstruction approach has also been applied to relate the time-dependent postsynaptic signal to the corresponding presynaptic input (Figure 7.6a). The results of these experiments have already been summarized earlier (see Section 7.2.2; Warzecha et al., 2003).

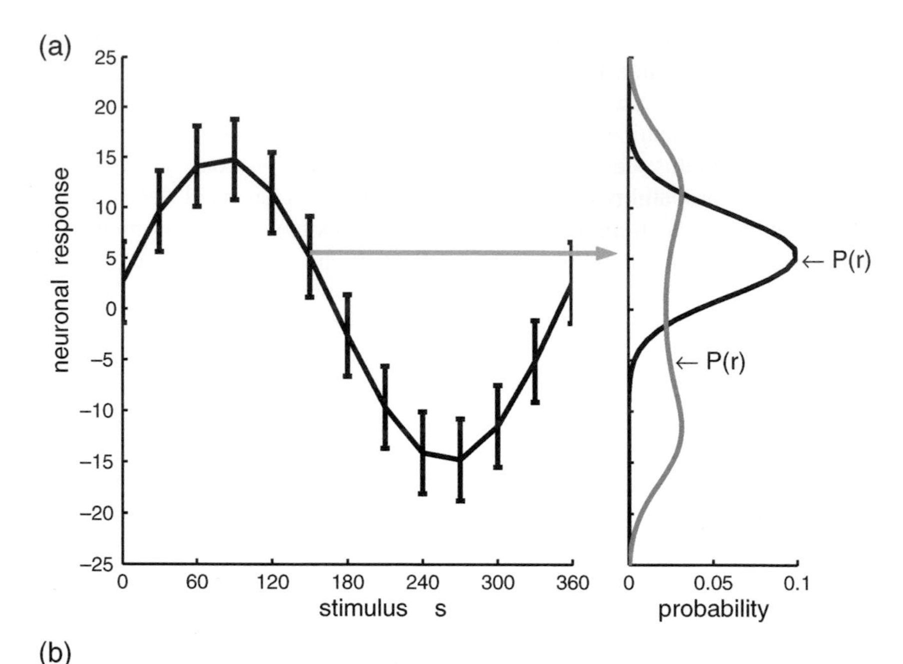

(b)

p(r)	probability distribution of the neural response to any stimulus
$p(s_X)$	probability that the stimulus takes the value s_X
$p(r\|s_X)$	probability of neural responses when the stimulus s_X is presented (conditional probability)

Information about a particular stimulus s_X

$$I_{s_X} = \Sigma\, p(r|s_X) \log_2 (p(r|s_X) / p(r))$$

Average mutual information obtained from all stimulus conditions:

$$I_S = \Sigma\Sigma\, p(s_X)\, p(r|s_X) \log_2 (p(r|s_X) / p(r))$$

FIGURE 7.5 Information transmitted by a neuron about an input signal (mutual information). **a**: Simulated tuning curve for a range of stimuli. The mean responses and standard deviations are shown as can be measured experimentally. The information conveyed by the neuron depends on the probability distribution of the response amplitudes. The more the probability distributions of responses to particular stimuli $p(r|s_x)$ deviate from the overall response distribution that is obtained for all stimuli p(r), the better the different stimuli can be distinguished on the basis of the neuronal responses and the higher the mutual information. **b**: Definition of mutual information and explanation of symbols. The mutual information between stimulus and response can be determined from the relationship of the probability distribution of response levels when a particular stimulus is presented (conditional probability) and the overall probability distribution of the different responses levels that is obtained for all stimulus conditions.

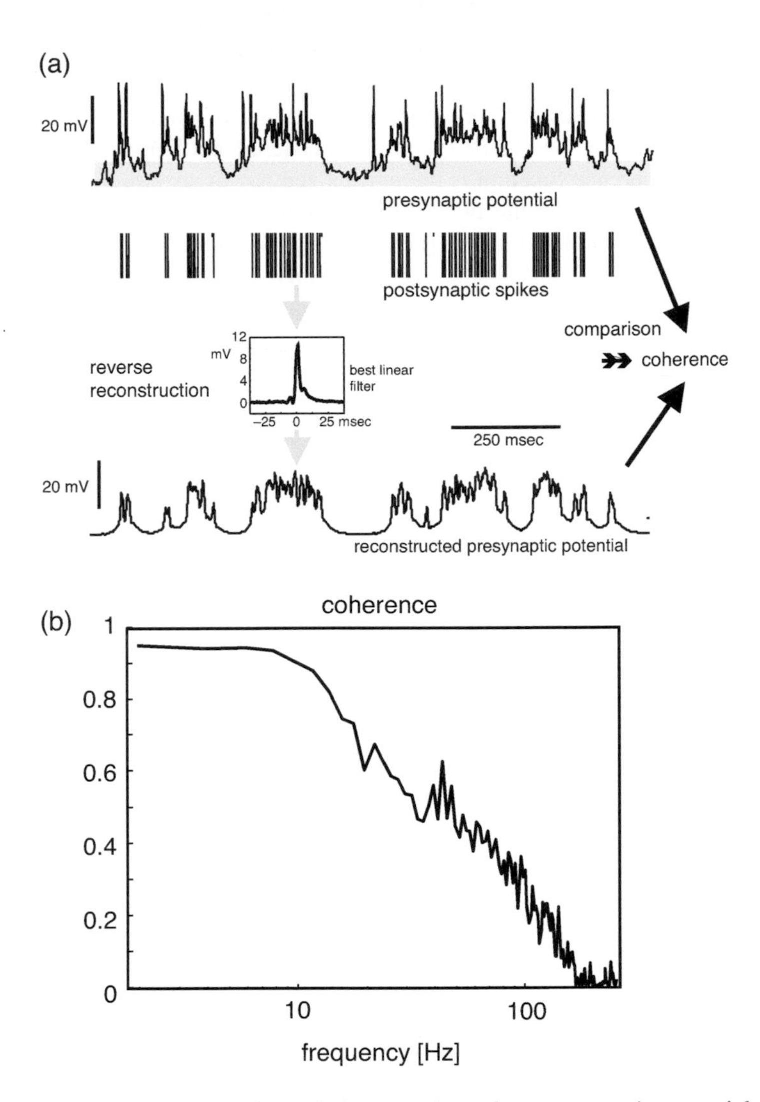

FIGURE 7.6 Reverse reconstruction of the time-dependent presynaptic potential from postsynaptic spike trains. **a**: Schematic outline of the procedure. The linear filter is determined that, when convolved with the postsynaptic spike train, leads to the best estimate of the presynaptic potential. Hyperpolarizations of the presynaptic neuron do not have much effect on the activity of the postsynaptic cell (due to the low postsynaptic resting activity). Therefore, the presynaptic potential was rectified at the resting potential of each response trace and was then used for the reconstruction. The part of the response that was rectified is marked by the shaded bar. The coherence function was determined as a measure of the similarity between the recorded and the estimated presynaptic membrane potential traces. **b**: Coherence determined for a cell pair analyzed with random velocity fluctuations. Coherence values close to 1 for frequencies up to 10 Hz indicate that the system can be regarded as very reliable and approximately linear in this frequency range. (Data from Warzecha, A.-K. et al., *Neurosci.* **119**:1103–1112, 2003.)

The similarity between the responses of the encoding model and the experimentally obtained responses can be determined on the basis of the so-called coherence function, which relates two time-dependent signals as a function of their frequency components (Bendat and Piersol, 2000; van Hateren and Snippe, 2001). In technical terms, the coherence function is the normalized power spectrum of the cross-correlation function between the two time-dependent signals. It varies between 0 (i.e., both signals are unrelated) and 1 (i.e., both signals can be linearly transformed into each other) (Figure 7.6b). There is one ambiguity with the coherence function, which is that a coherence value smaller than 1 can arise from two nonexclusive sources. First, it can be due to inherent noise in the neural pathway and, thus, arise from the variability in the neural responses. Second, the neural responses and the responses predicted on the basis of the encoding model are not linearly related. To account for the noise in encoding the stimulus, the so-called expected coherence can be assessed (Haag and Borst, 1998; van Hateren and Snippe, 2001). The expected coherence is determined by calculating the coherence function between the stimulus-induced response component (i.e., the ensemble average of a sufficiently large number of individual responses to the same input) and the individual responses. The expected coherence is related to the signal-to-noise ratio of the neuron (Haag and Borst, 1998; van Hateren and Snippe, 2001). The expected coherence represents the upper limit, given neuronal variability, of what can be maximally expected to be encoded even if a perfect encoding model were available.

The reliability of neural encoding has recently been addressed on the basis of another type of approach to assess the similarity of neural responses. In the context of blowfly motion vision, this approach has been applied to pinpointing the dominating noise source in the visual motion pathway. Here, the similarity of neural responses to a given stimulus is related to the similarity of response traces evoked by different stimuli. The similarity of spike trains can be determined in two ways. The first approach compares spike trains by calculating the minimal costs of transforming one spike train into another (Victor and Purpura, 1996). The transformation is done by either deleting, inserting, or temporally shifting single spikes. Each of these procedures is linked to defined costs (Figure 7.7a,b). By varying the costs for shifting a spike relative to inserting or deleting one, the temporal resolution of the procedure is changed. The second approach to determine similarity calculates the distance between two spike trains that are smoothed by temporal filtering. The distance is given by the square root of the squared differences between the smoothed spike trains (Kretzberg et al., 2001b). Changing the filter width leads to a varying temporal resolution of the procedure.

Both these approaches to determine the similarity of neural responses were used in a blowfly TC to test whether the timing of spikes elicited by visual motion is determined by photon noise, as was proposed in earlier studies (Borst and Haag, 2001; Lewen et al., 2001), or by noise inherent in the nervous system. Simulating photon noise by random brightness fluctuations of moving dots allowed us to show that the reliability of spike timing is dominated by noise intrinsic to the nervous system (Figure 7.7c) (Grewe et al., 2003).

This result is in accordance with findings of earlier studies based on dual recordings of TCs with largely overlapping receptive fields and a correlation analysis of their spike activity. Although the first spike after a rapid change in the direction

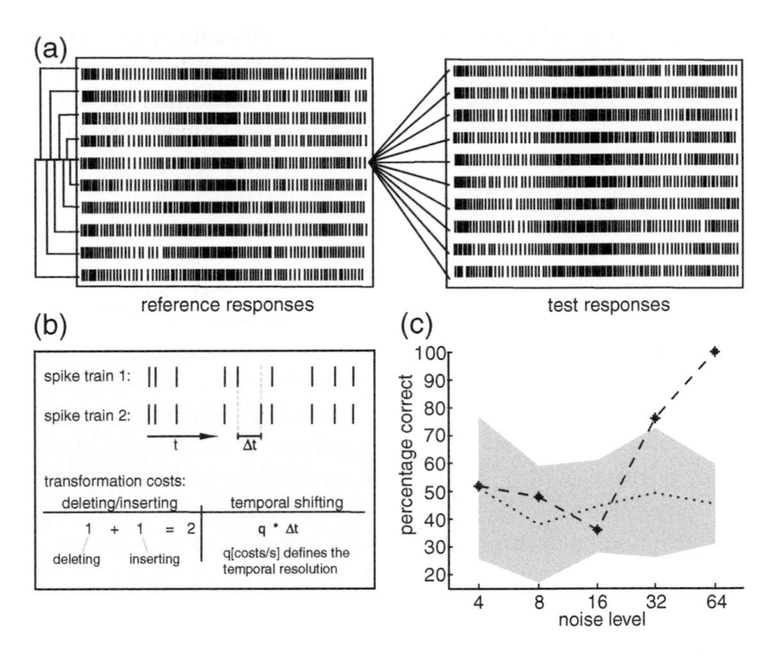

FIGURE 7.7 Similarity of spike trains as evoked by motion stimuli superimposed by variable noise. **a**: The raster plot shows sections out of the spike responses of a blowfly TC (the H1 cell) to a reference and a test stimulus. Both stimuli consisted of the same pattern of ten randomly positioned dots that were moved at a constant velocity in the cell's preferred direction. The dots were superimposed by random luminance fluctuations (noise). These were the same for each presentation of the reference stimulus, but differed, although they had the same statistical properties, for each presentation of the test stimulus (for details, see Grewe et al., 2003). A common spike pattern can be seen in all responses, despite differences in timing of individual spikes. The way of analyzing whether the temporals structures of the reference responses are more similar to each other than to the test responses is sketched by the lines connecting the highlighted reference response and all other reference and test responses. **b**: The similarity between two spike trains is defined as the inverse of the minimal costs of transforming one spike train into another one (Victor and Purpura, 1996). The transformation is done by either deleting, inserting, or temporally shifting single spikes. Deleting or inserting single spikes has the cost of 1. The cost of a temporal shift (*q* per second) is variable and determines the temporal resolution of the measure. For example, a *q* value of 200 equals a temporal resolution of 10 msec: for this time-scale, spikes in two response traces are considered nearly coincident if a given spike in one of the spike trains is shifted by less than ±10 msec with respect to the corresponding spike in the other spike train. As long as this condition is met, it is "cheaper" to adjust the spikes temporally in the two spike trains by shifting than it is by deleting or inserting one of them. **c**: The pairs of mean similarities of each reference response to all test responses, and of each reference response to all other reference responses are attributed to either the reference or the test stimulus. Assuming that the reference responses are more similar to each other than to the test responses, the larger similarity value was assigned to the reference response. The percentage of correct decisions is plotted as a function of added noise level. The shaded area represents the domain of uncertainty (see Grewe et al., 2003, for details). Discrimination performances falling into this range are likely to be a consequence of chance. A significant effect of the added noise on the responses can be assumed if the actual percentage-correct value is outside the domain of uncertainty. The added noise affects the responses, only at much higher levels than would be expected from photon noise. (From Grewe, J. et al., *J. Neurosci.* **23**:10776–10783, 2003. With permission.)

of motion may be precisely time-locked to the stimulus, the timing of most spikes, even during dynamical motion stimulation, is determined by the refractory period of the neuron and by random membrane potential fluctuations that are not directly related to the stimulus (Warzecha et al., 1998; Kretzberg et al., 2001a).

7.3.2 Coding of Motion Information by Neural Populations

The encoding of stimuli by neuronal populations is a basic operating principle of nervous systems. The number of neurons that constitute such a neuronal population may be very large. However, the population of TCs in the blowfly motion vision system comprises only a relatively small number of 50 to 60 cells in each half of the brain. Each of these cells is thought to encode different aspects of optic flow (see Section 7.2.1; reviews: Hausen and Egelhaaf, 1989; Krapp, 2000; Borst and Haag, 2002; Egelhaaf et al., 2002). The representation of sensory information by neuronal populations raises the question of the mechanism that eventually decodes this distributed information and uses it to guide behavior. In particular, it is important to assess how well relevant stimulus parameters are preserved in the population response and how specifically these parameters can be extracted by a readout mechanism. Although there are only a few experimental studies addressing this problem, the constraints that must be taken into account when interpreting neuronal population responses are analyzed theoretically or by model simulations in a wide variety of studies (reviews: Oram et al., 1998; Pouget et al., 1998, 2003;Dayan and Abbott, 2001).

Population coding is most frequently analyzed for a bank of sensory neurons encoding a single stimulus parameter. The optimal stimulus of different neurons varies along the stimulus axis. The tuning of an individual cell with respect to this stimulus parameter is usually described by a Gaussian function; that is, the response first increases with changing stimulus parameter, reaches an optimum, and then decreases again. Alternatively, for periodic stimuli axes (e.g., for orientation tuning), harmonic functions, such as cosine functions or truncated cosine functions, are employed to describe tuning curves.

Neural population coding has recently been studied in a subpopulation of blowfly TCs, the so-called VS cells. The ten VS cells in each half of the brain are thought to signal rotations of the head around different axes lying in the equatorial plane of the fly's eye (Krapp et al., 1998). However, VS cell responses are highly ambiguous because they are also activated during upward lift movements of the animal (Karmeier et al., 2003). Even if we assume VS cells to be perfect detectors for self-rotation, their responses are ambiguous due to their cosine-shaped tuning curves and the variability of their responses (Figure 7.8a,d). Assuming a single stimulus dimension (i.e., the orientation of the axis of self-rotation), encoding accuracy of the population of VS cells was determined (Figure 7.8b,f). For each stimulus condition (i.e., for each rotation axis), the activity distribution of all ten VS cells is determined. Given these response probabilities, the likelihood of a particular rotation axis given a certain response — and, thus, the most likely rotation axis — can be estimated using Bayes's theorem (Dayan and Abbott, 2001). The encoding error is determined from the estimated and the real rotation axis (Figure 7.8c,e,f).

This approach has been applied to VS cells for different rotational velocities and different time windows after the onset of motion. Despite the considerable neural noise, the rotation axis can be estimated within 25 msec with an accuracy of less than 2°. This result is surprising, given the ambiguous responses of individual sensory neurons (Figure 7.8a,b), and it thus stresses the significance of population coding. The decoding performance of VS cells does not improve for larger time windows but deteriorates considerably if only three instead of the ten VS cells are taken into account (Figure 7.8d,e). The finding that a very small time window is sufficient for decoding the orientation axis from the population response has important functional implications for flies, because they perform rapid acrobatic flight maneuvers, which require fast and accurate sensory control signals for visual flight and gaze stabilization (see Chapter 4).

Theoretical studies demonstrate that the accuracy of stimulus encoding depends on a variety of other aspects, such as noise correlation between single elements of the population, the width of the tuning curves, and the number of stimulus dimensions to be encoded (e.g., Abbott and Dayan, 1999; Pouget et al., 1999, 2003; Zhang and Sejnowski, 1999; Wilke and Eurich, 2001). These features must be taken into account although it is far from trivial to obtain the large amount of experimental data required to understand fully the encoding of sensory information by neuronal populations. All these analyses are only first steps toward understanding the encoding of complex natural motion stimuli as experienced by the animals in normal behavioral situations (see next section). This will require the development of novel theoretical approaches to analyze neuronal population data (see also Chapter 5).

7.4 APPROACHES TO INVESTIGATE THE ENCODING OF NATURAL VISUAL STIMULI

Information processing within a neural circuit is traditionally analyzed with stimuli that are much simpler with respect to their spatial and dynamical features than the input an animal encounters in behavioral situations. Because visual systems evolved in specific environments and behavioral contexts, the functional significance of the information being processed can only be assessed by analyzing neuronal performance under conditions that come close to natural situations.

One important aspect applies to the dynamical properties of the optic flow that is encountered during behavior. These are determined by the dynamics of the animal's self-motion and the three-dimensional layout of the environment. The characteristics of the optic flow may differ greatly for different species and in different behavioral situations. Some insects, such as hoverflies, dragonflies, and hawkmoths, are able to hover in midair in front of a flower (Collett and Land, 1975; Farina et al., 1994; Kern and Varjú, 1998). From their current position in space, these insects can accelerate rapidly and dart off at high velocities. Blowflies usually change the direction of self-motion rapidly by saccadic turns during flight (van Hateren and Schilstra, 1999) or, one order of magnitude more slowly, while walking (Horn and

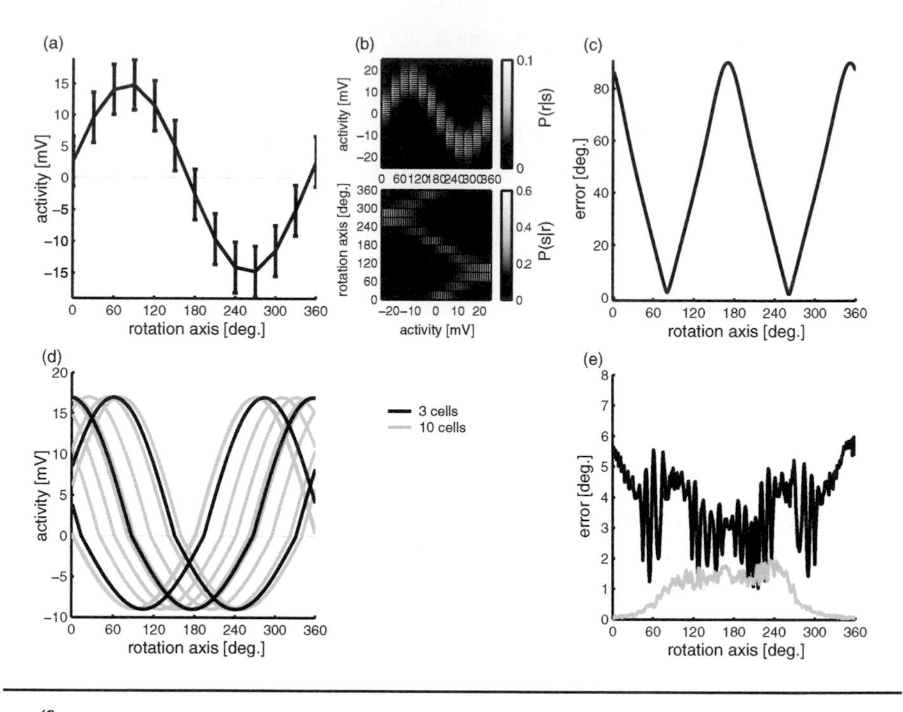

(f)

$\vec{r}$	response vector describing the population response
$p(\vec{r}\vert s_X)$	*can be measured*: probability distribution of the population response to the stimulus s_X
$p(s_X\vert\vec{r})$	*to be determined* probability of the stimulus s_X given the response vector $\vec{r}$

Applying Bayes Theorem allows determination of $p(s_X|\vec{r})$ from measurable data:

$$p(s_X|\vec{r}) = \frac{p(\vec{r}|s_X)\, p(s_X)}{p(\vec{r})}$$

Estimated stimulus for a given population responser $\quad s_{est} = \sum_{s_X} p(s_X|\vec{r})\, s_X \quad$ Error $\sqrt{(s - s_{est})^2}$

FIGURE 7.8 Accuracy of population coding illustrated for a class of TCs, the ten VS cells. **a**: The response of each VS cell depends on the orientation of an horizontally aligned axis about which a panoramic wide-field pattern is rotated. The tuning can be approximated by a distorted cosine function. **b**: The rotation axis can be estimated from the measured response distributions obtained for the different rotation axes $p(r|s_i)$ (compare Figure 7.5). By applying Bayes's theorem (see **f**), the probability distribution of rotation axes can be determined for a given response $p(s_i|r)$. **c**: The coding error is given by the standard deviation between the estimated orientation axis and the real orientation axis (for formal details, see **f**). If the response of only one VS cell is taken into account, the coding error varies strongly with the rotation axis and may assume very large values. **d**: If the responses of more VS cells are taken into account, the coding error decreases to very small values, even if only a 5 msec response interval is used for decoding. Tuning curves of all ten VS cells (gray lines) and a subgroup of three VS cells (black lines). **e**: Corresponding coding errors. **f**: Formal explanation of the decoding procedure.

Mittag, 1980; Kern et al., 2001a). Because the optic flow pattern and its dynamics are species and context specific, it is reasonable to assume that the mechanisms extracting motion information are adapted to behaviorally relevant conditions. Because it is hard to record from visual interneurons in freely moving insects, more indirect approaches have recently been used to determine the responses of motion-sensitive neurons to a variety of approximations to behaviorally generated optic flow.

Recordings of the H1 cell have been made from the brains of blowflies that were oscillated with a turntable with dynamics that mimic the rotational component of flight trajectories of unrestrained small houseflies (Lewen et al., 2001). For technical reasons, the most distinctive feature of optic flow of free-flying blowflies (i.e., the succession of saccadic turns and stable gaze) (van Hateren and Schilstra, 1999) was not taken into account and the angular velocity of the turntable was only half that of the flight trajectory.

In an alternative approach, neural responses were recorded from the brain of tethered moths flying in a flight simulator in which the animal can influence its visual input similarly to free-flight conditions (Gray et al., 2002). So far, this elegant approach is restricted to relatively large insect species, such as locusts or moths, changing their direction of flight relatively slowly.

In another approach, the optic flow experienced by flying blowflies was reconstructed and replayed to the animal during nerve cell recordings. This approach has been employed for various behavioral situations during tethered flight in a flight simulator (Warzecha and Egelhaaf, 1996, 1997; Kimmerle and Egelhaaf, 2000b), during unrestrained walking in a three-dimensional environment (Kern et al., 2000, 2001a), and recently during rapid free-flight maneuvers in a three-dimensional environment (Figure 7.9a) (Kern et al., 2001a, 2004; Lindemann et al., 2003).

The simulation of free flight has become possible thanks to the development of sophisticated techniques. First, free-flight behavior can be monitored by means of magnetic sensor coils mounted on the head and thorax of the animal (Schilstra and van Hateren, 1999; van Hateren and Schilstra, 1999) or by high-speed digital cameras (Oddos et al., 2003). Second, a panoramic visual stimulator (FliMax) for presentation of optic flow has been designed that is sufficiently fast for visual stimuli as experienced by free-flying insects (Figure 7.9a). FliMax is a special-purpose panoramic VGA output device generating image frames at a frequency of 370 Hz (Lindemann et al., 2003). FliMax is composed of printed circuit boards shaped like equilateral triangles (side: 30 cm), assembled to form 14 of the 20 sides of an icosahedron (radius of inscribed sphere: 22.4 cm). Each of these boards supports 512 regularly spaced, round light-emitting diodes, or LEDs (#WU-2-53GD, 5 mm, emitting wavelength: 567 nm, effective viewing angle: 25°)[33]. All 7168 LEDs are controlled individually via a computer equipped with a standard VGA graphics card and customized software. The luminance of each LED, adjustable to eight intensity levels, is kept constant between updates by sample-and-hold circuits. FliMax is open toward the back to mount an animal in its center and to make recordings from the blowfly's brain (Lindemann et al., 2003).

Although the analysis of how natural visual stimuli are processed has started only recently, it is already safe to conclude that the neuronal responses to complex optic flow as experienced under behavioral conditions can be understood only partly

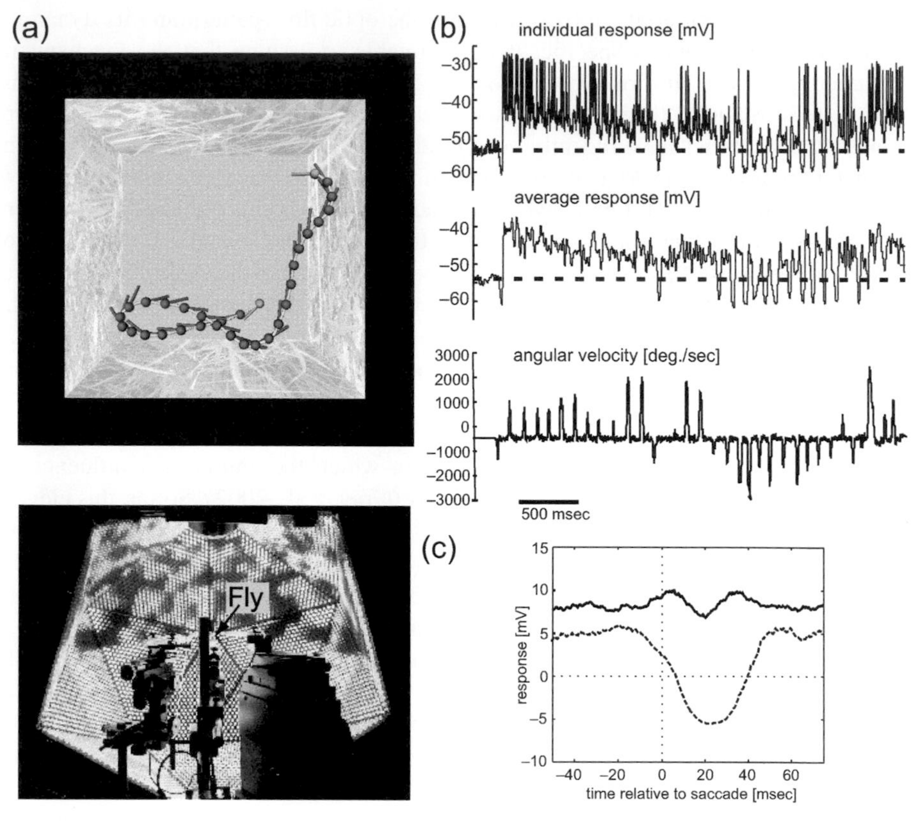

FIGURE 7.9 Responses of the HSE cell of a blowfly to behaviorally generated optic flow as experienced during a free-flight maneuver. **a**: Upper diagram: Flight trajectory as seen from above monitored in a cubic cage ($0.4 \times 0.4 \times 0.4$ m^3) covered on its side walls with images of herbage (for clarity shown only at low contrast). The position of the fly is displayed by small spheres every 10 msec. The position and orientation of the head are shown every 130 msec. Bottom diagram: Photograph of FliMax from behind. In the foreground, the micromanipulators can be seen by which recording electrodes are inserted into the fly's brain. **b**: Responses of an HSE cell in the right half of the visual system to behaviorally generated stimuli. Top trace: Individual response; cell responds to motion with graded de- and hyperpolarizations; spikelike depolarizations superpose the graded potential changes. Second trace: Average response (n = 7). Third trace: Angular velocity of the fly's head. Sharp angular velocity peaks corresponding to saccade-like turns of the fly dominate the time-dependent angular velocity profile. Positive (negative) values denote counterclockwise (clockwise) turns of the head in a head-centered coordinate system. In contrast to expectations based on the input organization of the HSE cell (Figure 7.1a), there are no obvious response peaks during preferred direction motion evoked by counterclockwise saccades. However, there are pronounced hyperpolarizations going along with clockwise saccades. Dotted horizontal lines indicate resting potential. **c**: Saccade-triggered average of the HSE responses. Counterclockwise saccades (solid line) go along with image motion in the HSE cell's preferred direction; clockwise saccades (dotted line) go along with image motion in the null direction. Zero time corresponds to the maximum angular velocity. The resting potential was subtracted before averaging. (Flight trajectory provided by JH van Hateren; Experimental data from Lindemann, J.P. et al., *Vision Res.* **43**:779–791, 2003. With permission.)

in terms of the concepts that were established on the basis of experiments done with conventional motion stimuli. This is illustrated in the following example.

In moving animals, retinal image flow is distinguished from conventional (experimenter-designed) visual stimuli by its characteristic dynamics, which are largely determined by the animal's own actions and reactions. During spontaneous flight, blowflies execute a series of saccadic turns where the head shows angular velocity peaks of up to several thousand degrees per second. Between saccades, the gaze is kept basically stable (Schilstra and van Hateren, 1999; van Hateren and Schilstra, 1999). The resulting retinal image flow was reconstructed and replayed to blowflies. The resulting complex time-dependent responses were related by reverse reconstruction approaches (see Section 7.3.1) to various self-motion parameters, such as yaw rotation, forward translation, and sideslip. How well the original self-motion parameters can be estimated from the neuronal responses is quantified by the coherence (see Section 7.3.1). Our results, obtained with natural optic flow on the so-called HSE neuron, do not match conclusions based on systems analysis of the neuron's properties with experimenter-designed stimuli (Kern et al., 2004b):

1. It was previously concluded that the neuron should mainly act as a detector of self-rotation of the animal around its vertical axis (see Figure 7.1). In contrast, our results with behaviorally generated optic flow show that the neuron fails to encode faithfully even the most prominent turns of the animal as found during saccades.
2. Although the cell experiences the largest optic flow during saccades, it may encode behaviorally relevant information, especially between saccades. Between saccades blowflies keep their gaze stable apart from small, broadband yaw rotations, so they may gather useful information about the outside world from the translational optic flow components that dominate at low frequencies in intersaccadic intervals. Indeed, between saccades, neural signals provide rich information about the spatial relation of the animal to its surroundings. It should be noted that distance is signaled only relative to the fly's own velocity, because retinal velocities evoked during translation are inversely proportional to distance and proportional to translation velocity. This implies that in walking flies, the visual surroundings should affect the responses of the HSE cell only when the fly is very close to environmental structures, just as has been found in electrophysiological experiments (Kern et al., 2001a, 2001b). This implicit scaling of distance information by the actual speed of the animal may be a parsimonious and advantageous way to extract from optic flow behaviorally relevant information about the outlay of the environment because, for instance, evasive actions evoked by obstacles in the path of locomotion need to be evoked at a smaller distance only when the animal moves more slowly.
3. Based on experimenter-designed motion stimuli, motion-sensitive neurons are frequently expected to encode stimulus velocity. Indeed, stimulus velocity can be reconstructed faithfully from the responses of blowfly motion-sensitive neurons, as long as the velocities and velocity changes are relatively small (Egelhaaf and Reichardt, 1987; Bialek et al., 1991;

> Haag and Borst, 1997). However, we could show that under behaviorally relevant stimulus conditions, the visual motion system of the blowfly operates far beyond this linear range for a considerable portion of time (Figure 7.9b,c). We concluded that, in contrast to previous views, the nonlinearities of the visual motion system may be essential for the HSE cell to encode information about the spatial relation of the animal to its environment. If the neuron encoded linearly the entire velocity range the system encounters in behavior, by far the largest responses would be generated during body saccades. This would leave only a very small response range for encoding optic flow in the intersaccadic interval. Given the noisiness of neuronal signals (see Section 7.2.1), it might be difficult to extract meaningful information about the spatial organization of the surroundings from the neuronal responses. Because angular velocities during saccades are far beyond the linear range of the motion detection system, the HSE cell appears to be able to encode useful information about translation and thus about the spatial relation of the animal to the outside world. This finding emphasizes that the significance of neuronal circuits can only be assessed if they are probed in the natural operating range.

So far, only the specificity of individual neurons has been analyzed in encoding of behaviorally relevant features from the natural optic flow. With the available techniques, the next step will be to understand how this specificity is increased by taking into account more of the relevant elements of the TC population. This analysis, however, will require the development of novel conceptual and technical approaches to handle the resulting complex time-dependent population data.

7.5 CONCLUSIONS AND OUTLOOK

Thanks to great methodological and conceptual developments, many aspects of the neural computations underlying visually guided orientation behavior in the blowfly have been elucidated in recent years. Analysis has ranged from the biophysical properties of neurons and their synaptic interactions, to the performance and reliability of neural populations in encoding behaviorally relevant motion sequences, to orientation behavior in flight simulators and under free-flight conditions. Accordingly, the employed methodological repertoire is very broad. On the one hand, sophisticated techniques of cellular physiology had to be adapted to *in vivo* conditions — to mention only the imaging of the spatially resolved intracellular activity patterns and, in particular, the time-dependent distribution of ions involved in intracellular information processing. Likewise, thanks to novel compounds such as photolabile calcium cages, targeted manipulation of intracellular ion concentrations could be performed. On the other hand, developments, for instance, in computer graphics, have recently allowed for the first time the presentation of natural motion stimuli in electrophysiological experiments, as they were encountered by freely behaving animals. The complex spatio-temporal properties of natural image sequences and the resulting complex time-dependent neuronal responses have made it necessary to establish and adapt sophisticated techniques for analysis of such complex data structures. Moreover, taking into account the variability of responses of individual neurons and the activity of whole populations of them has made another set of novel approaches to data analysis necessary and neurobiology an increasingly interdisciplinary effort.

Insect model systems, such as the visual motion pathway of the blowfly, are particularly advantageous for this multifaceted interdisciplinary effort. Experimental analysis down to the biophysical level can be done *in vivo,* where the system can be probed with behaviorally generated input and, thus, under its natural operating conditions. Because populations of neurons are relatively small compared with those of mammals, it appears to be possible to boil down visually guided behavior to the computational properties of neural networks of identified neurons. Despite the small size of insects' brains, these computations are far from trivial, given the fact that many insects, such as blowflies, are able to perform extraordinary things — at least when compared with autonomous manmade machines. Hence, disclosing computational design principles of insect brains down to the level of neurons and neural networks is not only one of the most fascinating missions of basic science, but also a strategic goal if robots with autonomous behavior are to be realized (see Chapter 8). To test the viability of biologically established computational principles, but also to translate these principles into a language that can be implemented in artificial systems, modeling is an indispensable tool. Indeed, in all laboratories investigating blowfly vision, modeling was intensively employed in parallel with experimental analysis at all levels, ranging from computations of single cells to overall behavioral performance (reviews: Borst, 2002; Egelhaaf et al., 2002).

Finally, it should be mentioned that because of the efficiency of visually guided orientation behavior in insects, there is great interest in applying principles of insect motion information processing to autonomous artificial systems (see Chapter 8). Although this has been successful for some behavioral components (Srinivasan et al., 1997, 2001; Huber et al., 1999; Harrison and Koch, 2000; Rind, 2002), biomorphic autonomous robots still appear to be dull compared with the original after which they are modeled. In contrast to man-made systems, natural vision systems have been tested and improved on a much longer timescale by many millions of years of evolution.

ACKNOWLEDGMENTS

The work in the authors' laboratory is supported by the Deutsche Forschungsgemeinschaft (DFG).

REFERENCES

Abbott LF, Dayan P (1999) The effect of correlated variability on the accuracy of a population code. *Neur. Comp.* **11**:91–101.

Barberini CL, Horwitz GD, Newsome WT (2000) A comparison of spiking statistics in motion sensing neurons of flies and monkeys. In: *Computational, Neural and Ecological Constraints of Visual Motion Processing*. Zanker JM, Zeil J, Eds., Springer, Heidelberg, pp. 307–320.

Bendat JS, Piersol AG (2000) *Random Data: Analysis and Measurement Procedures.* Wiley-Interscience, New York.

Bialek W, Rieke F, de Ruyter van Steveninck R, Warland D (1991) Reading a neural code. *Science* **252**:1854–1857.

Boeddeker N, Egelhaaf M (2003a) Steering a model fly: simulations on visual pursuit in blowflies. *Proc. R. Soc. Lond. B* **270**:1971–1978.

Boeddeker N, Egelhaaf M (2003b) Chasing behaviour of blowflies: a smooth pursuit system generates saccades. (submitted)

Boeddeker N, Kern R, Egelhaaf M (2003) Chasing a dummy target: smooth pursuit and velocity control in male blowflies. *Proc. Roy. Soc. Lond. B* **270**:393–399.

Borst A, Haag J (2001) Effects of mean firing on neural information rate. *J. Comput. Neurosci.* **10**:213–221.

Borst A, Haag J (2002) Neural networks in the cockpit of the fly. *J. Comp. Physiol. A* **188**:419–437.

Borst A, Theunissen FE (1999) Information theory and neural coding. *Natu. Neurosci.* **2**:947–957.

Collett TS (1971) Visual neurones for tracking moving targets. *Nature* **232**:127–130.

Collett TS, Land MF (1975) Visual control of flight behaviour in the hoverfly *Syritta pipiens* L. *J. Comp. Physiol.* **99**:1–66.

Dayan P, Abbott LF (2001) *Theoretical Neuroscience: Computational and Mathematical Modeling of Neural Systems.* MIT Press, Cambridge, MA.

Egelhaaf M (1985a) On the neuronal basis of figure–ground discrimination by relative motion in the visual system of the fly. II. Figure-detection cells, a new class of visual interneurones. *Biol. Cybern.* **52**:195–209.

Egelhaaf M (1985b) On the neuronal basis of figure–ground discrimination by relative motion in the visual system of the fly. III. Possible input circuitries and behavioural significance of the FD-cells. *Biol. Cybern.* **52**:267–280.

Egelhaaf M, Kern R (2002) Vision in flying insects. *Curr. Opin. Neurobiol.* **12**:699–706.

Egelhaaf M, Reichardt W (1987) Dynamic response properties of movement detectors: theoretical analysis and electrophysiological investigation in the visual system of the fly. *Biol. Cybern.* **56**:69–87.

Egelhaaf M, Warzecha A-K (1999) Encoding of motion in real time by the fly visual system. *Curr. Opin. Neurobiol.* **9**:454–460.

Egelhaaf M, Borst A, Warzecha A-K, Flecks S, Wildemann A (1993) Neural circuit tuning fly visual interneurons to motion of small objects. II. Input organization of inhibitory circuit elements by electrophysiological and optical recording techniques. *J. Neurophysiol.* **69**:340–351.

Egelhaaf M, Kern R, Kurtz R, Krapp HG, Kretzberg J, Warzecha A-K (2002) Neural encoding of behaviourally relevant motion information in the fly. *Trends Neurosci.* **25**:96–102.

Esch HE, Burns JM (1996) Distance estimation by foraging honeybees. *J. Exp. Biol.* **199**:155–162.

Esch HE, Zhang S, Srinivasan MV, Tautz J (2001) Honeybee dances communicate distances measured by optic flow. *Nature* **411**:581–583.

Farina WM, Varjú D, Zhou Y (1994) The regulation of distance to dummy flowers during hovering flight in the hawk moth *Macroglossum stellatarum*. *J. Comp. Physiol.* **174**:239–247.

Farrow K, Haag J, Borst A (2003) Input organization of multifunctional motion-sensitive neurons in the blowfly. *J. Neurosci.* **29**:9805–9811.

Fortune ES, Rose GJ (2001) Short-term synaptic plasticity as a temporal filter. *Trends Neurosci.* **24:**381–385.

Gauck V, Borst A (1999) Spatial response properties of contralateral inhibited lobula plate tangential cells in the fly visual system. *J. Comp. Neurol.* **406**:51–71.

Gilbert C, Strausfeld NJ (1991) The functional organization of male-specific visual neurons in flies. *J. Comp. Physiol. A* **169**:395–411.

Gray JR, Pawlowski VM, Willis MA (2002) A method for recording behavior and multineuronal CNS activity from tethered insects flying in virtual space. *J. Neurosci. Meth.* **120**:211–223.

Grewe J, Kretzberg J, Warzecha A-K, Egelhaaf M (2003) Impact of photon-noise on the reliability of a motion-sensitive neuron in the fly's visual system. *J. Neurosci.* **23**:10776–10783.

Haag J, Borst A (1997) Encoding of visual motion information and reliability in spiking and graded potential neurons. *J. Neurosci.* **17**:4809–4819.

Haag J, Borst A (1998) Active membrane properties and signal encoding in graded potential neurons. *J. Neurosci.* **18**:7972–7986.

Haag J, Borst A (2001) Recurrent network interactions underlying flow-field selectivity of visual interneurons. *J. Neurosci.* **21**:5685–5692.

Haag J, Borst A (2002) Dendro-dendritic interactions between motion-sensitive large-field neurons in the fly. *J. Neurosci.* **22**:3227–3233.

Haag J, Borst A (2003) Orientation tuning of motion-sensitive neurons shaped by vertical-horizontal network interactions. *J. Comp. Physiol. A* **189**:363–370.

Harrison RR, Koch C (2000) A silicon implementation of the fly's optomotor control system. *Neu. Comput.* **12**:2291–2304.

Hateren JH van, Schilstra C (1999) Blowfly flight and optic flow. II. Head movements during flight. *J. Exp. Biol.* **202**:1491–1500.

Hateren JH van, Snippe HP (2001) Information theoretical evaluation of parametric models of gain control in blowfly photoreceptor cells. *Vision Res.* **41**:1851–1865.

Hausen K (1981) Monocular and binocular computation of motion in the lobula plate of the fly. *Verh. Dtsch. Zool. Ges.* **74**:49–70.

Hausen K, Egelhaaf M (1989) Neural mechanisms of visual course control in insects. In: *Facets of Vision.* Stavenga D, Hardie RC, Eds., Springer, Berlin, pp. 391–424.

Hengstenberg R (1982) Common visual response properties of giant vertical cells in the lobula plate of the blowfly *Calliphora. J. Comp. Physiol.* **149**:179–193.

Horn E, Mittag J (1980) Body movements and retinal pattern displacements while approaching a stationary object in the walking fly, *Calliphora erythrocephala. Biol. Cybern.* **39**:67–77.

Horstmann W, Egelhaaf M, Warzecha A-K (2000) Synaptic interactions increase optic flow specificity. *Eur. J. Neurosci.* **12**:2157–2165.

Huber SA, Franz MO, Bülthoff HH (1999) On robots and flies: modeling the visual orientation behavior of flies. *Robotics Autono. Sys.* **29**:227–242.

Juusola M, French AS, Uusitalo RO, Weckström M (1996) Information processing by graded-potential transmission through tonically active synapses. *Trends Neurosci.* **19**:292–297.

Kalb J, Nielsen T, Fricke N, Egelhaaf M, Kurtz R (2004) *In vivo* two-photon laser-scanning microscopy of Ca^{2+} dynamics in visual motion-sensitive neurons. *Biochem. Biophys. Res. Communic.* **316**:341–347.

Karmeier K, Egelhaaf M, Krapp HG (2001) Early visual experience and receptive field organization of the optic flow processing interneurons in the fly motion pathway. *Vis. Neurosci.* **18**:1–8.

Karmeier K, Krapp HG, Egelhaaf M (2003) Robustness of the tuning of fly visual interneurons to rotatory optic flow. *J. Neurophysiol.* **90**:1626–1634.

Kern R, Varjú D (1998) Visual position stabilization in the hummingbird hawk moth, *Macroglossum stellatarum* L.: I. Behavioural analysis. *J. Comp. Physiol. A* **182**:225–237.

Kern R, Lutterklas M, Egelhaaf M (2000) Neural representation of optic flow experienced by unilaterally blinded flies on their mean walking trajectories. *J. Comp. Physiol. A* **186**:467–479.

Kern R, Petereit C, Egelhaaf M (2001a) Neural processing of naturalistic optic flow. *J. Neurosci.* **21**:1–5.

Kern R, Lutterklas M, Petereit C, Lindemann JP, Egelhaaf M (2001b) Neuronal processing of behaviourally generated optic flow: experiments and model simulations. *Network: Comp. Neur. Sys.* **12**:351–369.

Kern R, van Hateren JH, Michaelis C, Lindemann JP, Egelhaaf M (submitted) Eye movements during natural flight shape the function of a blowfly motion sensitive neuron.

Kimmerle B, Egelhaaf M (2000a) Detection of object motion by a fly neuron during simulated translatory flight. *J. Comp. Physiol. A* **186**:21–31.

Kimmerle B, Egelhaaf M (2000b) Performance of fly visual interneurons during object fixation. *J. Neurosci.* **20**:6256–6266.

Koenderink JJ (1986) Optic flow. *Vis. Res.* **26**:161–180.

Krapp HG (2000) Neuronal matched filters for optic flow processing in flying insects. In: *Neuronal Processing of Optic Flow.* Lappe M, Ed., Academic Press, San Diego, CA, pp. 93–120.

Krapp HG, Hengstenberg R (1996) Estimation of self-motion by optic flow processing in single visual interneurons. *Nature* **384**:463–466.

Krapp HG, Hengstenberg B, Hengstenberg R (1998) Dendritic structure and receptive-field organization of optic flow processing interneurons in the fly. *J. Neurophysiol.* **79**:1902–1917.

Krapp HG, Hengstenberg R, Egelhaaf M (2001) Binocular contribution to optic flow processing in the fly visual system. *J. Neurophysiol.* **85**:724–734.

Kretzberg J, Egelhaaf M, Warzecha A-K (2001a) Membrane potential fluctuations determine the precision of spike timing and synchronous activity: a model study. *J. Comput. Neurosci.* **10**:79–97.

Kretzberg J, Warzecha A-K, Egelhaaf M (2001b) Neural coding with graded membrane potential changes and spikes. *J. Comput. Neurosci.* **11**:153–164.

Kurtz R (2004) Ca^{2+} clearance in visual motion-sensitive neurons of the fly studied *in vivo* by sensory stimulation and UV photolysis of caged Ca^{2+}. *J. Neurophysiol.* **92**:458–467.

Kurtz R, Egelhaaf M (2003) Natural patterns of neural activity. *Molecular Neurobiol.* **27**:1–19.

Kurtz R, Warzecha A-K, Egelhaaf M (2001) Transfer of visual information via graded synapses operates linearly in the natural activity range. *J. Neurosci.* **21**:6957–6966.

Land MF, Collett TS (1974) Chasing behaviour of houseflies (*Fannia canicularis*): a description and analysis. *J. Comp. Physiol.* **89**:331–357.

Lappe M (2000) *Neuronal Processing of Optic Flow.* Academic Press, San Diego, CA.

Lewen GD, Bialek W, de Ruyter van Steveninck R (2001) Neural coding of naturalistic stimuli. *Network: Comput. Neural Syst.* **12**:317–329.

Lindemann JP, Kern R, Michaelis C, Meyer P, van Hateren JH, Egelhaaf M (2003) FliMax, a novel stimulus device for panoramic and high-speed presentation of behaviourally generated optic flow. *Vision Res.* **43**:779–791.

Marmarelis PZ, Marmarelis VZ (1978) *Analysis of Physiological Systems.* Plenum Press, New York.

Miller JP, Selverston AI (1979) Rapid killing of single neurons by irradiation of intracellularly injected dye. *Science* **206**:702–704.

Neher E (1998) Vesicle pools and Ca^{2+} microdomains: new tools for understanding their roles in neurotransmitter release. *Neuron* **20**:389–399.

Oddos F, Kern R, Boeddeker N, Egelhaaf M (2003) Flight performance modified by environmental changes in the blowfly Lucilia. In: *Göttingen Neurobiology Report.* Elsner N, Ed., Georg Thieme, Stuttgart, p. 444.

Olberg RM (1981) Object- and self-movement detectors in the ventral nerve cord of the dragonfly. *J. Comp. Physiol.* **141**:327–334.

Olberg RM (1986) Identified target-selective visual interneurons descending from the dragonfly brain. *J. Comp. Physiol.* **159**:827–840.

Olberg RM, Pinter RB (1990) The effect of mean luminance on the size selectivity of identified target interneurons in the dragonfly. *J. Comp. Physiol. A* **166**:851–856.

Oram MW, Földiák P, Perrett DI, Sengpiel F (1998) The "ideal homunculus": decoding neural population signals. *Trends Neurosci.* **21**:259–265.

Paulsen O, Sejnowski TJ (2000) Natural patterns of activity and long-term synaptic plasticity. *Curr. Opin. Neurobiol.* **10**:172–179.

Pouget A, Dayan P, Zemel RS (2003) Inference and computation with population codes. *Ann. Rev. Neurosci.* **26**:381–410.

Pouget A, Deneve S, Ducom J-C, Latham PE (1999) Narrow versus wide tuning curves: what's best for a population code? *Neural Comput.* **11**:85–90.

Pouget A, Zhang K, Deneve S, Latham PE (1998) Statistically efficient estimation using population coding. *Neural Comput.* **10**:373–401.

Rieke F, Warland D, de Ruyter van Steveninck R, Bialek W (1997) *Spikes: Exploring the Neural Code.* MIT Press, Cambridge, MA.

Rind FC (2002) Motion detectors in the locust visual system: from biology to robot sensors. *Microsc. Res. Tech.* **56**:256–269.

Rind FC, Simmons PJ (1999) Seeing what is coming: building collision-sensitive neurones. *Trends Neurosci.* **22**:215–220.

Ruyter van Steveninck R de, Lewen GD, Strong SP, Koberle R, Bialek W (1997) Reproducibility and variability in neural spike trains. *Science* **275**:1805–1808.

Sabatini BL, Regehr WG (1999) Timing of synaptic transmission. *Annu. Rev. Physiol.* **61**:521–542.

Schilstra C, van Hateren JH (1999) Blowfly flight and optic flow. I. Thorax kinematics and flight dynamics. *J. Exp. Biol.* **202**:1481–1490.

Simmons, PJ (2002) Signal processing in a simple visual system: the locust ocellar system and its synapses. *Microsc. Res. Tech.* **56**:270–280.

Srinivasan MV, Chahl JS, Zhang SW (1997) Embodying natural vision into machines. In: *From Living Eyes to Seeing Machines.* Srinivasan MV, Venkatesh S, Eds., Oxford University Press, Oxford, pp. 249–265.

Srinivasan MV, Zhang S, Altwein M, Tautz J (2000) Honeybee navigation: nature and calibration of the "odometer." *Science* **287**:851–853.

Srinivasan MV, Zhang S, Chahl JS (2001) Landing strategies in honeybees, and possible applications to autonomous airborne vehicles. *Biol. Bull.* **200**:216–221.

Thomson AM (2000) Facilitation, augmentation and potentiation at central synapses. *Trends Neurosci.* **23**:305–312.

Victor JD, Purpura K (1996) Nature and precision of temporal coding in visual cortex: a metric-space analysis. *J. Neurophysiol.* **76**:1310–1326.

Wagner H (1986) Flight performance and visual control of the flight of the free-flying housefly (*Musca domestica*). II. Pursuit of targets. *Phil. Trans. Roy. Soc. Lond. B* **312**:553–579.

Warzecha A-K, Egelhaaf M (1996) Intrinsic properties of biological motion detectors prevent the optomotor control system from getting unstable. *Phil. Trans. Roy. Soc. Lond. B* **351**:1579–1591.

Warzecha A-K, Egelhaaf M (1997) How reliably does a neuron in the visual motion pathway of the fly encode behaviourally relevant information? *Europ. J. Neurosci.* **9**:1365–1374.

Warzecha A-K, Egelhaaf M (1999) Variability in spike trains during constant and dynamic stimulation. *Science* **283**:1927–1930.

Warzecha A-K, Egelhaaf M, Borst A (1993) Neural circuit tuning fly visual interneurons to motion of small objects. I. Dissection of the circuit by pharmacological and photoinactivation techniques. *J. Neurophysiol.* **69**:329–339.

Warzecha A-K, Kretzberg J, Egelhaaf M (1998) Temporal precision of the encoding of motion information by visual interneurons. *Curr. Biol.* **8**:359–368.

Warzecha A-K, Kretzberg J, Egelhaaf M (2000) Reliability of a fly motion-sensitive neuron depends on stimulus parameters. *J. Neurosci.* **20**:8886–8896.

Warzecha A-K, Kurtz R, Egelhaaf M (2003) Synaptic transfer of dynamical motion information between identified neurons in the visual system of the blowfly. *Neurosci.* **119**:1103–1112.

Wehner R (1981) Spatial vision in arthropods. In: *Handbook of Sensory Physiology. Vol. VII/6C: Comparative Physiology and Evolution of Vision in Invertebrates.* Autrum H, Ed., Springer, Berlin, pp. 287–616.

Wilke SD, Eurich CW (2001) Representational accuracy of stochastic neural populations. *Neural Comput.* **14**:155–189.

Zhang K, Sejnowski TJ (1999) Neuronal tuning: to sharpen or broaden? *Neural Comput.* **11**:75–84.

8 Bioinspired Sensors: From Insect Eyes to Robot Vision

F. Claire Rind

CONTENTS

8.1 INTRODUCTION

An underlying theme in biology is that evolution favors survival of the fittest. Many insects rely on vision for predator evasion, prey capture, and mate selection, so their

0-8493-2024-0/05/$0.00+$1.50

visual systems are under strong selective pressure to perform well in these tasks. Anyone who has tried to swat a bothersome fly on a hot day knows how quick their reactions are. This is partly due to their vision which, although spatially at a lower resolution than our own, is often very fast, with a time resolution of up to 400 frames per second in some day-flying flies and bees (Autrum, 1952). Although each visual system is adapted to solve particular challenges faced by that species (see also Chapter 2), the challenges faced and overcome are often ones that exercise the designers of artificial vision systems and can provide us with a rich source of inspiration. Locusts, for example, fly in swarms many millions strong without colliding, and they evade bird capture on the ground and in the air. Could we exploit the locust visual system for collision warnings in cars on a crowded city street or in aircraft? Insects are attractive sources of inspiration for artificial visual systems because they have a rich repertoire of visual behaviors controlled by a compact nervous system containing as few as a million neurons (a locust brain, for example, weighs a mere 0.0037 g).

The precision engineering underlying the compound eye has fascinated biologists and engineers alike since the 1890s (Exner, 1891). Recently, though, it has been in the area of motion detection where advances in our understanding of insect vision have been made, mainly because motion-detecting neurons are some of the largest in the insect visual system and therefore are easier to record from. Pioneering investigations into correlation motion detectors and their suitability for robot vision were made by Nicholas Franceschini and colleagues from Marseilles University (Franceschini et al., 1989, 1992; Franceschini, 2003), who exploited the elementary motion-detecting neurons in the fly eye (Figure 8.1a,b) for navigation and obstacle avoidance in a purpose-designed robot (Figure 8.1c). The robot was equipped with analog electronic circuitry emulating the operation of a ring of elementary motion-detecting neurons (Figure 8.1d). To any rule there are exceptions, and polarization vision is the exception to the dominance of motion-detecting pathways for artificial vision. Rudiger Wehner and colleagues found that the ant *Cataglyphis* can use the patterns of polarization in the sky to explore new territory and to return directly to its nest after foraging over the arid Saharan desert (Müller and Wehner, 1988). With Rolf Pfeifer he built a robot (Lambrinos et al., 1997, 2000) that incorporated the behavior of the ant and also circuitry underlying the polarization sensitive interneurons of the cricket (Labhart, 1988) — the ant brain being too small to record from routinely. The robot was able to use the pattern of polarization to navigate back to its base after a period of exploration, so long as it performed a calibration procedure before setting out.

Insect vision has contributed to advances in three main types of artificial vision system:

1. Bioinspired circuits embedded in the control structure of mobile robots (locust LGMD; Blanchard et al., 2000), cricket polarization-analyzing neurons (Labhart, 1988), and fly motion detectors (Franceshini et al., 1992)
2. Neuromorphic chips consisting of very large silicon integrated circuits based on processing within directionally selective correlation motion

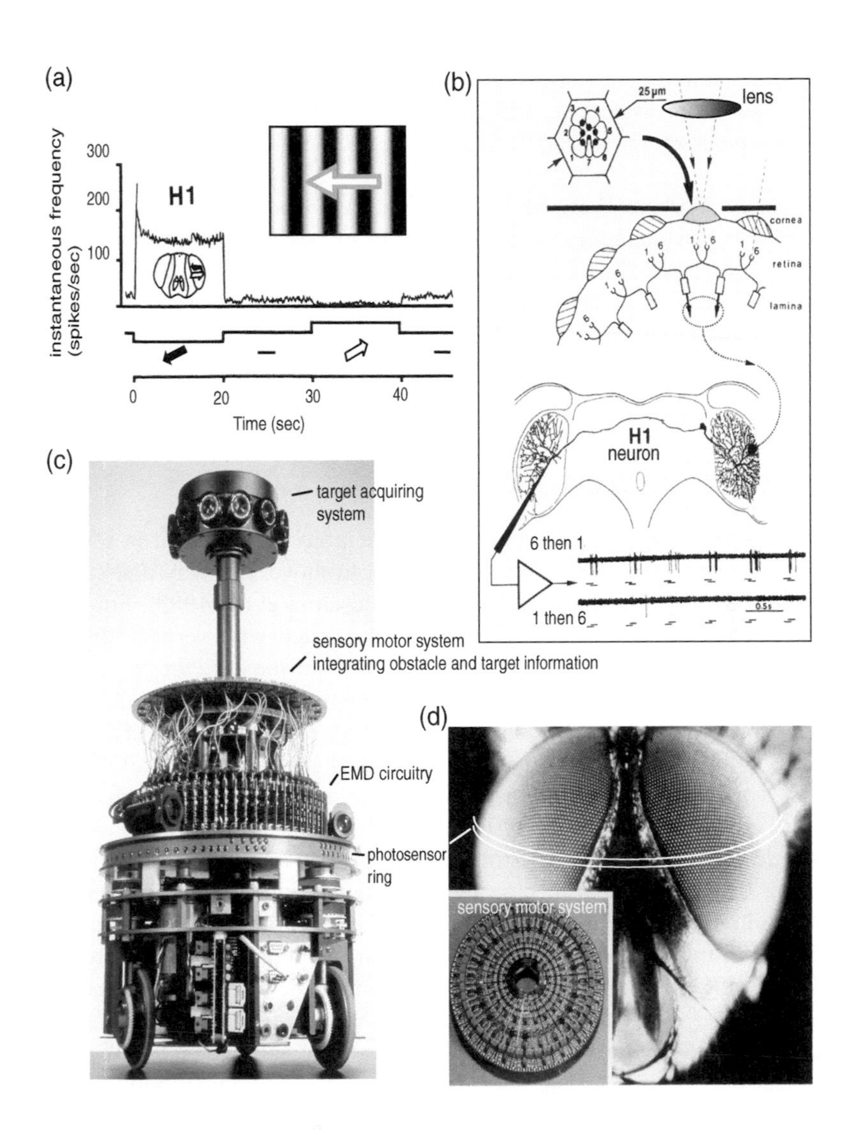

FIGURE 8.1 Autonomous robot-fly inspired by optomotor circuitry in the fly eye. **a**: Response of optomotor neuron H1 from the fly visual system to a wide-field drifting grating. **b**: Investigation of the elementary motion-detecting (EMD) circuitry of the fly. Light spots were moved either from photoreceptor 1 to 6 or from 6 to 1, and spikes produced by H1 were recorded in response to each motion direction. Only stimulation in one direction leads to spikes in H1. **c**: Robot-fly built by incorporating circuit details of the elementary motion detectors to control locomotion and obstacle avoidance. The robot moved in saccadelike straight paths toward a light target (see text for a detailed description). **d**: Ring of analog electronic circuits inspired by an equatorial ring of EMDs from the fly compound eye. (Adapted from Franceschini, N., in *Sensors and Sensing in Biology and Engineering*, Springer-Verlag, Vienna, 2003. With permission.)

detectors of the fly eye (Harrison, 2000), including VLSI retinal circuits (Liu, 2000; Liu et al., 2001)
3. Bioinspired behavioral strategies, such as wall following and landing in bees (Srinivasan et al., 1997, 1999, 2000, 2001) and ant navigation based on polarization sensors (Lambrinos et al., 1997)

Recently, Barbara Webb and coworkers have also combined an insect vision chip with auditory circuitry on a single robot to assess the contribution of motion detectors to performance of an auditory task such as phonotaxis (Webb and Harrison, 2000). The research advances described in this chapter have led both to the design of better artificial vision systems and to an increased understanding of insect vision.

8.2 LESSONS FROM THE HOUSEFLY

8.2.1 Motion-Detecting Circuits

In the fly's visual system, motion-detecting neurons are involved in the control of optomotor responses, which cause the animal to move in a way that stabilizes the image of the environment on the retina (Franceschini et al., 1989). For example, if a fly is suspended in the center of a drum decorated with vertical stripes, it will attempt to turn to follow the stripes when they are moved clockwise or counterclockwise. A particular group of neurons reacts to movements of this type in the way expected if they are to control the fly's turning behavior. The neurons are sensitive to the direction of movement of the background the fly sees (Figure 8.1a), and this sensitivity arises from the way that the small circuits of neurons that drive them are organized (Figure 8.1b). These small circuits are referred to as *elementary motion detectors*; each has a restricted field of view and is excited when an edge moves in one particular direction across it, but is inhibited by movement in the opposite direction (Figure 8.1b, bottom panel). This response to movement is achieved by correlating the time at which two different photoreceptors are stimulated by the moving edge, and this type of movement detector is, therefore, called a *correlation-type* elementary motion detector. A fundamental feature of this type of motion detector is that it is sensitive to the frequency with which moving edges pass over local parts of the retina. It cannot, therefore, distinguish between a field of narrow stripes moving slowly and a field of broad stripes moving more quickly. This characteristic of confounding spatial frequency with velocity was first determined in behavioral experiments by Reichardt and coworkers in the 1950s (Hassenstein and Reichardt, 1956) and is shared by many of the wide-field neurons in the fly visual system (Figure 8.1, H1 neuron), which is the main evidence that they are involved in optomotor responses.

Nicolas Franceschini and coworkers built an autonomous robot based on the design of H1 in the fly eye and tested its effectiveness for collision avoidance. They exploited analog, parallel processing to design the robot fly incorporating a single ring of photoreceptors (Figure 8.1c) with elementary motion-detecting (EMD) circuitry that correlated the outputs of neighboring receptors. Obstacle avoidance relied on the fact that during translation at a given speed, the velocity of motion of an object gives a measure of its distance, with nearer objects generating a greater signal. The robot fly moved at a constant 50 cm/sec, and the output of the EMDs was used to generate sensory motor commands if an obstacle was identified by the perturbations it produced in the flow field. This robot was extremely successful at avoiding collisions, even in novel environments. Motion was initiated by identification of an illuminated target using a ring of ordinary lenses mounted on top of the robot. This work led to an understanding of the need for parallel, analog processing for artificial vision sensors, and it proved the effectiveness of correlation-based motion-detection systems for the visual control of navigation.

With Nicolas Franceschini, Thomas Netter then adapted the fly EMD circuitry for height control of a small (850 g) unmanned aerial vehicle (UAV, Figure 8.2a,b) over uneven ground. The aircraft flies at 2 to 3 m/sec on a whirling arm, and the output of the EMDs is fed to the rotor thrust. The neuromorphic eye is a 19-pixel linear array and an aspheric lens (focal length 24 mm) set at only 13 mm from the array. The retinal image of the array is defocused because this improves its field of view and improves the measurement of the optic flow field. Each EMD produces a pulse whose voltage depends on the time delay between the photoreceptor activations; short delays give higher voltages so that the EMD signal is roughly proportional to flow-field velocity. The EMD outputs are aggregated, weighted, and compared with a reference measurement to provide a signal that is used to modulate thrust and, therefore, UAV height. This measurement is a feature-matching proce-

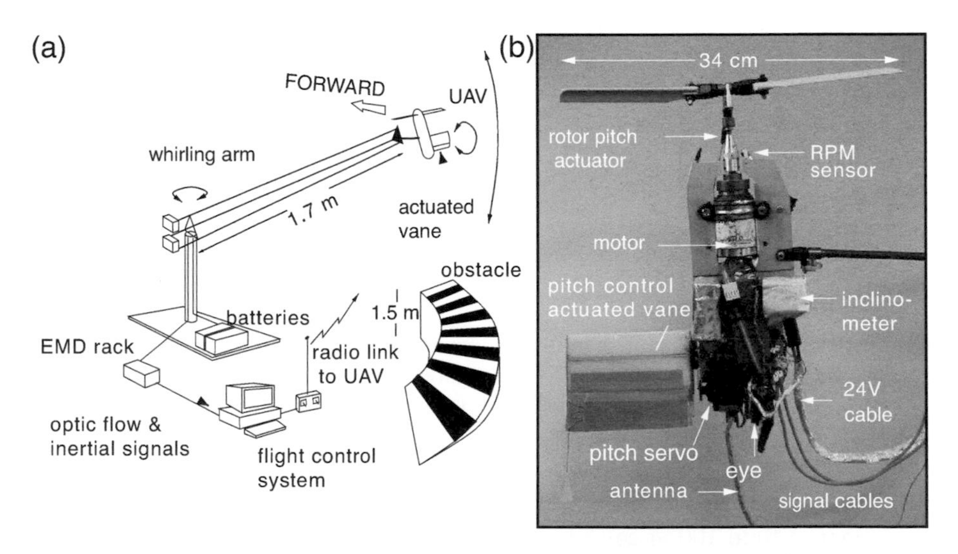

FIGURE 8.2 Rotorcraft guided by flow-field motion detectors based on fly elementary motion detectors. **a**: The height above the ground is modulated using an airborne photoreceptive array (eye) that measures optic flow over uneven terrain and feeds back to control the rotor thrust. **b**: Close-up view of the rotorcraft. The flight speed is maintained at a steady 2 to 3 m/sec and an inclinometer linked to an aerodynamic vane is used to control aircraft pitch. (Modified from Netter, T., and Franceschini, N., *SPIE Robotics and Machine Perception Newsletter*, 2003. With permission.)

dure, and the responses of this velocity sensor were not as sensitive to contrast as those of the correlation-based EMDs used in the robot-fly.

In recent experiments (Ruffier and Franceschini, 2003), the neuromorphic eye and the UAV have been miniaturized so that the eye weighs only 0.8 g and the whole rotorcraft only 100 g. In terrain-following trials, the miniature vehicle reliably followed the variable terrain at a variety of flight speeds. As flight speed increased, it maintained a greater height above the terrain, even though no explicit knowledge about ground speed or the local altitude was available onboard the vehicle. In both crafts, as a safe landing strategy, the flight speed is decreased voluntarily while retaining the reference optic flow, a strategy recently found to operate in insects (Srinivasan et al., 2000).

8.2.2 Designing Fly Eyes Using Silicon

The extreme precision of the fly eye, combined with a detailed knowledge of the motion-detecting circuitry behind neurons such as H1, has attracted circuit designers wanting to make fast, robust, lightweight, low-power vision systems. Analog very large scale silicon integrated (VLSI) circuit devices, with their transistors working in the subthreshold operating range, calculate like the neurons in the fly's eye and are able to process large numbers of inputs in parallel. This makes analog VLSI a natural choice for implementing and understanding insect vision. Recently, Harrison (Harrison and Koch, 1998; Harrison, 2000) described an analog VLSI circuit based on the input organization of the H1 neuron in the fly (Figure 8.1, Figure 8.3) and has used one to control a small mobile robot (Figure 8.4). The circuit he successfully used to guide his robot on a straight path operates in the voltage mode and has a 6×24 2D array of elementary motion detectors (EMDs), each consisting of two half-EMDs (Figure 8.3a). In turn, each half-EMD is composed of two subunits, 1 and 2. The outputs of the two half-EMDs are subtracted from one another to increase the directional selectivity of the array.

8.2.2.1 Subunit 1: The Adaptive Photoreceptor plus Temporal High-Pass Filter

A photodiode senses changes in light intensity and provides input to a five-transistor adaptive circuit followed by a two-transistor source follower and a five-transistor temporal high-pass filter (Delbrück and Mead, 1996). The effect of this is to provide a continuous output that has a low gain for static signals (including circuit mismatches) and high gain for transient signals. The response of the adaptive photoreceptor increases logarithmically with increasing illumination and can have a time constant of 10 to 100 msec.

8.2.2.2 Subunit 2: The Temporal Low-Pass Filter plus Multiplier

The output of subunit 1 is fed through a temporal low-pass filter followed by a Gilbert multiplier (Xs in Figure 8.3a). The first circuit's phase lag acts as a delay and the Gilbert multiplier then multiplies the delayed photoreceptor signal with the nondelayed photoreceptor signal from the neighboring photoreceptor. The output

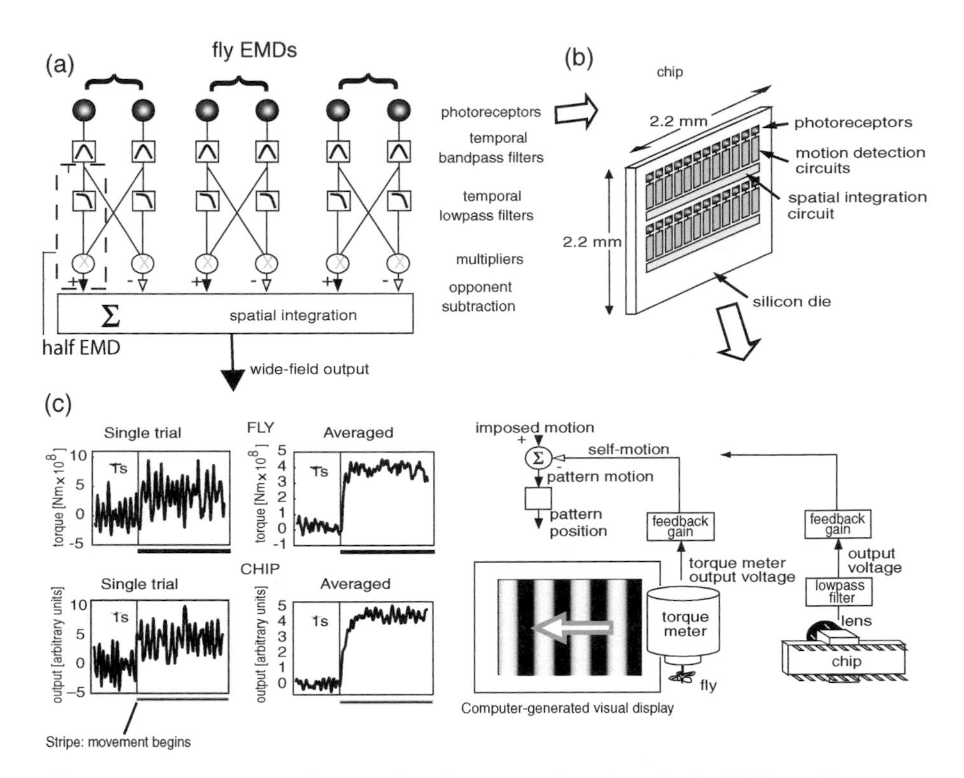

FIGURE 8.3 VLSI design for a motion flow sensor based on the fly EMD, and its use as a guidance system for a mobile robot. **a**: Block design of the fly EMDs. **b**: Schematic illustration of the layout of EMD circuitry on the silicon chip. **c**: Comparison of the responses of the fly optomotor pathway (measured as the fly's turning tendency or torque) with that of the silicon EMD chip to a wide-field drifting grating. Responses of the chip and the fly's optomotor responses were closely matched over single trials and on average (within the range of image-drift speeds illustrated). (Modified from Harrison, R., Ph.D. thesis, 2000. With permission.)

from the Gilbert multiplier is a current that can be summed with that from neighboring multipliers (in Figure 8.3a). The motion computation in this circuit was based on the Reichardt correlation detector (Hassenstein and Reichardt, 1956). If the velocity of the moving edges matches that of the delay line, then the signal on the delay line is reinforced. The delay line architecture generates motion information over an extended spatiotemporal range, with the result that the detectors are sensitive to motion over a range of spatial frequencies (1 to 2 orders of magnitude). Directional selectivity is generated by the polarity of the delay lines with a delay on the leftward arm. The preferred direction is to the right, so the hardwiring of the connection leads to directional tuning.

In this chip, one complete motion detector circuit measures 61 × 199 μm in a standard 1.2 μm process and consumes 50 nW of power under normal indoor illumination of 10 cd/m². The photocurrent contributes significantly to the overall power consumption, which is very low. Reid Harrison (2000) estimates that a 100 × 100 array of these 2D Reichardt motion sensors would consume less than 5 mW,

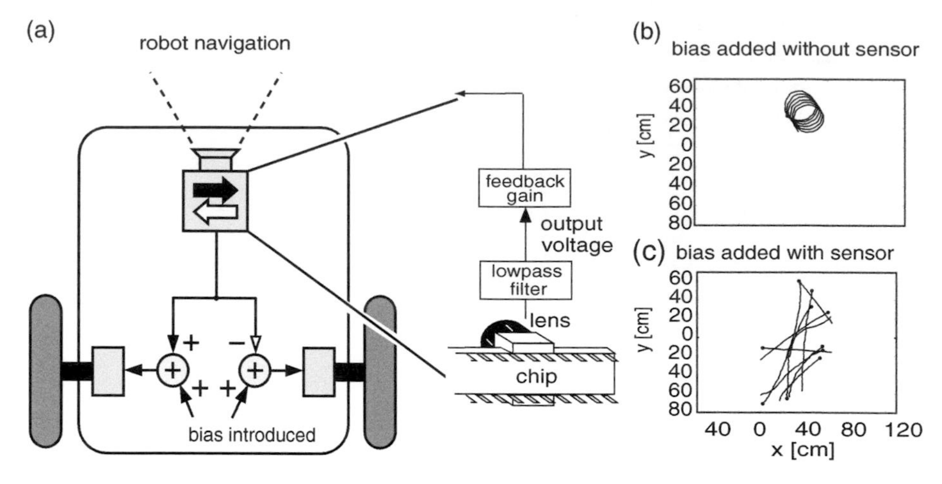

FIGURE 8.4 VLSI design for a motion flow sensor based on the fly EMD, and its use as a guidance system for a mobile robot. **a**: The EMD was mounted on a small robot with independent drive motors to its two rear wheels. The output of the chip was used to counteract any drift to the left or right, which was sensed by the chip during forward motion. **b**,**c**: The motion of the robot was plotted in two graphs, one showing the robot's path with no output from the chip **(b)**, the other with a bias introduced to the motor drives to its two wheels **(c)**, causing the robot to turn in one direction. The robot could not compensate for the bias, but even with a bias introduced to the motor drives to both wheels, it was still able to move forward in a straight path. The EMD chip is used to stabilize the course of the robot against unwanted perturbation introduced by biases in its motor system. (Modified from Harrison, R., Ph.D. thesis, 2000. With permission.)

which is very low. The chip matches the optomotor turning (torque) performance of the fly in response to a drifting striped stimulus, over a range of drift speeds (Figure 8.3c). Compared with the fly, the chip has a more limited range of drift velocities to which it gives a directional response, because, unlike the fly, the chip does not adapt to different velocities of image motion.

Reid Harrison and coworkers then used their wide-field motion-sensing chip to control the movement of a small mobile robot. The experiments were performed indoors in a cluttered laboratory environment while the robot's trajectories were recorded. The team introduced a bias in the motor drives to two wheels of the robot, so the robot tended to move in circles (top graph in Figure 8.4b). The optic flow feedback from the chip proved capable of nearly eliminating the effect of the bias in the drives to the wheels, although some glitches still occurred (bottom graph in Figure 8.4b). When moving in straight lines, the robot traveled at a speed of approximately 20 cm/sec, and objects or walls were typically 0.2 to 1.5 m from the robot.

Traditional imaging and image processing is expensive in terms of time, size, and power. Biologically inspired analog VLSI approaches to image processing can bring down the cost and can make robot vision more practical. On the down side, Reid Harrison noted that one consequence of building an integrated motion sensor that includes photoreceptors and motion processing on the same chip is low resolution. However, although high-resolution imagers may be necessary for some complex tasks, such as face recognition, low-resolution imagers are sufficient for many sophisticated visually guided behaviors. The visual acuity of flies, for example, is 100 to 500 times lower than that of humans, yet flies are capable of extraordinary maneuvers as they navigate through a dynamic, unstructured environment. Fly correlator-based motion detection may not be the only system that would benefit from the VLSI approach.

8.3 LOCUST LOOMING DETECTORS FOR COLLISION AVOIDANCE

Locusts and other orthopteroids are unique among insects in possessing a system of large and accessible collision-detecting neurons, the lobula giant movement detector (LGMD; O'Shea and Williams, 1974) and its post synaptic partner, the descending contralateral movement detector (DCMD; Rowell, 1971). In our laboratory (Rind and Simmons, 1992; Simmons and Rind, 1992; Rind and Bramwell, 1996; Rind and Simmons, 1997, 1998; Blanchard et al., 2000; Rind, 2002), we have used modeling and electrophysiological techniques to understand how these identifiable neurons detect collisions (Figure 8.5a–c) and have incorporated elements of the locust system into the control structure of a mobile robot (Figure 8.6). The robot extracts looming cues from its visual environment and avoids potential collisions, and the LGMD model has recently been elaborated into a design for a collision-warning sensor on a chip for cars (Figure 8.6b,c).

A locust compound eye may possess up to 8000 separate ommatidia (see also Chapter 6), and the lens of each ommatidium focuses light onto eight radially arranged photoreceptors or retinula cells (Nowel and Shelton, 1981). All eight photoreceptors are concerned with the same region of the locust's visual field and send axons into the same cartridge in the underlying optic neuropil. This pattern of connectivity means that just as the faceted surface of the compound eye divides up the world, the retinotopic connections beneath preserve this arrangement. As such, visual information is processed within separate, parallel channels throughout the optic lobe. The optic lobe itself is divided into three neuropil regions: the lamina, the medulla, and the lobula. In the lobula, large fan-shaped neurons are found, one such cell being the LGMD (O'Shea and Williams, 1974; Rind, 1996). The LGMD and its postsynaptic partner, the descending contralateral movement detector (DCMD), respond to approaching objects with a rapid train of spikes (Figure 8.4a; Rind and Simmons, 1992; Simmons and Rind, 1992; Hatsopoulos et al., 1995; Gabbiani et al., 1999). Both the frequency and number of these spikes are maximized if the looming object is approaching on a direct collision trajectory (Judge and Rind, 1997). Deviations of just 1.8° from such a course will halve the response in terms of mean peak spike frequency in either cell (Judge and Rind, 1997).

In order to discover exactly how these cells are able to respond to looming objects, they were challenged with a variety of stimuli, from clips from the *Star Wars* movie to real and computer-generated images. It seems that selectivity for approach is achieved by responding to the increasing extent of the image, signaled

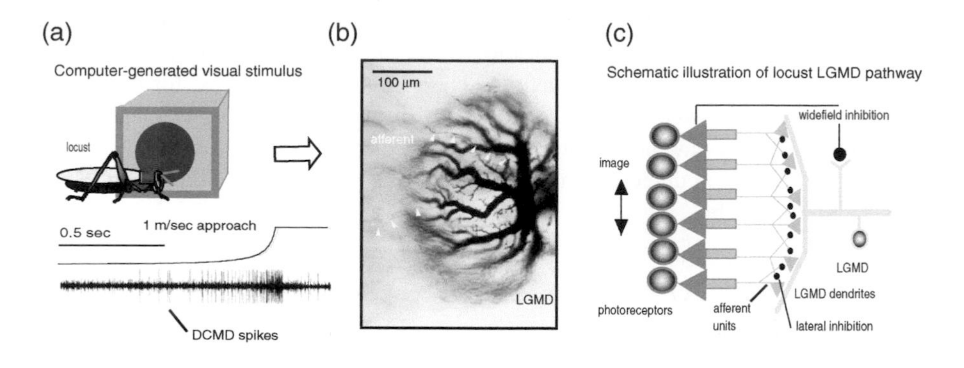

FIGURE 8.5 Circuits based on the input organization of the locust lobula giant movement detector (LGMD) neuron for the detection of looming objects. **a**: Responses of the LGMD monitored through its postsynaptic partner, the descending contralateral movement detector (DCMD), as a locust views the computer-generated image of an approaching object (80 mm diameter). **b**: Handlike structure of the input dendrites of the looming detector, the LGMD, in the optic lobe. Several visual afferent neurons have also been stained with cobalt hexammine (white arrow heads). **c**: Schematic illustration of the input organization of the LGMD neuron, showing the small-field afferent units exciting an LGMD dendrite and laterally inhibiting one another presynaptically to the LGMD. A wide-field inhibitory pathway also extends onto the LGMD dendrites, causing inhibitory potentials in the LGMD in response to rapid or whole-field motion.

by the growth of its edges and its increasing rate of expansion on the retina (Simmons and Rind, 1992). Using techniques ranging from electrophysiology to immunohistochemistry, it has been possible to determine the arrangement of the LGMD's inputs that allow it to respond to such cues: the LGMD receives excitatory inputs from a retinotopic array of small-field, spiking neurons excited transiently by changes in illumination (O'Shea and Rowell, 1976). The LGMD gathers information from the entire visual field of the compound eye (Figure 8.5b). With repetitive stimulation, there is a decrement in synaptic transmission between these afferents and the LGMD (O'Shea and Rowell, 1976). Evidence also exists for strong lateral inhibitory interactions between these afferents that are presynaptic to the LGMD (O'Shea and Rowell, 1976; Pinter, 1977; Rind and Simmons, 1998; Rind and Leitinger, 2000), and it seems that conduction delays occur as this inhibition spreads laterally (Pinter, 1977). Direct inhibitory input to the LGMD is also apparent, resulting from sudden or intense large-field stimuli such as a flash of light (Figure 8.5c; Rind, 1996). These inhibitory inputs have longer latencies than the excitatory postsynaptic potentials (EPSPs) evoked by the same instantaneous stimulus.

Although these features have been known by neurobiologists for some time, they were not initially understood in the context of the LGMD and DCMD neuron's selectivity for approaching objects. For this reason, the known circuitry was incorporated into a computational model to assess whether these alone were sufficient to explain the neuron's ability to detect looming stimuli. Claire Rind and David Bramwell (1996) constructed a four-layered neural network that was thought to represent ade-

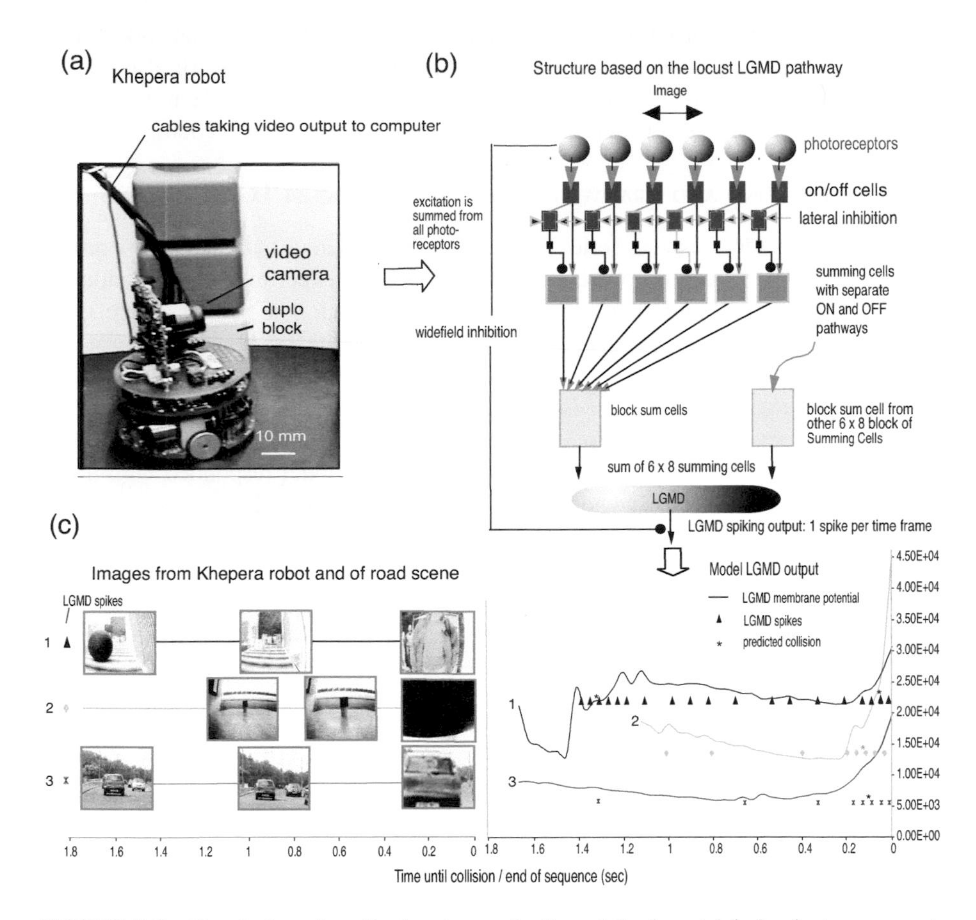

FIGURE 8.6 Circuits based on the input organization of the locust lobula giant movement detector (LGMD) neuron for the detection of looming objects. **a**: Small, mobile Khepera robot that incorporates the computational model of the input organization of the LGMD in its control structure. A small video camera mounted on the robot provides the model with black-and-white input images. Output from the model LGMD is used to control the movement of the robot and can be monitored continuously as membrane potential or used to generate a spiking output. **b**: Adaptation of the LGMD input organization for collision avoidance in complex environments. The light ON and OFF pathways onto the LGMD have been separated, and an intermediate level of signal integration, referred to as the summing cells, has been introduced before information is passed to the LGMD unit. **c**: The model LGMD signals impending collision. Video images provide input to the model LGMD in sequences 1–3: (1) Robot approaching a scaled cut-out of a person with a complex background (equivalent 80 km/h). (2) Robot approaching black box (equivalent 80 km/h). (3) Driving on a road with a car a constant distance in front (80 km/h) until 0.42 seconds when the images are manipulated to assume car in front has stopped. During each simulation or experiment, the level of excitation of the LGMD unit in the model can be monitored (solid lines 1–3) and the output of the model LGMD converted to a spiking output, with a maximum of one spike per timestep (25 spikes/sec max). Collision is indicated when a LGMD spike occurs in three consecutive time frames (*). In these videos, collision is predicted 50 to 100 msec before impact with the pedestrian, dark block, or car.

quately the input organization of the locust LGMD. The ability of the network to detect looming objects became quickly apparent, but studies of the network also revealed exactly how the LGMD is able to detect collision. A description of the model is given in the next section, but for more details, see also Blanchard et al. (2000).

8.3.1 The Rind and Bramwell Model of the Locust LGMD

Input to this model was via a series of computer-generated virtual images of a moving object. These movements were controllable by the experimenter, thus allowing velocity, trajectory, and even object size to be altered. The virtual image of the stimulus object was mapped onto the first layer of the LGMD network at each millisecond of simulated time. The first layer consisted of 250 P-units, representative of the locust's photoreceptors. Each viewed a specific region of the visual field, separated from its neighbor by 3.3°. This angle was chosen to spread the field of view of all 250 P-units in space. Each P-unit in the model responded with a brief (1 msec) excitation to a change in the level of illumination. In truth, these P-units are more accurately described as a composite between a photoreceptor and one or more postsynaptic neurons in which responses to light ON and OFF are processed to give the same signals. The excitation of these P-units was extremely transient and marked the passage of an edge with great precision. Interestingly, similar response time-courses have been observed in "transient cells" in the locust medulla responding to light increments or decrements (James and Osorio, 1996).

As in the locust optic lobe, processing is divided into channels, each concerned with a specific region of the visual field and each processed separately from, but in an identical fashion to, its neighbors. This is a crucial element in the wiring of any visual system and preserves the topographic map seen by the array of photoreceptors. Account was taken of this in the computational model as each P-unit excited a single E-unit in the same retinotopic position as itself in the second layer of the network. As with other excitatory connections in this model, conduction delays were set to 0 msec. Thus, excitation of any layer-2 E-unit would follow excitation of the P-unit feeding it, unless it was within a refractory period following excitation.

Layer 2 of the network also included a second class of neuronal unit. These are the inhibitory or I-units. As with the previously described E-units, each I-unit was excited by the P-unit in the same retinotopic position as itself. However, in contrast to the E-units, these "cells" passed inhibition to the 6 neighboring and 12 next-neighboring processing channels with a conduction delay which could be set to between 1 and 4 msec. This inhibition was weighted so that each neighboring channel would receive one-sixth of the total inhibition, and each next-neighboring channel one-twelfth.

These excitatory and inhibitory inputs were collected and summed in layer 3 of the network by a series of retinotopically arranged S-units, one per processing channel. The E- and I-unit inputs were summed linearly until a given threshold level of excitation was reached and a spike was produced. Immediately following the spike, voltage declined exponentially with time and was followed by a refractory period.

The final layer of the network (layer 4) consisted of a single LGMD-unit, which received and linearly summed excitation from all active S-units and inhibition from a further cell, the inhibitory F-unit. This sum was expressed as a voltage. The single F-unit constituted a feed-forward pathway, bypassing layer 3 of the network. The F-unit was activated when a set number of P-units (normally 50) were simultaneously active. In response to activation, the F-unit passed inhibition, delayed by 2 to 5 msec, directly to the LGMD-unit. Thus, IPSPs in the LGMD could be triggered by whole-field stimuli. During each simulation, a graphical display of activity in each layer of the network was available for analysis.

8.3.2 The "Critical Race" Hypothesis

The "critical race" hypothesis was originally suggested by Simmons and Rind (1992) in their analysis of the LGMD's input connections. Studies of the Rind and Bramwell (1996) computational network highlighted the importance of this hypothesis. Simply put, as an object approaches the eye, it appears to expand. This rate of expansion increases as the theoretical time of collision approaches. As this object expands, it excites more and more P-units, which in turn activate their respective processing channels. These maintain the retinotopic map of the image as viewed by the P-units and transfer it to the S-units in layer 3. At the same time, however, the I-units from each excited channel are inhibiting their neighbors. As a result, just as excitation expands over the array of S-units, so too does inhibition. Crucially, inhibition from the I-units is limited by its synaptic delay (1 to 4 msec), whereas excitation is limited only by the rate of image expansion. Thus, at the early stages of a loom, the stimulus object expands rather slowly. As a result, excitation also expands slowly over the array of S-units and is effectively canceled out by spreading lateral inhibition from the I-units. As a result, very little excitation is passed to the LGMD.

Conversely, as the time of collision with the object draws near, the excitation it generates expands more rapidly over the array of 250 S-units. Because inhibition may travel only as fast as its synaptic delay permits, there eventually comes a point that it can no longer keep up with the spreading excitation. At this stage, the S-units are strongly excited without the restrictive effects of I-unit mediated lateral inhibition, and the LGMD is strongly activated.

8.3.3 Feed-Forward Inhibition

It also appears that the F-unit plays a similarly crucial role in distinguishing approaching from receding objects. Because this cell is activated only by a large number of transiently excited P-units being active simultaneously, the perceived image must change very rapidly in order to activate it. Such changes occur during whole-field light flashes but also at the end of a very rapid loom or at the initiation of an object recession. With approaching objects greater than 100 mm in diameter, this pathway becomes activated increasingly early in a loom (Rind and Santer, 2004). As an object recedes, it appears to shrink in size most rapidly at the start of movement (when it is closest to the observer). As a result, these early stages of movement activate the F-unit, which in turn shuts down the LGMD-unit. Consequently, the model LGMD, like its biological inspiration, gives only a brief response to a receding stimulus.

The F-unit is also able to shut down the LGMD following a rapid loom. This may help to sharpen the cell's response so that it is only active while movement is occurring. This may prevent unnecessary avoidance behaviors from being triggered.

8.3.4 A Locust-Inspired Collision Sensor

The previously described computational techniques have revealed the Rind and Bramwell model of the LGMD (Rind and Bramwell, 1996) as a suitable starting place for building a looming detector. However, the major challenge is to produce a sensor capable of functioning in the real world. The real visual environment is dynamic in many respects. Light intensities change, shadows form, clouds move, all of which are not accurately represented by artificial looming stimuli. If indeed the locust circuitry were to be incorporated into a machine vision system, it must first be tested in this real-world environment. In order to meet these aims, the previously described model was first realized using a software system, $IQR_4 21$ (Verschure, 1997). This software package was originally developed by Paul Verschure at the Institute of Neuroinformatics in Zurich. $IQR_4 21$ is a computer program that allows the user to model neural networks precisely. Its chief advantage over earlier work is that it allows cells with realistic neuronal properties to be modeled, and these can be easily altered by the experimenter. It also permits the model network to be interfaced with a Khepera mobile robot[(20)] (Figure 8.6a). Input to the model was via a miniature black-and-white CCD camera with a 68° field of view mounted on the Khepera robot (see Figure 8.6a). Images from this camera were grabbed at each simulation timestep and relayed via serial port to the model network operating on a nearby PC. At each timestep, the camera's raw image was preprocessed to give a motion image. This was achieved by simply subtracting the previous timestep's image from the current one to give an *absolute difference* image. This processed image was then mapped onto an array of 400 P-units (the number of units in each processing layer was increased in this incarnation of the model to give a better resolution of the visual environment).

In addition, rather than simply recording the model LGMD's output, it was incorporated into the control structure of the mobile robot. Thus, in the absence of LGMD activity, the $IQR_4 21$ program would instruct the robot to perform a simple exploratory behavior (a forward movement). However, activity in the LGMD above a threshold spike rate would override this exploratory behavior and instigate a simple avoidance reaction (a counterclockwise turn). These commands were relayed to the Khepera robot via the PC's serial port. The robot was therefore able to detect and avoid potential collision using its LGMD system. The robot's behavior was tested in a circular arena with a Stonehenge arrangement of large Duplo bricks around its periphery (Figure 8.6a). The robot moved at a variety of speeds ranging from 1.5 to 12.5 cm/sec. Although very much slower than speeds encountered by the real animal, these were the highest practicable in the arena. The robot was able to avoid collision on 91% of occasions at speeds of 2.5 cm/sec, 81% at 5 cm/sec, and 88% at 10 cm/sec. The effectiveness of control via the LGMD was significantly better than 50% at every speed tested (Blanchard et al., 2000).

8.3.5 First Steps toward a Locust-Inspired Collision Detector on a Chip

In later experiments, the IQR_421 was replaced by a MATLAB program and the structure of the model was modified to that shown in Figure 8.6b. The program ran at 25 timesteps per second, and input could be provided by the video mounted on the robot (Figure 8.6c, traces 1,2) or from images stored as avi files with a frame rate of 25 frames per second (Figure 8.6c, trace 3). The model differed from that described in Blanchard et al. (2000) to make the responses of the model more robust to cluttered natural scenes. First, none of the input afferents produced spikes; all were simple graded potential neurons, more suitable for implementation in analog VLSI chips. The light ON and OFF pathways onto the LGMD were separated, and an intermediate level of signal integration (referred to as the *summing cells*) was introduced before information was passed to the LGMD unit. During each simulation or experiment, the level of excitation of the LGMD unit in the model could be monitored (solid lines 1–3 in Figure 8.6c) and the output of the LGMD could be converted to a spiking output, with a maximum of one spike per timestep (25 spikes per second, maximum). This was well below the 600 Hz maximum of the locust LGMD, but it was still possible to signal a collision reliably. Collision was indicated when an LGMD spike occurred in three consecutive time frames (* in Figure 8.6c). In these videos, collision was predicted 50 to 100 msec before impact with a pedestrian, a dark block, and a car. The robot also passed very close to the black ball seen in the first frame of video sequence #1, and a collision warning was given. In the future, the LGMD circuitry could be used to design effective sensors to signal collision with a looming object.

8.4 HONEYBEE BEHAVIORAL STRATEGIES FOR ESTIMATING RANGE AND DISTANCE TRAVELED

The visual abilities of honeybees have fascinated biologists for hundreds of years. Mandyam Srinivasan and coworkers (Srinivasan et al., 1999) have been designing unmanned surveillance vehicles that use visual strategies taken straight from the bee (Figure 8.7; Si et al., 2003; Srinivasan et al., 2000). The first strategy uses optic flow to measure distance flown. Bees communicate the distance to a food source to other bees in the hive by their dance. For short distances of less than 30 m, they perform a round dance, but if the food is farther away, they produce a waggle dance instead. Srinivasan used this change to see how the bee's odometer worked. He fooled the bee into thinking it had flown farther than it had by giving it stimuli that induced strong motion cues (vertical stripes; Figure 8.7b). Conversely, he could convince the bee that it had not flown far by giving it stimuli with few motion cues (horizontal stripes; Figure 8.7b). He was also able to calibrate the bee's odometer (17.7° of image motion per millisecond of waggle) and found that for a given route, angular motion of the image is a robust measure of distance traveled, regardless of how rapidly the bee flies. A similar strategy can be used for a reconnaissance vehicle making repeated sorties over the same terrain.

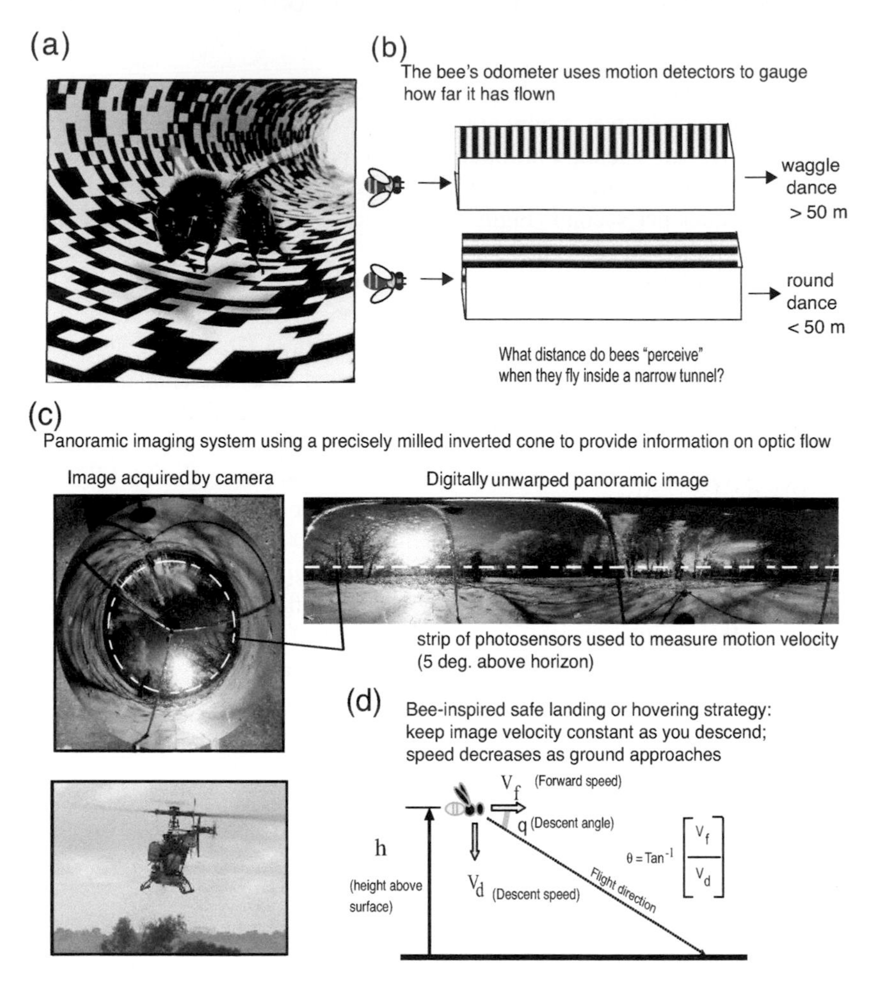

FIGURE 8.7 Honeybee-inspired strategies for the visual control of unmanned navigation vehicles. **a**: Experiments show the bee monitors optic flow to determine the distance flown. Here, a bee flies down a tunnel with a 10 mm pixel random-dot pattern. (Modified from Srinivasan, M.V. and Zhang, S.W., *Science* **287:**851–853, 2000. With permission.) **b**: Bees made to fly down a vertically striped tunnel overestimate the distance to a food source and do a waggle dance when they have not flown over 50 m, whereas those flying along a horizontally striped one underestimate the distance and perform a dance to indicate a distance under 50 m, when in fact it was farther away. **c**: The panoramic imaging system developed by Srinivasan and colleagues to provide a precise measure of optic flow over a narrow strip of photosensors, 5 degrees above the horizon around a 360-degree view. The panoramic lens has been incorporated into an autonomous helicopter and used to control its height above the ground and perform a safe landing. **d**: Safe landing strategy inspired by the bee (see text for details). (Modified from Srinivasan, M.V. et al., *Biol. Cybern.* **83:**171–183, 2000. With permission.)

Srinivasan and coworkers developed a smooth, cone-shaped panoramic imaging system so he could accurately measure motion over a strip 5° above the horizon. This was mounted on an unmanned helicopter (Figure 8.7c), which used angular motion to maintain a constant height above the ground (Barrows et al., 2003). When a bee lands, it slows to a stop as it approaches the ground (Figure 8.7d). Keeping image velocity constant as you lose height is a safe strategy for landing in bees and in a helicopter (Srinvasan et al., 2001). When a bee estimates distance to a wall, it is not using its correlation-based EMDs (described in Section 8.1). Rather, it is using a motion detector that is neither direction-selective nor sensitive to the temporal frequency of the stimulus (Srinivasan et al., 1999).

8.5 SENSOR FUSION LESSONS USING VISION AND PHONOTAXIS

How does information from multiple modalities combine to allow the performance of a specific task? Are there useful lessons here for artificial systems? Barbara Webb and coworkers are exploring how simple neural interactions at the motor-circuit level might be used to reproduce some of the observed behavioral effects of interacting modalities in the cricket (see Webb, 2002, for a recent review). In particular, optomotor responses from an EMD-based silicon chip (Harrison, 2000; described in Section 8.2.2) were used to stabilize a route toward a sound source (phonotaxis). Female crickets can locate conspecific males by moving toward the species-specific calling song produced by the males. Barbara Webb and Tom Scutt (2000) describe a robot that combines vision with audition to guide its navigation. They used a Koala; a six-wheeled base vehicle measuring approximately 40 × 30 × 20 cm with two drive motors and equipped with biologically based hardware and software for processing sound and visual signals. The auditory system mimicked the pressure-difference receiver properties of the cricket ear (Michelsen et al., 1994) to produce a highly directional signal for a specific frequency range. The visual system used was the analog VLSI implementation of the Hassenstein–Reichardt model of elementary motion detectors, developed by Reid Harrison and Christof Koch (1998) and described in Section 8.2.2. The output of the chip was a directional response indicating rightward or leftward horizontal motion. The chip summed the output of an array of elementary motion detectors from across the entire chip, providing a signal for full-field motion, like that of the H1 neuron in the fly. Neurons with similar responses to visual rotation have been found in the cricket.

8.5.1 NEURAL MODEL OF CRICKET PHONOTAXIS AND LOCOMOTION

In these experiments, robot hardware was interfaced with a spiking neural network simulation (Webb and Scutt, 2000), and it was shown that the auditory circuitry mediating phonotaxis could be captured in a four-neuron circuit that exploited spike timing and dynamic synaptic properties to be selective for the temporal pattern of signals resembling the female cricket (Webb, 1995; Webb and Reeve, 2003). There is little direct data on the neural connectivity from the brain to the motor control circuits for sound localization in the cricket, however. Staudacher and Schildberger

(1998) describe some 200 descending brain neurons (BN), some of which show a response to sound (for example, BNC2 in Figure 8.8a), with their response often "gated" by whether or not the animal is walking. One of these neurons has a firing rate that correlates with the walking direction of the animal.

Current models (e.g., Cruse et al., 1998) of motor control for six-legged walking in other insects suggest several features to include in the circuit. Motor patterns tend to be self-sustaining through local feedback or central pattern generator activity, so only a trigger signal is required to start the movement. Steering appeared to be modulated by fairly simple "turn" signals from the brain interacting with this pattern generator so as to appropriately modify limb movements. The neural circuitry devised by Barbara Webb and Richard Reeve (2003) in their robot model of cricket phonataxis behavior is illustrated in Figure 8.8a. The paired burst generators (BG), when initiated by an incoming spike, mutually activate each other to produce a continuous burst of spikes that go to right and left forward (RF and LF) neurons and drive the robot forward. The length of the burst is limited by the eventual activation of the STOP neuron, which inhibits the BG neurons (not shown). One trigger for movement is a spike in the left or right descending brain neuron, BNC2, which represents the premotor output of the auditory circuit. These act via a right or left turn (RT and LT) neuron to modulate the forward velocity additionally by appropriate excitatory and inhibitory connections to RF and LF.

8.5.2 Integration of Optomotor Response with an Analog VLSI Chip

The optomotor sensor (the analog VLSI chip) produced four spike trains: two for each direction of motion. These are summed in two optomotor interneurons (Figure 8.8a; opt-clockwise and opt-counterclockwise) that mutually inhibit one another, which effectively smoothes the signal. The output from the interneurons steers the robot in the appropriate direction by excitatory inputs to the left or right forward neurons (Figure 8.8a). Unless the robot is already moving (due to light or sound inputs), the optomotor input is not enough to move the robot on its own (i.e., it only modulates movement).

Figure 8.8b–d shows how the optomotor response can correct for randomly imposed turns (as might be produced by environmental disturbance such as uneven ground surface or, for a flying insect, wind gusts) to produce almost straight paths. The track is more direct and the robot moves more quickly when the optomotor responses are integrated with phonotaxis. In this circuit, Barbara Webb suggests that when the robot wants to make a turn to the right to follow a sound source, due to activity of the right BNC2 neuron, for example, then the optomotor correction of the turn should be suppressed. In this later work (Webb and Reeve, 2003), the suppression was direction specific (as hypothesized for the interaction of tracking and optomotor behavior in the fly by Heisenberg and Wolf, 1988). The suppression was the result of shunting inhibition, which counteracted any optomotor excitation up to the strength of the predicted input, but might still enable a response if the signal, even though in the expected direction, is much larger than expected. The mechanism implemented in this study reveals the stabilizing role optomotor

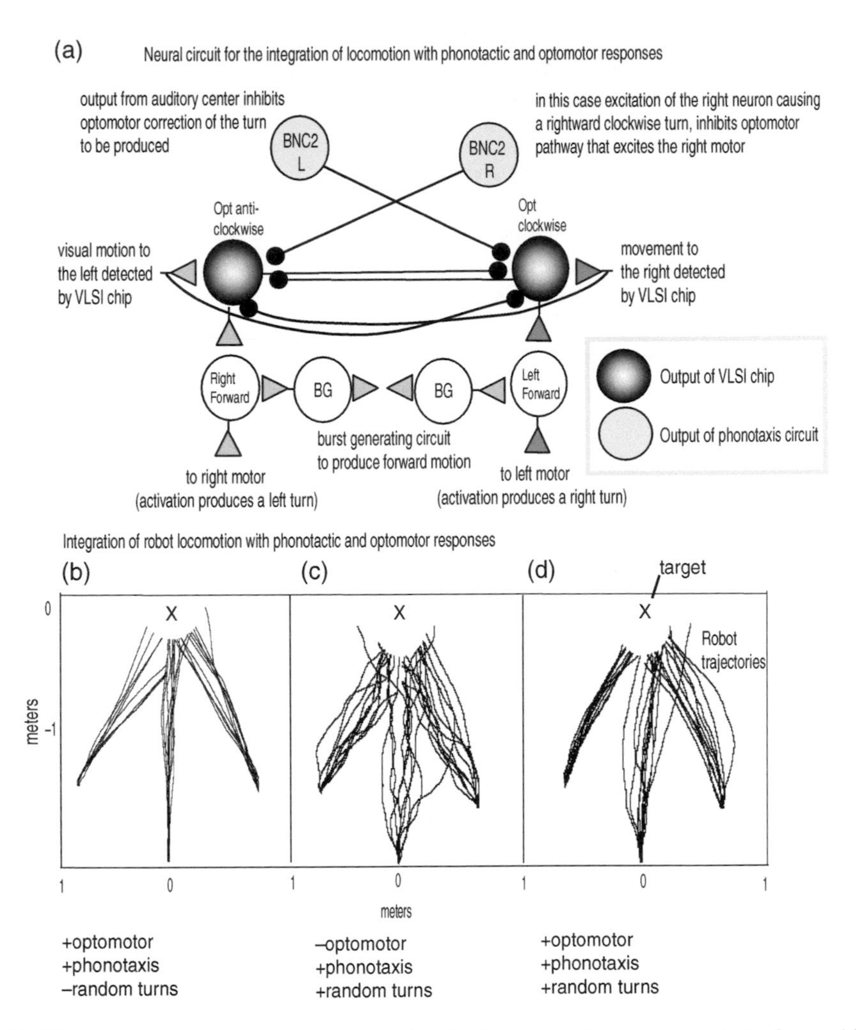

FIGURE 8.8 Integration of auditory and visual information for motor control of a mobile robot. Phonotaxis and optomotor circuitry from the cricket and the fly have been combined in a single robot. **a**: Schematic illustration to show how the output of the auditory pathway mediating phonotaxis (BNC2) and a fly motion sensor (VLSI circuit EMD from Harrison, R., Ph.D. thesis, 2000) interact in the control of a Koala robot. In the neural circuit for the optomotor response, left or right visual movement excites opt-clockwise (OC) or opt-anticlockwise (OA) interneurons to produce an appropriate turn, but the response is inhibited by phonotaxis turn signals from BNC2. The OC and OA interneurons mutually inhibit one another. The output from the interneurons steers the robot in the appropriate direction by excitatory inputs to the left or right forward neurons. **b–d**: The optomotor response can correct for randomly imposed turns (as might be produced by an environmental disturbance such as an uneven ground surface or, for a flying insect, wind gusts) to produce almost straight paths. Paths of the robot with randomly imposed turns **(c)** and the same conditions with the random turning almost fully corrected by the optomotor reflex **(d)** are shown. (Modified from Webb, B. and Harrison, R.R., *Sensor Fusion and Decentralised Control in Robotic Systems III,* 2000. With permission.)

responses could play during phonotaxis, especially if the guidance of the motor system is less than perfect, as is the case with many robots — particularly legged ones (Quinn and Ritzmann, 1998).

8.6 CONCLUSIONS AND OUTLOOK

Insects reveal a range of visual behaviors controlled by a tractable amount of neural machinery: a locust brain, for example, is about 3 mm from retina to retina and weighs 0.004 g. Insects possess sophisticated and rapid image-processing systems that are providing unique insights into the neuronal interactions that create a selective tuning to image motion, image velocity, and looming objects. They also reveal that embodying a model in a robot is an important way of understanding the system and being able to use it for real-world applications.

We have some distance to go in this endeavor. Artificial visual systems, even those like the VLSI fly eye incorporating analog components instead of digital circuitry (Harrison and Koch, 1998), still reveal a remarkable inflexibility compared with the biological system that inspired them. The silicon retina based on the wide-field, direction-selective fly neuron H1 (Harrison and Koch, 1998; Franceschini et al., 1989), for example, has only a narrow range of image velocities that it can signal, largely because the comparison between neighboring elements that underlies the selective response has a fixed interval over which the interaction can take place. Fly motion detectors get over this problem in a clever way — but not by changing the time constants of the neurons involved (Harris et al., 1999). Locust collision-sensing neurons have been successfully incorporated into the control structure of a mobile robot and the system performs robustly in a world more complex than that normally encountered by a locust. Insects still have much to reveal, in particular the way motion detectors adapt to steady velocities and signal changes, the economical use of scarce neuronal resources for more than one task, and the ability to make behavioral judgments quickly.

ACKNOWLEDGMENTS

I would like to thank Nicolas Franceschini, Mandyam Srinivasan, and Reid Harrison for providing good copies of their figures for inclusion in this chapter. This work was supported by the Gatsby Charitable Foundation and a European Community grant, IST 2001-38097-LOCUST.

REFERENCES

Autrum H (1952) Über zeitliches Auflösungsvermörgen und Primärvorgänge im Insektenauge. *Naturwissenschaften*. **39:**290–297.

Barrows GL, Chahl JS, Srinivasan MV (2003) Biologically inspired visual sensing and flight control. *Aeronaut. J.* **107:**159–168.

Blanchard M, Rind FC, Vershure PFM (2000) Collision avoidance using a model of the locust LGMD neuron. *J. Robot. Auto. Syst.* **30**:16–38.

Cruse H, Kindermann T, Schumm M, Dean J, Schmitz J (1998) Walknet: a biologically inspired network to control six-legged walking. *Neur. Net.* **11:**1435–1447.

Delbrück T, Mead CA (1996) Analog VLSI phototransduction by continuous-time, adaptive, logarithmic photoreceptor circuits. *CNS Memo 30*, California Institute of Technology.

Exner S (1891) *Die Physiologie der facettirten Augen von Krebsen und Insecten.* Deuticke, Leipzig.

Franceschini N (2003) From fly vision to robot vision: reconstruction as a mode of discovery. In: *Sensors and Sensing in Biology and Engineering.* Barth FG, Humphrey EJ, Secomb TW, Eds., Springer-Verlag, Vienna, pp. 237–250.

Franceschini N, Riehle A, Le Nestour A (1989) Directionally selective motion detection by insect neurons. In: *Facets of Vision.* Stavenga D, Hardie R, Eds., Springer-Verlag, Berlin, pp. 360–390.

Franceschini N, Pichon JM, Blanes C (1992) From insect vision to robot vision. *Phil. Trans. Roy. Soc. Lond. B* **337**:283–294.

Gabbiani F, Krapp HG, Laurent G, (1999) Computation of object approach by a wide-field, motion-sensitive neuron. *J. Neurosci.* **19:**1122–1141.

Harris RA, O'Carroll DC, Laughlin SB (1999) Adaptation and the temporal delay filter of fly motion detectors. *Vision Res.* **39:**2603–2613.

Harrison RR (2000) *An analog VLSI motion sensor based on the fly visual system.* Ph.D. thesis, California Institute of Technology, Pasadena, CA.

Harrison RR, Koch C (1998) An analog VLSI model of the fly elementary motion detector. In: *Advances in Neural Information Processing Systems 10.* Jordan MI, Kearns MJ, Solla SA, Eds., MIT Press Cambridge, MA, pp. 880–886.

Hassenstein B, Reichardt W (1956) Systemtheoretische Analyse der Zeit-, Reihenfolgen-, und Vorzeichenauswertung bei der Bewungsperzeption des Rüsselkäfers *Chlorophanus. Z. Naturforsch.* **11:** 513–524.

Hatsopoulos N, Gabbiani F, Laurent G (1995) Elementary computation of object approach by a wide-field visual neuron. *Science* **270:**1000–1003.

Heisenberg M, Wolf R (1988) Reafferent control of optomotor yaw torque in *Drosophila melanogaster. J. Comp. Physiol. A* **163:**373–388.

James AC, Osorio D (1996) Characterisation of columnar neurons and visual signal processing in the medulla of the locust optic lobe by system identification techniques. *J. Comp. Physiol. A* **178:**183–199.

Judge SJ, Rind FC (1997) The locust DCMD: a movement detecting neurone tightly tuned to collision trajectories. *J. Exp. Biol.* **200:**2209–2216.

Labhart T (1988) Polarization-opponent interneurons in the insect visual system. *Nature* **331:**435–437.

Lambrinos D, Maris M, Kobayashi H, Labhart T, Pfeifer R, Wehner R (1997) An autonomous agent navigating with a polarized light compass. *Adapt. Behav.* **6**:131–161.

Lambrinos D, Moller R, Labhart T, Pfeifer R, Wehner R (2000) A mobile robot employing insect strategies for navigation. *Robot. Auton. Syst.* **30:**39–64.

Liu S-C (2000) A neuromorphic VLSI model of global motion processing in the fly. *IEEE Trans. Circ. Syst. II* **47**:1458–1467.

Liu S-C, Usseglio-Viretta A (2001) Fly-like visuomotor responses of a robot using VLSI motion-sensitive chips. *Biol. Cybern.* **85**:449–457.

Michelsen A, Popov AV, Lewis B (1994) Physics of directional hearing in the cricket *Gryllus bimaculatus. J. Comp. Physiol. A,* **175:**153–164.

Muller M, Wehner R (1988) Path integration in desert ants, *Cataglyphis*-fortis, *P. Natl. Acad. Sci. USA* **85**: 5287–5290.

Netter T, Franceschini N (1999) Towards UAV nap-of-the-Earth flight using optical flow. *Adv. Artific. Life Proc.*, European Conference on Artificial Life (ECAL), Lausanne, Switzerland, 334–338.

Netter T, Franceschini N (2003). A robot that flies with a neuromorphic eye. *SPIE Robot. Mach. Percept. Newsl.*, **12:**.

Nowel MS, Shelton PM (1981) A Golgi-electron-microscopical study of the structure and development of the lamina ganglionaris of the locust optic lobe. *Cell Tiss. Res.* **216**: 377–401.

O'Shea M, Rowell CHF (1976) The neuronal basis of a sensory analyzer, the acridid movement detector system. II. Response decrement, convergence and the nature of the excitatory afferents to the fan-like dendrites of the LGMD. *J. Exp. Biol.* **65**: 289–308.

O'Shea M, Williams JLD (1974) The anatomy and output connections of a locust visual interneurone; the lobula giant movement detector (LGMD) neurone. *J. Comp. Physiol.* **91**:257–266.

Pinter RB (1977) Visual discrimination between small objects and large textured backgrounds. *Nature* **270:** 429–431.

Quinn RD, Ritzmann RE (1998) Construction of a hexapod robot with cockroach cinematics benefits both robotics and biology. *Connect. Sci.* **10:**239–254.

Rind FC (1996) Intracellular characterization of neurons in the locust brain signalling impending collision. *J. Neurophysiol.* **75:**986–995.

Rind FC (2002) Motion detectors in the locust visual system: from biology to robot sensors. *Microsc. Res. Tech.* **56**:256–269.

Rind FC, Bramwell DI (1996) A neural network based on the input organisation of an identified neuron signalling impending collision. *J. Neurophysiol.* **75:**967–985.

Rind FC, Leitinger G (2000) Immunocytochemical evidence that collision sensing neurons in the locust visual system contain acetylcholine. *J. Comp. Neurol.* **423:**389–401.

Rind FC, Santer RD (2004) Collision avoidance and a looming sensitive neuron: size matters but biggest is not necessarily best. *Proc. Roy. Soc. Lond. B* (Suppl.) **271**:S27–S29.

Rind FC, Simmons PJ (1992) Orthopteran DCMD neuron: a re-evaluation of responses to moving objects. I. Selective responses to approaching objects. *J. Neurophysiol.* **68:**1654–1666.

Rind FC, Simmons PJ (1997) Signalling of object approach by the DCMD neuron of the locust. *J. Neurophysiol.* **77:**1029–1033.

Rind FC, Simmons PJ (1998) Local circuit for the computation of object approach by an identified visual neuron in the locust. *J. Comp. Neurol.* **395**: 405–415.

Rind FC, Simmons PJ (1999) Seeing what is coming: building collision sensitive neurones. *Trends Neurosci.* **22**:215–220.

Rowell CHF (1971) The orthopteran descending movement detector (DMD) neurones: a characterisation and review. *Zeit. Vergl. Physiol.* **73:**167–194.

Ruffier F, Franceschini N (2003) Octave, a bioinspired visuo-motor control system for the guidance of micro-air-vehicles. In: *Bioengineered and Bioinspired Systems.* Rodriguez-Varquez A, Abbott D, Carmona R, Eds. *Proc. SPIE 5119*, pp.1–12.

Si A, Srinivasan MV, Zhang Z (2003) Honeybee navigation: properties of the visually driven "odometer." *J. Exp. Biol.* **206:**1255–1273.

Simmons PJ, Rind FC (1992) Orthopteran DCMD neuron: a re-evaluation of responses to moving objects. II. Critical cues for detecting approaching objects. *J. Neurophysiol.* **68:**1667–1682.

Srinivasan MV, Zhang SW (2000) Honeybee navigation: nature and calibration of the "odometer." *Science* **287**: 851–853.

Srinivasan MV, Chahl JS, Zhang SW (1997) Robot navigation by visual dead-reckoning: inspiration from insects. *Int. J. Pattern Recognit. Artif. Intel.* **11**:35–47.

Srinivasan MV, Chahl JS, Weber K, Venkatesh S, Nagle MG, Zhang SW (1999) Robot navigation inspired by principles of insect vision. *Robot. Auton. Syst.* **26**:203–216.

Srinivasan MV, Zhang SW, Chahl JS, Barth E, Venkatesh S (2000). How honeybees make grazing landings on flat surfaces. *Biol. Cybern.* **83**:171–183.

Srinivasan MV, Zhang SW, Chahl JS (2001) Landing strategies in honeybees, and possible applications to autonomous airborne vehicles. *Biol. Bull.* **200**:216–221.

Staudacher S, Schildberger K (1998) Grating of sensory responses of descending brain neurones during walking in crickets *J. Exp. Biol.* **201**:559–572.

Verschure PFMJ, Xmorph: a software tool for the synthesis and analysis of neural systems Technical Report, Institute of Neuroinformatics. ETH University Zürch, June 1997.

Webb B (1995) Using robots to model animals: a cricket test. *Robot. Auton. Syst.* **16:**117–134.

Webb B (2001) Can robots make good models of biological behaviour? *Behav. Brain Sci.* **24:**1033–1094.

Webb B (2002) Robots in invertebrate neuroscience. *Nature* **417:**359–363.

Webb B, Harrison RR (2000) Integrating sensorimotor systems in a robot model of cricket behavior. In: *Sensor Fusion and Decentralised Control in Robotic Systems III* McKee GT, a Schenker PS (Eds.) Boston, Nov. 6–8: SPIE.

Webb B, Scutt T (2000) A simple latency dependent spiking neuron model of cricket phonotaxis. *Biol. Cybernet.* **82:**247–269.

Webb B, Reeve R (2003) Reafferent or redundant: integration of phonotaxis and optomotor behavior in crickets and robots. *Adapt. Behav.* **11:**137–158.

Section IV

Molecular Characterization of Chemosensory Systems

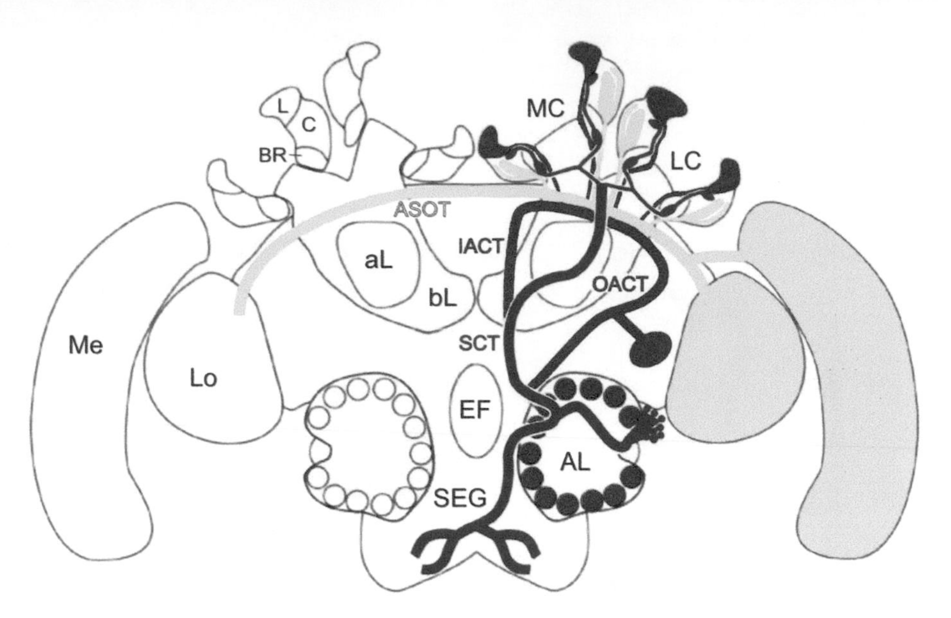

FIGURE 1.2 Topographic segregation of multisensory input to the calyces of the mushroom body in the honeybee brain. (See page 8 for details.)

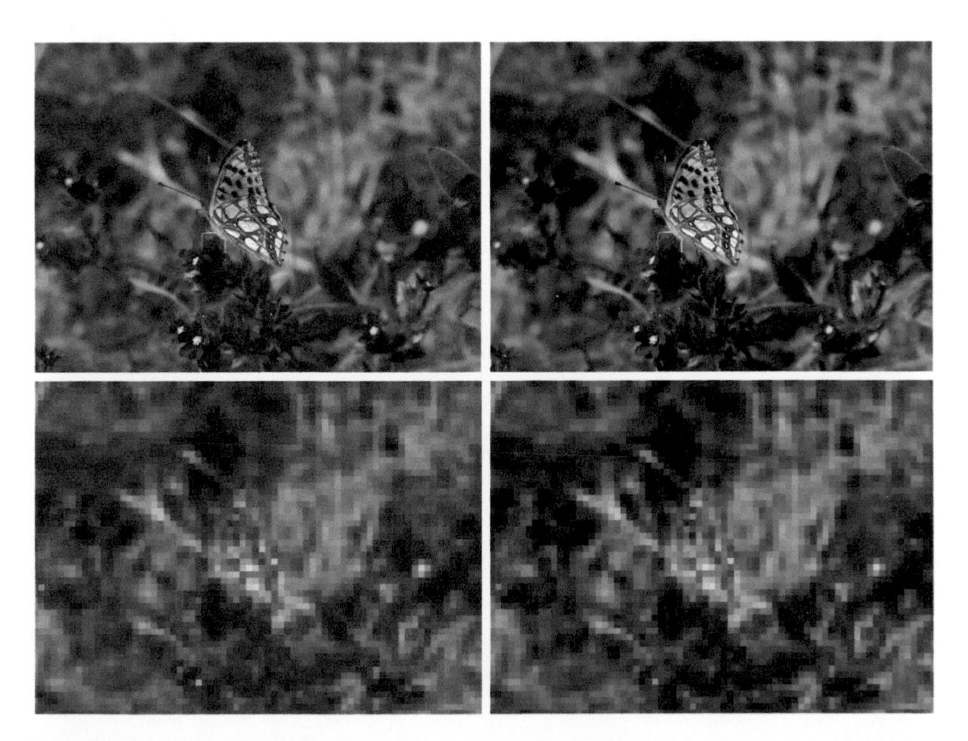

FIGURE 2.4 These four photos of a Queen-of-Spain fritillary (*Issoria lathonia*) feeding on flowers of alkanet (*Anchusa officinalis*) illustrate the potential advantages of color vision. (See page 41 for details.)

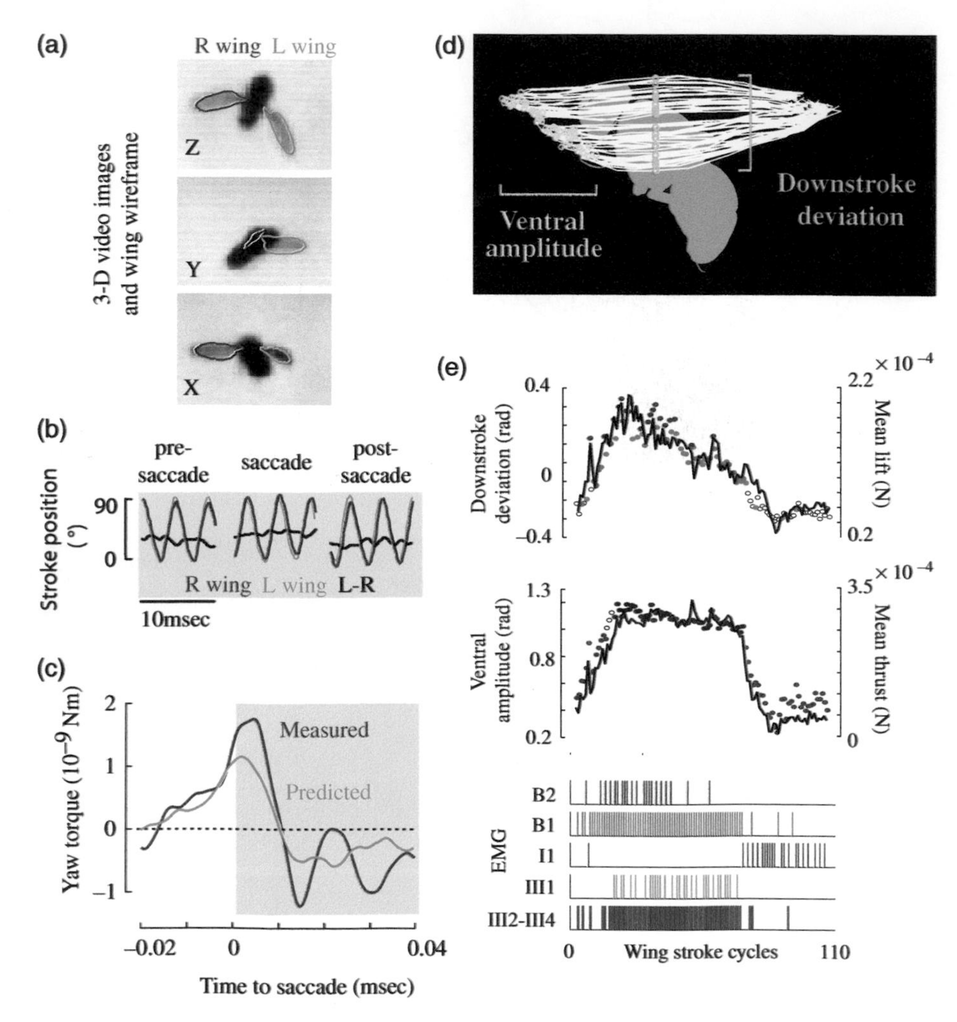

FIGURE 4.3 Neuromuscular control of wing kinematics and flight forces in the fruit fly *Drosophila*. (See page 113 for details.)

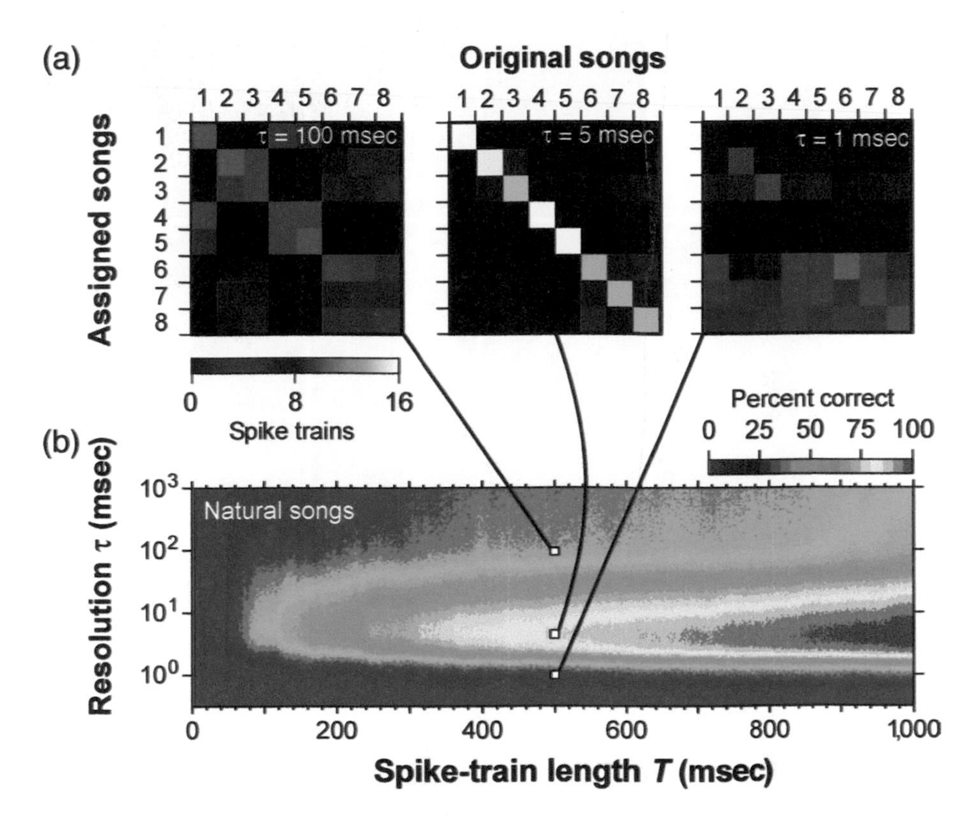

FIGURE 5.9 Song discrimination in the grasshopper auditory system. (See page 150 for details.)

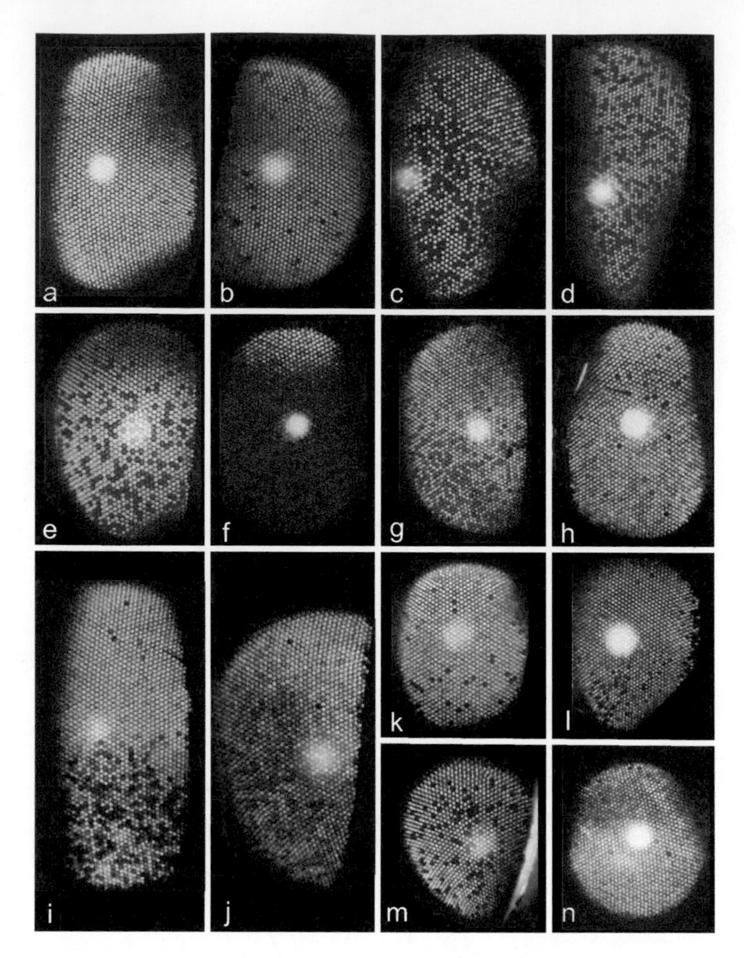

FIGURE 6.8 Butterfly eye shine photographed with a large aperture epi-illumination system. (See page 177 for details.)

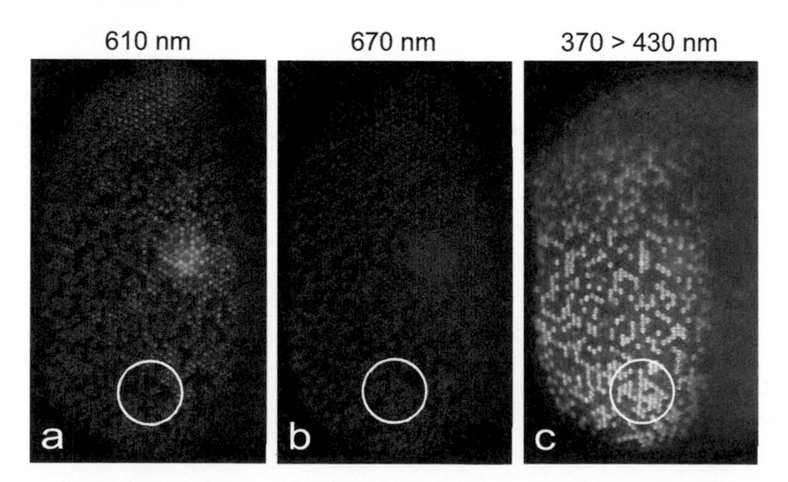

FIGURE 6.9 Eye shine of a male small white butterfly, *Pieris rapae crucivora*, created by reflection (a, b) and fluorescence (c), using the extended setup of Figure 6.7. (See page 178 for details.)

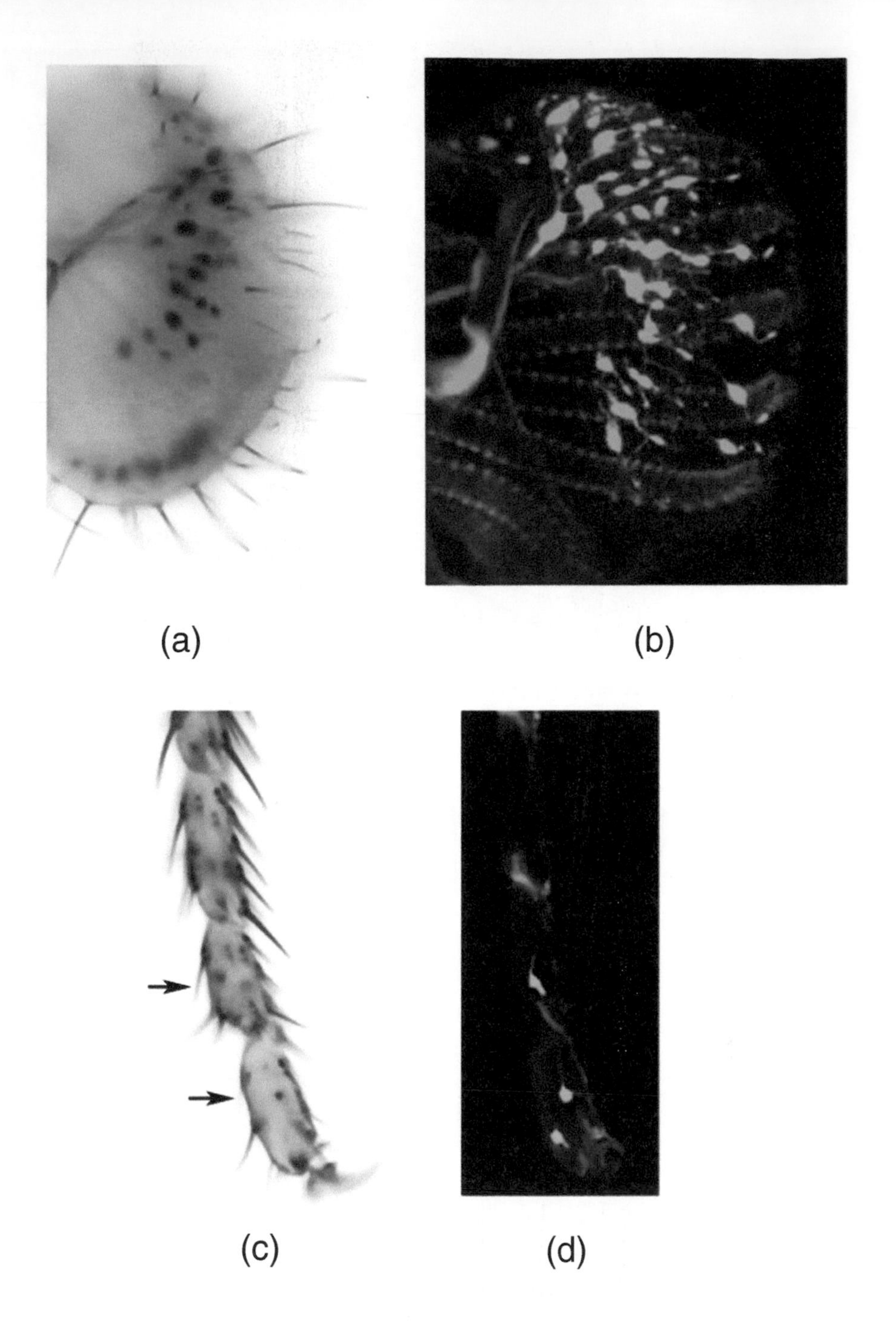

FIGURE 9.2 Expression patterns of the gene Gr5a in the taste neurons in the labellum (a,b) and distal segments of the leg (c,d) in the fruit fly *Drosophila*. (See page 245 for details.)

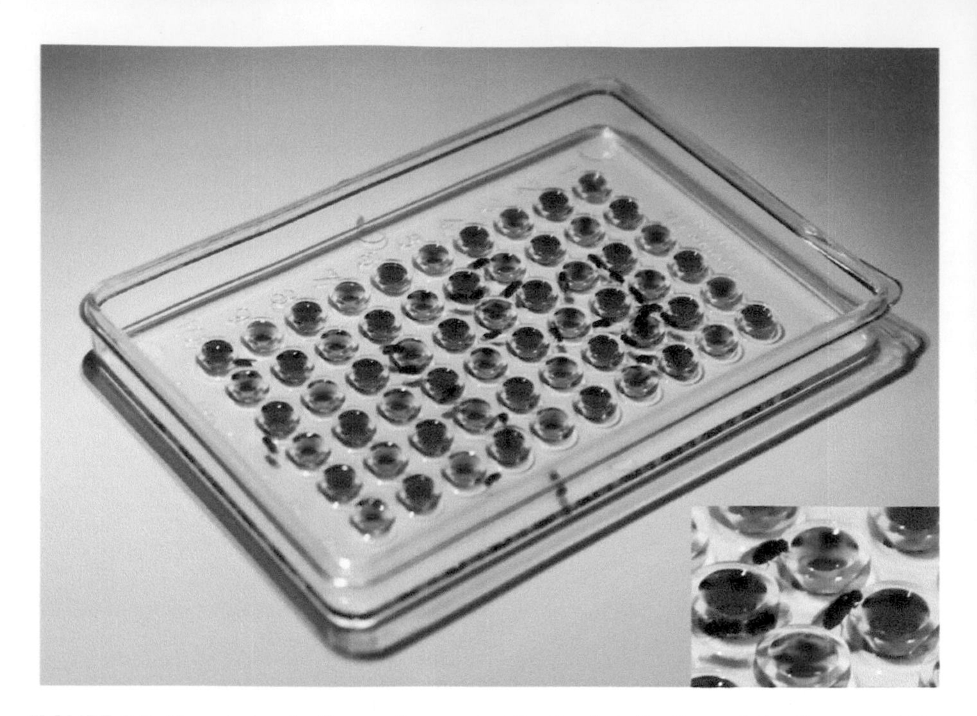

FIGURE 9.3 Two-choice feeding preference test for the fruit fly *Drosophila*. (See page 248 for details.)

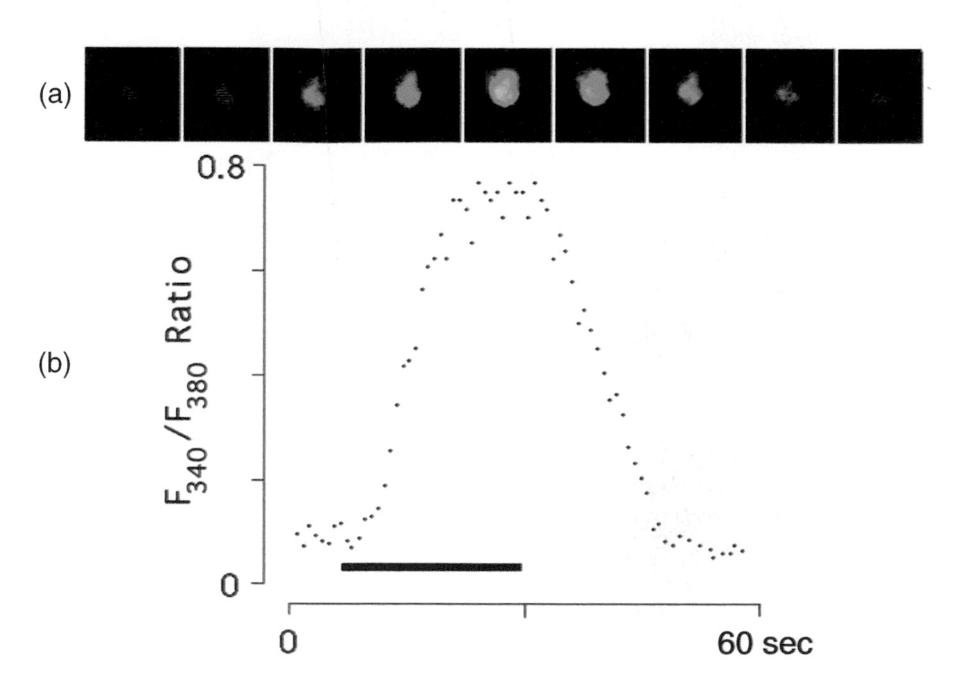

FIGURE 9.5 Calcium response to trehalose in S2-Gr5a cells in *Drosophila*. (See page 258 for details.)

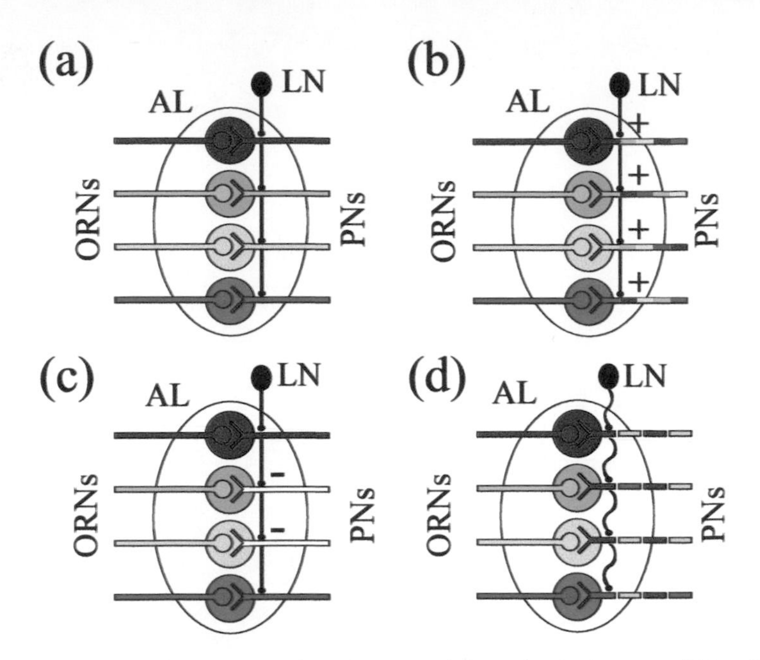

FIGURE 12.3 Models for transformation of olfactory information in the antennal lobe. (See page 331 for details.)

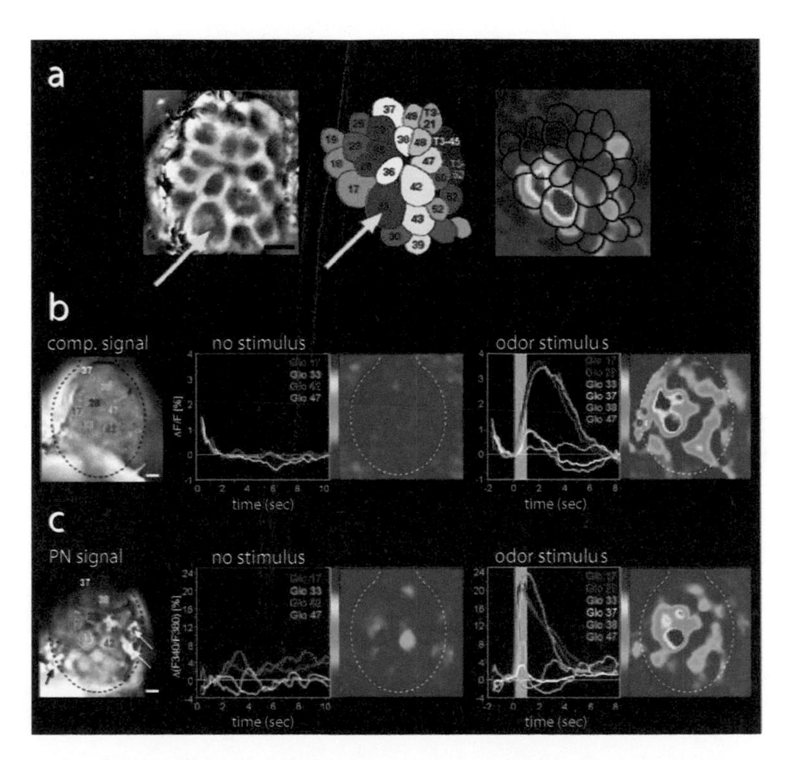

FIGURE 13.3 Odor responses visualized with calcium imaging in the honeybee. (See page 368 for details.)

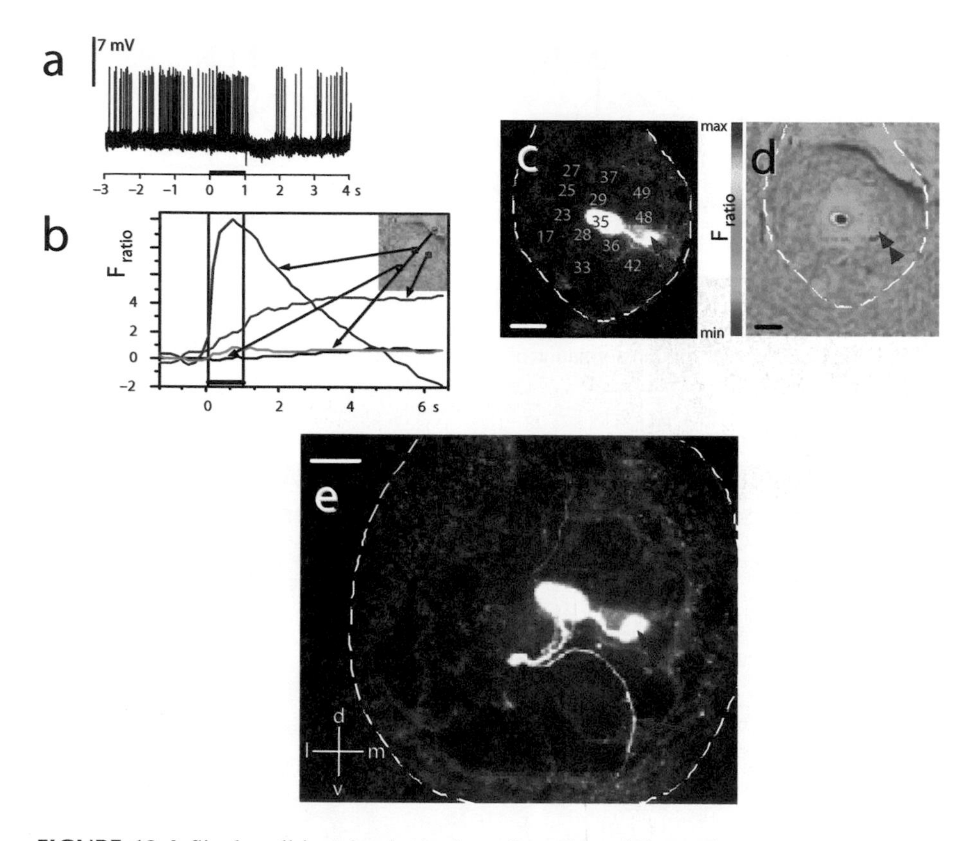

FIGURE 13.4 Single-cell imaging in the honeybee of a uniglomerular PN filled with Fura by iontophoretic injection. (See page 370 for details.)

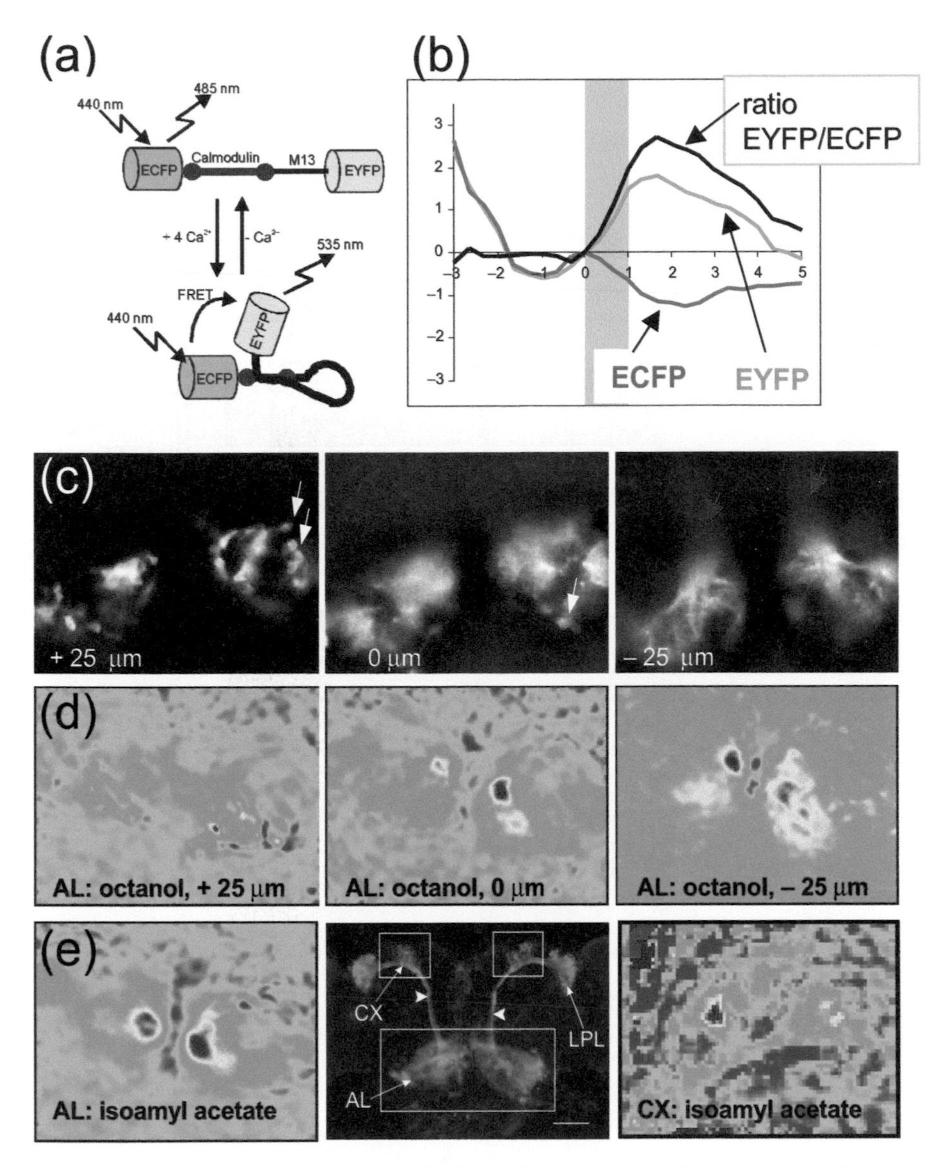

FIGURE 13.5 Odor responses recorded in *Drosophila* using a genetically engineered calcium reporter, cameleon. (See page 371 for details.)

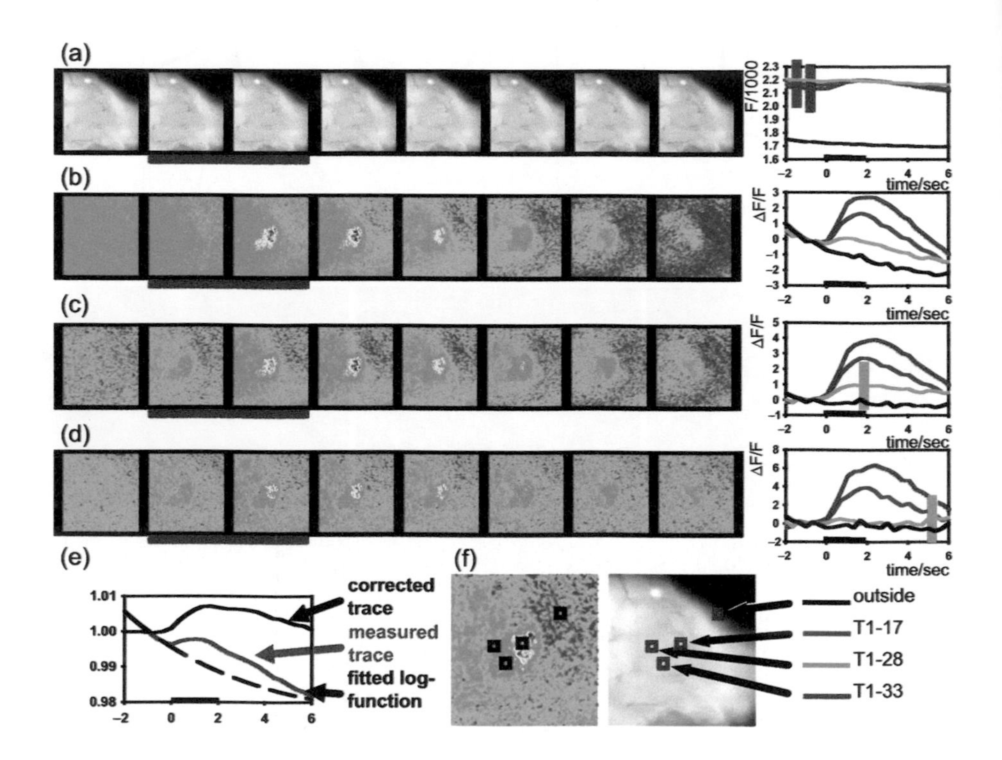

FIGURE 13.7 Scattered light and bleach correction in calcium imaging. (See page 380 for details.)

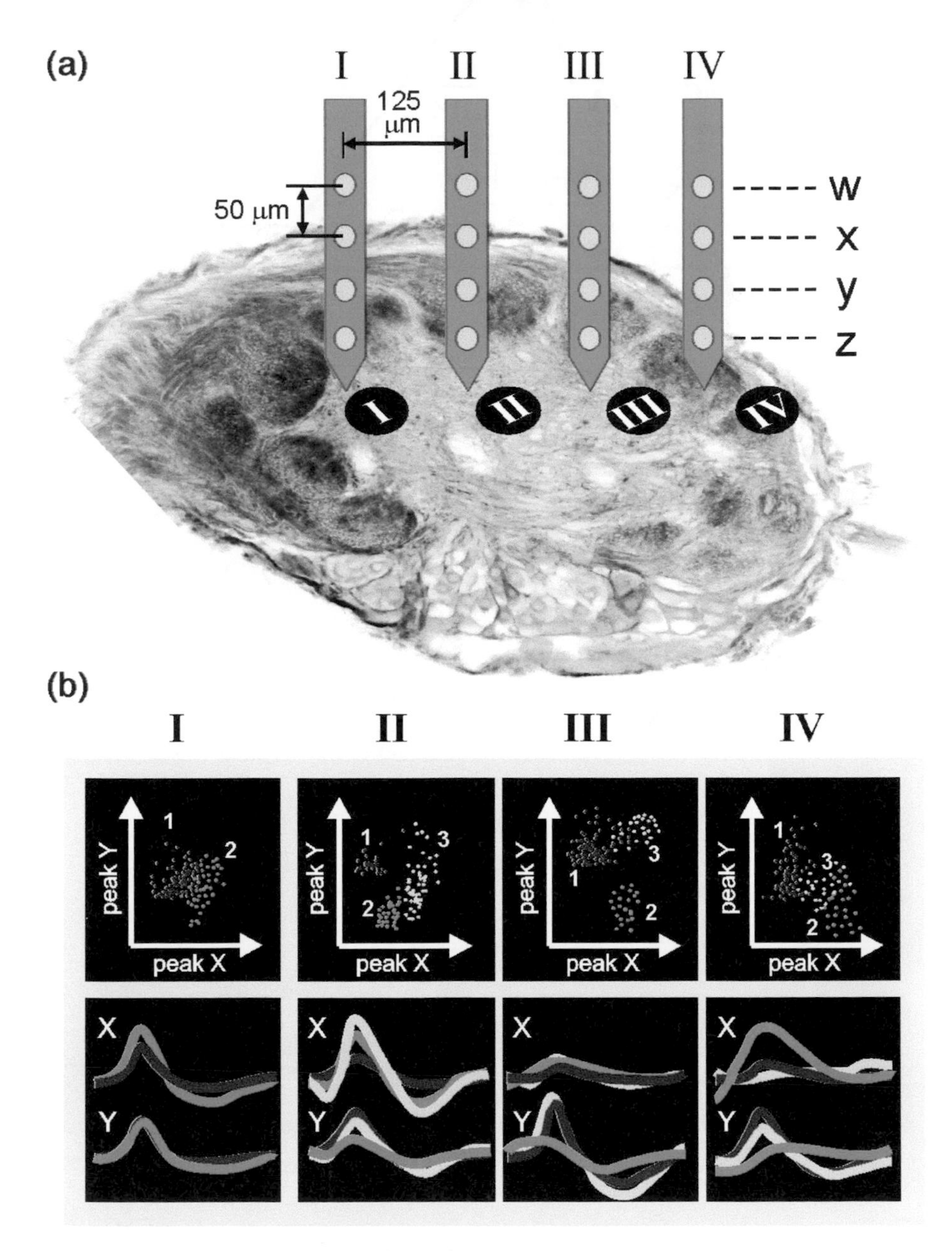

FIGURE 14.1 Diagram of the silicon multichannel recording probes (www.engin.umich.edu/center/cnct) we currently use in the characterization of olfactory networks in the moth *Manduca*. (See page 396 for details.)

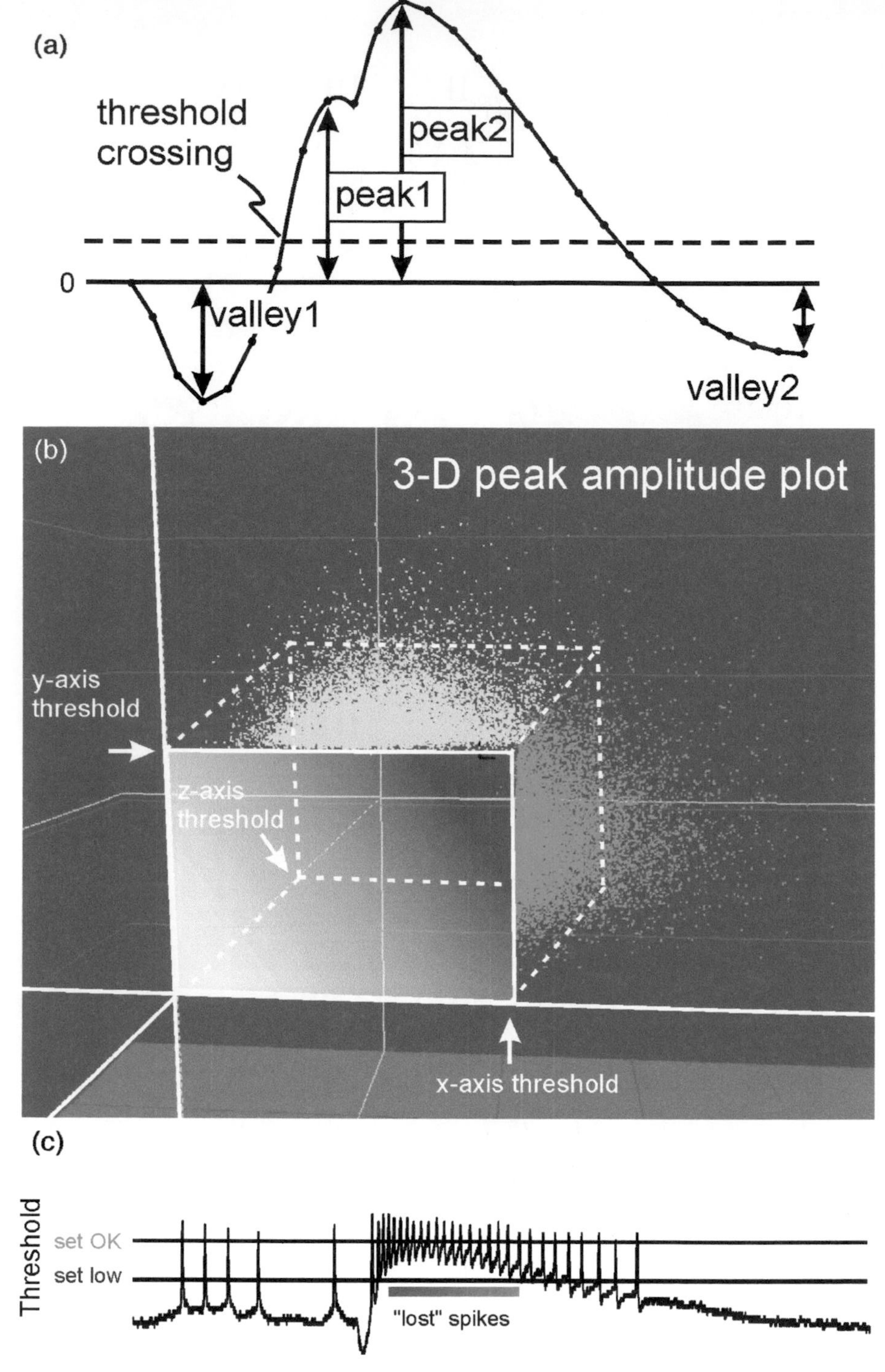

FIGURE 14.3 Some pitfalls of simple fixed-voltage thresholding in cluster cutting. (See page 399 for details.)

9 Analysis of Taste Receptors in *Drosophila*

Jennifer J. Perry, Anupama Dahanukar, and John R. Carlson

CONTENTS

9.1 INTRODUCTION

For most insects, including *Drosophila*, the ability to detect chemical cues from the environment is vital for survival. Gustation, or the sense of taste, is crucial to flies for finding high-quality food sources and avoiding poisonous compounds (Dethier, 1976). Evidence suggests that taste is also essential in fruit flies for pheromone detection during courtship behaviors and possibly for oviposition (Jallon, 1984; Tompkins, 1984; Hall, 1994; Stocker, 1994). Despite its importance, little is known

0-8493-2024-0/05/$0.00+$1.50

about the basic principles of taste perception. Major questions concerning how chemical cues are converted to electrical impulses, how gustatory information is encoded, and how central processing of gustatory input leads to appropriate behaviors remain unresolved.

Anatomical studies of taste organs in flies have revealed that taste receptor neurons are located in hairlike structures called *sensilla* (Falk et al., 1976; Nayak and Singh, 1983; Stocker, 1994). As in vertebrates, reception of taste stimuli in insects appears to be mediated by a large, divergent family of G protein–coupled receptors (Clyne et al., 2000; Dahanukar et al., 2001; Dunipace et al., 2001; Scott et al., 2001; Robertson et al., 2003). Although the gustatory system in vertebrates is numerically more complex at both the peripheral and central levels, some parallels can be drawn between the organization of the taste systems in vertebrates and insects, making *Drosophila* an excellent model system in which to study taste. Fruit flies do not have taste buds *per se*, but they have dedicated taste sensilla that house chemosensory neurons, perhaps tuned to specific taste modalities (Hodgson et al., 1955; Dethier, 1976; Rodrigues and Siddiqi, 1978; Meunier et al., 2003). Axons of these chemosensory neurons project to the subesophageal ganglion (see Chapter 1) in the brain of the fly (Stocker, 1994), which is the equivalent of the cortical gustatory association areas in mammals. The value of *Drosophila* as a model system for the study of taste is enhanced by the relative simplicity of its taste system, the power of its genetics, and the convenience with which taste function can be measured *in vivo*, either physiologically or behaviorally.

In this chapter, we review current approaches used to study *Drosophila* taste receptors. We begin by describing how candidate taste receptor genes were identified in *Drosophila melanogaster* and subsequently in other insects. We then discuss methods for examining the spatial and temporal expression patterns of gustatory receptors in *Drosophila,* which is critical to our understanding of the molecular organization and development of the taste system. Next, we examine methods for behavioral and electrophysiological analysis of taste receptor function, which can provide a map of the functional organization of the gustatory system. Subsequently, we describe the isolation of taste mutants and approaches for manipulating levels of taste receptor gene expression, essential for the functional analysis of taste receptors *in vivo*. Finally, we examine a heterologous expression system that provides a tool with which to examine the biochemical and pharmacological properties of individual taste receptors.

9.2 ISOLATION OF TASTE RECEPTOR GENES

9.2.1 Bioinformatic Approaches

After many years of unsuccessful efforts to isolate insect taste receptors by biochemical, genetic, and molecular means, a bioinformatic approach proved to be successful in the identification of taste receptors in *Drosophila* (Clyne et al., 2000). This approach began with the assumption that taste receptors were *G protein–coupled receptors* (GPCRs), which are widely divergent in sequence but which contain a common structure consisting of seven transmembrane domains. A computer algorithm was designed to search the *Drosophila* genome sequence for proteins on the basis of their structural features rather than their primary sequence (Clyne et al., 1999; Kim et al., 2000). Screens of an early release of the genome sequence with this algorithm led to the identification of two large families encoding predicted GPCRs that were putative odorant (*Or*) and gustatory receptors (*Gr*) (Clyne et al., 1999, 2000). Upon completion of the *Drosophila* genome sequencing project, an additional 25 transcripts of the *Gr* gene family were identified and annotated using basic local alignment search tool (BLAST) sequence comparisons (Dunipace et al., 2001; Scott et al., 2001; Robertson et al., 2003; see Chapter 10 for comparisons with olfactory studies). By these criteria, the *Gr* gene family comprises 60 genes that encode 68 predicted receptors (Figure 9.1; Robertson et al., 2003).

The *Gr* proteins are extremely divergent, sharing as little as 8% amino acid identity (Robertson et al., 2003). The *Gr* genes are distributed throughout the genome, suggesting that the family is ancient. In many cases, the genes are found in clusters with intergenic regions ranging from 150 to 450 base pairs. In particular, there are two large clusters containing six genes each. In addition, some of these loci have been shown to generate multiple alternatively spliced transcripts. *Gr39a*, for example, encodes at least four transcripts, each of which has a distinct 5′ exon, encoding six transmembrane domains, that is spliced to three common downstream exons, encoding the seventh transmembrane domain and the C terminus (Clyne et al., 2000).

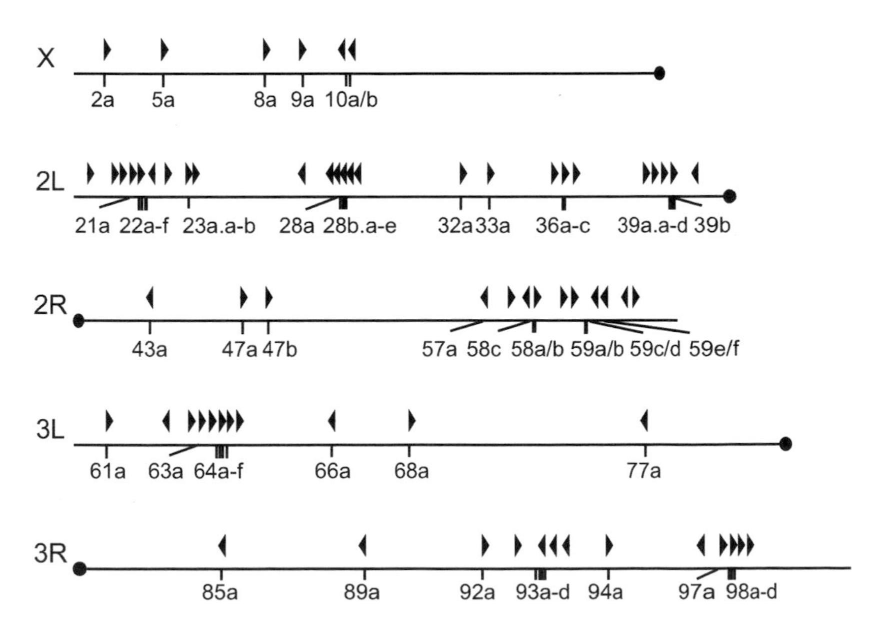

FIGURE 9.1 Organization of *Gr* genes. Arrows indicate location and transcription orientation of individual *Gr* transcripts. Chromosome arms (not drawn to scale) are labeled to the left. (Adapted from Robertson, H.M. et al., *Proc. Natl. Acad. Sci. USA* **100:**14537–14542, 2003.)

9.2.2 Comparative Genomic Analysis

A bioinformatics approach was subsequently used to harvest information from the *Anopheles gambiae* genome, a draft sequence for which was released in 2002 (Holt et al., 2002; see also Chapter 10). *D. melanogaster* and *A. gambiae* diverged about 250 million years ago, and roughly half the genes in the genome of each species can be defined as orthologs, sharing approximately 56% sequence identity (Zdobnov et al., 2002). Genomic analysis identified 276 GPCRs in *A. gambiae*, 76 of which were identified as *Gr* genes (Hill et al., 2002). Comparison of the *Gr* genes of *D. melanogaster* and *A. gambiae* revealed seven possible orthologous pairs, some of which display as much as 68% identity, such as *AgGr22/DmGr21a* and *AgGr24/DmGr63a* (Hill et al., 2002). Conserved receptors between species share common sequences that may be responsible for recognizing similar ligands. Ongoing efforts are directed toward the identification of *Gr* genes in honeybees (*Apis mellifera*), mosquitoes (*Aedes aegypti*), and other fruit fly species (*D. yakuba* and *D. simulans*) (Robertson HM, personal communication).

The identification of *Gr* genes in other *Drosophila* species provides tools for studying the evolutionary divergence of this family of genes and their role in genetic isolation or speciation. The genome of *D. pseudoobscura* has been completely sequenced (Human Genome Sequencing Center, 2003), and the task of sequencing the genomes of at least ten other *Drosophila* species is underway (National Human Genome Research Institute, 2003). A number of the *D. melanogaster Gr* genes have orthologs in *D. pseudoobscura* (Human Genome Sequencing Center, 2003; Robertson et al., 2003), a species that diverged from a common ancestor 21 to 46 million years ago (Beverley and Wilson, 1984; Thomas and Hunt, 1991; Beckenbach et al., 1993). These orthologs have been identified not only on the basis of their sequence similarity, but also in many instances on the basis of their conserved chromosomal locations.

It will be interesting to determine whether receptors involved in recognition of common food sources are highly conserved, whereas those involved in reception of more exclusive ligands (such as species-specific contact pheromones or components of narrow ecological niches) are more divergent. Comparison of genomic positions may also shed light on evolutionary relationships among taste receptors.

9.3 EXPRESSION PATTERNS OF *DROSOPHILA* TASTE RECEPTORS

Initially, expression in taste tissue was an important criterion used to identify the *Gr* genes as a candidate taste receptor family. For 18 of the first 19 identified *Gr* transcripts, expression was detected in the *labellum*, the primary taste organ of the fly. Moreover, expression was specific in that only 1 of the 19 transcripts was detected in heads depleted of taste organs, and expression in a variety of other tissues was also limited to a small fraction of genes (Clyne et al., 2000). To test the hypothesis that the *Gr* genes are expressed in taste neurons, expression was also investigated in a mutant, *pox-neuro* (Awasaki and Kimura, 1997), in which taste neurons are eliminated. Expression of *Gr* genes in the labellum of this mutant was eliminated for 18 of 19 *Gr* transcripts tested (Clyne et al., 2000).

More detailed expression analysis (Dunipace et al., 2001; Scott et al., 2001; Bray and Amrein, 2003; Chyb et al., 2003) confirmed the expression of *Gr* genes in taste neurons and was important for determining the spatial distribution of a number of taste receptors, their tissue specificity, and the number and location of cells within different taste tissues. This analysis is critical for an understanding of the functional organization of the gustatory system.

9.3.1 RT-PCR

The most expeditious method to analyze expression of *Gr* genes has been the *reverse transcription–polymerase chain reaction* (RT-PCR). RNA can be collected from different taste tissues, such as legs, labella, or even microdissected tissue from the labral sense organ (LSO; a taste organ located in the pharynx) to screen for the presence of *Gr* gene RNAs as described in Clyne et al. (2000). This method can be extended to study the onset and duration of expression by extracting RNA from different larval stages, pupae, or appropriately aged adults, to investigate sexual dimorphisms in *Gr* gene expression and to examine the effects of epigenetic factors such as diet on the expression of taste receptors. An olfactory receptor gene in mosquitoes has been shown to be down-regulated after a blood meal using this assay (Fox et al., 2001).

Although the advantages of this method are clear, a significant drawback is that this approach provides limited information about the spatial distribution of these receptors in a specific taste tissue. The labellum, for example, has about 250 taste neurons housed in taste bristles and pegs (Nayak and Singh, 1983; Stocker, 1994), and the use of other techniques (see Section 9.3.3) has demonstrated that these are not equivalent with respect to receptor gene expression.

9.3.2 *In Situ* Hybridization

Although RT-PCR was invaluable in identifying the *Gr* genes as candidate taste receptors, two other methods have been used to describe their spatial expression patterns in more detail. The first, described here, is *in situ* hybridization, which is used to detect the presence of the specific RNA sequences in fixed tissue. The hybridization can be performed on either cryosections or whole-mount samples of tissue and has the advantage that the morphology of the sample is preserved and the cells can be observed in their native environment. However, Scott et al. (2001) are the only investigators to have reported any success with this technique, likely due to low levels of *Gr* gene expression. Of 56 *Gr* genes tested, expression of six *Gr* genes, *Gr28b*, *Gr32a*, *Gr33a*, *Gr47a*, *Gr66a*, and *Gr98b*, was detected in the proboscis using this technique; three others, *Gr10a*, *Gr21a*, and *Gr63a*, were detected in the third antennal segment, the main olfactory organ of the adult fly.

RNA probes, designed from candidate taste receptor sequences, were labeled with digoxigenin in the study of Scott et al. (2001). Probes were then hybridized to sectioned tissue, which was subsequently treated with antidigoxigenin antibody and enzymatic detection procedures. However, because most *Gr* gene messages were not detected in the labellum using this method (Clyne et al., 2000; Scott et al., 2001), there is ample reason to try other techniques. For the *Or* genes, fluorescent probes appear to be more sensitive (A. Goldman and J. Carlson, unpublished results) and can be used in conjunction with signal amplification systems, such as tyramide with alkaline phosphatase (Yang et al., 1999). Further attempts at hybridization experiments to detect *Gr* gene mRNAs *in situ* are warranted in order to investigate the coexpression of *Gr* genes and to examine the patterns of *Gr* gene expression with respect to those of genes that label individual neurons in single taste sensilla.

9.3.3 GAL4-UAS Expression System

In part due to the limited success of *in situ* hybridization experiments, another approach has been applied to examine expression patterns of *Gr* genes: the bipartite GAL4/UAS system (Brand and Perrimon, 1993). This technique exploits the activity of a yeast transcription factor, GAL4, and its identified *cis*-regulatory binding sites (UAS, upstream activation sequences) that when present upstream of a gene, result in its transcriptional activation by GAL4. Expression of the GAL4 protein can be driven by a variety of tissue-specific promoters and can be visualized using a variety of reporter transgenes, such as *lacZ* or GFP (green fluorescent protein).

Expression of at least 12 *Gr* genes has been reported using this method, revealing an assortment of expression patterns (Dunipace et al., 2001; Scott et al., 2001; Bray and Amrein, 2003; Chyb et al., 2003). Some *Gr* receptors are exclusive to a particular taste tissue, such as *Gr22f*, which is expressed in the labellum (Dunipace et al., 2001; Scott et al., 2001), and *Gr68a*, which is expressed in male legs (Bray and Amrein, 2003), whereas others appear to be more widely distributed in multiple taste organs. There is also considerable variation in the number of taste neurons in which individual *Gr* genes are expressed. *Gr22f* is only expressed in four to eight of the labellar taste neurons (Dunipace et al., 2001; Scott et al., 2001), whereas *Gr5a* is expressed in more than 50 (Figure 9.2; Chyb et al., 2003). Finally, in some instances, expression has been detected in other stages of the life cycle. *Gr2a*, for example, is detected in larval gustatory neurons (Scott et al., 2001).

The advantages of this method are its spatial resolution, its sensitivity, and its versatility; a limitation is the uncertainty as to whether the pattern of GAL4-driven expression reflects the expression pattern of the endogenous *Gr* gene. Typically, up to ~10 Kb regions of DNA upstream of *Gr* genes have been used to drive expression of GAL4, but many such regions have failed to express detectable levels of GAL4 protein. Moreover, even in the cases where reporter gene expression has been observed, in no instance has the expression pattern been confirmed at high resolution by *in situ* hybridization or other means to recapitulate that of the endogenous *Gr* gene.

In addition to driving expression of reporter genes such as GFP or *lacZ*, the promoter-GAL4 lines can also be used to express proteins that affect neuronal function, such as ricin or tetanus toxin (Moffat et al., 1992; Sweeney et al., 1995; Bray and Amrein, 2003), or proteins that affect cell viability, such as *Reaper*, *Grim*, or *Hid* (Grether et al., 1995; Chen et al., 1996; White et al., 1996). One can use

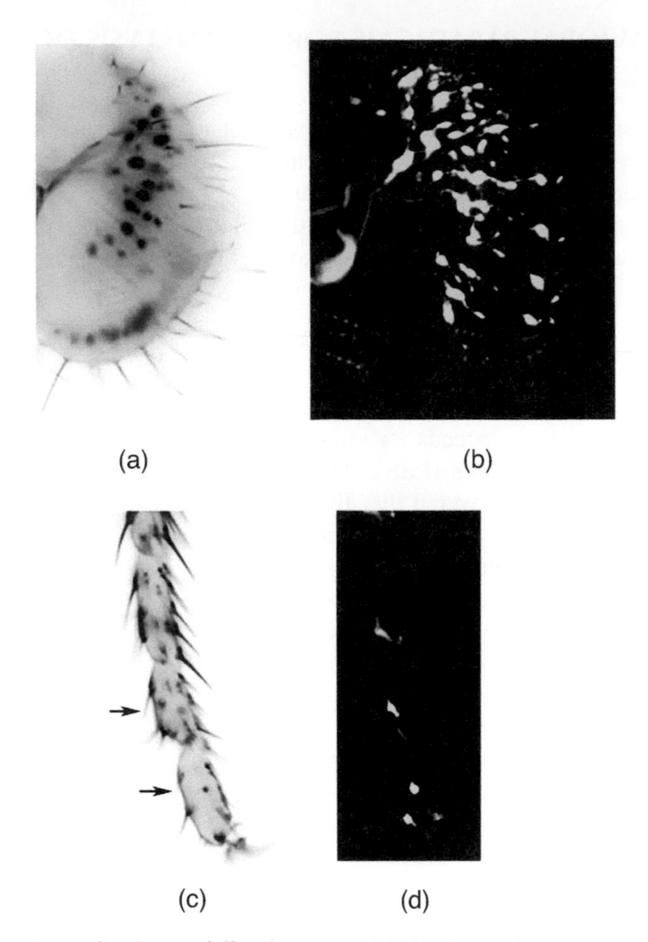

FIGURE 9.2 (See color insert following page 202.) Expression patterns of *Gr5a* in taste neurons in the labellum **(a,b)** and distal segments in the leg **(c,d)**. Shown here are whole mount samples of labella or tarsi. Genotypes examined were as follows: *Gr5a-GAL4*; *UAS-lacZ* **(a,c)** and *Gr5a-GAL4/UAS-mCD8:GFP*; *Gr5a-GAL4/UAS-mCD8:GFP* **(b,d)**. Shown in **b** is a composite of a series of confocal images. Reporter gene expression was observed in ~30 cells in each half of the labellum and two cells in each of the two to three distal-most segments of the leg. (Reprinted from Chyb, S. et al., *Proc. Natl. Acad. Sci. USA* **100:**14526–14530, 2003. With permission.)

GAL4 drivers to suppress specific neurons by expressing a modified Shaker K^+ channel that inhibits firing of the cell (White et al., 2001). The levels of *Gr* proteins can be decreased by expression of RNAi, which knocks down expression (Bray and Amrein, 2003), or increased by expression of full-length cDNA (A. Dahanukar and J. Carlson, unpublished results). Expression of a GFP fusion protein, such as synaptobrevin-GFP (Ito et al., 1998) is also being used to trace the projections of these neurons to the CNS (Scott et al., 2001).

9.4 BEHAVIORAL PARADIGMS FOR THE ANALYSIS OF TASTE RECEPTOR FUNCTION

Drosophila exhibits a variety of taste-driven behaviors that are stereotypical in response to controlled stimuli and can be easily measured under experimental conditions. In particular, there are two well-studied responses in the adult: one that is based on an all-or-none response and another that requires the animal to make a choice between two stimuli. Gustatory behaviors have also been described in larvae and will be discussed at the end of this section.

9.4.1 Proboscis Extension Response

When a fruit fly is presented with a sugar solution, by application to either the labellum or the tarsi, it extends its mouthparts in a predictable reflex called the *proboscis extension response* (PER). The study of gustatory reception in insects began when Minnich discovered that the legs of *Lepidoptera* were sensitive to a sugar solution (Minnich, 1921). The classical PER was shown to be evoked by stimulation of either the labellum or leg of *Phormia*, *Calliphora,* and *Drosophila* (Minnich, 1926, 1929; Vaysse and Médioni, 1973; Tompkins et al., 1979; Arora et al., 1987). For a specific stimulus, the results of a single trial are scored as either positive (the proboscis extends) or negative (the proboscis does not extend). The results can be quantified by measuring the proportion of responding animals for a given stimulus. This reflex occurs only in the case of attractive stimuli. Repellent substances, such as bitter compounds or salts, can be tested for their ability to inhibit the response to sugar when presented in a mixture (Falk and Atidia, 1975; Arora et al., 1987). The PER is dependent on the concentration of the stimulus, as is suppression of the PER, which is dependent on the dose of the inhibitory stimulus (Falk and Atidia, 1975; Arora et al., 1987). In most cases, individual animals have been tested, but groups of flies can also be tested at the same time (Vargo and Hirsch, 1982), and the paradigm can be automated (Vaysse and Médioni, 1973).

To present the stimulus, the flies are usually immobilized in pipette tips (Vargo and Hirsch, 1982) or secured on slides or needle using either wax (Tompkins et al., 1979) or adhesive (Deak, 1976). A different technique was developed more recently in order to circumvent the need to immobilize the flies. This involves coating the inside of a 1-ml pipette with a thin layer of agarose containing the stimulus and allowing the fly to navigate through the length of the pipette (Cheng et al., 1992). In this case, the response is measured as the number of proboscis extensions observed within a predetermined time frame.

Age, sex, and starvation have been shown to affect the PER. The PER threshold for sucrose increases with age (Brigui et al., 1990), and accordingly, most studies use flies less than one week old for the PER assay (Devaud, 2003). Sex also has significant influence on the PER. In males, the PER threshold with tarsal stimulation is ten times lower than in females (Vaysse and Médioni, 1973). A period of starvation is also necessary to motivate flies during the PER assay. Excessive starvation before testing, however, can lead to unhealthy flies and variable results. Minnich furthermore showed that starvation of *Calliphora* increased the sensitivity of the tarsal sensilla to sugar (Minnich, 1929).

9.4.2 Two-Choice Feeding Preference Assay

The second commonly used behavioral test requires the flies to make a choice between two taste stimuli. That fruit flies can evaluate two sets of stimuli and choose between them was demonstrated in an experiment where flies were presented with a choice between a sugar medium that contained a bitter compound (quinine) and one that contained sugar alone (Tompkins et al., 1979). The two-choice assay developed by Tanimura et al. (1982) measures this behavior in response to two stimuli that are in alternating wells of a microtiter dish with different dyes present in the alternating wells (Figure 9.3). Flies are allowed to feed on these dishes in the dark for a certain time interval and scored for their feeding preference by the color of their abdomens. In contrast to the proboscis extension test, this assay is carried out on populations of flies and the results are interpolated as a preference index indicating the proportion of flies that prefer one of the stimuli as compared with the other. Thus, flies have to make a choice to be included in the calculations; those that do not feed, for any reason at all, are excluded from the data set.

This is a simple but powerful and exquisitely sensitive assay. In most of these experiments, the standard stimulus is chosen as 2 mM sucrose, and flies can distinguish between this concentration and 1 mM sucrose. However, physiological responses from individual labellar sugar neurons cannot easily be detected at such low concentrations, illustrating the sensitivity of this behavioral paradigm. Increasing the concentration of the standard sugar solution to 10 mM sucrose reduces the variability among trials and increases the proportion of flies that feed on the media (A. Dahanukar and J. Carlson, unpublished results).

9.4.3 Larval Chemotaxis Assay

Larvae also show a preferential response when presented with two different stimuli. Using a partitioned Petri dish or Plexiglas box, different compounds dissolved in agarose are placed into different partitions. Larvae are placed in the center of the dish or box and are allowed to explore for a predetermined time, after which time the larvae in each partition are counted (Tompkins, 1979; Miyakawa et al., 1980; Miyakawa, 1981; Lilly and Carlson, 1990). Using this assay, Miyakawa analyzed responses to several sugars and demonstrated that sucrose and fructose had a response threshold ten times lower than glucose, maltose, and fucose (Miyakawa et al., 1980). As do adults, larvae avoid a high concentration of salt or quinine (Tompkins, 1979; Miyakawa, 1981). This method allows convenient screening of large numbers of larvae to attractive or aversive stimuli.

9.5 SINGLE-UNIT ELECTROPHYSIOLOGICAL ANALYSIS

Another powerful tool for studying the function of taste receptors is physiological recording from individual sensilla that house taste neurons. Taste behaviors result from the integration of information from the ensemble of peripheral neurons, as well as being influenced by central processes such as learning and memory. Single-unit physiological recordings allow the researcher to assess the events occurring within

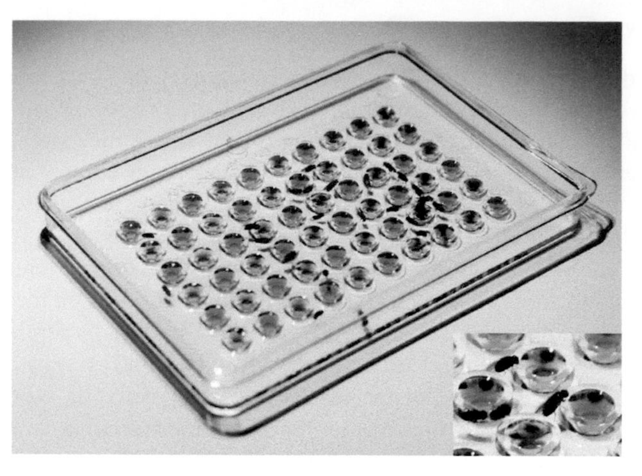

FIGURE 9.3 (See color insert following page 202.) Two-choice feeding preference test. In alternating wells of a microtiter dish flies are presented with two stimuli that also contain different dyes.

individual taste sensilla in live animals. Hodgson and his colleagues first described the tip-recording technique in 1955 (Hodgson et al., 1955). A glass electrode is filled with a solution that contains the tastant and an electrolyte, thus serving both to present the stimulus and to record the response (Figure 9.4). This electrode is guided toward the sensillum using a micromanipulator until it just encompasses the tip, so that the solution is in contact with the sensillar lymph via the pore that is present at the tip. The tastant can diffuse through the lymph and stimulate the dendrites, and the electrolyte provides a means for conducting electrical activity from the sensillum. A reference electrode is inserted into the fly head. The signal from the recording electrode is fed into a high input impedance nonblocking amplifier and transferred via an A/D converter to a computer with software for analyzing output (AutoSpike, Syntech)[(29)]. Action potentials (Figure 9.4) are recorded and are generally reported as the number of spikes in a given window of time after onset of the stimulus.

Taste sensilla in *Drosophila* usually house a single mechanosensory neuron and up to four gustatory neurons (Falk et al., 1976; Nayak and Singh, 1983). Based on their best stimuli, these four chemosensory neurons have been classified as a sugar cell (S), a water cell (W), a highly sensitive salt cell (L1), and a salt cell of lower sensitivity (L2) (Hodgson et al., 1955; Rodrigues and Siddiqi, 1978). The L2 cell has also been called the fifth cell, and recent studies suggest that it responds to bitter compounds (Meunier et al., 2003; Ozaki et al., 2003). In spite of the presence of multiple neurons in each sensillum, the activities of the different neurons can be distinguished by their differing spike amplitudes (for more on spike sorting, see Chapter 14). Moreover, spike sorting is aided by the fact that these neurons are apparently silent when the appropriate stimulus is absent. Only the L1 and L2 cells respond to a stimulus of 100 mM NaCl, for example, which elicits no action potentials from the S cell. Responses of the sugar cell can also be measured using

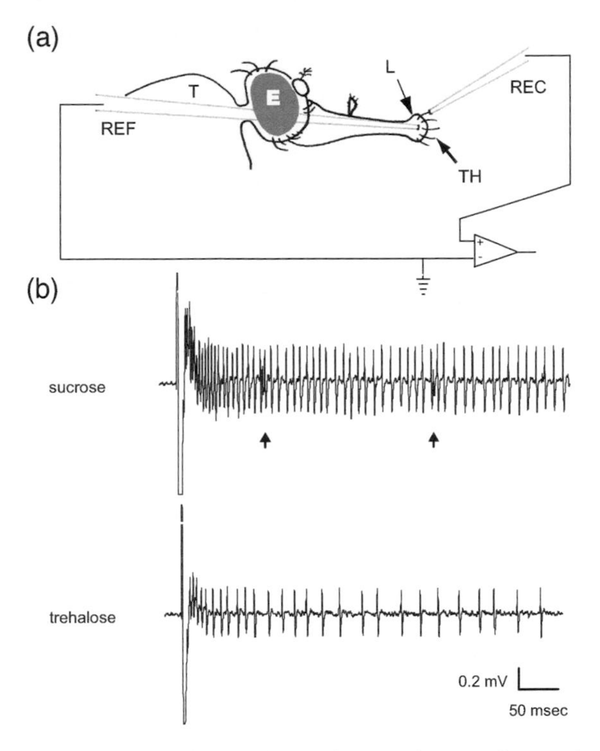

FIGURE 9.4 Electrophysiological recordings using the tip-recording method. **a**: Schematic representation of the tip-recording method. T, thorax; E, eye; L, labellum; TH, taste hair; REF, reference electrode; REC, recording electrode. **b**: Action potential recordings from an M-class sensillum in response to 100 mM sugar solutions (as indicated). Large spikes are those of the sugar cell. Spikes of the water cell are indicated by arrows below the trace.

tricholine citrate as the electrolyte, which inhibits the response from the water cell and does not evoke a response from the salt cell (Wieczorek and Wolff, 1989).

Although most taste sensilla contain four chemosensory neurons, some contain only one or two. Initial characterization of these sensilla was based on morphology and location, rather than on function (Ray et al., 1993; Shanbhag et al., 2001). Whether each morphological type is functionally identical or can be subdivided into functional classes (as for the olfactory sensilla; De Bruyne et al., 2001) awaits further investigation. A study in the blowfly, *Protophormia*, showed that responses to sugars vary between morphological types but not within, suggesting that sensilla from the same morphological type are functionally identical (Wieczorek and Koppel, 1982). A more recent study in *Protophormia* suggests that sensilla of a given morphological type vary in their responses depending upon their location on the labellum (Liscia et al., 1998). In *Drosophila*, the only published survey on comparisons of responses between labellar taste sensilla was unable to document qualitative differences in sugar responses between morphologically different sensilla, but showed that the dose response curves varied (Hiroi et al., 2002). In all these studies, however, a limited panel of stimuli was tested. It will be interesting to test a wider and more diverse choice of tastants in order to establish a detailed functional map of the taste sensilla.

The tarsal sensilla have not been studied as extensively as their labellar counterparts, but studies show that the morphological types can comprise more than one functional class in *Phormia* as well as in *Musca* (McCutchan, 1969; Shiraishi and Tanabe, 1974; Hansen et al., 1998). Tarsal neurons may also vary in their sensitivities as compared with labellar sensilla (Schnuch and Seebauer, 1998). In *Drosophila*, males have more tarsal sensilla than females, and their responses to nonpheromonal substances, such as salts, sugar, and water, are also sexually dimorphic (Meunier et al., 2000). Meunier et al. (2000) divided the tarsal sensilla into three groups, based on their responses to sucrose, NaCl, and water, and found that one type of sugar-sensitive sensillum was absent in males (Meunier et al., 2000). Although there is no behavioral evidence for sexual differences in response to nonpheromonal substances, this data shows that a dimorphism does exist and may be significant for oviposition site selection. A difference between *Drosophila* tarsal sensilla and those of either *Phormia* or *Musca* is that the water cell may be present in *Drosophila* but not in the other flies. This is controversial, as Van der Starre reported that two tarsal cells in *Phormia* and *Calliphora* were responsive to water (Den Otter and Van der Starre, 1967; Van der Starre, 1970).

An obvious limitation of the tip-recording method is that the stimulus must be dissolved in the electrolyte solution. This means that compounds that are not water-soluble cannot be tested conveniently. A modification of this technique was developed to accommodate the testing of hydrophobic or nonpolar compounds (Morita and Yamashita, 1959) by placing an independent recording electrode in direct contact with the sensillar lymph via a crack made in the side of the sensillum. In these recordings, called *side-wall recordings*, any kind of stimulus can then be presented to the tip of the sensillum. So far, this method has been employed in a very limited manner in *Drosophila*, mainly due to the small size of the taste sensilla. However, it may become invaluable for the study of tarsal sensilla, at least some of which are suggested to play a role in the reception of hydrophobic cuticular pheromones (Nayak and Singh, 1983; Stocker, 1994; Bray and Amrein, 2003).

Electrophysiological recordings have been used to assess not only qualitative differences in response to different stimuli, but also quantitative differences in response to different concentrations of the same stimulus. Furthermore, responses can be either excitatory or inhibitory. Bitter compounds, which are perhaps the most diverse class of taste stimuli, were reported not to elicit action potentials in *Drosophila* taste neurons (Morita, 1959; Siddiqi and Rodrigues, 1980). The response to a bitter stimulus can be measured by the extent of inhibition of a sugar or salt response when the two stimuli are presented in a mixture (Dethier and Bowdan, 1992; Glendinning et al., 2000). However, in some insects such as *Manduca* caterpillars, bitter stimuli do evoke an excitatory response (Glendinning and Hills, 1997), and recent evidence suggests that this may also be the case for the fifth cell in labellar sensilla of *Phormia* (Liscia et al., 1998; Ozaki et al., 2003) and *Drosophila* (Meunier et al., 2003).

Another feature of the response that can be investigated with this method is adaptation. This has not been investigated in *Drosophila* but has been studied in other insects, particularly in the *Manduca sexta* caterpillar (Glendinning and Hills, 1997). More recent studies examined the phenomenon of cross-adaptation among three bitter compounds that stimulate the same cell in the caterpillar, and the results provide evidence that two transduction pathways operate within this cell (Glendinning et al., 2002).

The effects of amino acids on gustatory cells are also complex. In the tsetse fly, *Glossina fuscipes fuscipes*, tarsal taste cells respond strongly to many amino acids, which are found in abundance in the sweat of their primary host, humans (Van der Goes van Naters and Den Otter, 1998). The labellar sensory responses of the fleshfly, *Boettcherisca peregrina*, and the blowfly, *Phormia regina*, to 19 L-amino acids were examined (Shiraishi and Kuwabara, 1970). Application of each amino acid to individual sensilla either had no effect or evoked one of the following three responses:

1. The amino acid inhibited all the taste cells within that sensillum.
2. It stimulated the salt cell.
3. It stimulated the sugar cell.

These flies apparently do not have a dedicated cell that responds solely to amino acids. There are few, if any, reports in the literature of amino acid responses in the fruit fly (Wieczorek and Wolff, 1989).

The characterization of response spectra of sensilla in wild-type flies will be invaluable for uncovering the functional organization and complexity of the gustatory system. Furthermore, in combination with the expression analysis of *Gr* receptors, it may provide a map of the molecular basis of gustatory coding. A comprehensive study of taste sensitivity is lacking in *Drosophila*, as the responses to only a handful of salts and bitter and sweet compounds have been analyzed (Hodgson et al., 1955; Isono and Kikuchi, 1974; Tanimura et al., 1982; Fujishiro et al., 1984; Arora et al., 1987; Hiroi et al., 2002; Meunier et al., 2003). Selection of appropriate test molecules is challenging in that there is a vast number of compounds found in nature from which to choose. Compounds selected from the habitat of *Drosophila melanogaster*, specifically food sources, are the most likely candidates for strong taste receptor ligands. As their name implies, fruit flies live on and around fruit, particularly rotting and fermenting fruits. In addition to sugars, sugar alcohols and fermentation products are likely ligands as they are found abundantly in fermenting fruit to which *Drosophila* are attracted. Cuticular pheromones may also be potential ligands for gustatory receptors, given the evidence that tarsal sensilla may be playing a role in gustatory-based courtship behavior (Jallon, 1984; Tompkins, 1984; Hall, 1994; Bray and Amrein, 2003).

9.6 GENETIC ANALYSIS OF TASTE RECEPTOR GENES

Prior to the identification of *Gr* genes using bioinformatics, several genetic approaches had been drawn on in attempts to isolate mutations of taste receptors and other taste genes (Table 9.1). These included many forward genetic screens, which will be described first, followed more recently by reverse genetic approaches that ultimately produced mutations of a *Drosophila* taste receptor gene.

TABLE 9.1
Summary of Taste Mutants

Mutant	Phenotype	Reference
126BO4	Decreased response to glucose and sucrose, but not fructose	a
Lot mutants	Increased tolerance to high concentrations of NaCl	b
***gus* mutants**		
gusA	Abnormal response to quinine and sucrose, but not NaCl; abnormal larval chemotaxis to quinine	c, d
gusB	Abnormal response to quinine, sucrose, and NaCl	c, d
gusC	Abnormal response to NaCl, but not quinine or sucrose	c, d
gusD	Abnormal response to both quinine and NaCl, but not sucrose	c, d
gusE	Abnormal response to quinine, sucrose and NaCl; normal larval chemotaxis to quinine	c, d
gusF	Abnormal response to NaCl, but not quinine or sucrose	c, d
***gust* mutants**		e
gustA	Reduced response to sucrose, glucose, and maltose, but not fructose	f
gustB	Increased tolerance to NaCl accompanied by salt-induced firing of the sugar neuron	g
gustC	Defects in responses to sugars, salts, and quinine, and increased labellar hair sensitivity to NaCl	h
gustD	Increased tolerance to quinine	h
gustE	Eliminated sodium, but not potassium, response of the L1 neurons	i
gustF	Reduced sensitivity to both NaCl and KCl	i
GustR	Reduced response to sucrose and increased attraction to NaCl, but not KCl; enhanced firing of the L1 neuron to NaCl	i
GustS	Increased attraction to NaCl and decreased response to sugars	i
Shaker	Reduced sensitivity to sucrose and increased tolerance to salts; electrical responses of labellar taste hairs to salts and sucrose were normal	j
shaking-B	Reduced response to sucrose and fructose and increased threshold of the salt response	j
malvolio	Reduced response to sugars and increased acceptance to salt; electrophysiological responses to sucrose were normal	k
dpr	Reduced inhibitory response to salt	l
Gr5a	Reduced response to trehalose, but not sucrose	m

Sources: **a:** Isono and Kikuchi, 1974; **b**: Falk and Atidia, 1975; **c:** Tompkins et al., 1979; **d:** Tompkins, 1979; **e:** Rodrigues and Siddiqi, 1978; **f:** Rodrigues and Siddiqi, 1981; **g:** Arora et al., 1987; **h:** Rodrigues et al., 1991; **i**: Siddiqi et al., 1989; **j:** Balakrishnan and Rodrigues, 1991; **k:** Rodrigues et al., 1995; **l:** Nakamura et al., 2002; **m:** Dahanukar et al., 2001

9.6.1 Genetic Screens for Taste Mutants

Initial screens for taste mutants employed various behavioral assays. Some of the first taste mutants reported were the X-linked *Lot* mutants, which were isolated based on their increased tolerance to high concentrations of salt (Falk and Atidia, 1975). The screen used a feeding assay in which salt-tolerant flies were identified visually after being allowed to feed on a red-dyed sugar solution that was laced with salt. Further characterization suggested that at least one of the *Lot* loci is involved in central processing, rather than peripheral recognition of salt.

Another feeding assay using four stimuli — water, fructose, glucose, and sucrose — was used to isolate the first EMS-induced mutant that exhibited a defect in sugar response (Isono and Kikuchi, 1974). This mutant, *126B04*, showed a decreased response to glucose and sucrose, but not to fructose, suggesting that reception of sugars is mediated through multiple receptors. This suggestion agreed with the results of previous biochemical and physiological experiments in which protease treatments were used to define pyranose and furanose receptor sites in the membrane (Tanimura and Shimada, 1981). Although the phenotype of this mutant was consistent with that expected for a mutation in a sugar receptor gene, it has since been lost.

A behavioral countercurrent paradigm, in which flies were required to make a series of binary choices between a sugar solution and a sugar solution containing either quinine or a high concentration of salt, was used to isolate the *gus* (*gustatory*) mutants (Tompkins et al., 1979). Wild-type flies are repelled by quinine or high concentrations of salt. Twelve *gus* mutants were identified and were broadly divided into "mistactic" mutants, which preferentially distributed to the quinine or salt substrates, and "atactic" mutants, which distributed equally between the two choices. These mutations mapped to six complementation groups: *gusA* through *gusF.* In the mistactic class, *gusA* mutants showed an abnormal response to quinine but not to NaCl. In contrast, *gusC* responded normally to quinine but abnormally to salt. The remaining mistactic mutants, *gusD* and *gusE,* responded abnormally to both quinine and NaCl. Of the atactic mutants, *gusB* responded abnormally to both quinine and NaCl; in contrast, *gusF* mutants responded normally to quinine but were atactic when tested with salt. The sugar responses of these mutants were further characterized using the proboscis extension assay (Tompkins et al., 1979), in which *gusA*, *gusB,* and *gusE* showed abnormal responses to sugar, whereas *gusC*, *gusD,* and *gusF* had normal responses to sugar. Further analysis implicated both *gusA* and *gusE* in developmental roles. *gusA* affects the behavior of both larvae and adults, and analysis of a temperature-sensitive allele showed that its temperature-sensitive period is during embryogenesis, consistent with a role in development. This is also the case for *gusE*, which does not affect the behavior of larvae, but its temperature-sensitive period occurs during the third larval instar when the adult nervous system differentiates (Tompkins, 1979).

Another study resulted in the isolation of an independent set of gustatory mutants using the proboscis extension assay and a Y-maze in which one arm was treated with quinine (Rodrigues and Siddiqi, 1978). These mutants included *gustA*, which did not respond to sucrose but responded normally to quinine and salt; *gustB*, which showed an increased tolerance to salt; and *gustC*, which did not respond to any tastant. Single-unit electrophysiology experiments showed that the specificity of the sugar cell in *gustB* labellar sensilla was altered to allow responses to salts, suggesting that this gene is involved in a developmental or regulatory role (Arora et al., 1987). *gustC* is also not a likely candidate for a receptor gene, but is involved in a more central role, because electrophysiological responses were normal (Rodrigues et al., 1991). In contrast, *gustA* seemed the best candidate for a G protein–coupled receptor (Rodrigues and Siddiqi, 1981) but has been lost.

Other mutants with taste phenotypes include *Shaker* and *shaking-B* (Balakrishnan and Rodrigues, 1991) and *malvolio* (Rodrigues et al., 1995). The *Shaker* gene encodes a potassium channel, and its alleles show behavioral defects that include a reduced sensitivity to sugar and an increased tolerance to salt. However, their peripheral electrophysiological responses are comparable to wild-type, suggesting a more central role (Balakrishnan and Rodrigues, 1991). Mutations in *shaking-B*, whose function is involved with gap junctions (Zhang et al., 1999), result in a reduced response to sugars and an increase in the threshold of the salt response (Balakrishnan and Rodrigues, 1991). As with the *Shaker* alleles, mutations in *shaking-B* do not appear to affect the peripheral gustatory response.

The *malvolio* mutant was isolated in a screen for mutants defective in sugar perception (Rodrigues et al., 1995). Mutant flies show a reduced response to sugars and other attractants and an increased acceptance to salt. The *malvolio* gene encodes a protein with striking similarity to a class of transporters named natural resistance–associated macrophage proteins (NRAMP), containing a consensus sequence found in a number of ATP-coupled transporters (Rodrigues et al., 1995). As with most of the other taste mutants identified so far, the electrophysiological responses are normal, suggesting that the aberrant behavioral responses are the result of a lesion at a more central level.

Recently, a new gustatory mutant called *defective proboscis response* (*dpr*) was isolated (Nakamura et al., 2002). This mutant has a phenotype similar to the *Lot* mutants and exhibits a proboscis extension response to a sugar solution also containing high salt. Such a mixture inhibits the PER in wild-type flies. No morphological defects were observed in the taste sensilla of *dpr* mutant animals. The *dpr* gene encodes a membrane-bound protein with two Ig-like domains. The *Drosophila Klingon* protein (Butler et al., 1997) and the mammalian protein LAMP (limbic system–associated membrane protein) (Pimenta et al., 1995), proteins with similar Ig domains, have been shown to function in the nervous system.

Although many of these genetic screens were successful in the isolation of mutants that affect taste behavior, none of the mutants, with the exception of *gustA* and *126B04*, showed specific peripheral phenotypes consistent with those of taste receptor mutations. There are several possible explanations for this paucity of candidate receptor mutants. One possibility is that there is some degree of functional redundancy among receptors and that the assays were not sufficiently sensitive to detect mutations of a single receptor gene. Another explanation is that the number of stimuli tested has been limited and likely does not include the strongest stimuli for most of the receptors.

9.6.2 The Trehalose Response Locus and Gr5a

Elegant behavioral and genetic analysis by Tanimura and colleagues identified a locus whose alleles, some naturally occurring, confer different levels of response to the disaccharide trehalose (Tanimura et al., 1982). The locus was mapped to cytogenetic position 5A on the X chromosome (Tanimura et al., 1988). A GPCR gene in the region was identified, characterized, named *Tre1*, and proposed to encode a trehalose receptor (Ishimoto et al., 2000). It has sequence similarity to vertebrate melatonin receptors but not to *Gr* proteins.

Following the identification of the *Gr* genes as candidate taste receptors, the genomic location of *Gr* genes was compared with those of loci implicated in taste function. One gene, *Gr5a*, mapped less than 900 base pairs from *Tre1*. A number of deletion mutations isolated by Ueno et al. that affect trehalose response were found to uncover the *Gr5a* gene or both *Gr5a* and *Tre1* (Dahanukar et al., 2001; Ueno et al., 2001). To test directly the possibility of a role for *Gr5a* in trehalose response, Dahanukar et al. (2001) engineered three transformation rescue constructs, each containing copies of *Tre1* and *Gr5a*. The first construct contained wild-type copies of both genes, the second construct was identical but contained a stop codon in *Tre1*, and the third contained a wild-type copy of *Tre1* but carried a stop codon in *Gr5a*. Rescue experiments showed that trehalose response depended on the presence of a wild-type *Gr5a* gene but not the *Tre1* gene, in either behavioral measurements or physiological recordings from taste neurons, and in either of two deletion mutant backgrounds. These results thus provide direct functional evidence for a role for *Gr5a* in trehalose response.

The results are consistent with those of Ueno et al. (2001), who examined multiple *Drosophila* strains and identified a number of polymorphisms in *Gr5a*. One, a substitution of threonine for alanine in the predicted second intracellular loop, segregated with a reduced behavioral response to trehalose. Further analysis of *Tre1* by others has recently shown that it is expressed in germ cells and is required for their transepithelial migration in embryos (Kunwar et al., 2003). Further evidence that *Gr5a* confers response to trehalose has been obtained from heterologous expression studies that are described in Section 9.7.

An interesting finding in the Dahanukar et al. (2001) study was that although elimination of *Gr5a* affected the physiological response of sugar-sensitive neurons to trehalose, the response of these neurons to sucrose was unaffected. The simplest interpretation of these results is that these neurons contain more than one taste receptor: one for trehalose and one for sucrose.

9.6.3 RNAi Analysis of Gr68a Function

Another means to dissect the function of individual *Gr* genes when a mutant is not available is to use molecular genetic techniques to reduce levels of *Gr* gene expression. Most of the *Gr* genes appear to be expressed in small subsets of taste neurons; *Gr* gene promoter–GAL4 fusion lines can be used to express double-stranded RNA to knock down the level of *Gr* gene expression by RNA-mediated interference (RNAi). Double-stranded RNA has been shown to inhibit gene expression in a sequence-specific manner in many different organisms such as worms, plants, and flies, and also in cultured cells (Fire et al., 1998; Furner et al., 1998; Kennerdell and Carthew, 1998; Bahramian and Zarbl, 1999). Fire and Mello showed that a mixture of sense and antisense strands of RNA effectively silenced expression of the targeted gene in *C. elegans* (Fire et al., 1998). Double-stranded RNA injected into *Drosophila* embryos reduced gene expression in the embryo but was not consistent in the adult (Kennerdell and Carthew, 1998). Subsequently, it was found that gene expression could be reduced in transgenic flies via constructs containing two cDNA sequences arranged in an inverted repeat; transcription of this DNA is predicted to lead to dsRNA (Kennerdell and Carthew, 2000). More recently, this technique has been modified to incorporate constructs comprising genomic DNA fused to inverted cDNA. These transgenes are also predicted to express dsRNA and have been more effective at reducing the expression of the targeted gene than the cDNA transgenes described in previous studies (Kalidas and Smith, 2002).

RNAi has been used to study the function of *Gr68a* (Bray and Amrein, 2003). *Gr68a* is expressed in a very restricted pattern in a subset of taste sensilla on male forelegs. The forelegs of male flies have significantly more taste sensilla than those of females and have been suggested to play a role in pheromone reception and courtship behaviors (Nayak and Singh, 1983; Jallon, 1984; Hall, 1994; Stocker, 1994; Bray and Amrein, 2003). Male transgenic flies expressing RNAi directed against the *Gr68a* gene had defects in courtship behavior consistent with a role in pheromone reception (Bray and Amrein, 2003). The authors also drove expression of tetanus toxin using a *Gr68a* promoter–GAL4 transgene to inactivate selectively the neurons expressing *Gr68a*. These males had a more severe courtship defect than the males expressing RNAi, suggesting either that the RNAi phenotype is a hypomorphic phenotype or that *Gr68a* is not the only gene involved in pheromone reception in these cells. Consistent with the second possibility is the observation that flies expressing RNAi for *Gr68a* as well as *Gr32a*, which is closely related to *Gr68a* and also expressed in the forelegs, have a more severe defect in courtship behavior than those that express *Gr68a* RNAi alone.

RNAi has not been attempted for any of the *Gr* genes that are primarily expressed in the labellum, but it provides a simple and effective method for silencing expression of specific genes in specific cells or tissues in *Drosophila*. The main disadvantage of knock-down approaches is that identification of the ligand may be quite challenging. A large number of taste stimuli may have to be tested in order to determine the function of a particular taste receptor. (See also Chapter 10 for a discussion of RNAi methods in olfaction.)

9.7 HETEROLOGOUS EXPRESSION OF TASTE RECEPTORS

There are two main advantages to studying *Gr* gene function in a heterologous system. First, this approach allows direct confirmation of the role of a *Gr* gene as a taste receptor. Second, heterologous expression provides a convenient method by which to characterize the response spectrum of an individual *Gr* gene in detail. It may provide an efficient approach for screening a large number of ligands for orphan *Gr* receptors.

The *Drosophila* Schneider 2 (S2) cell line has been used for heterologous protein expression, with the aid of a number of different promoters. Plasmid vectors that allow constitutive expression, as well as those that can be induced to achieve a tightly regulated pattern of expression, are available (Bunch et al., 1988; Angelichio et al., 1991). S2 cells are loosely adherent cells that have a doubling time of approximately

24 hours. They can be transfected by calcium phosphate–DNA coprecipitation, similar to the standard techniques used to transfect mammalian cells (Cherbas et al., 1994). Other reagents include commercially available transfection agents, many of which are liposome formulations that form lipid–DNA complexes upon incubation with the plasmid and cause efficient transfection (e.g., Cellfectin)[(10)]. Stably transfected S2 cells may contain up to several thousand copies of the plasmid DNA (Cherbas et al., 1994).

The TRP and TRPL cation channels are functional when expressed in S2 cells (Hardie et al., 1997; Reuss et al., 1997; Chyb et al., 1999). Their activity is enhanced by coexpression of activated G_q subunits (Obukhov et al., 1996) and addition of IP3 (Dong et al., 1995), suggesting that they couple with the G_q alpha subunit present in S2 cells (Hardie et al., 1997; Chyb et al., 1999), as believed to be the case *in vivo* (Reuss et al., 1997; Raghu et al., 2000). Activation of the G_q alpha subunit results in the accumulation of the IP3 second messenger and release of Ca^{2+} from intracellular stores (Macrez-Leprêtre et al., 1997; Cabrera-Vera et al., 2003). Changes in the level of cytosolic Ca^{2+} can be measured using calcium indicator dyes, such as FURA-2, which is a fluorescent ratiometric indicator (Tsien et al., 1985). FURA-2 and its complex with Ca^{2+} fluoresce strongly at different wavelengths, so changes in the level of free Ca^{2+} may be measured by an assessment of the ratio of the fluorescence of the dye–Ca^{2+} complex to that of the free dye (see also Chapter 13). Upon presentation of a stimulus, responding cells are defined as those with fluorescence ratios above a predetermined threshold, and results are usually quantified as a percentage of responding cells.

In vivo experiments showed that the *Gr5a* gene is required in taste neurons for response to trehalose (Dahanukar et al., 2001). These results suggested that Gr5a is a taste receptor for trehalose, but the suggestion was tested directly by expressing *Gr5a* in S2 cells and analyzing the response of transfected S2 cells to trehalose by Ca^{2+} imaging (Chyb et al., 2003). Cells transfected with *Gr5a*, but not with the vector alone, showed a transient increase in the level of intracellular Ca^{2+} after application of the stimulus. The kinetics and dose responses were characterized in detail (Figure 9.5. This system also allows for a number of ligands to be tested. Examination of the responses of Gr5a to a number of sugars closely related to trehalose suggests that the Gr5a receptor shows a high degree of specificity for trehalose.

One of the disadvantages of this system is that S2 cells are notoriously difficult to grow under conditions of low density, making the cloning of transfected cells a challenge (Graziano et al., 1998). It may be possible to use mammalian cell lines for these types of experiments if *Drosophila* G protein–coupled receptors can be efficiently targeted to the membrane in these cells. Mammalian bitter, sweet, and amino acid taste receptors have been successfully expressed in cell culture (Chandrashekar et al., 2000; Nelson et al., 2001, 2002). In one case, however, receptors were expressed as fusion proteins attached to the N-terminus of rhodopsin to achieve correct localization to the membrane (Chandrashekar et al., 2000).

Gr5a is only the first *Gr* gene to be expressed; additional *Gr* genes await testing in this system. Heterologous expression studies provide a powerful tool with which to study biochemical properties of individual receptors, such as the

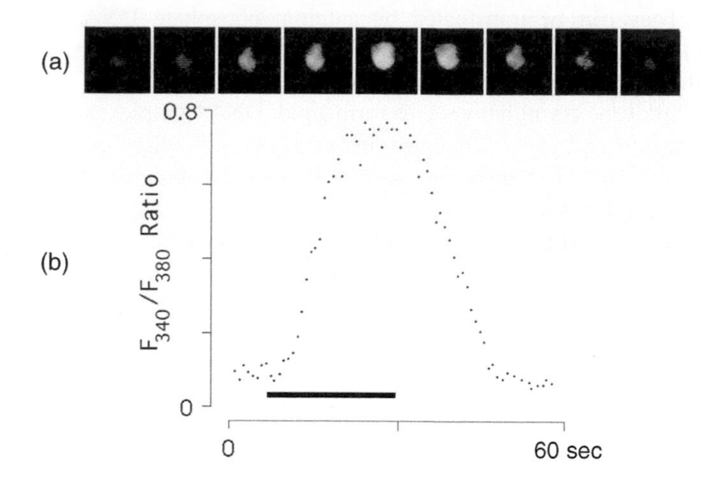

FIGURE 9.5 (See color insert following page 202.) Calcium response to trehalose in S2-*Gr5a* cells. **a**: A series of images of a single Fura 2-loaded S2-*Gr5a* cell, taken at 5 sec intervals. The first image is taken 5 sec before the application of 100 mM trehalose. **b**: A quantitative representation of the calcium response in the same cell. Bar indicates stimulus period. (Reprinted from Chyb, S. et al., *Proc. Natl. Acad. Sci. USA* **100:**14526–14530, 2003. With permission.)

tuning specificity and kinetics of ligand binding. Because multiple receptors appear to be expressed in individual taste neurons (Dahanukar et al., 2001), the study of these properties *in vivo* presents a challenge. Another advantage of the heterologous expression system is that expression of chimeric receptors, or those with specific point mutations, can be used to test perturbations in the structure of individual receptors (A. Dahanukar, J. Perry, and J. Carlson, unpublished results). Finally, receptors may also be tested in combinations to address whether they can function as heterodimers, as has been described in vertebrates (Nelson et al., 2001, 2002), a property that could be another facet of combinatorial gustatory coding.

9.8 CONCLUSIONS AND OUTLOOK

In this chapter, we have reviewed current approaches used to study *Drosophila* taste receptors. Our basic understanding of the molecular mechanisms underlying taste perception is limited, but recent advances in the field have begun to elucidate these mechanisms. We have attempted to provide a guide to the tools used for studying taste receptors in *Drosophila* by examining recent discoveries and current techniques. We have shown how taste receptors were identified and describe how their expression patterns are being analyzed in *Drosophila.* We also have surveyed behavioral and electrophysiological approaches used to study taste receptor function *in vivo*. We have reviewed the isolation of classical taste mutants and illustrate the methods that are currently being used to analyze the function of taste receptor genes *in vivo*. Lastly, we have examined a method for screening numerous candidate ligands and for studying properties of individual taste receptors using a heterologous expression system. With the identification of candidate taste receptors and the current techniques in place, advances in this field should come rapidly.

ACKNOWLEDGMENTS

This work was supported by NIH research grants (to J. Carlson) and an NRSA (to A. Dahanukar). We thank Aaron Goldman for photography.

REFERENCES

Angelichio ML, Beck JA, Johansen H, Ivey-Hoyle M (1991) Comparison of several promoters and polyadenylation signals for use in heterologous gene expression in cultured *Drosophila* cells. *Nucleic Acids Res.* **19:**5037–5043.

Arora K, Rodrigues V, Joshi S, Shanbhag S, Siddiqi O (1987) A gene affecting the specificity of the chemosensory neurons of *Drosophila. Nature* **330:**62–63.

Awasaki T, Kimura K (1997) *pox-neuro* is required for development of chemosensory bristles in *Drosophila. J. Neurobiol.* **32:**707–721.

Bahramian MB, Zarbl H (1999) Transcriptional and posttranscriptional silencing of rodent alpha1(I) collagen by a homologous transcriptionally self-silenced transgene. *Mol. Cell. Biol.* **19:**274–283.

Balakrishnan R, Rodrigues V (1991) The *shaker* and *shaking-B* genes specify elements in the processing of gustatory information in *Drosophila melanogaster. J. Exp. Biol.* **157:**161–181.

Beckenbach AT, Wei YW, Liu H (1993) Relationships in the *Drosophila obscura* species group, inferred from mitochondrial cytochrome oxidase II sequences. *Mol. Biol. Evol.* **10:**619–634.

Beverley SM, Wilson AC (1984) Molecular evolution in *Drosophila* and the higher Diptera II. A time scale for fly evolution. *J. Mol. Evol.* **21:**1–13.

Brand AH, Perrimon N (1993) Targeted gene expression as a means of altering cell fates and generating dominant phenotypes. *Development* **118:**401–415.

Bray S, Amrein H (2003) A putative *Drosophila* pheromone receptor expressed in male-specific taste neurons is required for efficient courtship. *Neuron* **39:**1019–1029.

Brigui N, Le Bourg E, Médioni J (1990) Conditioned suppression of the proboscis-extension response in young, middle-aged, and old *Drosophila melanogaster* flies: acquisition and extinction. *J. Comp. Psychol.* **104:**289–296.

Bunch TA, Grinblat Y, Goldstein LS (1988) Characterization and use of the *Drosophila* metallothionein promoter in cultured *Drosophila melanogaster* cells. *Nucleic Acids Res.* **16:**1043–1061.

Butler SJ, Ray S, Hiromi Y (1997) *klingon*, a novel member of the *Drosophila* immunoglobulin superfamily, is required for the development of the R7 photoreceptor neuron. *Development* **124:**781–792.

Cabrera-Vera TM, Vanhauwe J, Thomas TO, Medkova M, Preininger A, Mazzoni MR, Hamm HE (2003) Insights into G protein structure, function, and regulation. *Endocr. Rev.* **24:**765–781.

Chandrashekar J, Mueller KL, Hoon MA, Adler E, Feng L, Guo W, Zuker CS, Ryba NJ (2000) T2Rs function as bitter taste receptors. *Cell* **100:**703–711.

Chen P, Nordstrom W, Gish B, Abrams JM (1996) *grim*, a novel cell death gene in *Drosophila. Genes Dev.* **10:**1773–1782.

Cheng E, Chungming C, Sun YH (1992) An improved proboscis extension assay. *Drosophila Information Newsletter* **6**.

Cherbas L, Moss R, Cherbas P (1994) Transformation techniques for *Drosophila* cell lines. *Methods Cell Biol.* **44:**161–179.

Chyb S, Raghu P, Hardie RC (1999) Polyunsaturated fatty acids activate the *Drosophila* light-sensitive channels TRP and TRPL. *Nature* **397:**255–259.

Chyb S, Dahanukar A, Wickens A, Carlson JR (2003) *Drosophila Gr5a* encodes a taste receptor tuned to trehalose. *Proc. Natl. Acad. Sci. USA* **100:**14526–14530.

Clyne PJ, Warr CG, Freeman MR, Lessing D, Kim J, Carlson JR (1999) A novel family of divergent seven-transmembrane proteins: candidate odorant receptors in *Drosophila. Neuron* **22:**327–338.

Clyne PJ, Warr CG, Carlson JR (2000) Candidate taste receptors in *Drosophila. Science* **287:**1830–1834.

Dahanukar A, Foster K, van der Goes van Naters WM, Carlson JR (2001) A *Gr* receptor is required for response to the sugar trehalose in taste neurons of *Drosophila. Nat. Neurosci.* **4:**1182–1186.

De Bruyne M, Foster K, Carlson JR (2001) Odor coding in the *Drosophila* antenna. *Neuron* **30:**537–552.

Deak, II (1976) Demonstration of sensory neurones in the ectopic cuticle of *spineless-aristapedia*, a homoeotic mutant of *Drosophila. Nature* **260:**252–254.

Den Otter CJ, Van der Starre H (1967) Responses of tarsal hairs of the bluebottle, *Calliphora erythrocephala* Meig., to sugar and water. *J. Insect Physiol.* **13:**1177–1188.

Dethier VG (1976) *The Hungry Fly.* Harvard University Press, Cambridge, MA.

Dethier VG, Bowdan E (1992) Effects of alkaloids on feeding by *Phormia regina* confirm the critical role of sensory inhibition. *Physiol. Entomol.* **17:**325–330.

Devaud JM (2003) Experimental studies of adult *Drosophila* chemosensory behaviour. *Behav. Processes* **64:**177–196.

Dong Y, Kunze DL, Vaca L, Schilling WP (1995) Ins(1,4,5)P3 activates *Drosophila* cation channel TRPL in recombinant baculovirus-infected Sf9 insect cells. *Am. J. Physiol.* **269:**C1332–1339.

Dunipace L, Meister S, McNealy C, Amrein H (2001) Spatially restricted expression of candidate taste receptors in the *Drosophila* gustatory system. *Curr. Biol.* **11:**822–835.

Falk R, Atidia J (1975) Mutation affecting taste perception in *Drosophila melanogaster. Nature* **254:**325–326.

Falk R, Bleiseravivi N, Atidia J (1976) Labellar taste organs of *Drosophila melanogaster. J. Morphol.* **150:**327–341.

Fire A, Xu S, Montgomery MK, Kostas SA, Driver SE, Mello CC (1998) Potent and specific genetic interference by double-stranded RNA in *Caenorhabditis elegans. Nature* **391:**806–811.

Fox AN, Pitts RJ, Robertson HM, Carlson JR, Zwiebel LJ (2001) Candidate odorant receptors from the malaria vector mosquito *Anopheles gambiae* and evidence of down-regulation in response to blood feeding. *Proc. Natl. Acad. Sci. USA* **98:**14693–14697.

Fujishiro N, Kijima H, Morita H (1984) Impulse frequency and action potential amplitude in labellar chemosensory neurones of *Drosophila melanogaster. J. Insect Physiol.* **30:**317–325.

Furner IJ, Sheikh MA, Collett CE (1998) Gene silencing and homology-dependent gene silencing in *Arabidopsis*: genetic modifiers and DNA methylation. *Genetics* **149:**651–662.

Glendinning JI, Nelson NM, Bernays EA (2000) How do inositol and glucose modulate feeding in *Manduca sexta* caterpillars? *J. Exp. Biol.* **203 Pt 8:**1299–1315.

Glendinning JI, Davis A, Ramaswamy S (2002) Contribution of different taste cells and signaling pathways to the discrimination of "bitter" taste stimuli by an insect. *J. Neurosci.* **22:**7281–7287.

Glendinning JI, Hills TT (1997) Electrophysiological evidence for two transduction pathways within a bitter-sensitive taste receptor. *J. Neurophysiol.* **78:**734–745.

Graziano M, Broderick D, Tota M (1998) Expression of G protein-coupled receptors in *Drosophila* Schneider 2 cells. In: *Identification and Expression of G Protein-Coupled Receptors*, Lynch K, Ed. Wiley-Liss, New York, pp. 181–195.

Grether ME, Abrams JM, Agapite J, White K, Steller H (1995) The head involution defective gene of *Drosophila melanogaster* functions in programmed cell death. *Genes Dev.* **9:**1694–1708.

Hall JC (1994) The mating of a fly. *Science* **264:**1702–1714.

Hansen K, Wacht S, Seebauer H, Schnuch M (1998) New aspects of chemoreception in flies. *Ann. N.Y. Acad. Sci.* **855:**143–147.

Hardie RC, Reuss H, Lansdell SJ, Millar NS (1997) Functional equivalence of native light-sensitive channels in the *Drosophila trp301* mutant and TRPL cation channels expressed in a stably transfected *Drosophila* cell line. *Cell Calcium* **21:**431–440.

Hill CA, Fox AN, Pitts RJ, Kent LB, Tan PL, Chrystal MA, Cravchik A, Collins FH, Robertson HM, Zwiebel LJ (2002) G protein-coupled receptors in *Anopheles gambiae. Science* **298:**176–178.

Hiroi M, Marion-Poll F, Tanimura T (2002) Differentiated response to sugars among labellar chemosensilla in *Drosophila. Zoolog. Sci.* **19:**1009–1018.

Hodgson ES, Lettvin JY, Roeder KD (1955) Physiology of a primary chemoreceptor unit. *Science* **122:**417–418.

Holt RA, Subramanian GM, Halpern A, Sutton GG, Charlab R, Nusskern DR, Wincker P, Clark AG, Ribeiro JM, Wides R, et al. (2002) The genome sequence of the malaria mosquito *Anopheles gambiae. Science* **298:**129–149.

Human Genome Sequencing Center at Baylor College of Medicine (2003) *Drosophila* Genome Project. http://www.hgsc.bcm.tmc.edu/projects/drosophila/index.html.

Ishimoto H, Matsumoto A, Tanimura T (2000) Molecular identification of a taste receptor gene for trehalose in *Drosophila. Science* **289:**116–119.

Isono K, Kikuchi T (1974) Autosomal recessive mutation in sugar response of *Drosophila. Nature* **248:**243–244.

Ito K, Suzuki K, Estes P, Ramaswami M, Yamamoto D, Strausfeld NJ (1998) The organization of extrinsic neurons and their implications in the functional roles of the mushroom bodies in *Drosophila melanogaster* Meigen. *Learn. Mem.* **5:**52–77.

Jallon JM (1984) A few chemical words exchanged by *Drosophila* during courtship and mating. *Behav. Genet.* **14:**441–478.

Kalidas S, Smith DP (2002) Novel genomic cDNA hybrids produce effective RNA interference in adult *Drosophila. Neuron* **33:**177–184.

Kennerdell JR, Carthew RW (1998) Use of dsRNA-mediated genetic interference to demonstrate that *frizzled* and *frizzled 2* act in the *wingless* pathway. *Cell* **95:**1017–1026.

Kennerdell JR, Carthew RW (2000) Heritable gene silencing in *Drosophila* using double-stranded RNA. *Nat. Biotechnol.* **18:**896–898.

Kim J, Moriyama EN, Warr CG, Clyne PJ, Carlson JR (2000) Identification of novel multi-transmembrane proteins from genomic databases using quasi-periodic structural properties. *Bioinformatics* **16:**767–775.

Kunwar PS, Starz-Gaiano M, Bainton RJ, Heberlein U, Lehmann R (2003) *Tre1*, a G protein-coupled receptor, directs transepithelial migration of *Drosophila* germ cells. *PLoS Biol.* **1:**E80.

Lilly M, Carlson J (1990) *smellblind*: a gene required for *Drosophila* olfaction. *Genetics* **124:**293–302.

Liscia A, Majone R, Solari P, Tomassini Barbarossa I, Crnjar R (1998) Sugar response differences related to sensillum type and location on the labella of *Protophormia terraenovae*: a contribution to spatial representation of the stimulus. *J. Insect Physiol.* **44:**471–481.

Macrez-Leprêtre N, Kalkbrenner F, Schultz G, Mironneau J (1997) Distinct functions of G_q and G11 proteins in coupling alpha1-adrenoreceptors to Ca^{2+} release and Ca^{2+} entry in rat portal vein myocytes. *J. Biol. Chem.* **272:**5261–5268.

McCutchan MC (1969) Responses of tarsal chemoreceptive hairs of the blowfly, *Phormia regina*. *J. Insect Physiol.* **15:**2059–2068.

Meunier N, Ferveur JF, Marion-Poll F (2000) Sex-specific non-pheromonal taste receptors in *Drosophila*. *Curr. Biol.* **10:**1583–1586.

Meunier N, Marion-Poll F, Rospars JP, Tanimura T (2003) Peripheral coding of bitter taste in *Drosophila*. *J. Neurobiol.* **56:**139–152.

Minnich DE (1921) An experimental study of the tarsal chemoreceptors of two nymphalid butterflies. *J. Exp. Zool.* **33:**173–203.

Minnich DE (1926) The chemical sensitivity of the tarsi of certain muscid flies. *Biol. Bull.* **51:**166–178.

Minnich DE (1929) The chemical sensitivity of the legs of the blowfly *Calliphora vomitoria* Linn., to various sugars. *Z. Vergl. Physiol.* **11:**1–55.

Miyakawa Y (1981) Bimodal response in a chemotactic behavior of *Drosophila* larvae to mono-valent salts. *J. Insect Physiol.* **27:**387–392.

Miyakawa Y, Fujishiro N, Kijima H, Morita H (1980) Differences of feeding response to sugars between adults and larvae in *Drosophila melanogaster*. *J. Insect Physiol.* **26:**685–688.

Moffat KG, Gould JH, Smith HK, O'Kane CJ (1992) Inducible cell ablation in *Drosophila* by cold-sensitive ricin A chain. *Development* **114:**681–687.

Morita H (1959) Initiation of spike potentials in contact chemosensory hairs of insects. *J. Cell. Compar. Physl.* **54:**189–204.

Morita H, Yamashita S (1959) Generator potential of insect chemoreceptor. *Science* **130:**922.

Nakamura M, Baldwin D, Hannaford S, Palka J, Montell C (2002) Defective proboscis extension response (DPR), a member of the Ig superfamily required for the gustatory response to salt. *J. Neurosci.* **22:**3463–3472.

National Human Genome Research Institute (2003) Genome Sequencing Proposals, Status of Organisms in the Prioritization Process for Genome Sequencing and their "White Paper" Proposals. http://genome.gov/page.cfm?pageID=10002154.

Nayak SV, Singh RN (1983) Sensilla on the tarsal segments and mouthparts of adult *Drosophila melanogaster* meigen (Diptera: Drosophilidae). *Int. J. Insect Morphol. Embryol.* **12:**273–291.

Nelson G, Hoon MA, Chandrashekar J, Zhang Y, Ryba NJ, Zuker CS (2001) Mammalian sweet taste receptors. *Cell* **106:**381–390.

Nelson G, Chandrashekar J, Hoon MA, Feng L, Zhao G, Ryba NJ, Zuker CS (2002) An amino-acid taste receptor. *Nature* **416:**199–202.

Obukhov AG, Harteneck C, Zobel A, Harhammer R, Kalkbrenner F, Leopoldt D, Luckhoff A, Nurnberg B, Schultz G (1996) Direct activation of *trpl* cation channels by G alpha11 subunits. *EMBO J.* **15:**5833–5838.

Ozaki M, Takahara T, Kawahara Y, Wada-Katsumata A, Seno K, Amakawa T, Yamaoka R, Nakamura T (2003) Perception of noxious compounds by contact chemoreceptors of the blowfly, *Phormia regina*: putative role of an odorant-binding protein. *Chem. Senses* **28:**349–359.

Pimenta AF, Zhukareva V, Barbe MF, Reinoso BS, Grimley C, Henzel W, Fischer I, Levitt P (1995) The limbic system-associated membrane protein is an Ig superfamily member that mediates selective neuronal growth and axon targeting. *Neuron* **15:**287–297.

Raghu P, Usher K, Jonas S, Chyb S, Polyanovsky A, Hardie RC (2000) Constitutive activity of the light-sensitive channels TRP and TRPL in the *Drosophila* diacylglycerol kinase mutant, rdgA. *Neuron* **26:**169–179.

Ray K, Hartenstein V, Rodrigues V (1993) Development of the taste bristles on the labellum of *Drosophila melanogaster. Dev. Biol.* **155:**26–37.

Reuss H, Mojet MH, Chyb S, Hardie RC (1997) *In vivo* analysis of the *Drosophila* light-sensitive channels, TRP and TRPL. *Neuron* **19:**1249–1259.

Robertson HM, Warr CG, Carlson JR (2003) Molecular evolution of the insect chemoreceptor gene superfamily in *Drosophila melanogaster. Proc. Natl. Acad. Sci. USA* **100:**14537–14542.

Rodrigues V, Siddiqi O (1978) Genetic analysis of chemosensory pathway. *P. Indian Acad. Sci. B* **87:**147–160.

Rodrigues V, Siddiqi O (1981) A gustatory mutant of *Drosophila* defective in pyranose receptors. *Mol. Gen. Genet.* **181:**406–408.

Rodrigues V, Sathe S, Pinto L, Balakrishnan R, Siddiqi O (1991) Closely linked lesions in a region of the X chromosome affect central and peripheral steps in gustatory processing in *Drosophila. Mol. Gen. Genet.* **226:**265–276.

Rodrigues V, Cheah PY, Ray K, Chia W (1995) *malvolio*, the *Drosophila* homologue of mouse NRAMP-1 (Bcg), is expressed in macrophages and in the nervous system and is required for normal taste behaviour. *EMBO J.* **14:**3007–3020.

Schnuch M, Seebauer H (1998) Sugar cell responses to lactose and sucrose in labellar and tarsal taste hairs of *Musca domestica. J. Comp. Physiol. A* **182:**767–775.

Scott K, Brady R, Jr., Cravchik A, Morozov P, Rzhetsky A, Zuker C, Axel R (2001) A chemosensory gene family encoding candidate gustatory and olfactory receptors in *Drosophila. Cell* **104:**661–673.

Shanbhag SR, Park SK, Pikielny CW, Steinbrecht RA (2001) Gustatory organs of *Drosophila melanogaster*: fine structure and expression of the putative odorant-binding protein PBPRP2. *Cell Tissue Res.* **304:**423–437.

Shiraishi A, Kuwabara M (1970) The effects of amino acids on the labellar hair chemosensory cells of the fly. *J. Gen. Physiol.* **56:**768–782.

Shiraishi A, Tanabe Y (1974) The proboscis extension response and tarsal and labellar chemosensory hairs in the blowfly. *J. Comp. Physiol.* **92:**161–179.

Siddiqi O, Rodrigues V (1980) Genetic analysis of a complex chemoreceptor. *Basic Life Sci.* **16:**347–359.

Siddiqi, O, Joshi S, Arora K, Rodrigues V (1989) Genetic investigation of salt reception in *Drosophila melanogaster. Genome* **31**:646–651.

Stocker RF (1994) The organization of the chemosensory system in *Drosophila melanogaster*: a review. *Cell Tissue Res.* **275:**3–26.

Sweeney ST, Broadie K, Keane J, Niemann H, O'Kane CJ (1995) Targeted expression of tetanus toxin light chain in *Drosophila* specifically eliminates synaptic transmission and causes behavioral defects. *Neuron* **14:**341–351.

Tanimura T, Shimada I (1981) Multiple receptor proteins for sweet taste in *Drosophila* discriminated by papain treatment. *J. Comp. Physiol.* **141:**265–269.

Tanimura T, Isono K, Takamura T, Shimada I (1982) Genetic dimorphism in the taste sensitivity to trehalose in *Drosophila melanogaster. J. Comp. Physiol.* **147:**433–437.

Tanimura T, Isono K, Yamamoto MT (1988) Taste sensitivity to trehalose and its alteration by gene dosage in *Drosophila melanogaster. Genetics* **119:**399–406.

Thomas RH, Hunt JA (1991) The molecular evolution of the alcohol dehydrogenase locus and the phylogeny of Hawaiian *Drosophila. Mol. Biol. Evol.* **8:**687–702.

Tompkins L (1979) Developmental analysis of two mutations affecting chemotactic behavior in *Drosophila melanogaster. Dev. Biol.* **73:**174–177.

Tompkins L (1984) Genetic analysis of sex appeal in *Drosophila. Behav. Genet.* **14:**411–440.

Tompkins L, Cardosa MJ, White FV, Sanders TG (1979) Isolation and analysis of chemosensory behavior mutants in *Drosophila melanogaster. Proc. Natl. Acad. Sci. USA* **76:**884–887.

Tsien RY, Rink TJ, Poenie M (1985) Measurement of cytosolic free Ca2+ in individual small cells using fluorescence microscopy with dual excitation wavelengths. *Cell Calcium* **6:**145–157.

Ueno K, Ohta M, Morita H, Mikuni Y, Nakajima S, Yamamoto K, Isono K (2001) Trehalose sensitivity in *Drosophila* correlates with mutations in and expression of the gustatory receptor gene *Gr5a. Curr. Biol.* **11:**1451–1455.

Van der Goes van Naters WM, Den Otter CJ (1998) Amino acids as taste stimuli for tsetse flies. *Physiol. Entomol.* **23:**278–284.

Van der Starre H (1970) Tarsal water receptor cell responses in the blowfly *Phormia regina. Netherlands J. Zool.* **20:**289–290.

Vargo M, Hirsch J (1982) Central excitation in the fruit fly (*Drosophila melanogaster*). *J. Comp. Physiol. Psychol.* **96:**452–459.

Vaysse G, Médioni J (1973) First experiments on tarsal gustation in *Drosophila melanogaster*: stimulation with sucrose. *C R Seances Soc. Biol. Fil.* **167:**560–564.

White BH, Osterwalder TP, Yoon KS, Joiner WJ, Whim MD, Kaczmarek LK, Keshishian H (2001) Targeted attenuation of electrical activity in *Drosophila* using a genetically modified K(+) channel. *Neuron* **31:**699–711.

White K, Tahaoglu E, Steller H (1996) Cell killing by the *Drosophila* gene *reaper. Science* **271:**805–807.

Wieczorek H, Koppel R (1982) Reaction spectra of sugar receptors in different taste hairs of the fly. *J. Comp. Physiol.* **149:**207–213.

Wieczorek H, Wolff G (1989) The labellar sugar receptor of *Drosophila. J. Comp. Physiol. A* **164:**825–834.

Yang H, Wanner IB, Roper SD, Chaudhari N (1999) An optimized method for *in situ* hybridization with signal amplification that allows the detection of rare mRNAs. *J. Histochem. Cytochem.* **47:**431–446.

Zdobnov EM, von Mering C, Letunic I, Torrents D, Suyama M, Copley RR, Christophides GK, Thomasova D, Holt RA, Subramanian GM, et al. (2002) Comparative genome and proteome analysis of *Anopheles gambiae* and *Drosophila melanogaster. Science* **298:**149–159.

Zhang Z, Curtin KD, Sun YA, Wyman RJ (1999) Nested transcripts of gap junction gene have distinct expression patterns. *J. Neurobiol.* **40:**288–301.

10 Combining Molecular, Physiological, and Behavioral Tools to Study Insect Olfactory Processing

Alan J. Nighorn and Lawrence J. Zwiebel

CONTENTS

0-8493-2024-0/05/$0.00+$1.50

10.1 INTRODUCTION

One of the most exciting developments in modern biology has been the vigorous and renewed interest in studying complex problems such as olfactory information processing by using approaches that integrate methods from molecular biology, cell culture, and physiology within the context of the normal behavior of the animal. The sequencing of the complete *Drosophila* genome in 2000 and its subsequent refinements have resulted in remarkable advances in our understanding of chemosensory processes in insects (Adams et al., 2000; Celniker and Rubin, 2003). In many instances, however, studies using model insect experimental systems other than *Drosophila melanogaster* can provide invaluable opportunities to dissect complex biological processes such as olfaction. Combining the results from multiple insect model systems is likely to yield critical insights into the function of chemosensory systems.

The groundbreaking studies on the organization and molecular genetics of olfaction in *D. melanogaster* have been extensively reviewed in this volume and elsewhere (Carlson, 1996, 2001; Keller and Vosshall, 2003; see also Chapter 9). Accordingly, we will make little attempt to recapitulate those efforts here, but instead, we will turn the bulk of our attention to the question of olfactory studies in non-Drosophilid insect systems. There are many reasons for studying non-Drosophilid insects. First and foremost is the appreciation that these insects themselves are often of critical importance from a medical or economic perspective. This statement is underscored by the dramatic effects on global public health of disease vectors, such as the Afrotropical malaria vector mosquito *Anopheles gambiae,* which is responsible for the annual transmission of human malaria to over 250 million individuals, resulting in over one million deaths (Greenwood and Mutabingwa, 2002). In addition to the profound impacts of other mosquito vectors, such as *Aedes aegypti* (dengue and yellow fever virus) and *Culex pipens* (West Nile virus), and other insect vectors, including *Glossina morians* (Trypanosomisis), a host of agricultural pest insects too numerous to list is responsible for the loss of a significant percentage of the world's food and export crops. Taken together, these criteria make it hard to overlook the relevance of studying these nonacademic insects.

A second reason to focus on non-Drosophilid insects is that they can often be more appropriate models for vertebrate and eventual human function in terms of their anatomical organization and regulation. For example, the olfactory system in *Manduca sexta,* with its long olfactory receptor neurons that project ipsilaterally to spherical glomeruli, is remarkably similar to the anatomical organization in vertebrates (Hildebrand and Shepherd, 1997; Christensen and White, 2000). The olfactory system of *Drosophila,* with relatively small olfactory receptor neurons (ORNs) that project both ipsilaterally and contralaterally over a much smaller distance, is subtly but significantly different, particularly for studies on olfactory axon guidance.

A third reason to study other insects is that the specific biology of a particular insect species may lend itself well to a particular type of analysis. For example, the accessibility and structure of developing locust motoneurons greatly facilitated a set of classic axon guidance studies (Doe et al., 1986). In a similar context, in cases where differential size can be an important advantage, the sheer dimensions of the olfactory apparatus in larger insects such as *Manduca* facilitates electrophysiological and biochemical analyses that would be difficult or impossible in smaller insects the size of *Drosophila*.

Finally, one of the most essential and compelling reasons to study non-Drosophilid insects is to take advantage of their incredibly rich biological diversity. This allows an investigator to study sensory processing in species that are specialized for a particular sensory modality and to take advantage of the detailed descriptive and biological studies that have already been performed in many of these species. These studies often describe the exact behavioral context for the sensory processing, and in many cases, an anatomical and cell biological context is also available (see other chapters in the book for detailed accounts specific to different modalities).

All of this information provides an ideal background for studies that also focus on a molecular analysis of function. The ultimate goal is an integrated study in which the function of specific molecules is measured in biochemical, behavioral, physiological, or other assays and evaluated in the context of the natural behavior of the organism. In many systems, the genetic or biochemical manipulation of molecules is very difficult but the contextual analysis of the effects of those manipulations is relatively straightforward. In others, most notably *D. melanogaster*, the genetic and molecular manipulations are relatively straightforward but the contextual analyses of function are more difficult. When applied considerations do not prevail, the choice of system to investigate a particular question is not always straightforward. Therefore, these considerations lead to the inevitable conclusion that there is enormous value in studying related questions using multiple species, either in the same lab or collaboratively within the scientific community as a whole. Too much emphasis on a single experimental system likely will produce a skewed and, at best, incomplete understanding of the question under study. Taken together, these considerations provide a compelling argument in favor of addressing such questions in more than a single "model" system.

In this chapter, we will first describe current methods for identifying molecular targets of interest for the processing of olfactory information in insects. We will then outline current methods for manipulation of specific genes, and finally we will briefly review methods for assaying potential functional implications of those manipulations.

10.2 IDENTIFICATION AND CLONING OF OLFACTORY GENES

Along with an increasingly refined set of molecular tools, the recent progress of insect genome sequencing projects is transforming the way in which olfactory genes of interest (GOI) are identified. This includes data mining from completed genomes, but it also includes those insects whose genomes have not yet been fully or partially sequenced. The identification of genes can be divided into two complementary approaches. The first is a candidate molecule approach. In this approach, one can identify candidate molecules that are hypothesized to be involved in olfactory processing and then clone them from the species of interest. This typically involves using as much information about that molecule as possible from what is known about its biochemistry or molecular characteristics in model species. The second avenue is a screen for elements of a particular pathway, without identifying specific molecules. This allows for the development of completely novel hypotheses.

10.2.1 Identifying Candidate Molecules

The identification of candidate molecules requires that we know something about the characteristics or likely sequence of the molecules being sought. Taking advantage of the ever growing accumulation of individual gene sequences into public databases, as well as the completion of several insect whole-genome projects, the first step in this process is done *in silico*.

10.2.1.1 Bioinformatics/Genomics

In this approach, one uses a directed set of software tools to search one or more databases for olfactory genes of interest. At present, a growing number of genetic databases are available that contain either the complete or partial genomic sequences for several insects, including *D. melanogaster*, *D. psuedoobscura*, *Apis meliferia*, *A. gambiae*, and *A. aegypti*, to name but a few. Indeed, there is every expectation that in the interim between the writing and the publication of this volume, there will several rounds of additions to these bioinformatic resources. Nevertheless, it is worthwhile to consider the reasonable approaches that one takes in order to maximize the utilization of these databases.

While each search entails its own distinct characteristics, several broad "rules" may nevertheless be stated. As an example, we will use the recent identification by one of us of the complete repertoire of G protein–coupled receptors (GPCRs) from the malaria vector mosquito *An. gambiae* (AgGPCRs) (Hill et al., 2002). Initially, this investigation involved a directed homology-based search of the raw and assembled sequences from the *An. gambiae* genome (Holt et al., 2002), using one or more software platforms. The most widely used homology-based database search engines belong to the Basic Local Alignment Search Tool (BLAST) family of applications (Altschul et al., 1990). These tools offer the user a robust set of programs for identifying localized primary sequence homology with wide default settings that are used to search for related nucleotide and protein databases against either a DNA or a protein sequence of interest. Furthermore, the investigator may adjust a number of the default characteristics of these search algorithms to provide an even more finely tuned search capability. In addition to BLAST-based algorithms, similar approaches based on both local and global sequence homology can be achieved with a number of other search platforms (Barton and Sternberg, 1990; Attwood et al., 1997).

Although of particular importance for the analysis of whole genomes, a nevertheless critical element in this process is the ability to increase the ultimate utility of these efforts by taking a reiterative approach, where the results of each successive search are used to facilitate the subsequent rounds. In this manner, both inter- and intraspecific criteria are represented in the overall search. For example, in the case of AgGPCR superfamily members, this strategy involved an initial focus on an interspecific BLAST screen using all previously identified GPCR family members from the complete genome sequence of *D. melanogaster*. Following several rounds of robust interspecific analyses, a set of intraspecific screens was undertaken in order to identify potential AgGPCRs that are more related to other AgGPCR family members than to GPCRs from *Drosophila*. In this manner, a number of potentially important species-specific gene expansions were uncovered in several GPCR subfamilies (Hill et al., 2002). Last, in order to address the conservation of amino acid physico-chemical properties rather than strict primary amino acid sequence homologies, a novel algorithm (Kim et al., 2000) that identified quasi-periodic properties such as hydropathy, polarity, pI, pKa, and overall amino acid composition was used to screen for novel AgGPCRs.

Overall, in order to maximize the use of bioinformatics-based resources, it is important to address both amino acid and nucleotide homologies from the perspective of both inter- and intraspecies relationships. Secondarily, the inclusion of additional bioinformatics screens based on conservation of parameters other than primary sequence characteristics provides an important capacity for addressing the conservation a wide range of physico-chemical properties that may be characteristic of the genes of interest. Furthermore, the maximal utilization of genomic and genetic data and other elements from *D. melanogaster* is a hallmark in helping to establish similar efforts in non-Drosophilid insects, where the ability to take advantage of a wide range of resources from this extraordinarily developed insect model system provides an essential vantage point from which other, less tractable, insect systems may be approached.

10.2.1.2 Expression Cloning

In this approach to olfactory GOI discovery, one can take advantage of the expected expression characteristics of a particular GOI in order to facilitate its cloning. Typically, this involves a large-scale survey of gene expression in which one can target a particular gene of interest by taking advantage of a previously known (or hypothesized) temporal or spatially restricted pattern of expression. In this manner, the screen can be carried out so as to maximally reduce the effects of signal to noise that would otherwise obscure the outcome of these screens. Although there are numerous approaches in this area, we will focus on the three most basic types of approach.

10.2.1.3 Random Sampling of cDNAs

In this method, one takes advantage of the availability of high-throughput DNA sequencing capabilities in order to survey a population of clones derived from a particular tissue type or developmental stage. Although there is considerable variation on the elements of the basic methodologies that may be involved, this process can be viewed on a basic level as harvesting representative populations of mRNAs from the tissue or stage of interest, faithfully converting them into clonable cDNA pools that give rise to large numbers of individual clones, known as *expressed sequence tags* (ESTs), which are randomly selected for characterization by DNA sequencing. Extreme care must be taken at each preparatory step to retain as much of the characteristics of the original mRNA population as possible. However, even though these methods have been instrumental in a number of systems, there are significant downsides that should be appreciated when utilizing this technology. The most important reservation is that this type of EST survey will largely be biased toward mRNAs that are highly expressed and that provide the most abundant templates for conversion to cDNA. Therefore, any attempts to catalog rare or underexpressed genes will suffer in this type of survey. Nevertheless, these strategies have led to the generation and screening of EST collections that can be randomly selected for DNA sequencing of partial-length cDNAs to yield sufficient information to facilitate the bioinformatics-based identification of particular transcripts. Several EST projects have been successfully carried out for non-Drosophilids, such as the honey bee *Apis mellifera* (Whitfield et al., 2002), the silk moth *Bombyx mori* (Mita et al., 2003), the pine engraver beetle *Ips pini* (Eigenheer et al., 2003), and the malaria vector mosquito, *An. gambiae* (Ribeiro, 2003).

10.2.1.4 Differential/Subtractive Screening

These approaches form the basis for much of the microarray-based mRNA profiling that will be elaborated on in subsequent sections. In essence, they are dependent on the availability of differential collections of cDNA probes or templates that can be used to screen duplicate cDNA libraries or can be screened by collections of oligonucleotide primers, respectively. In the first instance, which has been widely applied to both vertebrate and invertebrate systems (Sargent, 1987), a cDNA or other type of DNA library, from which a GOI is expected to be present, is typically constructed from a collection of cells, a specific tissue, or perhaps a targeted developmental or temporal stage in the life cycle of an organism. The next step is to prepare two or more differential sets of molecular probes that represent cDNA pools generated from biological material where the GOI is expected to be present (positive) and, most importantly, where it is expected not to be expressed (negative). With these reagents in hand, it is then possible to conduct duplicate hybridization-based screens in order to identify cDNAs that are differentially represented in one complex probe relative to another. Although the precise format of this screen has changed over time such that "dot" blots have given way to microarray-based assays, the concept is still essentially the same. Accordingly, there are similar caveats to this approach that must be taken into account.

The inherent complexity of mRNA pools that are used for library or probe construction present an extraordinary challenge insofar as their ability to be represented faithfully at each step of these protocols. Indeed, although each particular differential screen has its own set of challenges, it is widely accepted that these approaches are limited to identifying only relatively large changes (about fivefold) in GOI abundance between positive and negative cDNA pools. Even more problematic is the question of relative levels of gene expression between target tissues, inasmuch as successful screens typically occur when the abundance of these desired mRNAs in at least one of the differential probe pools is higher than 0.05% (Sargent, 1987) of the total population of transcripts. However, although these concerns clearly pose significant problems for the identification of rarely expressed olfactory GOIs, such as candidate odorant receptors, the use of high throughput-based approaches, such as microarrays screened with chromophore-labeled probes, has dramatically fueled the design of increasingly more sensitive differential screens.

Another highly popular differential hybridization–based method, known as *differential display* (DD-PCR), was invented by Liang and Pardee (Liang et al., 1994). This approach relies on the use of reverse transcriptase–coupled polymerase chain reactions (RT-PCR) to amplify cDNAs that correspond to the 3′ ends mRNAs and the ability to display these products by high-resolution gel electrophoresis. Although the widespread utility of DD-PCR has been questioned due to critical issues, such as the ability to screen for rare transcripts and a high rate of false positives (Debouck, 1995), it has seen considerable use in non-Drosophilid insects, where it has been used to identify abundant and moderately abundant mRNAs (Rogers et al., 1999). The original methodology relied on the use of upstream oligonucleotide primers that consisted of randomly designed 10mers, and the downstream primers included 11 thymidines ahead of the two terminal nucleotides with complementarity to the ultimate and penultimate nucleotides of the mRNA. More recently, in order to improve its utility (especially with regard to its utility for the identification of rare mRNAs), several modifications to this approach have been reported. One such DD-PCR improvement is known as *amplification of double-stranded (ds) cDNA end restriction fragments* (ADDER) (Kornmann et al., 2001) and entails the amplification of ds cDNA restriction fragments complementary to the 3′ terminal domains of mRNAs. ADDER has been used successfully to characterize extremely rare circadian-controlled genes in hepatocytes (Damiola et al., 2000).

Subtractive screening (Sagerstrom et al., 1997) enables the investigator to remove selectively (subtract) the highly redundant mRNA component corresponding to "housekeeping genes" that is in common between two complex transcript populations. This effectively increases the concentration of the subset of mRNAs that are enhanced in the target (tester) mRNA population relative to a population of sequences that are not expected to contain them (driver). This would effectively raise the concentration of specific GOIs that may be subsequently identified using one of the methods already discussed. In these protocols, the driver is used in vast molar excess relative to the tester in the hybridization reaction, leading to the formation of duplex structures containing either RNA-DNA hybrids or simple cDNA duplexes that can be selectively removed by one of several alternative techniques. In the original scheme, this involved differential adsorption of single-stranded nucleic acids and double-stranded nuclei acids to hydroxyapatite (Davis et al., 1984) and now includes improvements involving the use of biotinylation and streptavidin for removal of both driver–driver homodimers and driver–tester hybrids. A particularly attractive adaptation of the subtraction concept is known as SABRE: selective amplification via biotin and restriction-mediated enrichment (Lavery et al., 1997). This combines the use of PCR amplification of cDNA pools to enrich for particularly rare mRNAs, together with the use of both streptavidin paramagnetic beads and restriction enzymes to select specifically for tester–tester homodimers. These methods have been adapted successfully to facilitate olfactory gene discovery studies in *An. gambiae* (Merrill et al., 2002).

10.2.1.5 Sequence Similarity-Based Cloning

Lacking a completed genome project or a reliable functional cloning assay is not necessarily an insurmountable barrier to using a candidate molecule approach. In the absence of specific sequence information, much information can be gleaned from homologous molecules from other species for which there is good sequence information. The main assumption with this set of methods is that there are regions of high sequence identity between the known and, presumably, homologous sequences and your candidate GOI. These areas of presumed high sequence identity may then be used as tools to identify and isolate full-length cDNAs of the candidate olfactory GOIs. However, this presumption of regions of sequence identity should not be lightly embraced. An illustrative cautionary tale is derived from the many researchers who have tried to use these methods to clone cDNAs encoding candidate odorant receptor (OR) proteins from many insect systems — only to discover that there is very little sequence conservation between ORs in different species. That being said, this method has been highly successful for the cloning of many other candidate olfactory GOIs.

There are two basic strategies for sequence similarity–based cloning approaches: low stringency hybridization and degenerate oligo RT-PCR. In low stringency hybridization, the first step is to identify the closest relative for which sequence information is known and a cDNA clone is available. This cDNA clone, or a fragment that includes the domains that are thought to be the most highly conserved, is then used as a template for probe generation and library screening. Although many different methods can be used for probe generation, including PCR and random priming, we have had the most success using *nick translation*. This method produces probes of widely varying lengths from all parts of the probe template cDNA. Once a probe is generated, the struggle is to identify hybridization conditions that will allow for probe binding without generating too much background. Because the overall levels of sequence identity are unknown, the appropriate stringency conditions (if they exist) must be determined empirically. To accomplish this, the candidate cDNA library or, alternatively, an appropriate blot is screened under a variety of different hybridization stringencies. The relevant conditions can be varied through changing the hybridization and wash temperatures or by a combination of different concentrations of formamide and/or salt in the hybridization buffer. The chances of success can often be significantly improved by employing an ordered set of hybridization conditions with a series of Northern or Southern blots and by screening a high-quality genomic library rather than a cDNA library. This latter step is worthwhile to eliminate any question about whether the target sequence is even present in the cDNA library. However, even with these caveats, insofar as all of the conditions have to be determined experimentally and library screening in the best of times can be fraught with problems of background, this method is usually very tedious and time-consuming.

A second approach to the cloning of candidate olfactory GOIs using sequence similarity is to use degenerate oligo RT-PCR. In this case, rather than count on a generalized level of sequence identity, small regions of very high sequence similarity are identified and used for the generation of PCR primers. These primers are then used to amplify a fragment of the candidate GOI using PCR. Once the fragment has been successfully amplified, cloned, and sequenced to confirm its identity, it is used as a probe template in a standard high stringency cDNA library screen to identify the full-length cDNA of the candidate molecule. The combination of relatively inexpensive PCR primers (state-of-the-art gradient PCR machines that allow the researcher to test a wide variety of annealing temperatures in a single experiment) and the availability of commercially available kits for rapid PCR cloning make it possible to test an incredible variety of conditions in a relatively short period of time. These include different primer combinations and cycling conditions. Overall, the high efficiency of these methods has optimized the ability to conduct multiple assays over a short period of time, resulting in the popularity of these approaches.

The key to success with this approach is the correct design of the degenerate oligonucleotide PCR primers. The first step is to identify a group of homologous amino acid sequences from other organisms and perform an amino acid alignment. Sequences from a wide variety of different species will help to identify absolutely conserved domains, whereas a more phylogenetically restricted subset of sequences can be used to identify domains that are more specific to a particular group. In either case, the goal is to identify at least two domains of at least five amino acids that are likely to be absolutely conserved in the target molecule. For efficient PCR, these domains should be at least 40 but not more than 300 amino acids apart. In addition, there should be as few gaps in the alignment as possible, so that the size of the appropriate PCR product can be accurately judged. Of course, because only the amino acids are known, the corresponding oligonucleotide primers must be degenerate in order to allow for the use of alternative codons. The primers with the best chance of success have a relatively high melting temperature and low degeneracy. There are many different ways of designing degenerate primer mixtures that minimize the overall degeneracy, including the inclusion of nonspecific nucleotides (such as inosine) that do not require precise complementarity to base pair and simply guessing some of the correct sequences based on the known codon usage in the insect of interest. In addition, a novel strategy that has been successful in one of our laboratories is the use of the CODEHOP algorithm (Rose et al., 2003) and the resulting CODEHOP primers, in which the 3′ end is degenerate while maintaining the 5′ end of the oligo as nondegenerate.

Unfortunately, whichever strategy is used for primer design, the exact melting temperature of the primers is impossible to determine from the degenerate sequence used for the synthesis because the exact variant that will bind to the target template is unknown. In this instance, a gradient PCR machine is invaluable because it allows one to test a wide range of melting temperatures in one experiment. In this way, it is possible to test a large series of PCR primer combinations quickly. All of the PCR products are examined on agarose gels, and products roughly corresponding to the predicted size of the target sequence are cut out of the gel, purified, cloned, and sequenced. Following bioinformatic analysis, those fragments with high similarity to the target sequence family can then be used as probes to screen a cDNA library. In addition, full-length cDNA sequences can often be obtained from incomplete fragments by a combination of 5′ and 3′ rapid amplification of cDNA end (RACE) technologies (Chenchik et al., 1996). See also Chapter 9 for application of these methods to the taste system.

10.2.2 Screening for New Genes

A complementary strategy to the candidate molecule approach is to screen for molecules that have the characteristics of the desired GOIs. Here, screening methods do not presuppose the involvement of a particular molecule or class of molecules; rather, they attempt to identify a particular characteristic. For example, this may comprise being expressed in the right tissue and at the right time to play an important role in the physiological process of interest. The two main ways to do this, in the absence of a facile genetic system to screen for mutants, are to use the differential cloning methods already discussed or microarray technologies.

10.2.2.1 Microarray Methods

The ever advancing state of knowledge about genomic sequences makes microarray-based experiments an increasingly popular way to screen for particular genes. Even in the absence of a completed genome project or an extensive EST library, however, it is possible to use microarrays to screen for new genes. The first step is to ensure that the particular question of interest is one that can be addressed by these methods. Because microarrays generally focus on mRNA populations, these methods are most effective at identifying genes that rapidly and dramatically change their levels of expression in response to a particular stimulus. The first step is to choose the samples to be placed on the microarray glass slide or "chip." If a genome project has been completed, it may be possible to represent the entire known genome on a small number of chips, depending on the size of the genome. However, the cost of doing this properly can be very high. It strongly depends on a properly annotated genome, so that all of the possible expressed genes are present. The chip also must be designed such that the results can be accurately interpreted. This usually includes having the appropriate control sequences and the test sequences present in triplicate (Stafford and Liu, 1993).

A second way to take advantage of microarray methods to screen for genes expressed in a particular physiological state is to use a naïve microarray strategy. In this case, the cDNAs plated on the microarray are taken from a library generated from the tissue in which the mRNA corresponding to an olfactory GOI is thought to be highly expressed. For example, if an investigator wanted to identify the genes expressed in the antennae of an ant in response to a hormone treatment, the first step would be to construct a cDNA library from ant antennae under the influence of that hormone. The next step would be to choose sequences with which to seed the microarray. In this case, however, the identities of those sequences are unknown. The key, therefore, is to normalize the library as well as possible to increase the possibility of detecting relatively rare cDNAs. There are many different ways to normalize cDNAs, each with its advantages and disadvantages (Bonaldo et al., 1996). Some of the most common are to use a subtraction technique, similar to those used for difference cloning described earlier, or to use negative selection. In negative selection, a cDNA library is screened with a probe generated from a nonnormalized sample of the cDNA from which the library was generated originally. Clones with inserts that do not hybridize strongly to this complex probe are then chosen to be seeded onto the microarray (Nelson et al., 1999).

Because the identities of the clones in a naïve microarray are, by definition, unknown, it is impossible to design a microarray with all of the sequences present in triplicate. Rather, the microarray is designed with some controls for known sequences to ensure that the overall hybridization process is complete. The resulting microarray is then probed with cDNA generated from antennae, with and without exposure to the hormone. Because the targets are not there in triplicate and because there is no way to ensure that all of the important sequences are present, the results from this type of array must be used as the first step in a series of experiments to identify the important molecules. Furthermore, the expression patterns of any positives must be confirmed independently with PCR; the identities of the positives can then be determined by sequencing the appropriately identified cDNA. Although naïve microarrays cannot be used to identify exhaustively every molecule important to a physiological process, they can be quite useful for identifying a subset of the important molecules without being limited by the constraints of any one assumption of what must be there. With a properly normalized library, they can also be used for any organism — regardless of the amount of knowledge (or lack of it) about its genome.

10.2.2.2 Manipulating Gene Expression

In order to bring these analyses to the next level, it is important that the investigator has tools available to test hypotheses about the function of particular, identified olfactory GOIs. Although function can be inferred from the expression patterns and biochemical examination of the molecules in question, the most convincing examination of function is either to interfere with the endogenous function of the molecule or to cause it to become more active and examine the effects of those manipulations on the physiological process of interest. Pharmacological tools exist for looking at many different enzyme classes, but often they are designed for use in vertebrate systems. Moreover, their effectiveness on the desired target and their specificity of effect are often less than desired, and there are many targets for which pharmacological tools are simply not available. Fortunately, considerable progress has recently been made in the methods available for manipulating gene expression in organisms that are not amenable to standard or conventional genetic manipulations. The most promising of these are RNA interference (RNAi) and transposon-mediated insertional mutagenesis.

10.2.2.3 RNAi

RNAi is a form of posttranscriptional gene silencing (PTGS) that was first discovered in plants, when it was often found that high copy numbers of transgenes paradoxically resulted in a specific decrease in mRNA levels from both the transgenes and the homologous endogenous genes (Jakowitsch et al., 1999). Interestingly, it was subsequently noted that PTGS was mediated by molecules that were soluble in the cytoplasm and, perhaps even more remarkably, could pass from cell to cell through the vascular system of the plant (Voinnet and Baulcombe, 1997). The first hint that this phenomenon was not limited to plants came from the serendipitous observation in *C. elegans* that injection of double-stranded RNA (dsRNA) resulted in the specific, targeted degradation of homologous mRNAs (Fire et al., 1998). One of the most striking features of this knockdown was its effectiveness. The dsRNA construct was clearly effective at lower-than-stoichiometric levels. Remarkably, dsRNA-mediated gene silencing could even be passed on to the next generation (Fire et al., 1998). Analysis of the mechanism in both plants and *C. elegans* showed that both processes were mediated by double-stranded RNA intermediates. Since that time, dsRNA has been shown to mediate degradation of specific mRNAs in a wide variety of organisms, including plants, flies (Kennerdell and Carthew, 1998), zebra fish (Wargelius et al., 1999), mosquitoes (Levashina et al., 2001), cockroaches (Marie et al., 2000), planaria (Sanchez Alvarado and Newmark, 1999), and hydra (Lohmann et al., 1999)

In most insect species examined to date, RNAi can be triggered by large fragments of dsRNA. This is a huge advantage over vertebrate systems in which RNAi has to be triggered using the siRNAs to avoid nonspecific effects. To date, published reports suggest that this method can be effective at knocking down gene expression in a wide range of insects, including the honeybee (Farooqui et al., 2003), the cockroach (Marie and Blagburn, 2003), and the moth (Rajagopal et al., 2002). In our experience, the use of large fragments, about 700 to 900 base pairs, has been the most effective and, presumably, is taken up by the cells more effectively or else there is an amplification step that occurs within the cell when the large dsRNA molecules are converted in siRNAs.

A second way in which RNAi can be used in insect cell systems is vector-based knockdown strategies. In this approach, a vector is created that contains fragments of the target sequence in an inverted repeat structure. When this plasmid is expressed in the organism, it generates an RNA that folds back onto itself and can act as a trigger for RNAi (Paddison et al., 2002). This approach eliminates concerns about how the dsRNA enters the cell. Unfortunately, it creates two additional concerns. First, your construct must include the appropriate promoter to drive expression of the snap-back construct in the cells of interest. This is often done by including promoters from generally highly expressed genes such as viral genes or actin. Alternatively, this could be used to direct expression within a restricted cell population if the appropriate promoter were used. Second, you must get the construct into the cells. In cell culture, there are many tools available to do this. Getting the construct into cells *in vivo* can be problematic. One promising approach is *in vivo* electroporation. In this method, the tissue is surrounded by a solution containing the plasmid. Electrodes are then placed around the tissue, and electric current is used to drive the plasmid into the cells of interest. This has been used successfully in the developing vertebrate spinal cord (Krull, 2004), and labs are adapting this method for use in the larger insects. Finally, methods have also been developed to create transgenic insects using transposon-mediated transformation (see Section 10.2.2.4). These could be used to generate stable lines expressing snap-back RNAi constructs under the control of tissue specific or conditional promoters. See also Chapter 9 for a discussion of RNAi methods in gustation.

10.2.2.4 Transposon-Mediated Germline Transformation

The ability to generate transgenic *Drosophila* using P-elements (Perrimon et al., 1991) has transformed the world of *Drosophila* genetics. Using transposable elements, it is possible to direct the expression (or overexpression) of particular genes in particular places, express snap-back-type RNAi constructs, or generate mutations through the insertion of the transposable element in the endogenous gene. It is now possible to take advantage of this powerful set of tools in other insects through the use of Hermes, piggyBac, and mariner transposable element families, among others (Wimmer, 2003).

The new generation of constructs takes advantage of three copies of the binding site for the evolutionary conserved Pax-6 transcription factor to drive expression of fluorescent marker proteins in the eyes of target animals. These "3xP3" vectors facilitate the selection of transgenic insects from the general population without the need for the maintenance of a specific mutant strain, such as white-eyed animals or balancer chromosomes. These transposable elements have been used to generate germline transformant strains of mosquitoes (Jasinskiene et al., 1998), moths (Imamura et al., 2003), and other insects (Allen et al., 2004). Basically, a plasmid expressing an engineered transposable element and a second plasmid expressing transposase are coinjected into embryos. The actions of the transposase allow the transposable element to insert semirandomly into the genome. Germline transformation can then be confirmed by screening the progeny of a mating with the injected animals and wild-type to look for the specific eye-color phenotype associated with the transposable element construct.

Germline transformation can be used to generate animals that manipulate gene expression in a number of different ways. Depending on the nature of the gene of interest, constructs can be generated to investigate the effects of overexpression of the protein or expression of dominant negative forms of the protein. Down-regulation of the levels of the gene of interest can be effected by the incorporation of snap-back RNAi constructs (see previous section). Finally, because the elements insert semirandomly, they can also be used in insertional mutagenesis screens, in which new mutants and new genes of interest can be identified through screening for particular olfactory phenotypes.

10.3 ASSAYING FUNCTION

Insect olfaction can be assayed using a variety of approaches that target several discrete points, ranging from the peripheral signal transduction cascade to the overall behavioral output at the organismal level. In contrast to the molecular studies (which in *Drosophila* take advantage of the robust availability of genetic and other resources developed over the last 100 years of research and which are considerably more challenged in non-Drosophilid systems), the functional studies discussed in this section are often more predisposed in nonmodel insect systems. Indeed, more often than not, this owes to size considerations; as previously discussed, non-Drosophilid insects often possess olfactory systems that are several orders of magnitude larger than *D. melanogaster* itself.

10.3.1 Electrophysiological Tools

10.3.1.1 Electroantennographic Methods

Insect ORNs are principally located on the antennae, along with several other peripheral sensory structures, including the maxillary palps. Within these tissues are specialized cuticular hairs known as sensilla that house the basic neuronal machinery used to detect chemical stimuli (Steinbrecht, 1996). Therefore, its not surprising that electrophysiological analyses at the level of whole antennae involving a technique known as the *electroantennogram* (EAG) have been used for almost half a century as a measure of the functional nature of insect olfactory sensitivity (Schneider, 1957; see also Chapter 2 and Chapter 12). Paradoxically, although there has been widespread use of the EAG and related methodologies (e.g., electropalpograms) in *Drosophila* and non-Drosophilid insects to monitor both the amplitude and the kinetics of olfactory signal transduction, the precise mechanism behind these signals is not understood. It is, however, generally accepted that the measured voltage fluctuations detected in EAGs reflect the overall voltage change that can be measured across an antennal preparation, essentially representing an array of individual voltage sources and resistors that are derived from multiple ORNs responsive to odorant stimulation in the insect antennae.

Nevertheless, these assays have provided a reliable empirical tool with which to survey olfactory sensitivity broadly. There is considerable variation in EAG procedures, which depend on the morphology and other relevant biological characteristics of the insect system to be assayed. Generally, EAG protocols can be divided into those that utilize fully excised or partially dissected olfactory tissues (antennae, palps, etc.), as opposed to those that rely on whole insect preparations. These are largely reflective of the relative size of the antennae and their ability to be reliably manipulated in an excised state. Typically, Lepidopterans such as *Bombyx mori* or *Antherea polyphemus,* which are characterized by large, easily accessible antennae, have been assayed in excised preparations (Kaissling, 1995), whereas smaller insects, such as *An. gambiae,* with more delicate olfactory structures have been studied using whole insect preparations (Cork and Park, 1996).

The absolute value of EAG signals vary through the microvolt (μV) to millivolt (mV) range, depending on the biological preparation and the concentration and nature of the stimulus. As with any electrophysiological measurement, the ratio between EAG signals and background noise represents the true output of these studies, and it is not surprising that extreme care should taken to reduce both biological and external electrical noise. Both these effects are usually minimized by the use of fresh biological material and experimental setups that are adequately grounded, incorporate electronic filters, and are often enclosed inside electromagnetic shielding (Faraday cages). At an elemental level, EAG instrumentation involves setting up a highly sensitive antennal input circuit attached to an inline amplifier that is linked to a recording device. Typically, modern EAG setups incorporate integrated circuits as operational amplifiers and use computer-based systems for both data recording and analysis. In addition to these detection and recording elements, a robust stimulus delivery system must be employed to provide a constant flow of filtered and humidified air into which test odorants can be added for specific pulse intervals. High-quality instrumentation can be built *ad hoc* or is commercially available.

10.3.1.2 Single-Unit Recordings

In an effort to supercede the analytical constraints inherent to EAG analyses, electrophysiological study of insect olfactory processes is also routinely carried out at the single-unit level. In insect systems, this more often than not reflects monitoring of stimulus-evoked responses from a single sensillum. Once again, although there are numerous variations on the theme, these studies typically are carried out using one of two broad experimental approaches. The first method is known as *cut-tip sensillum recording* (Van der Pers and den Otter, 1978; Kaissling, 1995) and involves the placement of a recording electrode containing sensillum lymph saline solution (Kaissling and Thorson, 1980) over the exposed and ablated end of a single sensillum. Cut-tip recordings are generally the method of choice for larger insects, such as the sphinx moth *Manduca sexta* (Shields and Hildebrand, 2001). The second approach, employed for the smaller sensilla found on the antennae and palps of insects such as mosquitoes (Meijerink and van Loon, 1999) is conducted using extracellular recordings mediated by the *surface-contact method* (den Otter et al., 1980), in which movement of a tungsten microelectrode along the cuticle at the base of an individual sensillum is used to establish contact with underlying ORNs. In this manner, the distinct electrophysiological responses of each ORN can be distinguished individually using one of several dedicated computer software packages such as Autospike[(29)] (Syntech, Hilversum, The Netherlands; also see Chapter 2 and Chapter 12 for further discussion of these methods).

10.3.1.3 Multiunit Recordings

Using silicon multielectrode arrays, it is now possible to record from and identify many different neurons in an ensemble (Christensen et al., 2000; see also Chapter 14). Because these methods are described in detail elsewhere in this book, we will only outline a few of the advantages here. First, these methods allow for the simultaneous analysis of multiple units. Although this seems self-evident at first, the impact of this ability is profound because it allows the investigator to watch relationships between units change as a function of odor, odor concentration, or olfactory experience. Second, these recordings are extremely stable, allowing for the long-term analysis of these units. This allows for the visualization of long-term changes, such as might occur during olfactory learning. It also allows time for many different pharmacological manipulations on the same set of units, allowing the experimenter to perform a dose response and washout of pharmacological agents in the same preparation. In total, both single-unit and multiple-unit recordings provide a rich substrate for investigating the function of specific molecules in an olfactory context.

10.3.2 Heterologous/Transgenic Systems

In an effort to understand further the mechanisms underlying insect olfaction, considerable attention has been placed on establishing experimental platforms that will allow investigators to study olfactory GOIs and proteins in a functional context. Until recently, these studies have been plagued by failure, as evidenced by the paucity of published studies on this subject. There have, however, been notable advances of late, suggesting that the experimental impediments that have hampered this field may have largely been circumvented. Interestingly enough, *Drosophila* has provided a large measure of the tools for these studies on non-Drosophilids.

The ability to remove specifically the endogenous components of the olfactory signal transduction apparatus in a defined *Drosophila* ORN and replace them with olfactory genes from other insect species has provided a powerful approach for the study of olfaction in non-Drosophilids. This has been accomplished by taking advantage of a novel olfactory neuron expression system that has been developed in *D. melanogaster* in the laboratory of John Carlson (Dobritsa et al., 2003; see also Chapter 9). In this system, the endogenous *Or22a* and *Or22b* genes that are normally expressed in the ab3A neuron have been removed as a consequence of a chromosomal deficiency that spans the 22a/b locus. The reintroduction of exogenous olfactory genes is then accomplished through the use of standard tools of *Drosophila* molecular genetics that utilize P-element transformation (Rubin and Spradling, 1982) and the Gal4-UAS gene driver system (Brand and Perrimon, 1993).

These techniques were recently utilized to functionally characterize two putative ORs from *An. gambiae*: AgOR1 and AgOR2s (Hallem et al., 2004). In addition to validating their roles as OR encoding genes, these studies demonstrate that the female-specific AgOR1 gene was specifically tuned to 4-methyl phenol, a component of human sweat to which *An. gambiae* adult females are particularly responsive (Cork and Park, 1996). These studies open the possibility of further dissecting the function of additional olfactory GOIs and their corresponding proteins, with particular emphasis on those family members that play a direct role in Anopheline's and other mosquitoes' ability to establish and maintain the insect's host preference, which is critical in its overall vectorial capacity.

10.3.3 Behavioral Tools for Testing Reverse Genetics

There is a wealth of data describing olfactory-driven insect behaviors which have been extensively studied in a number of biological contexts (Hartleib and Anderson, 1999). Indeed, it would be pointless here to attempt to catalog the range of insect behaviors rooted by olfactory inputs. Rather than revisit that well-beaten path, we will instead focus on the more specialized literature that describes the behavioral assays developed to examine the underlying basis of several olfactory-driven behaviors in non-Drosophilids. These include laboratory- and field-based protocols that are directed toward mating, oviposition cues, and host preference behaviors of mosquitoes (Takken and Knols, 1999), as well as a number of similar contexts for moths, parasitoid wasps, and tsetse flies (Card, 1996).

Behavioral responses of mosquitoes toward chemical cues have been assayed in the field using odor-baited trapping systems principally employed for vector surveillance and control (Kline et al., 1990). Similarly, dual-port systems have been employed to assay olfactory responses of natural populations of *An. gambiae* mosquitoes (Costantini et al., 1993). In a laboratory context, olfactometers have been developed that consist of dual-port (Eiars and Jepson, 1991) and dual-choice (Y-tube) (Geier et al., 1996) wind tunnels that have been used to provide fine-level analyses of olfactory responses for *Aedes aegypti* and other mosquitoes. Trap-based olfactometers have also been used extensively with *An. gambiae* (Knols et al., 1994), where they have been used to study differential behavior toward complex odors (Dekker and Takken, 1998; Pates et al., 2001), to identify human-signifying kairimones (Dekker et al., 2002), and to monitor the temporal dynamics of host-seeking responses following blood feeding (Takken et al., 2001). Furthermore, wind tunnel–based olfactometry has facilitated studies of the effect of host-odor plume structure in *An. gambiae* and *A. aegypti* (Dekker et al., 2001).

In moths, both wind tunnel and field studies have been used to investigate the subtleties of both pheromone-driven flight behavior in males (Vickers et al., 2001) and host-plant selection behavior in females (Cunningham et al., 2004). These include studies of the regulation of upwind flight and parsing out the relative behavioral salience of specific pheromone- or plant-derived compounds (Fraser et al., 2003) and odor vs. visual cues (Raguso and Willis, 2002). These behavioral assays are well worked out, and the specific behavioral responses to particular odorants have been investigated in the context of both wind tunnel and field studies. Behavioral studies can, therefore, provide an important context for understanding the function of genes of interest after pharmacological or molecular manipulations.

10.4 CONCLUSIONS AND OUTLOOK

Taken together, the tools outlined in this chapter provide an invaluable range of resources that have resulted in a significant appreciation of olfactory processes in *D. melanogaster* and in non-Drosophilid insects. In many cases, the studies in non-Drosophilids have extended studies that were originally carried out in *D. melanogaster*, where the genetic and molecular resources have provided important entry points. In addition to augmenting such efforts, we have tried to argue that olfactory studies revolving around non-Drosophilid insects have considerable "legs" of their own. Indeed, these systems have an inherent ability, in an experimental context, to go where *Drosophila* cannot. As such, a number of critical insights regarding olfactory processing and odor coding have been achieved and are discussed in several chapters throughout this volume.

Over and above these spectacular advances in our appreciation of the processes underlying insect olfaction is the potential for dramatically reducing the negative impact of numerous insect agricultural pests and vectors for a host of globally important human diseases. For example, because several behaviors that critically affect vectorial capacity in mosquitoes depend on the insect's olfactory system (Takken and Knols, 1999), these studies will likely provide insight into the mechanisms by which these insects act as vectors for maladies such as malaria, filarisis, West Nile virus, encephalitis, and dengue and yellow fevers. Inasmuch as the combined ill effects of these diseases affect hundreds of millions of individuals each year throughout both the developed and the developing worlds, this constitutes a significant context for scientific focus.

By further integrating molecular olfactory studies with physiological and behavioral investigations, it will be possible to address critical questions concerning olfactory behavior in these systems. These may ultimately lead to novel disease-prevention strategies, not only in mosquitoes but also in other disease-transmitting arthropods. These will include the identification and intelligent design of a new generation of novel and economically synthesized attractants and repellents. Such compounds may eventually be used not only to reduce vector–human contact, but to otherwise enable innovative insect- and disease-control strategies that focus on disrupting vector–host interactions.

ACKNOWLEDGMENTS

The authors would like to thank Dr. Thomas Christensen for many helpful changes and editorial advice. Financial support was provided from the UNDP/World Bank/WHO Special Programme for Research and Training in Tropical Diseases (TDR V30/181/208, to L.J.Z.) and the NIH (DCO4292 to A.J.N. and DC04692/AI56402 to L.J.Z).

REFERENCES

Adams MD et al. (2000) The genome sequence of *Drosophila melanogaster. Science* **287**:2185–2195.

Allen ML, Handler AM, Berkebile DR, Skoda SR (2004) piggyBac transformation of the New World screwworm, *Cochliomyia hominivorax*, produces multiple distinct mutant strains. *Med. Vet. Entomol.* **18**:1–9.

Altschul SF, Gish W, Miller W, Myers EW, Lipman DJ (1990) Basic local alignment search tool. *J. Mol. Biol.* **215**:403–410.

Attwood TK, Avison H, Beck ME, Bewley M, Bleasby AJ, Brewster F, Cooper P, Degtyarenko K, Geddes AJ, Flower DR, Kelly MP, Lott S, Measures KM, Parry-Smith DJ, Perkins DN, Scordis P, Scott D, Worledge C (1997) The PRINTS database of protein fingerprints: a novel information resource for computational molecular biology. *J. Chem. Inf. Comput. Sci.* **37**:417–424.

Barton GJ, Sternberg MJ (1990) Flexible protein sequence patterns. A sensitive method to detect weak structural similarities. *J. Mol. Biol.* **212**:389–402.

Bonaldo MF, Lennon G, Soares MB (1996) Normalization and subtraction: two approaches to facilitate gene discovery. *Genome Res.* **6**:791–806.

Brand A, Perrimon N (1993) Targeted gene expression as a means of altering cell fates and generating dominant phenotypes. *Development* **118**:401–415.

Card RT (1996) Odour plumes and odour-mediated flight in insects. *Ciba Found. Symp.* **200**: 54–66; discussion 66–70.

Carlson J (1996) Olfaction in *Drosophila*: from odor to behavior. *Trends Genet.* **12**:175–180.

Carlson JR (2001) Viewing odors in the mushroom body of the fly. *Trends Neurosci.* **24**:497–498.

Celniker SE, Rubin GM (2003) The *Drosophila melanogaster* genome. *Annu. Rev. Genomics Hum. Genet.* **4**:89–117.

Chenchik A, Diachenko L, Moqadam F, Tarabykin V, Lukyanov S, Siebert PD (1996) Full-length cDNA cloning and determination of mRNA 5' and 3' ends by amplification of adaptor-ligated cDNA. *Biotechniques* **21**:526–534.

Christensen TA, White J (2000) Representation of olfactory information in the brain. In: *The Neurobiology of Taste and Smell,* Finger TE, Silver WL, Restrepo D, Eds., Wiley-Liss Inc., New York, pp. 201–232.

Christensen TA, Pawlowski VM, Kei H, Hildebrand JG (2000) Multi-unit recordings reveal context-dependent modulation of synchrony in odor-specific neural ensembles. *Nat. Neurosci.* **3**:927–931.

Cork A, Park KC (1996) Identification of electrophysiologically-active compounds for the malaria mosquito, *Anopheles gambiae*, in human sweat extracts. *Med. Vet. Entomol.* **10**:269–276.

Costantini C, Gibson G, Brady J, Merzagora L, Coluzzi M (1993) A new odour-baited trap to collect host-seeking mosquitoes. *Parassitologia* **35**:5–9.

Cunningham JP, Moore CJ, Zalucki MP, West SA (2004) Learning, odour preference and flower foraging in moths. *J. Exp. Biol.* **207**:87–94.

Damiola F, LeMinh N, Preitner N, Kornmann B, Fleury-Olela F, Schibler U (2000) Restricted feeding uncouples circadian oscillators in peripheral tissues from the central pacemaker in the suprachiasmatic nucleus. *Genes Dev.* **14**:2950–2961.

Davis MM, Cohen DI, Nielsen EA, Steinmetz M, Paul WE, Hood L (1984) Cell-type-specific cDNA probes and the murine I region: the localization and orientation of Ad alpha. *Proc. Natl. Acad. Sci. USA* **81**:2194–2198.

Debouck C (1995) Differential display or differential dismay. *Curr. Opin. Biotech.* **6**:597–599.

Dekker T, Takken W (1998) Differential responses of mosquito sibling species *Anopheles arabiensis* and *An. quadriannulatus* to carbon dioxide, a man or a calf. *Med. Vet. Entomol.* **12**:136–140.

Dekker T, Steib B, Cardé RT, Geier M (2002) L-lactic acid: a human-signifying host cue for the anthropophilic mosquito *Anopheles gambiae. Med. Vet. Entomol.* **16**:91–98.

Dekker T, Takken W, Cardé RT (2001) Structure of host-odour plumes influences catch of *Anopheles gambiae* and *Aedes aegypti* in a dual choice olfactometer. *Physiol. Entomol.* **26**:124–134.

den Otter CJ, Behan M, Maes FW (1980) Single cell responses in female *Pieris brassicae* (Lepidoptera: Pieidaae) to plant volitiles and conspecific egg odours. *J. Insect Physiol.* **26**:465–472.

Dobritsa AA, van der Goes van Naters W, Warr CG, Steinbrecht RA, Carlson JR (2003) Integrating the molecular and cellular basis of odor coding in the *Drosophila* antenna. *Neuron* **37**:827–841.

Doe CQ, Bastiani MJ, Goodman CS (1986) Guidance of neuronal growth cones in the grasshopper embryo. IV. Temporal delay experiments. *J. Neurosci.* **6**:3552–3563.

Eiars AE, Jepson PC (1991) Host location by *Aedes aegypti* (Diptera: Culicidae) a wind tunnel study of chemical cues. *Bull. Entomol. Res.* **81**:151–160.

Eigenheer AL, Keeling CI, Young S, Tittiger C (2003) Comparison of gene representation in midguts from two phytophagous insects, *Bombyx mori* and *Ips pini*, using expressed sequence tags. *Gene* **316**:127–136.

Farooqui T, Robinson K, Vaessin H, Smith BH (2003) Modulation of early olfactory processing by an octopaminergic reinforcement pathway in the honeybee. *J. Neurosci.* **23**:5370–5380.

Fire A, Xu S, Montgomery MK, Kostas SA, Driver SE, Mello CC (1998) Potent and specific genetic interference by double-stranded RNA in *Caenorhabditis elegans. Nature* **391**:806–811.

Fraser AM, Mechaber WL, Hildebrand JG (2003) Electroantennographic and behavioral responses of the sphinx moth *Manduca sexta* to host plant headspace volatiles. *J. Chem. Ecol.* **29**:1813–1833.

Geier M, Sass H, Boeckh J (1996) A search for components in human body odour that attract females of *Aedes aegypti. Ciba Found. Symp.* **200**:132–144; discussion 144–148, 178–183.

Greenwood B, Mutabingwa T (2002) Malaria in 2002. *Nature* **415**:670–672.

Hallem E, Fox AN, Zwiebel LJ, Carlson JR (2004) A mosquito odorant receptor tuned to a component of human sweat. *Nature* **427**:212–213.

Hartleib E, Anderson P (1999) Olfactory-released behaviors. In: *Insect Olfaction.* Hansson BS, Ed., Springer-Verlag, Berlin, pp. 315–350.

Hildebrand JG, Shepherd GM (1997) Mechanisms of olfactory discrimination: converging evidence for common principles across phyla. *Annu. Rev. Neurosci.* **20**:595–631.

Hill CA, Fox AN, Pitts RJ, Kent LB, Tan PL, Chrystal MA, Cravchik A, Collins FH, Robertson HM, Zwiebel LJ (2002) G protein-coupled receptors in *Anopheles gambiae. Science* **298**:176–178.

Holt RA, et al. (2002) The genome sequence of the malaria mosquito *Anopheles gambiae. Science* **298**:129–149.

Imamura M, Nakai J, Inoue S, Quan GX, Kanda T, Tamura T (2003) Targeted gene expression using the GAL4/UAS system in the silkworm *Bombyx mori. Genetics* **165**:1329–1340.

Jakowitsch J, Papp I, Moscone EA, van der Winden J, Matzke M, Matzke AJ (1999) Molecular and cytogenetic characterization of a transgene locus that induces silencing and methylation of homologous promoters *in trans. Plant J.* **17**:131–140.

Jasinskiene N, Coates CJ, Benedict MQ, Cornel AJ, Rafferty CS, James AA, Collins FH (1998) Stable transformation of the yellow fever mosquito, *Aedes aegypti,* with the Hermes element from the housefly. *Proc. Natl. Acad. Sci. USA* **95**:3743–3747.

Kaissling K-E (1995) Single unit and electroanntennogram recordings in insect olfactory organs. In: *Experimental Cell Biology of Taste and Olfaction. Current Techniques and Protocols,* Kaissling K-E, Ed., CRC Press, Boca Raton, FL.

Kaissling K-E, Thorson J (1980) Insect olfactory sensilla: structure, chemical and electrical components of the functional organization. In: *Receptors for Neurotransmittors, Hormones and Pheromones in Insects.* Sattelle DB, Hall JM, Hildebrand JG, Eds., Elsevier/North-Holland, Amsterdam, pp. 261–282.

Keller A, Vosshall LB (2003) Decoding olfaction in *Drosophila. Curr. Opin. Neurobiol.* **13**:103–110.

Kennerdell JR, Carthew RW (1998) Use of dsRNA-mediated genetic interference to demonstrate that frizzled and frizzled 2 act in the wingless pathway. *Cell* **95**:1017–1026.

Kim J, Moriyama, EN, Warr CG, Clyne PJ, Carlson JR (2000) Identification of novel multi-transmembrane proteins from genomic databases using quasi-periodic structural properties. *Bioinformatics* **16**:767–775.

Kline DL, Takken W, Wood JR, Carlson DA (1990). Field studies on the potential of butanone, carbon dioxide, honey extract, 1-octen-3-ol, L-lactic acid and phenols as attractants for mosquitoes. *Med. Vet. Entomol.* **4**:383–391.

Knols BG, de Jong R, Takken W (1994) Trapping system for testing olfactory responses of the malaria mosquito *Anopheles gambiae* in a wind tunnel. *Med. Vet. Entomol.* **8**:386–388.

Kornmann B, Preitner N, Rifat D, Fleury-Olela F, Schibler U (2001) Analysis of circadian liver gene expression by ADDER, a highly sensitive method for the display of differentially expressed mRNAs. *Nucleic Acids Res.* **29**:E51–1.

Krull CE (2004) A primer on using *in ovo* electroporation to analyze gene function. *Dev. Dyn.* **229**:433–439.

Lavery DJ, Lopez-Molina L, Fleury-Olela F, Schibler U (1997) Selective amplification via biotin- and restriction-mediated enrichment (SABRE), a novel selective amplification procedure for detection of differentially expressed mRNAs. *Proc. Natl. Acad. Sci. USA* **94**:6831–6836.

Levashina EA, Moita LF, Blandin S, Vriend G, Lagueux M, Kafatos FC (2001) Conserved role of a complement-like protein in phagocytosis revealed by dsRNA knockout in cultured cells of the mosquito, *Anopheles gambiae*. *Cell* **104**:709–718.

Liang P, Zhu W, Zhang X, Guo Z, O'Connell RP, Averboukh L, Wang F, Pardee AB (1994) Differential display using one-base anchored oligo-dT primers. *Nucleic Acids Res.* **22**:5763–5764.

Lohmann JU, Endl I, Bosch TC (1999) Silencing of developmental genes in Hydra. *Dev. Biol.* **214**:211–214.

Marie B, Blagburn JM (2003) Differential roles of engrailed paralogs in determining sensory axon guidance and synaptic target recognition. *J. Neurosci.* **23**:7854–7862.

Marie B, Bacon JP, Blagburn JM (2000) Double-stranded RNA interference shows that engrailed controls the synaptic specificity of identified sensory neurons. *Curr. Biol.* **10**:289–292.

Meijerink J, van Loon JJA (1999) Sensitivities of antennal olfactory neurons of the malaria mosquito, *Anopheles gambiae*, to carboxylic acids. *J. Insect Physiol.* **45**:365–373.

Merrill CE, Riesgo-Escovar J, Pitts RJ, Kafatos FC, Carlson JR, Zwiebel LJ (2002) Visual arrestins in olfactory pathways of *Drosophila* and the malaria vector mosquito *Anopheles gambiae*. *Proc. Natl. Acad. Sci. USA* **99**:1633–1638.

Mita K, Morimyo M, Okano K, Koike Y, Nohata J, Kawasaki H, Kadono-Okuda K, Yamamoto K, Suzuki MG, Shimada T, Goldsmith MR, Maeda S (2003) The construction of an EST database for *Bombyx mori* and its application. *Proc. Natl. Acad. Sci. USA* **100**:14121–14126.

Nelson PS, Hawkins V, Schummer M, Bumgarner R, Ng WL, Ideker T, Ferguson C, Hood L (1999) Negative selection: a method for obtaining low-abundance cDNAs using high-density cDNA clone arrays. *Genet. Anal.* **15**:209–215.

Paddison PJ, Caudy AA, Bernstein E, Hannon GJ, Conklin DS (2002) Short hairpin RNAs (shRNAs) induce sequence-specific silencing in mammalian cells. *Genes Dev.* **16**:948–958.

Pates HV, Takken W, Stuke K, Curtis CF (2001) Differential behaviour of *Anopheles gambiae* sensu stricto (Diptera: Culicidae) to human and cow odours in the laboratory. *Bull. Entomol. Res.* **91**:289–296.

Perrimon N, Noll E, McCall K, Brand A (1991) Generating lineage-specific markers to study *Drosophila* development. *Dev. Genet.* **12**:238–252.

Raguso RA, Willis MA (2002) Synergy between visual and olfactory cues in nectar feeding by naive hawkmoths, *Manduca sexta*. *Anim. Behav.* **64**:685–695.

Rajagopal R, Sivakumar S, Agrawal N, Malhotra P, Bhatnagar RK (2002) Silencing of midgut aminopeptidase N of *Spodoptera litura* by double-stranded RNA establishes its role as *Bacillus thuringiensis* toxin receptor. *J. Biol. Chem.* **277**:46849–46851.

Ribeiro JM (2003) A catalogue of *Anopheles gambiae* transcripts significantly more or less expressed following a blood meal. *Insect Biochem. Mol. Biol.* **33**:865–882.

Rogers ME, Jani MK, Vogt RG (1999) An olfactory-specific glutathione-S-transferase in the sphinx moth *Manduca sexta*. *J. Exp. Biol.* **202**:1625–1637.

Rose TM, Henikoff JG, Henikoff S (2003). CODEHOP (COnsensus-DEgenerate Hybrid Oligonucleotide Primer) PCR primer design. *Nucleic Acids Res.* **31**:3763–3766.

Rubin GM, Spradling AC (1982) Genetic transformation of *Drosophila* with transposable element vectors. *Science* **218**:348–353.

Sagerstrom CG, Sun BI, Sive HL (1997) Subtractive cloning: past, present, and future. *Annu. Rev. Biochem.* **66**:751–783.

Sanchez Alvarado A, Newmark PA (1999) Double-stranded RNA specifically disrupts gene expression during planarian regeneration. *Proc. Natl. Acad. Sci. USA* **96**:5049–5054.

Sargent TD (1987) Isolation of differentially expressed genes. *Methods Enzymol.* **152**:423–432.

Schneider D (1957) Elektrophysiologische Untersuchungen von Chemo un Mechanorecepo-ren de Anetnne des Seidenspinners *Bombyx mori* L. *Z. Vergl. Physiol.* **40:**8–41.

Shields VD, Hildebrand JG (2001) Responses of a population of antennal olfactory receptor cells in the female moth *Manduca sexta* to plant-associated volatile organic compounds. *J. Comp. Physiol. A* **186**:1135–1151.

Stafford P, Liu X (1993) Microarray technology comparison, statistical methods and experimental design. In: *Microarray Methods and Applications,* Hardiman G, Ed., DNA Press, Eagleville, PA, pp. 273–324.

Steinbrecht RA (1996) Structure and function of insect olfactory sensilla. *Ciba Found. Symp.* **200**:158–174.

Takken W, Knols BG (1999) Odor-mediated behavior of Afrotropical malaria mosquitoes. *Annu. Rev. Entomol.* **44**:131–157.

Takken W, van Loon JJ, Adam W (2001) Inhibition of host-seeking response and olfactory responsiveness in *Anopheles gambiae* following blood feeding. *J. Insect Physiol.* **47**:303–310.

Van der Pers J, den Otter CJ (1978) Single cell responses from olfactory receptors of small ermine moths, *Yponomeuta* spp. (Lepidotera: Yponomeutidae). *Int. J. Insect Morphol. Embryol.* **9**:15–23.

Vickers NJ, Christensen TA, Baker TC, Hildebrand JG (2001) Odour-plume dynamics influence the brain's olfactory code. *Nature* **410**:466–470.

Voinnet O, Baulcombe DC (1997) Systemic signaling in gene silencing. *Nature* **389**:553.

Wargelius A, Ellingsen S, Fjose A (1999) Double-stranded RNA induces specific developmental defects in zebrafish embryos. *Biochem. Biophys. Res. Commun.* **263**:156–161.

Whitfield CW, Band MR, Bonaldo MF, Kumar CG, Liu L, Pardinas JR, Robertson HM, Soares MB Robinson GE (2002) Annotated expressed sequence tags and cDNA microarrays for studies of brain and behavior in the honey bee. *Genome Res.* **12**:555–566.

Wimmer EA (2003) Innovations: applications of insect transgenesis. *Nat. Rev. Genet.* **4**:225–232.

Section V

Population Analysis of Sensory Systems

11 Taste Processing in the Insect Nervous System

Philip L. Newland

CONTENTS

11.1 INTRODUCTION

Finding the right kinds of food, and in sufficient quantity, is one of the most basic behaviors an animal must accomplish to survive. The sense of taste is a vital component in that process: for all animals, chemicals must be detected, encoded in the central nervous system, and processed, and finally, the neural signals must be acted upon. At higher levels, tastes must be remembered and recognized to prevent us from accidentally eating something harmful or unpleasant. For insects, the sense of taste underlies a range of specific behavior patterns, such as egg-laying and avoidance movements. Yet despite its critical importance in the everyday life of all animals, we know surprisingly little of how different chemosensory cues are encoded within the central nervous system.

In insects, the sense of taste (or *contact chemoreception*, as it is commonly known) plays a crucial role in the sampling and selection of food (Dethier, 1976). Desert locusts (Figure 11.1), for example, will regulate their intake of protein and carbohydrate when offered a choice of diet (Simpson et al., 2002), and the composition of that diet

0-8493-2024-0/05/$0.00+$1.50

FIGURE 11.1 The desert locust, *Schistocerca gregaria* (Forskål). **a:** Photograph of the head of a gregarious adult showing the mouthparts that bear many fields of contact chemoreceptors. **b:** A gregarious fifth-instar nymph. **c:** Multiple fields of contact chemoreceptors are distributed over the mouthparts, legs, and abdomen of adult locusts.

can influence how their taste receptors respond to subsequent chemical cues in the environment (Abisgold and Simpson, 1988; Amakawa, 2001). Moreover, diet can also influence development time and survivorship (Simpson and Raubenheimer, 1993, 1996). Taste is central to all of these processes, as it is in humans, where we associate the sense of taste with food and dietary regulation. In insects, however, taste also plays a crucial role in a range of other behaviors associated with interactions with plants, such as host-plant selection (Chapman and Bernays, 1989; Chapman and Sword, 1993), and in egg laying (Woodrow, 1965; Ma and Schoonhoven, 1973; Städler et al., 1995; Städler, 2002).

In the last few years, there has been a significant advance in our understanding of taste and the processing of taste signals in the central nervous systems of insects, and in particular in the locust (Rogers and Newland, 2002). Although it is unlikely that insects use the same categorization of tastes as humans, they still must solve the same basic problem of taste processing faced by all animals. This involves detecting the many different chemicals at various concentrations and often in complex mixtures, and then categorizing, processing, and ultimately translating that information into an appropriate behavioral response.

A good starting point in this review would be to ask: What value can there be in studying taste in insects if our ultimate goal is to understand how tastes are encoded in the human brain? Perhaps the most important reason is that even after 40 years of research into vertebrate taste, we still do not clearly understand how different gustatory chemicals are encoded in the brain. This lack of understanding has been highlighted by a number of recent reviews (Smith and St. John, 1999; Scott and Giza, 2000; Smith et al., 2000) in which there has been considerable debate of two contrasting models of taste coding that have been proposed. Pfaffmann (1959) postulated that individual sensory neurons might be broadly tuned to respond to a range of chemical stimuli. In this *across-fiber pattern code* or *population code*, the range of chemicals that an individual chemosensory neuron responds to varies among members of a population. It follows that the identity of a chemical must be decoded in the central nervous system by comparing the response of the entire population of chemosensory neurons. Alternatively, the *labeled-line code* theory suggests that individual chemosensory neurons can be grouped into specific classes that each code for a specific taste quality, with no overlap in selectivity (Pfaffmann, 1974; Pfaffmann et al., 1976). More recent studies at a central level, however, suggest that gustatory neurons are broadly tuned to a range of taste stimuli (Pfaffmann et al., 1976; Spector, 2000). Even though these two theories are not mutually exclusive, a similar debate has also occurred in the field of olfaction (discussed in Christensen and White, 2000).

The reason that this debate has continued for so long perhaps resides in our lack of knowledge of the behavioral significance of different gustatory qualities in influencing feeding decisions. The insect nervous system, however, is amenable to rigorous analysis, and it has provided significant insights into how the vertebrate central nervous system functions through studies on the control of reflex movements (Burrows, 1996), locomotion, and vision and, in recent years, through studies on the olfactory system (in locusts, see for example Laurent et al., 2001). Similarly, the insect central nervous system provides an ideal system in which to investigate gustation.

Not only can the insect nervous system provide insights into the general principles underlying taste coding in higher vertebrates, but studies on taste in insects are also crucial in their own right. Many insects are pest species that are responsible for immense economic damage to crops. Locusts, for example, devastate large areas of our planet by consuming vast quantities of vegetation. A single locust can consume its own body weight (approximately 2 g) of food each day; given that there can be up to 50 million locusts/km^2 in a single swarm and that swarms cover many kilometers (Uvarov, 1977), it is clear that they can have a major impact on the agriculture of affected countries. Taste plays a crucial role in dietary selection, and thus by understanding what drives locusts to eat what they do, we have the potential to develop alternative control measures. For example, can we treat crops with a chemical that we know is nontoxic, but through studies on taste we know is unpleasant to, or rejected by, locusts? Moreover, globally we currently spend (on average) millions of dollars each year on control with toxic pesticides, from the United States to Africa, and from Asia to Australia. Again, understanding how locusts detect and respond to tastes could allow the future development of alternative control measures.

In recent years, there have been many studies of olfactory coding in insects (Laurent, 1996, 1997; Galizia et al., 1998, 1999; Christensen et al., 2000, 2003; Laurent et al., 2001; Carlsson and Hansson, 2003) that have addressed how different odors are encoded in the antennal lobes and at higher levels in the brain (see also Chapter 12 through Chapter 14). In insects, however, our knowledge of where and how taste stimuli are processed and coded in the central nervous system comes from only two key studies. The first pioneering study by Mitchell and Itagaki (1992) used intracellular recordings from interneurons in the suboesophageal ganglion to analyze the central coding of taste qualities in blowflies. Their results suggested that particular interneurons in the ganglion responded well to sugars but not to salts, whereas other interneurons showed the opposite coding properties, responding to salts but not to sugars. Such responses are characteristic of the labeled-line coding model proposed by Pfaffmann (1974). In contrast, however, Rogers and Simpson (1999) found by studying the responses of interneurons in the suboesophageal ganglion of locusts to chemical stimuli applied to taste receptors on the mouthparts, that they differed in the latency of response. Rogers and Simpson (1999) showed that some interneurons responded with the shortest latency to wheat leaves applied to the mouthparts, whereas other interneurons responded with the shortest latencies to cabbage leaves.

Although these coding theories are not mutually exclusive, it is not clear how the huge numbers of chemicals in the environment can be encoded by either or both mechanisms. From a neuroethological standpoint, it is important that if we want to understand the neural basis of taste, to relate the properties of neurons and networks back to behavior. One way to resolve our lack of understanding about how different qualities of tastes are encoded in the central nervous system is to adopt a multidisciplinary approach to understand contact chemoreception, in which the behavior of the locust is related to the responses of central neurons and to the organization of the neuronal pathways and neurons. What makes this task possible in the desert locust is the detailed information we already have available of the sensory systems, the local circuits that control leg movements, how they are organized centrally, which types of interneurons are involved, and how they control the motor pattern. Much of this information has been obtained through the detailed studies carried out in Malcolm Burrows's laboratory in Cambridge, England (described in detail in Burrows, 1996).

11.2 BEHAVIORAL PARADIGMS FOR TASTE RESEARCH IN INSECTS

Before undertaking such an analysis, it is worth considering some of the key taste-related reflex behaviors exhibited by insects. This section discusses how detailed studies of these behaviors have been crucial in the analysis of taste coding strategies and have facilitated our understanding of the neural circuits that ultimately control these behaviors.

11.2.1 Proboscis Extension Reflex in Flies

One of the best-described taste-related behaviors of insects is the proboscis extension reflex of flies, first described by Minnich (1926). The classic work of Vincent Dethier (1976), described eloquently in his book *The Hungry Fly*, showed that stimulation of taste receptors on the tarsi of the forelegs of flies initiates an extension of the proboscis. When the proboscis comes into contact with the same food source, the stimulation of further taste receptors on the labellum initiates drinking. Getting (1971) showed that in hungry flies, the stimulation of single taste receptors on the labellum is sufficient to evoke activity in the proboscis muscles involved in feeding. He was also able to demonstrate that ingestion was dependent on an initial chemosensory sampling. This simple feeding reflex has been used as a classic phenotype in the isolation of *Drosophila* taste mutants (see Chapter 9) and also forms the basis of classical conditioning reflexes in studies of learning and memory in bees and moths (Menzel, 2001). Although we already know a little of the motor responses controlling movements of the proboscis, we do not yet understand the neuronal circuits in sufficient detail to attempt a detailed intracellular analysis of gustatory coding. As yet, we do not know what interneuron types are involved in generating and controlling the complex movements of the proboscis itself. Taken together, these factors make the task of relating taste-evoked responses of neurons to movement of the proboscis a difficult one.

11.2.2 Oviposition Behavior in Locusts

Another behavior in which contact chemoreception plays an important role is in the oviposition (or egg-laying) behavior of locusts and their choice of egg-laying sites. Locusts normally lay their eggs in damp soil, and the physical and chemical characteristics of the substrate influence both the selection of an appropriate site in which to lay eggs and the amount of time spent egg laying or digging (Uvarov, 1977). Egg-laying behavior is easily monitored using time-lapse video recording, although the analysis of the behavior can be somewhat time consuming (Figure 11.2a). The role of contact chemoreception in the selection of egg-laying sites can be assessed by offering locusts a simple choice of either damp (control) sand or sand that is chemically treated. Behavioral analyses show that locusts preferentially choose to lay eggs in control sand (Figure 11.2b), avoiding higher concentrations of test chemicals, and also spend proportionately less time egg laying as the concentration of a chemical added to the test sand is increased (Figure 11.2c) (Woodrow, 1965; Yates and Newland, 2003).

Clearly, intracellular recording from central interneurons during free movement is an almost impossible task, and the choice behavior itself is too complex to use in taste-coding analyses. A much more relevant behavioral response, however, and one that can be studied under the microscope, is the rhythmic movement of the abdomen produced during the digging phase of egg-laying behavior (Figure 11.3) (Thompson, 1986a). For locusts to lay eggs, they must dig down into the sand using two pairs of hard sclerotized ovipositor valves (the dorsal and ventral valves) at the end of the abdomen, in order to deposit eggs at depths where they are less likely to be exposed to desiccation. During this behavior, the locusts will gradually extend their abdomens by up to twice their resting lengths to dig to depths of 6 to 8 cm.

Isolating the abdomen from the thorax leads to rhythmic movements of the ovipositor valves (Thompson, 1986a) as the circuits responsible for generating the rhythmic motor pattern are released from descending inhibition from higher centers (Lange et al., 1984; Thompson, 1986b; Facciponte and Lange, 1996). The rhythmic digging movements of the valves take the form of a triphasic motor pattern consisting of a protraction phase, an opening phase, and a simultaneous retraction and closing phase (Figure 11.3a).

Contact chemoreception influences the digging movements of the ovipositor valves (Yates and Newland, 2003) in such a way that when they are moving rhyth-

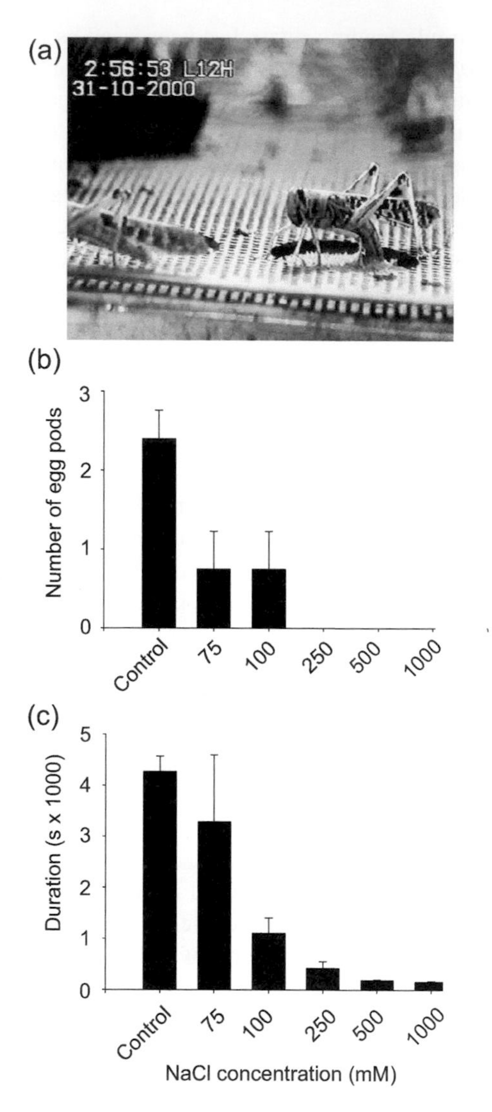

FIGURE 11.2 The chemical composition of the substrate influences the choice of oviposition sites. **a:** Female locusts lay their eggs in damp soil/sand in clusters of up to 100 in cylindrical pods. **b,c:** The number of egg pods laid (**b**) and the duration spent in oviposition (**c**) are dependent on the chemical concentration of the soil. For NaCl, the greater the concentration, the fewer egg pods are laid (250 mM NaCl completely abolishes egg laying) and the greater is the time spent attempting to lay eggs.

mically and chemicals are applied in droplet fashion to the tip of the abdomen, the valves are immediately closed and retracted (Figure 11.3b). The duration of valve retraction is dependent on the concentration of the chemical in the droplet. For example, application of a solution containing 2.5 mM nicotine hydrogen tartrate (NHT), a secondary plant chemical that is a known feeding deterrent, inhibits the digging rhythm for a longer period than either a low NHT concentration or water

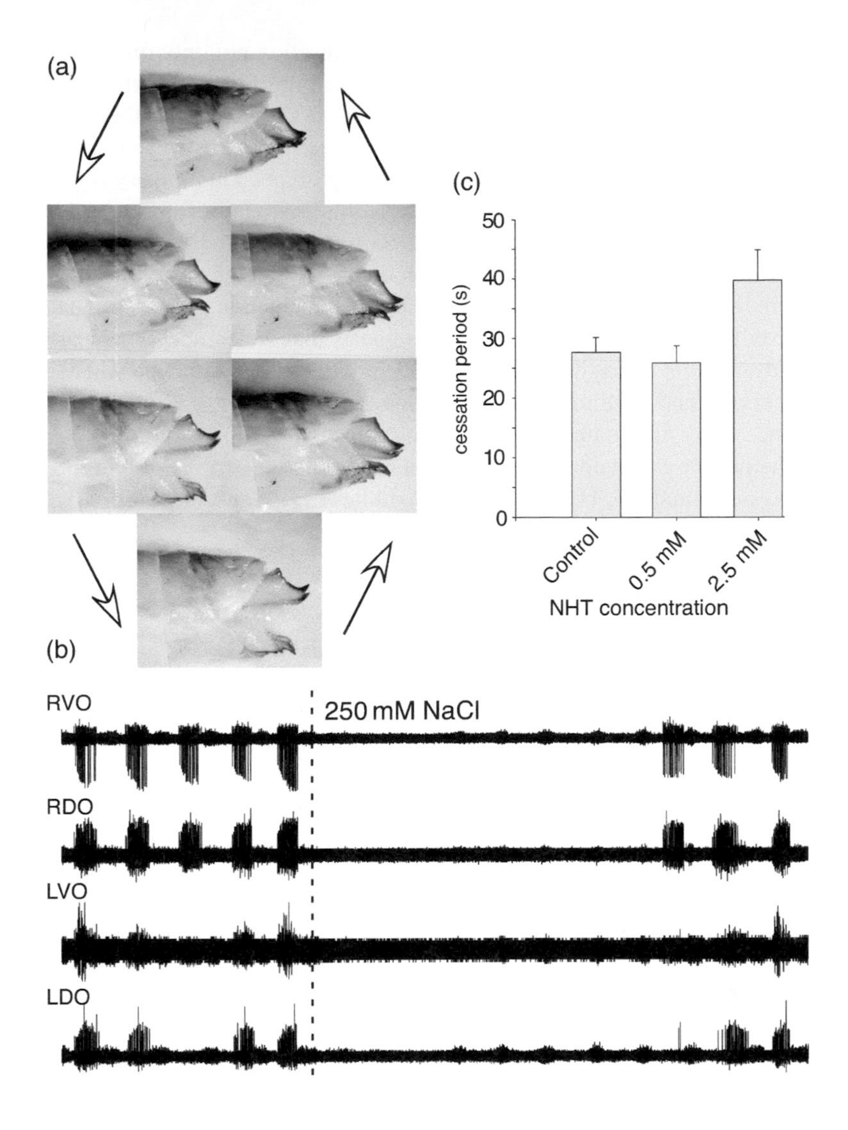

FIGURE 11.3 Rhythmic abdominal digging movements are influenced by gustatory cues. Female locusts use rhythmic digging movements of hard scleretized ovipositor valves on the last abdominal segment to dig to depths of up to 8 cm to lay eggs. **a:** Images of the ovipositor valves during a single cycle of movement. The paired dorsal and ventral valves are protracted and opened, followed by closing and retraction. **b:** Electromyogram recordings from the opener muscles of each of the right and left (R and L) dorsal and ventral (D and V) valves. Application of 250 mM NaCl to the ventral valves abolishes the rhythmic activity of the opener muscles. **c:** The duration for which the rhythm ceases is dependent on chemical concentration. The greater the concentration of the secondary plant compound nicotine hydrogen tartrate (NHT), the greater is the time the rhythm is suppressed.

applied as a control (Figure 11.3c). Thus, chemosensation not only underlies behavioral choice of egg-laying sites but also influences rhythmic motor behavior. Recent studies by Kalogianni (1996), Kalogianni and Burrows (1996), and others (Belanger and Orchard, 1992) have begun to describe in detail the neuronal circuitry within the terminal ganglion responsible for the processing of mechanosensory signals from tactile hairs on the valves. Moreover, recent anatomical studies (Tousson and Hustert, 2000) have described the organization of the central projections of the chemosensory neurons that innervate the taste sensillae found on the ovipositor valves. The increasing knowledge we now have of the central nervous system that produces and controls these movements of the ovipositor valves suggests that we can begin to learn much from studying this behavior. For example, are there parallels between contact chemoreception and rhythmic digging movements in insects, and human taste and its influence on rhythmic chewing and biting movements? Only further studies will reveal whether we can gain insights into these interactions from studying "simple" nervous system function, but if these early studies are of any indication, the future is bright.

11.2.3 Feeding and Food Sampling Behavior

Feeding is perhaps the most obvious behavior in which taste plays a major role, and it is arguable that this is where we should be focusing our research efforts. Indeed, the few studies that have been carried out on central taste coding in insects have focused in this area (Mitchell and Itagaki, 1992; Rogers and Simpson, 1999). The problems with working on feeding, however, are threefold:

1. The movements of the different parts of the mouth and head are numerous and complex. This means that determining their respective roles during feeding is a difficult and immensely time-consuming task. These movements include head movements, antennal movement, palp movements that occur at high speed as the locust samples food, and mouthpart movements.
2. Access to the appropriate part of the central nervous system is not straightforward, although these previous studies have shown that it is possible (Simpson, 1992). Accessibility, however, remains a negative factor when analyzing feeding behavior.
3. Arguably most importantly, we know less about the region of the central nervous system that controls and moves the mouthparts, the suboesophageal ganglion, than most other parts of the insect nervous system (Burrows, 1996).

All of these factors contrive to make feeding and food sampling difficult to understand in sufficient detail to relate the responses of individual neurons to specific movements of the mouthparts and palps. Thus, if our goal is to understand the neurobiology of taste, it is arguable that it is necessary to analyze the equally important chemosensory-based nonfeeding behaviors before we undertake an analysis of feeding in greater detail.

11.2.4 Leg Avoidance Behavior

The legs are often the first part of the animal to come into contact with a plant, and the taste receptors on the legs play an important role in the assessment of food quality and in food rejection (Ma and Schoonhoven, 1973; Szentesi and Bernays, 1984; Chapman et al., 1987; White and Chapman, 1990). Movements of the legs are produced in response to chemical stimulation, and these take the form of avoidance movements that move the leg away from the point of stimulation. Such chemosensory-evoked leg movements of the grasshopper were first described by Slifer (1954) in response to odors of specific heavy oils, such as eucalyptus or clove oil. This early work was concerned exclusively with movements of the grasshopper limb and did not examine the responses of contact chemoreceptors on the leg. Hence, the detectors responsible for initiating these movements were unknown. Recent repetition of these studies on the locust, but with the head and antennae covered with a cap to prevent odors from activating more rostral olfactory receptors, failed to result in leg movements (Newland, 1998), and electrophysiological analyses failed to show evoked responses of the contact chemosensory neurons innervating central chemoreceptors on the leg. Nevertheless, odors of acids did activate the chemosensory neurons and evoked a withdrawal movement of the leg. Dethier (1976) had also shown that odors of specific acids activate the chemosensory neurons of taste sensilla on the legs and mouthparts of blowflies. We still have not discovered how these acidic odors activate the chemosensory neurons or whether this response is of any true biological function. It is possible, however, that these sensilla have evolved to detect the acidic odors of ants that use formic acid for defense. The true ecological and evolutionary benefits of such responses have yet to be addressed and remain of particular interest.

White and Chapman (1990) showed that locusts will move their legs away from the source of nonvolatile chemicals. Contact with the secondary plant compound, nicotine hydrogen tartrate, or sodium chloride evokes rapid withdrawal movements similar in form to those evoked by acid-odor stimulation (cf. Newland, 1998) and also to those evoked by tactile stimulation of exteroceptive hairs on the locust leg (Pflüger 1980; Siegler and Burrows, 1986). Droplets of chemicals applied to the hind leg lead to a rapid movement of the leg away from that chemical. These movements are easily quantifiable because they occur in a single plane and thus are easy to monitor and record using standard video equipment, followed by single-frame video analysis (Figure 11.4a).

Importantly, in our search for a behavior that we can relate to neuronal responses, we know in exquisite detail the circuits of neurons that produce and control the movements of the legs, the types of neurons involved in processing sensory signals (apart from chemosensory signals), the transmitters and neuromodulators involved, and the types of interneurons controlling the output of the motor neurons to the leg musculature. These limb movements represent an ideal behavior that we can relate back to the response properties of interneurons in the local circuits to attempt to understand how chemical cues are coded in the central nervous system. Recent detailed studies have analyzed these leg movements and asked how the legs move in response to a range of different chemical and in a range of concentrations. The

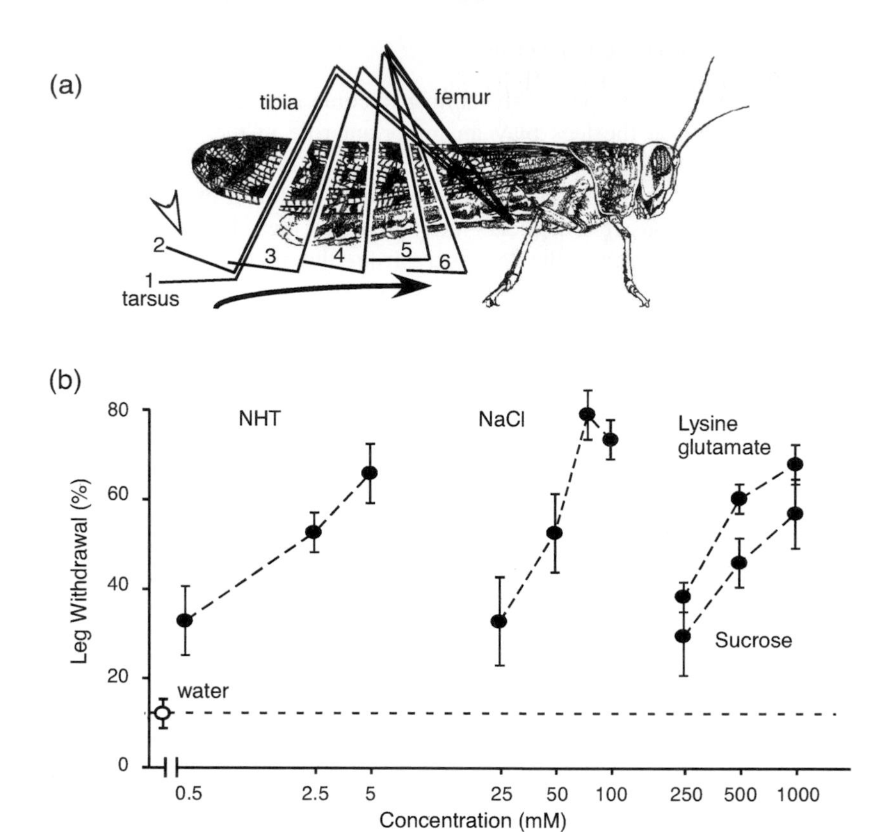

FIGURE 11.4 Local avoidance movements of the leg are evoked by chemical stimulation. **a:** A droplet of an aqueous solution of a chemical applied to the hind leg (arrowhead) causes a rapid withdrawal movement of the leg (in less than 1 sec), in which the femur is levated, the tibia flexed, and the tarsus levated. The hind leg is shown as stick diagrams of the position of the leg segments on subsequent video frames (interframe interval of 40 msec). **B:** All chemicals applied to the leg evoke a withdrawal movement, the frequency of which is dependent on chemical type and concentration. NHT is the most potent chemical, with 2.5 mM NHT causing withdrawal in approximately 50% of trials, whereas sucrose is less potent, with concentrations of approximately 1000 mM being required to evoke a similar frequency of withdrawal. (From Rogers, S.M. and Newland, P.L., *J. Exp. Biol.* **203**:423–433, 2000. With permission.)

effects of five chemicals (NaCl, sucrose, the antifeedant nicotine hydrogen tartrate (NHT), a lysine–glutamate salt, and water) have been analyzed at different concentrations for their effects on the frequency of leg withdrawal (Rogers and Newland, 2000). These chemicals were chosen as representative of a number of dietary requirements and known toxins. The probability of withdrawal increased for all chemicals in a concentration-dependent fashion (Figure 11.4b). Interestingly, any chemical, when present at an appropriate concentration, will give rise to an avoidance movement. This suggests that all chemicals are processed in the same way within the thoracic local circuits, so that there is no separate categorization of these chemical stimuli. All chemicals are treated as aversive and act to evoke an avoidance movement of the leg if they are present in high concentrations. Such responses are not surprising if we consider our own behavioral responses to a range of chemicals. If we put food in our mouths, some chemicals, such as salt and sucrose, can be considered pleasant at low concentrations but are actively rejected at higher concentration.

Taste, or contact chemoreception, is therefore important for a number of different insect behaviors. Feeding is not the only insect behavior for which the sense of taste is important, however. We already know much about the nervous control of limb movements, and that makes this system ideal to study the general principles underlying taste coding.

11.3 TASTE RECEPTORS AND PERIPHERAL CODING

The taste receptors responsible for detecting chemicals are housed in tiny hairs known as *basiconic sensilla* (singular: *sensillum*) and are found not only on the mouthparts, where they occur at their highest densities, but also on most parts of the body and legs (Chapman, 1982). On the leg of the locust, the sensilla range in length from 30 to 50 μm (Figure 11.5a,b), and each has a pore in its tip leading to a fluid-filled lumen containing the dendrites of the sensory neurons. The cell bodies of these sensory neurons lie just below the receptor (Figure 11.5d,e), and they send their axons back to the central nervous system. It is worth pointing out that basiconic sensilla are bimodal and, on the locust leg, are innervated by four chemosensory neurons that respond to contact with chemicals (Figure 11.5c) and one mechanosensory neuron that responds to touch (Newland and Burrows, 1994).

One key factor that makes the insect nervous system suitable for analyses of gustation is accessibility to these sensilla in the periphery, and our understanding of gustatory processing in insects has been based almost entirely on analyses of the activity patterns of the chemosensory neurons that innervate the sensilla. Until recently however, very little has been known of the destinations of the chemosensory neurons in the central nervous system and the central processing of the gustatory information they carry. Again, this stands in stark contrast to the wealth of studies that have focused on the central processing of olfactory cues (see Anton and Homberg, 1999).

The decision an insect makes to initiate feeding or reject food is based largely on the information it receives from its basiconic sensilla. How, then, is chemosensory information encoded by these receptors? Most recordings from basiconic sensilla have been made using extracellular recording methods obtained with either the tip-recording technique (Hodgson et al., 1955) or with sidewall or basal recording obtained by driving sharpened tungsten or glass microelectrodes through the cuticle adjacent to or within the walls of a sensillum (see also Chapter 1 through Chapter 3). Much information has been gathered from different types of gustatory sensilla from many insects using the tip-recording method (Schoonhoven and van Loon, 2002). In tip recording, all test chemicals must be in an aqueous solution, and in fluid feeding insects, these solutions may provide a reasonably close approximation of the substances on which they feed. It is unclear, however, how these solutions equate to the chemosensory environment found on the surfaces of solid matter, such as a leaf surface. In order to detect chemicals from such surfaces, it is thought that

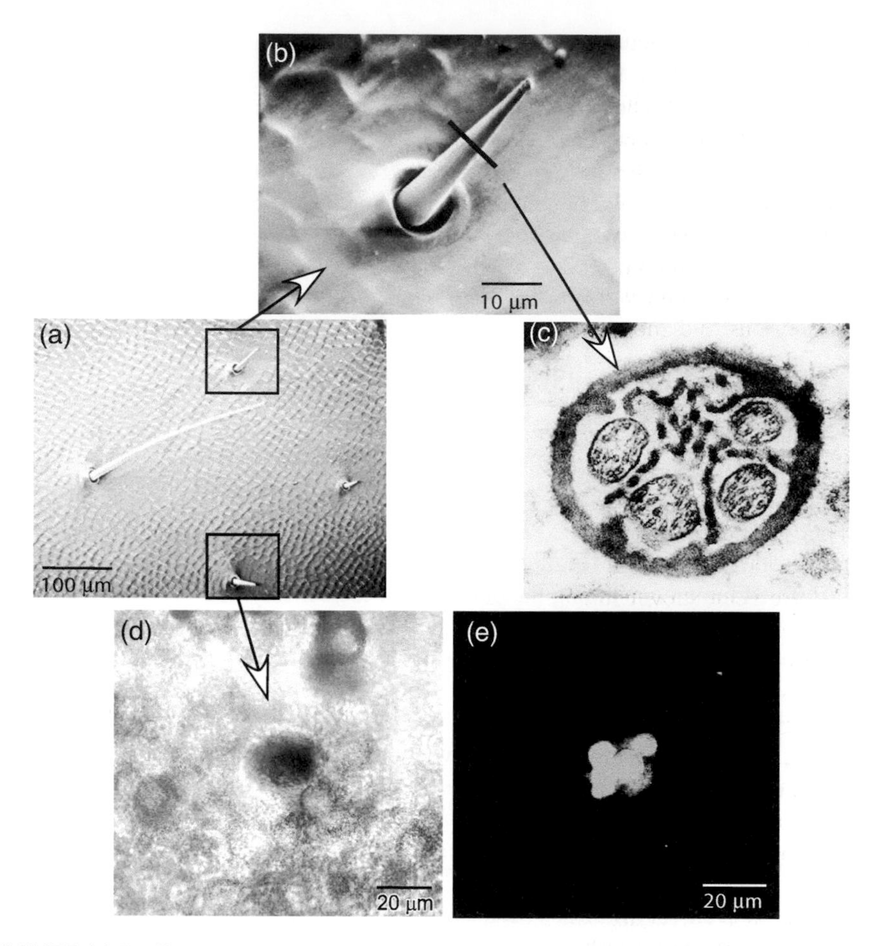

FIGURE 11.5 Contact chemoreceptors on the locust leg. **a:** Scanning electron micrograph of a region on the dorsal femur showing a long tactile hair and three short basiconic sensillae, the contact chemoreceptors. **b:** Higher-power view of a basiconic sensillum. **c:** A section through a sensillum at a level similar to that indicated in **b**, showing sections of four chemosensory dendrites within the shaft of the sensillum. A fifth, mechanosensory dendrite inserts at the base of the sensillum, out of the plane of focus. **d,e:** Confocal images of the internal surface of the cuticle indicating the site of a sensillum in **d** and a fluorescent image (**e**) of the cell bodies of the sensory neurons stained by application of Texas Red to the cut receptor shaft.

chemicals from the leaves are taken up in the receptor lymph that occasionally exudes from the terminal pore of a sensillum (Zacharuk, 1985).

In his recent review of contact chemoreception of phytophagous insects, Chapman (2003) described in detail how plant chemicals are monitored by the gustatory systems of insects. Due to the timeliness of this excellent review, a further review here of the same subject matter would add little. It is worth reiterating, however, the recent suggestion of Bernays and Chapman (2001) that an alternative terminology for chemosensory neurons should be based not on what chemicals they respond to, but instead on their function. In other words, chemosensory neurons could be classified as being deterrent or phagostimulatory, depending on a chemical's action on behavior, rather than being sugar cells or salt cells. Studies on many insect species now clearly show that chemosensory neurons may respond not just to one chemical, but instead to more than one, indicating that a population code is utilized (Mitchell and Gregory, 1979; Bernays et al., 2000). Indeed, this argument is supported by Rogers and Newland (2002) who analyzed the central coding of taste in locusts (see Section 11.4.3) and who showed that all chemicals have the potential to activate interneurons and motor neurons in gustatory pathways and that their effects are dependent on the aversiveness of the chemical stimulus (see Section 11.4.3).

11.4 LOCAL CIRCUITS CONTROLLING LIMB MOVEMENTS

The locust central nervous system has been analyzed in great detail (Burrows, 1996), and the local circuits of neurons that produce and control the movements of a hind leg have been shown to be located within the ipsilateral half of the metathoracic ganglion. From these studies we know that sensory receptors on and in just one hind leg alone send some 10,000 sensory axons to one-half of the metathoracic ganglion. These sensory neurons are from different types of receptors that provide mechanosensory information about stress on the cuticle (campaniform sensilla; Zill and Moran, 1981), movement about joints (chordotonal organs; Bräunig et al., 1981; see Chapter 3), and touch (tactile hairs; Newland, 1991a), as well as chemosensory information (Newland, 1998). Of these 10,000 sensory neurons, it is estimated that approximately 80% come from the taste receptors (Newland and Burrows, 1997). On the other hand, fewer than 100 motor neurons are responsible for the complex patterns of movements that are required for adjusting the posture of a leg or for locomotion (Burrows, 1996).

Thus, there is a huge convergence and integration of sensory signals (100:1 ratio), and this occurs locally within the ipsilateral half of the ganglion. Here, there are fewer than 1000 interneurons, just over half of which have no axons and have branches restricted entirely to this ganglion. On this basis, they are called local interneurons and are of two physiological types:

1. The first are *spiking local interneurons,* which communicate by means of all-or-nothing action potentials or spikes. They play a key role in the processing of exteroceptive (Burrows and Siegler, 1982, 1984; Siegler and Burrows, 1983; Burrows, 1992) and proprioceptive signals (Burrows, 1987), in addition to signals from the campaniform sensilla (Newland and Emptage, 1996) and, as shown more recently, to chemical stimuli (Newland, 1999).
2. The second type of interneuron is the *nonspiking local interneuron,* which communicates without action potentials and instead utilizes graded electrical signals (Burrows, 1996). A key feature of the nonspiking interneu-

rons that makes them important components in motor control is their divergent patterns of connection with different leg motor neurons. Each nonspiking interneuron can make synaptic connections with a number of motor neurons; in turn, each motor neuron receives convergent input from a number of nonspiking interneurons (Burrows, 1980). This organization ensures that the movements of a leg are generated from the coordinated action of many interneurons and motor neurons.

The remaining interneurons have axons that project to adjacent ganglia in the segmental chain and are called *intersegmental* or *projection interneurons* (Newland, 1990). They receive sensory information in one ganglion and project to another where, by virtue of their patterns of outputs, they are able to regulate the action of local circuits in different ganglia (Laurent and Burrows, 1989) and are, thus, involved in interleg coordination. So far, three populations of intersegmental interneurons with ascending or descending axons have been described in the metathoracic ganglion (Laurent and Burrows, 1989; Newland, 1990).

Together, these local, intersegmental, and motor neurons provide the neuronal framework within which the motor actions of the leg are organized, and studies of their actions and roles in local circuits provide the detailed background information that is needed to attempt to understand the central coding of tastes.

11.4.1 Anatomical Organization of Sensory Neurons

Recent studies have analyzed chemosensory processing in local circuits at three levels:

1. The organization of the chemosensory neurons within the central nervous system
2. The identification of spiking local interneurons in the gustatory pathways and their responses to chemical stimulation
3. The responses of leg motor neurons to chemical stimuli applied to the legs

If we want to understand where taste signals are processed in the central nervous system, then anatomical cues are essential. The problem is that there appears to be no clear anatomical structure that is clearly involved in taste processing, unlike those that occur in the olfactory system, in which numerous, dense projections form distinct glomeruli in the antennal lobes of almost all insect species (Anton and Homberg, 1999). The problem, then, is where to look in a central nervous system that lacks structure related to taste. Recent advances in our understanding of where and how taste stimuli may be processed come from studies that have addressed where chemosensory neurons send their central projections, how sensory neurons segregate in the central nervous system, and what the fine structure of their central projections is.

For many years, it has been clear that determining the destination and branching patterns of sensory neurons within the central nervous system has a major contribution to make to our understanding of how signals from different sensory modalities are processed. Once we know where sensory neurons project, we then have guides to what central neurons may process their signals. For example, sensory neurons commonly have branches in specific regions of the central nervous system, depending on the sensory modality to which they respond. Proprioceptive sensory neurons from receptors that monitor movements about limb joints have branches of fine neurites predominantly in dorso-lateral regions of the ganglion, whereas sensory neurons that innervate exteroceptive tactile hairs project into ventral regions of the ganglion (Pflüger et al., 1981, 1988). This type of organization is common across most insect species so far analyzed.

Another major feature of sensory systems is that the projections of their sensory neurons form orderly maps in the central nervous system. These orderly maps can form consistent and predictable representations of the location of sensory receptors on or in the body within the central nervous system (i.e., somatosensory maps, Johnson and Murphey, 1985; Newland, 1991b), or they may be organized according to particular coding properties of the sensory neurons themselves, such as auditory tonotopic maps (Oldfield, 1982; Römer, 1983; Römer et al., 1988). The somatotopic maps of sensory neurons that innervate tactile hairs on the legs have been analyzed in great detail, as has the structure of interneurons related to these maps. Sensory neurons from tactile hairs form a three-dimensional map in which the spatial location of the receptor on the leg is retained in the spatial organization of the branches of individual sensory neurons within the central nervous system (Newland 1991b; Newland et al., 2000). This is illustrated in Figure 11.6, which shows the projections of mechanosensory neurons from tactile hairs along just one axis of the mesothoracic leg, the proximo-distal axis (Figure 11.6a–c). A mechanosensory neuron that innervates a hair on the femur projects toward the midline in the ganglion, one from a tibial hair projects centrally, and one from a tarsal hair projects laterally. This means that the location of a hair on a proximo-distal axis on a leg is represented by its projections along a medio-lateral axis in the ganglion. Sensory neurons that innervate hairs located dorsally on the leg send projections dorsally in the ganglion, whereas those from hairs located ventrally project ventrally in the ganglion (Newland, 1991b). Finally, those that innervate hairs located on the anterior surface of the leg send projections laterally in the ganglion, whereas those from hairs on the posterior surface of the leg send their projections posteriorly (Newland, 1991b). Thus, the central projections of the mechanosensory neurons form a complete central three-dimensional map of the spatial location of the tactile hairs on the leg. Spatial information is, therefore, preserved in the specific patterns of projections that sensory neurons make in the central nervous system: a key organizational principle of invertebrate nervous systems (Newland and Burrows, 1997).

If we want to understand more about the gustatory system, then the obvious question to ask is: How are the chemosensory neurons organized within the central nervous system? Can we make any predictions based on what we already know about how different sensory systems are organized? Since the basiconic sensilla on the leg are bimodal and innervated by one mechanosensory neuron responding to touch and four responding to taste, we can make predictions on the basis of what we know of the anatomy of the mechanosensory systems.

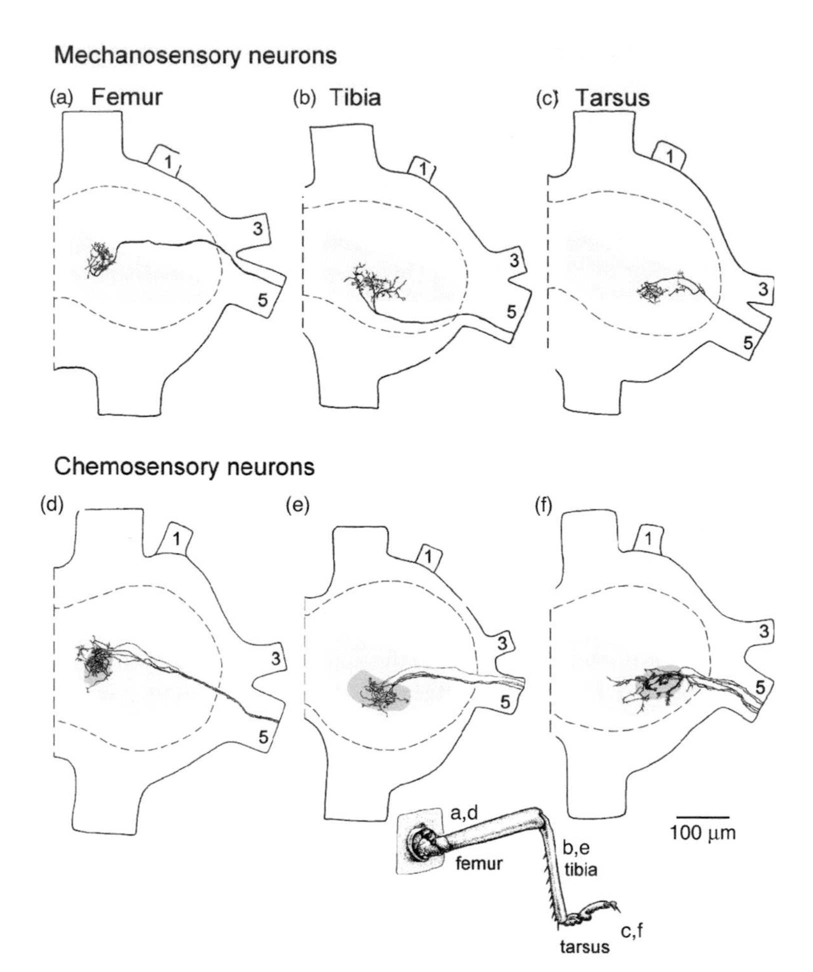

FIGURE 11.6 Parallel somatoptopic maps of mechanosensory and gustatory neurons in the central nervous system. **a–c:** The single mechanosensory neuron from a tactile hair on the femur of the middle leg projects to a region close to the midline of the mesothoracic ganglion (**a**). A mechanosensory neuron from a hair on the tibia projects more centrally (**b**), and that from a hair on the tibia projects laterally (**c**). Thus, the position of a hair along the proximo-distal leg axis is represented by the projections of its sensory neurons along a medio-lateral axis within the ganglion. The light stippling represents the total area occupied by the mechanosensory neurons from tactile hairs on the dorsal surface of the leg. **d–f:** Central projection of sensory neurons from basiconic sensillae from locations similar to those of the tactile hairs in **a–c**. **d:** Sensory neurons from a basiconic sensillum project near the midline and overlap with the projections of the tactile hair shown in **a** (dark stippling). **e:** Sensory neurons from a sensillum on the tibia project more centrally and overlap with the projections of the mechanosensory afferent shown in **b**. **f:** Sensory neurons from a sensillum on the tarsus project most laterally and overlap with a mechanosensory afferent from a similar location (dark stippling). The light stippling represents the total area occupied by chemosensory projections from basiconic sensilla on the dorsal surface of each segment of the leg. Inset shows the locations of the tactile hairs and basiconic sensillae on the leg. (Modified from Newland, P.L., Rogers, S.M., Gaaboub, I., Matheson, T., *J. Comp. Neurol.* **425**:82–96, 2000. With permission.)

1. We might expect a modality-specific segregation of the different sensory neurons into taste and touch regions.
2. We might expect that the mechanosensory afferents from the basiconic sensilla would project to similar regions as the sensory neurons that innervate the tactile hairs.
3. We might expect that the mechanosensory afferents of the basiconic sensilla would form somatotopic maps in the thoracic ganglia.
4. We might expect that chemosensory afferents would be organized in some way related to taste quality.

By staining the sensory neurons of basiconic sensilla, it is possible to compare their projections with those that innervate tactile hairs from a similar location on the legs. Figure 11.6d–f shows the central projection of chemosensory neurons from basiconic sensilla along the proximo-distal axis of the mesothoracic leg. Our first prediction is that we would expect to see a modality-specific segregation of the different sensory neurons, with the mechanosensory afferents projecting ventrally and the chemosensory afferents possibly projecting to another area. Interestingly, this pattern is never observed. Instead, all afferents from a single sensillum project to similar ventral regions of the mesothoracic ganglion. It is clear that although the axons of the different neurons sometimes take slightly different routes through the central nervous system, their terminal arborizations all project into the same area of neuropil without apparent segregation (Newland et al., 2000).

Our next two predictions suggest that the mechanosensory afferents from the basiconic sensilla would project to similar regions as the sensory neurons from tactile hairs and that they would also form somatotopic maps. These predictions do hold true, but surprisingly, not just for the one mechanosensory afferent but for all sensory neurons from a given sensillum, including all of the chemosensory neurons. This is clear in Figure 11.6d–f, in which the sensory neurons from a basiconic sensillum on the femur project toward the midline in the ganglion and overlap with the tactile hair projections shown by the dark stippling. The sensory neurons from a basiconic sensillum on the tibia overlap with the central projections from a neighboring tactile hair. Similarly, the projections of sensory neurons from a tarsal sensillum project laterally in the central nervous system, but again overlapping with mechanosensory hair projections. Thus, the proximo-distal leg axis is represented along a medio-lateral axis in the ganglion, in a similar manner to that of the tactile hairs. Finally, our fourth prediction was that the chemosensory afferents would be organized in some way related to taste quality, possibly in a fashion analogous to the chemotopy demonstrated in the olfactory system. There is, however, no evidence to suggest any clear segregation in branching patterns from any of the chemosensory neurons.

More recent studies have also concluded the same in other regions of the locust central nervous system from taste receptors that serve different roles in the gustatory behaviors described earlier. Rogers (1998), for example, found that there was no modality-dependent segregation of projections from sensory neurons that innervate basiconic sensilla on the palps of the mouthparts, either when stained individually or when many of the entire 400 at the end of a palp are stained *en masse.* Similarly, Tousson and Hustert (2000) analyzed the central projections of chemosensory neurons from basiconic sensilla on the abdomen and ovipositor and found no modality-specific segregation of their projections within the terminal abdominal ganglion.

Clearly, there is a close link between chemoreception and mechanoreception in the central nervous system of insects. This parallels what is also found in vertebrates, where the cranial nerve that innervates taste buds produces a map in the central nervous system that overlaps with a similar map from the mechanosensory trigeminal nerve. The maps also appear to be spatially organized, each projection position within a map corresponding to specific areas of taste buds in the mouth (Smith and Shepherd, 1999). For humans, the behavioral function appears to be in the ability to locate edible food particles in the mouth. For the locust, the implication is that there may well be a similar function that allows the locust to localize chemicals on different parts of the leg.

11.4.2 Gustatory Pathways in Local Circuits

Armed with knowledge of where chemosensory neurons terminate in the central nervous system, the key question that needs to be addressed now is: Where and how are the taste signals processed and encoded in the central nervous system? Anatomical studies give us clues about where to start such analyses, since they indicate that the central projections of the chemosensory afferents project to similar regions of the central nervous system as tactile afferents (Newland et al., 2000). This is where detailed knowledge of the insect central nervous system is very useful; previous studies have shown that the tactile afferents make synaptic connections with spiking local interneurons that have their branches in the same ventral region of neuropil (Burrows and Newland, 1994). Does this imply that the chemosensory neurons make connections with similar classes of spiking local interneuron?

Intracellular recordings from this class of interneuron (Figure 11.7a) reveal that they do receive convergent inputs from both touch- and taste-sensitive neurons (Newland, 1999). Touching tactile hairs on the leg evokes large depolarizations and spikes in an interneuron (Figure 11.7b). Applying droplets of chemicals to the same leg also leads to a response in the same interneuron, the duration of which depends on the identity of a chemical and its concentration. It is clear in the example shown in Figure 11.7c that the duration of response to NaCl is longer and of higher frequency than to water alone. A major problem with this type of analysis is that there is an ambiguity in the response of an interneuron because each interneuron has bimodal properties responding to both taste and touch. If we want to establish what the chemosensory pathways are and how they respond to taste stimuli, this bimodal property of interneurons is more than an inconvenience. Recent studies, however, have shown that the basiconic sensilla and interneurons respond to odors of specific acids, such as formic and acetic (Newland, 1999). When an acid odor stimulus is held under the tarsus, it evokes a response in interneurons and a withdrawal of the leg (Figure 11.7d). The beauty of this stimulus is that it activates only the chemosensory neurons and not the mechanosensory neurons within the same sensilla, providing us with the means to establish which neurons are exclusively responsible for the processing of taste signals.

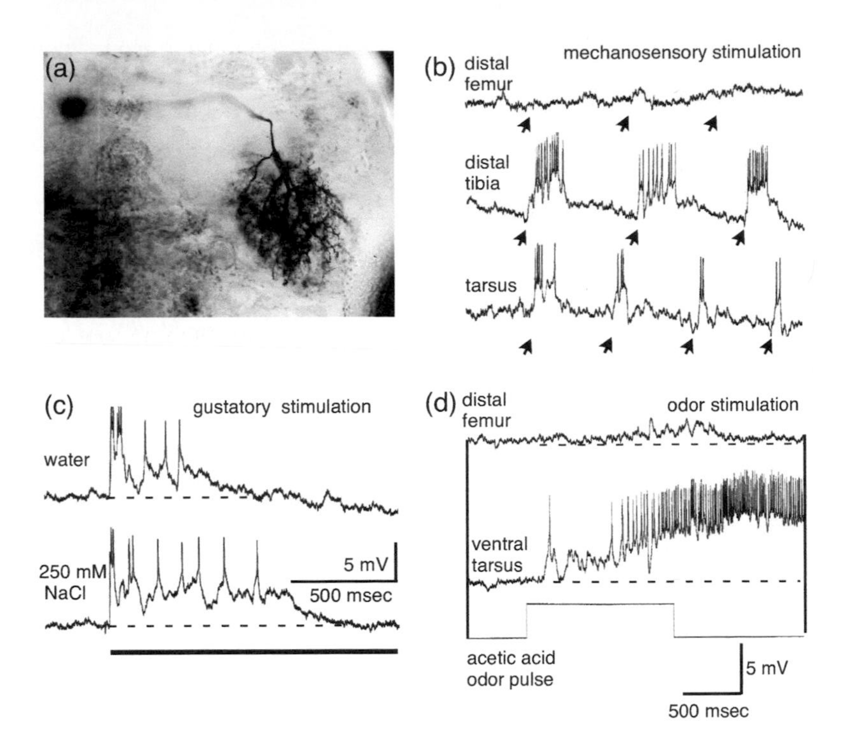

FIGURE 11.7 Spiking local interneurons act as summing points for mechano- and chemosensory inputs. **a:** Cobalt hexammine stain of the ventral field of branches of a midline spiking local interneuron, with its cell body on the contralateral side, just out of the plane of focus. **b:** Mechanical deflection of tactile hairs on a hind leg evokes excitatory postsynaptic potentials (EPSPs) and spikes in another local interneuron. **c:** A droplet of water applied to the hind leg evokes a long-lasting depolarization in the same interneuron. Application of a droplet of 250 mM NaCl evokes a longer-lasting depolarization in the same interneuron. **d:** The odor of acetic acid directed over the hind leg also evokes a depolarization and spikes. (Modified from Newland, P.L., *J. Neurophysiol.* **82**:3149–3159, 1999. With permission.)

Analyses of the receptive field of the spiking local interneurons bring together recent anatomical and physiological studies of gustatory processing. The receptive field of an interneuron is simply the area on the leg that when stimulated, evokes a response in an interneuron. Using the odor method of stimulation, it is possible to compare the receptive fields of interneurons in response to mechano- and chemosensory stimulation. The interneuron shown in Figure 11.8a has a mechanosensory receptive field restricted to the tarsus and only responds when hairs on the tarsus are stimulated. Similarly, by applying odors to different parts of the leg, it is possible to define the chemosensory receptive field on the tarsus (Figure 11.8b). By contrast, other interneurons have different receptive fields. The neuron shown in Figure 11.8c has a mechanoreceptive field restricted to the femur, which overlaps with a similar chemoreceptive field restricted to the femur (Figure 11.8d). Different interneurons have different receptive fields, and across the entire population they map the surface

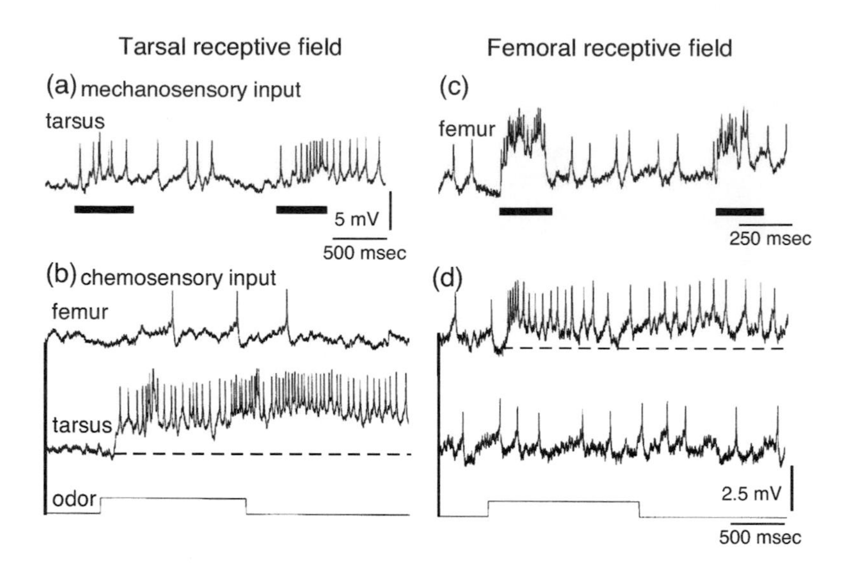

FIGURE 11.8 Mechano- and chemosensory receptive fields of spiking local interneurons. **a:** An interneuron is depolarized and spikes when a tactile stimulus is applied to tactile hairs on the tarsus. **b:** The same interneuron is activated when an odor is directed over the tarsus, but not when directed over the femur. **c:** A different interneuron is depolarized and spikes when a tactile stimulus is applied to hairs on the femur. **d:** The same interneuron is activated by the odor of acetic acid only when it is directed at the femur. Thus, there is an overlap of chemo- and mechanoreceptive fields. (Modified from Newland, P.L., *J. Neurophysiol.* **82**:3149–3159, 1999. With permission.)

of the leg so that spatial information is preserved. This organization of receptive fields could be produced only if there is convergence of chemosensory afferents to the same regions within the central nervous system as the mechanosensory afferents. We have also recently found that the chemo- and mechanosensory afferents make convergent monosynaptic connections with these spiking local interneurons (Newland, 1999).

11.4.3 Gustatory Coding by Interneurons and Motor Neurons

Once we have established what interneurons are involved in processing chemosensory signals, the obvious next step is to ask how they encode different chemicals. The procedure is relatively simple but involves making long-term intracellular recordings from the spiking local interneurons while applying a range of chemicals, in a range of concentrations, to the leg contact chemoreceptors (Rogers and Newland, 2002). Figure 11.9 shows the recordings of four different interneurons. Each interneuron responds more strongly as chemical concentration is increased for all of the chemicals tested. Thus, there is a significant correlation between chemical concentration and the duration of response (and the number of action potentials) produced by any one of the four chemicals.

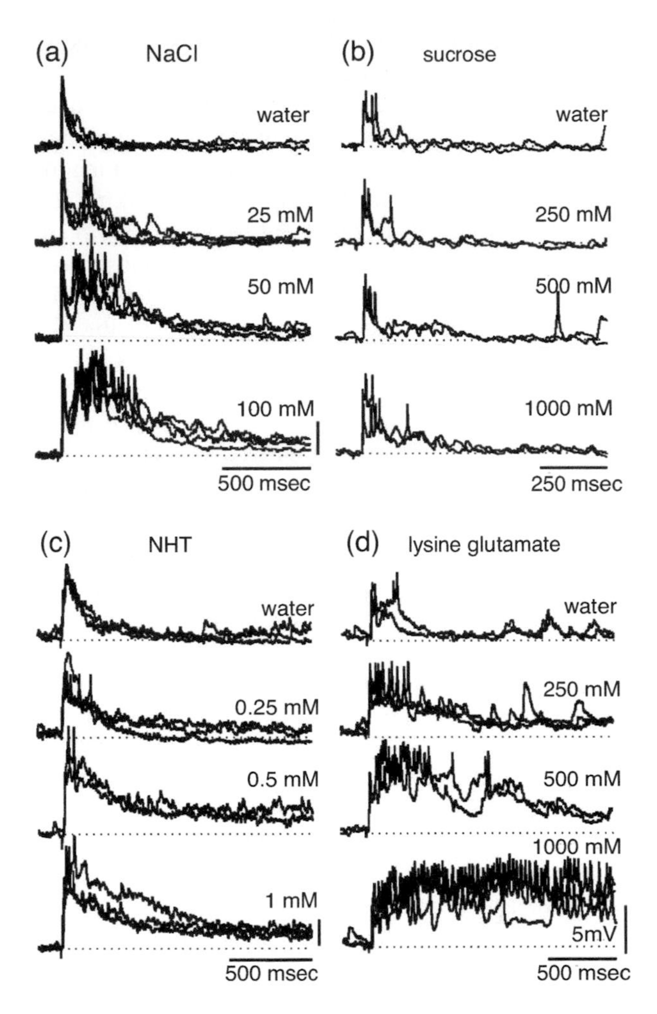

FIGURE 11.9 Responses of spiking local interneurons to chemosensory stimulation. **a:** Responses to NaCl; **b:** responses to sucrose; **c:** responses to NHT; **d:** responses to lysine glutamate. For each chemical, the greater the concentration, the greater are the depolarization and number of spikes evoked in an interneuron. The concentration at which a chemical is effective varies, as do its effects on the avoidance behavior shown in Figure 11.5. (Based on Rogers, S.M., Newland, P.L., *J. Neurosci.* **22**:8324–8333, 2002. With permission.)

It has also been shown that there is a convergence of different taste qualities onto a given interneuron; in other words, an interneuron can respond to different chemicals such as salt and sucrose. Local interneurons, therefore, differ only in their overall sensitivity to the various chemicals, and there is no evidence that different interneurons are differentially sensitive to only one chemical. Locust chemosensory neurons do show some differentiation in response to different chemicals within a basiconic sensillum (Blaney, 1975; White and Chapman, 1990; Chapman et al., 1991), in much the same way as typical vertebrate gustatory neurons do. Thus, chemosensory neurons respond to a range of chemical stimuli.

Although it is possible that there are other chemosensory processing pathways in the locust where more segregated chemosensory information is used or that the responses of populations of interneurons could distinguish one chemical from another (Varkevisser et al., 2001), within the metathoracic ganglion many different chemosensory neurons appear to synapse onto the same spiking local interneurons (see Section 11.4.2). Recent results showing that the spiking local interneurons all have similar coding properties (Rogers and Newland, 2002) are also supported by previous studies showing that the chemo- and mechanosensory afferents from basiconic sensilla on the same regions of the leg send projections to the same regions of neuropil in the metathoracic ganglion, with no visible spatial segregation of different modalities or sensitivities (Newland et al., 2000). Such convergence of modalities is also found in the taste pathways of mammals, where specific taste-sensitive neurons also respond to mechanosensory and temperature-dependent inputs (Roper, 1989; Smith and Frank, 1993; Barry, 1999).

Recordings from the output stage of the local circuits, for example, from the flexor tibiae motor neurons that flex the tibia relative to the femur, reveal that the pattern of neural sensitivity is maintained throughout the neuronal network. The size of the synaptic inputs onto leg motor neurons also depends on chemical identity and concentration (Figure 11.10), just as in the behavioral model and the physiological responses of the spiking local interneurons (see Section 11.2.4). The antagonistic extensor tibiae motor neurons receive an increasing inhibitory drive that mirrors the levels of excitation received by the flexor tibiae motor neurons (Rogers and Newland, 2002).

What does this all mean in terms of central coding of chemical cues? How can a locust tell the difference between a solution of salt or sugar in contact with the leg? Spiking local interneurons are broadly tuned, as are the gustatory neurons in the nucleus of the solitary tract (Smith and Travers, 1979; Smith and St. John, 1999), and the duration of their responses depends on both chemical identity and concentration. It is, therefore, unlikely that they are able to code for chemical identity *per se*. For example, a given response duration or specific number of spikes in an interneuron could be evoked by a weak solution of NHT or a much stronger solution of sucrose (see, for example, Figure 11.11). The duration of response to the test solutions, however, does provide a measure of one chemical quality, that of aversiveness, as the relative size of responses in spiking local interneurons and motor neurons parallels the likelihood of the locust's removing its leg from a droplet of a particular chemical at a given concentration (Rogers and Newland, 2000) (Figure 11.11). Behavioral studies have shown that different chemicals become aversive at different concentrations and that aversiveness is a function of both chemical identity and chemical concentration, both of which are key determinants of the size of response in the spiking local interneurons and motor neurons. Thus, there is a very close correspondence between behavior and neural responses. This organization contrasts with that of the olfactory system in insects, where there is a high degree of anatomical chemotopy with different odorants, activating a wide variety of combinations of interneurons (Hildebrand and Shepherd, 1997; Laurent, 1997; Hansson and Anton, 2000).

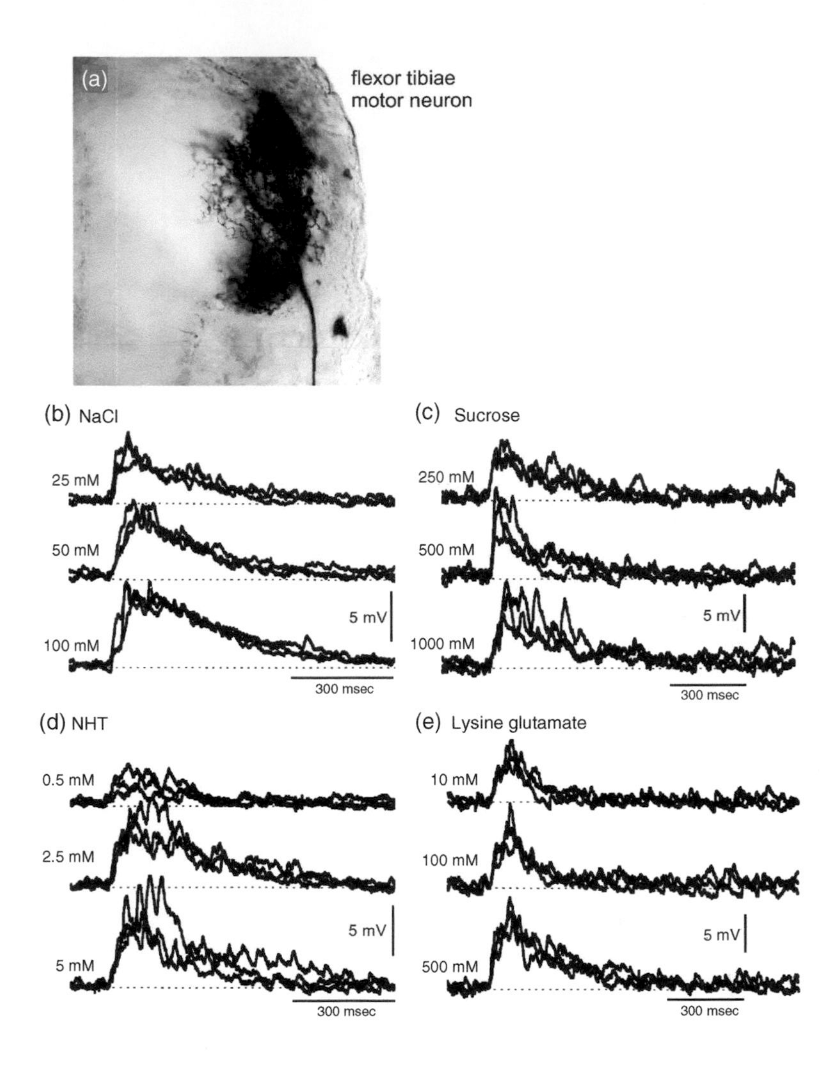

FIGURE 11.10 Responses of a flexor tibiae motor neuron to chemosensory stimulation. **a:** An example of a flexor motor neuron. **b:** Responses to NaCl; **c:** responses to sucrose; **d:** response to NHT; **e:** responses to lysine glutamate. The responses of the motor neurons show a dose-dependent relationship with chemical concentration: the greater the concentration, the greater is the depolarization of a motor neuron. All flexor tibiae motor neurons show similar responses. (Based on Rogers, S.M., Newland, P.L., *J. Neurosci.* **22**:8324–8333, 2002. With permission.)

11.5 CONCLUSIONS AND OUTLOOK

Until recently, most of what we knew about the coding of taste in insects has come from many studies addressing peripheral encoding by chemosensory neurons inner-

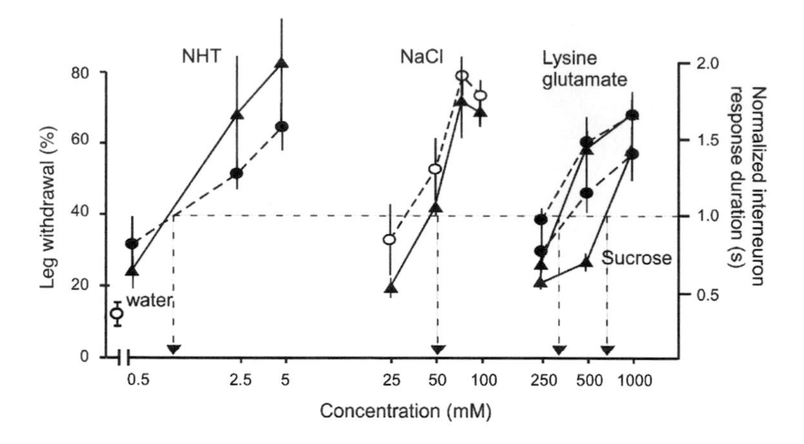

FIGURE 11.11 Correlation between interneuron responses and behavioral avoidance. The mean response durations of spiking local interneurons correspond closely with the behavioral effectiveness of the different chemicals. The solid lines represent the normalized response durations of the interneuron, whereas the dashed lines represent the frequency of leg withdrawal movements. (Based on Rogers, S.M., Newland, P.L., *J. Neurosci.* **22**:8324–8333, 2002. With permission.)

vating basiconic sensilla. Using a multidisciplinary approach, recent studies have, however, revealed the pathways responsible for taste processing and have asked if the responses of central interneurons and motor neurons can be related to the behavior of an insect. Central to these studies is the question of what chemosensory information is used for. The obvious answer is in the selection and rejection of food, which can involve fields of contact chemoreceptors on different parts of the body. Frequently, the initial contact an animal makes with a potential food or noxious chemical is through the contact chemoreceptors on the leg. This means that the first assessment can be made by these chemoreceptors (Dethier, 1976). For the locust and many other insects, what is important appears to be whether the chemical in contact with the legs is phagostimulatory or aversive. If it is aversive, then the animal takes its leg away; if it is not or is phagostimulatory, then an animal may go on and sample the food with its mouthparts, the palps (Bernays and Simpson, 1982). If again the chemical stimulant is considered to be aversive, it can be rejected, or if a phagostimulant, it can be bitten and chewed. At this third level, decisions can again be made on whether to ingest the food. The chemosensory receptors involved in this later stage of sampling are located on the inner surface of the labrum and cibarial cavity (Chapman, 1982). This chain of rejection at different stages suggests a hierarchical organization of behaviors, but this is not always the case: studies have shown that locusts will not lay eggs in sand containing more than 100 mM NaCl, yet they will lower their heads and eat the treated sand (Yates and Newland, 2003).

Clearly, rejection forms an important part of an animal's response to chemosensory stimulation, as it does in humans. It is, therefore, not so surprising to find that what makes a chemical aversive is a combination of its identity and its concentration. This has clearly been demonstrated in recent studies of locust behavior to gustatory cues (Rogers and Newland, 2000). What is perhaps most surprising from recent studies is the lack of segregation of the mechanosensory and chemosensory neurons in the central nervous system that innervate the contact chemoreceptors. This lack of segregation implies that within the central nervous system what may be important is, again, not the identity of a chemical *per se*, but instead its influence on particular local circuits. In the case of the contact chemoreceptors on the leg, what is crucial is their action on local circuits that control limb movement. For those on the palps, it is their action on local circuits controlling the palps and mouthparts. Once the local circuits responsible for gustatory processing have been identified, then the task of analysis is made easier because we know the locations and types of interneuron present, their arborization patterns, and the locations of their somata. Spiking local interneurons appear to play a pivotal role in gustatory processing, but, again, at the level of these neurons, the identity of the chemical does not appear to be the only important signal. Instead, these interneurons have broad responses and convergent inputs from a range of chemosensory qualities. This means that their responses are dependent on both identity and concentration, which we have established through behavioral studies as a single quality of aversiveness.

Why insects should be highly sensitive to a toxic secondary plant compound such as NHT is easily understandable, but it is less clear why they find high concentrations of nutrients aversive. Recent studies of the role of taste in dietary regulation in locusts and other animals (Simpson and Raubenheimer, 1996, 2000) provide a clue as they show that animals do not have an open-ended appetitive desire for nutrient chemicals, but instead nutrients must be consumed in the right quantities and proportions to achieve a balanced diet. In such circumstances, it is not surprising that solutions of highly concentrated nutrients may be actively rejected at the gustatory sampling stage of feeding, depending on ongoing nutritional requirements.

Recently, a model with similar features has been developed for taste processing in vertebrates, in which interactions between the basic taste qualities provide overarching measures of two opposing taste qualities: nutritional suitability or toxicity (Scott and Giza, 2000; Scott and Verhagen, 2000). Recent work on taste processing and coding in the locust may provide a corollary of the food selection process, but in the insect, the criterion is rejection rather than acceptance, with the consequence that high concentrations of single nutrients presented to the leg evoke large neural responses that cue withdrawal behavior.

ACKNOWLEDGMENTS

This work was supported by grants from the Biotechnology and Biological Sciences Research Council. I would like to thank the many people who have contributed to the work reported here, in particular Drs. Ibrahim Gaaboub, Tom Matheson, Stephen Rogers, and Professor Malcolm Burrows. I am grateful to Paul Yates for allowing me to use his unpublished data on oviposition behavior of locusts. I am also grateful to Hans Schuppe and Paul Yates for their valuable comments on an earlier draft of this manuscript.

REFERENCES

Abisgold JD, Simpson SJ (1988) The effect of dietary protein levels and haemolymph composition on the sensitivity of the maxillary palp chemoreceptors of locusts. *J. Exp. Biol.* **135**:215–229.

Amakawa T (2001) Effects of age and blood sugar levels on the proboscis extension of the blow fly *Phormia regina*. *J. Insect Physiol.* **47**:195–203.

Anton S, Homberg U (1999) Antennal lobe structure. In: *Insect Olfaction.* Hansson BS, Ed., Springer-Verlag, Berlin, pp. 95–124.

Barry MA (1999) Recovery of functional response in the nucleus of the solitary tract after peripheral gustatory nerve crush and regeneration. *J. Neurophysiol.* **82**:237–247.

Belanger JH, Orchard I (1992) The role of sensory input in maintaining output from the locust oviposition digging central pattern generator. *J. Comp. Physiol. A* **171**:495–503.

Bernays EA, Chapman RF (2001) Electrophysiological responses of taste cells to nutrient mixtures in the polyphagous caterpillar of *Grammia geneura*. *J. Comp. Physiol. A* **187**:205–213.

Bernays EA, Simpson SJ (1982) The control of food intake. *Adv. Insect Physiol.* **19**:59–102.

Bernays EA, Chapman RF, Singer MS (2000) Sensitivity to chemically diverse phagostimulants in a single gustatory neuron of a polyphagous caterpillar. *J. Comp. Physiol. A* **186**:13–19.

Blaney WM (1975) Behavioural and electrophysiological studies of taste discrimination by the maxillary palps of larvae of *Locusta migratoria* (L.). *J. Exp. Biol.* **62**:555–569.

Bräunig P, Hustert R, Pflüger H-J (1981) Distribution and specific central projections of mechanoreceptors in the thorax and proximal leg joints of locusts. I. Morphology, location and innervation of internal proprioceptors of pro- and metathorax and their central projections. *Cell Tiss. Res.* **216**:57–77.

Burrows M (1980) The control of sets of motor neurones by local interneurones in the locust. *J. Physiol. Lond.* **298**:213–233.

Burrows M (1987) Parallel processing of proprioceptive signals by spiking local interneurons and motor neurons in the locust. *J. Neurosci.* **7**:1064–1080.

Burrows M (1992) Reliability and effectiveness of transmission from exteroceptive sensory neurones to spiking local interneurones in the locust. *J. Neurosci.* **12**:1477–1489.

Burrows M (1996) *The Neurobiology of an Insect Brain.* Oxford University Press, Oxford, U.K.

Burrows M, Newland PL (1994) Correlation between the receptive fields of locust interneurons their dendritic morphology, and the central projections of mechanosensory neurons. *J. Comp. Neurol.* **329**:412–426.

Burrows M Newland PL (1994) Convergence of mechanosensory afferents from different classes of exteroceptors onto spiking local interneurons in the locust. *J. Neurosci.* **14**:3341–3350.

Burrows M, Siegler MVS (1982) Spiking local interneurones mediate local reflexes. *Science* **217**:650–652.

Burrows M, Siegler MVS (1984) The diversity and receptive fields of spiking local interneurones in the locust metathoracic ganglion. *J. Comp. Physiol. A* **224**:483–508.

Carlsson MA, Hansson BS (2003) Dose-response characteristics of glomerular activity in the moth antennal lobe. *Chem. Senses* **28**:269–278.

Chapman RF (1982) Chemoreception: the significance of sensillum numbers. *Adv. Insect Physiol.* **16**:247–356.

Chapman RF (2003) Contact chemoreception in feeding by phytophagous insects. *Annu. Rev. Entomol.* **48**:455–484.

Chapman RF, Bernays EA (1989) Insect behaviour at the leaf surface and learning aspects of host plant selection. *Experientia* **45**:215–222.

Chapman RF, Sword G (1993) The importance of palpation in food selection by a polyphagous grasshopper (Orthoptera: Acrididae). *J. Insect Physiol.* **6**:79–91.

Chapman RF, Bernays EA, Wyatt T (1987) Chemical aspects of host-plant specificity in three Larrea-feeding grasshoppers. *J. Chem. Ecol.* **14**:557–575.

Chapman RF, Ascoli-Christensen A, White PR (1991) Sensory coding for feeding deterrence in the grasshopper *Schistocerca gregaria. J. Exp. Biol.* **158**:241–259.

Christensen TA, White JE (2000) Representation of olfactory information in the brain. In: *Neurobiology of Taste and Smell,* Vol. II, Finger TE, Silver WL, Restrepo D, Eds., John Wiley & Sons, New York.

Christensen TA, Pawlowski VM, Lei H, Hildebrand JG (2000) Multi-unit recordings reveal context-dependent modulation of synchrony in odor-specific neural ensembles. *Nature Neurosci.* **3**:927–931.

Christensen TA, Lei H, Hildebrand JG (2003) Coordination of central odor representations through transient non-oscillatory synchronization of glomerular output neurons. *Proc. Nat. Acad. Sci. (USA)* **100**:11076–11081.

Dethier VG (1976) *The Hungry Fly.* Harvard University Press, Cambridge, MA.

Facciponte G, Lange AB (1996) Control of the motor pattern generator in the VIIth abdominal ganglion of locusts: descending neural inhibition and coordination with the oviposition hole digging central pattern generator. *J. Insect Physiol.* **42**:791–798.

Galizia CG, Nagler K, Hölldobler B, Menzel R (1998) Odour coding is bilaterally symmetrical in the antennal lobes of honeybees (*Apis mellifera*). *Europ. J. Neurosci.* **10**:2964–2974.

Galizia CG, Sachse S, Rappert A, Menzel R (1999) The glomerular code for odor representation is species specific in the honeybee *Apis mellifera. Nat. Neurosci.* **2**:473–478.

Getting PA (1971) The sensory control of motor output in fly proboscis extension. *Z. Vergl. Physiol.* **74**:103–120.

Hansson BS, Anton S (2000) Function and morphology of the antennal lobes: new developments. *Annu. Rev. Entomol.* **45**:203–231.

Hildebrand JG, Shepherd GM (1997) Mechanisms of olfactory discrimination: converging evidence for common principles across phyla. *Annu. Rev. Neurosci.* **20**:595–631.

Hodgson ES, Lettvin JY, Roeder KD (1955) Physiology of a primary chemoreceptor unit. *Science* **122**:417–418.

Johnson SE, Murphey RK (1985) The afferent projection of mesothoracic bristle hairs in the cricket, *Acheta domesticus. J. Comp. Physiol. A* **156**:369–379.

Kalogianni E (1996) Morphology and physiology of abdominal projection interneurones in the locust with mechanosensory inputs from ovipositor hair receptors. *J. Comp. Neurol.* **366**:656–673.

Kalogianni E, Burrows M (1996) Parallel processing of mechanosensory inputs from the locust ovipositor by intersegmental and local interneurones. *J. Comp. Physiol. A* **178**:735–748.

Lange AB, Orchard I, Loughton BG (1984) Neural inhibition of egg-laying in the locust *Locusta migratoria. J. Insect. Physiol.* **30**:271–278.

Laurent G (1996) Dynamical representation of odors by oscillating and evolving neural assemblies. *Trends Neurosci.* **19**:489–496.

Laurent G (1997) Olfactory processing: maps, time and codes. *Curr. Opin. Neurobiol.* **7**:547–553.

Laurent G, Burrows M (1989) Intersegmental interneurons can control the gain of reflexes in adjacent segments of the locust by their action on nonspiking local interneurons. *J. Neurosci.* **9**:3030–3039.

Laurent G, Stopfer M, Friedrich RW, Rabinovich MI, Volkovskii A, Abarbanel HDI (2001) Odor encoding as an active, dynamical process: experiments, computation, and theory. *Annu. Rev. Neurosci.* **24**:263–297.

Ma W-C, Schoonhoven LM (1973) Tarsal contact chemosensory hairs of the large white butterfly *Pieris brassicae* and their possible role in oviposition behaviour. *Entomol. Exp. Appl.* **16**:343–357.

Menzel R (2001) Searching for the memory trace in a mini-brain, the honeybee. *Learn. Mem.* **8**:53–62.

Minnich DE (1926) The organs of taste on the proboscis of the blowfly, *Phormia regina*, Meigen. *Anat. Rec.* **34**:126.

Mitchell BK, Gregory P (1979) Physiology of the maxillary sugar sensitive cell in the red turnip beetle, *Entomoscelis americana. J. Comp. Physiol. A* **132**:167–178.

Mitchell BK, Itagaki H (1992) Interneurons of the subesophageal ganglion of *Sarcophaga bullata* responding to gustatory and mechanosensory stimuli. *J. Comp. Physiol. A* **171**:213–230.

Newland PL (1990) The morphology of a population of ascending interneurones in the metathoracic ganglion of the locust. *J. Comp. Neurol.* **299**:242–260.

Newland PL (1991a) Physiological properties of afferents from tactile hairs on the hindlegs of the locust. *J. Exp. Biol.* **155**:487–503.

Newland PL (1991b) Morphology and somatotopic organisation of the central projections of afferents from tactile hairs on the hind leg of the locust. *J. Comp. Neurol.* **311**:1–16.

Newland PL (1998) Avoidance reflexes mediated by contact chemoreceptors on the legs of locusts. *J. Comp. Physiol. A* **183**:313–324.

Newland PL (1999) Processing of gustatory information by spiking local interneurones in the locust. *J. Neurophysiol.* **82**:3149–3159.

Newland PL, Burrows M (1994) Processing of mechanosensory information from gustatory receptors on a hind leg of the locust. *J. Comp. Physiol. A* **174**:399–410.

Newland PL, Burrows M (1997) Processing of tactile information in neuronal networks controlling leg movements of the locust. *J. Insect Physiol.* **43**:107–128.

Newland PL, Emptage NJ (1996) The central connection and actions during walking of tibial campaniform sensilla in the locust. *J. Comp. Physiol. A* **178**:749–763.

Newland PL, Rogers SM, Gaaboub I, Matheson T (2000) Parallel somatotopic maps of gustatory and mechanosensory neurons in the central nervous system of an insect. *J. Comp. Neurol.* **425**:82–96.

Oldfield BP (1982) Tonotopic organisation of auditory receptors in Tettigoniidae (Orthoptera: Ensifera). *J. Comp. Physiol. A* **147**:461–469.

Pfaffmann C (1959) The afferent code for sensory quality. *Am. Psychol.* **14**:226–232.

Pfaffmann C (1974) Specificity of the sweet receptors of the squirrel monkey. *Chem. Sens. Flav.* **1**:62–67.

Pfaffmann C, Frank M, Bartoshuk LM, Snell TC (1976) Coding gustatory information in the squirrel monkey chorda tympani. In: *Progress in Psychobiology and Physiological Psychology* Vol. 6, Sprague JM, Epstein AN, Eds., Academic Press, New York, pp. 1–27.

Pflüger HJ (1980) The function of hair sensilla on the locust's leg: the role of tibial hairs. *J. Exp. Biol.* **87**:163–175.

Pflüger HJ, Bräunig P, Hustert R (1981) Distribution and specific central projections of mechanoreceptors in the thorax and proximal leg joints of locusts. II. The external mechanoreceptors: hair plates and tactile hairs. *Cell Tiss. Res.* **216**:79–96.

Pflüger HJ, Bräunig P, Hustert R (1988) The organisation of mechanosensory neuropiles in locust thoracic ganglia. *Phil. Trans. Roy. Soc. Lond.* **321**:1–26.

Rogers SM (1998) Chemosensory development and integration in *Locusta migratoria*. Ph.D.. thesis, University of Oxford, U.K.

Rogers SM, Newland PL (2000) Local movements evoked by chemical stimulation of the hind leg in the locust *Schistocerca gregaria. J. Exp. Biol.* **203**:423–433.

Rogers SM, Newland PL (2002) Gustatory processing in thoracic local circuits of locusts. *J. Neurosci.* **22**:8324–8333.

Rogers SM, Simpson SJ (1999) Chemo-discriminatory neurones in the sub-oesophageal ganglion of *Locusta migratoria. Entomol. Exp. Appl.* **91**:19–28.

Römer H (1983) Tonotopic organization of the auditory neuropile in the bushcricket *Tettigonia viridissima. Nature* **306**:60–62.

Römer H, Marquart V, Hardt M (1988) Organization of a sensory neuropile in the auditory pathway of two groups of Orthoptera. *J. Comp. Neurol.* **275**:201–215.

Roper SD (1989) The cell biology of vertebrate taste receptors. *Annu. Rev. Neurosci.* **12**:329–353.

Schoonhoven LM, van Loon, JJA (2002) An inventory of taste in caterpillars: each species its own key. *Acta. Zool. Acad. Sci. H.* **48**:215–263.

Scott TR, Giza BK (2000) Issues of neural coding: where they stand today. *Physiol. Behav.* **69**:65–76.

Scott TR, Verhagen JV (2000) Taste as a factor in the management of nutrition. *Nutrition* **16**:874–885.

Siegler MVS, Burrows M (1983) Spiking local interneurons as primary integrators of mechanosensory information in the locust. *J. Neurophysiol.* **50**:1281–1295.

Siegler MVS, Burrows M (1986) Receptive fields of motor neurons underlying tactile reflexes in the locust. *J. Neurosci.* **6**:507–512.

Simpson SJ (1992) Mechanoresponsive neurones in the suboesophageal ganglion of the locust. *Physiol. Entomol.* **17**:351–369.

Simpson SJ, Raubenheimer D (1993) A multi-level analysis of feeding behaviour: the geometry of nutritional decisions. *Phil. Trans. Roy. Soc. Lond. B* **3542**:381–402.

Simpson SJ, Raubenheimer D (1996) Feeding behaviour, sensory physiology and nutrient feedback: a unifying model. *Entomol. Exp. Appl.* **80**:55–64.

Simpson SJ, Raubenheimer D (2000) The hungry locust. *Adv. Study Behav.* **29**:1–44.

Simpson SJ, Raubenheimer D, Behmer ST, Whitworth A, Wright GA (2002) A comparison of nutritional regulation in solitarious- and gregarious-phase nymphs of the desert locust *Schistocerca gregaria. J. Exp. Biol.* **205**:121–129.

Slifer EH (1954) The reaction of a grasshopper to an odorous material held near one of its feet (Orthoptera: Acrididae). *Proc. Roy. Ent. Soc. Lond. A* **29**:177–179.

Smith DV, Frank ME (1993) Sensory coding by peripheral fibers. In: *Mechanisms of Taste Transduction.* Simon SA, Roper SD, Eds., CRC Press, Boca Raton, FL, pp. 295–338.

Smith DV, St. John SJ (1999) Neural coding of gustatory information. *Curr. Opin. Neurobiol.* **9**:427–435.

Smith DV, Shepherd, GM (1999) Chemical senses: taste and olfaction. In: *Fundamental Neuroscience*, Zigmond, MJ, Bloom FE, Landis SC, Roberts JL, Squire LR, Eds., Academic Press, San Diego, CA, pp. 719–759.

Smith DV, Travers JB (1979) A metric for the breadth of tuning of gustatory neurons. *Chem. Sens. Flavor* **4**:215–229.

Smith DV, St. John SJ, Boughter Jr. JD (2000) Neuronal cell types and taste quality coding. *Physiol. Behav.* **68**:77–85.

Spector AC (2000) Linking gustatory neurobiology to behaviour in vertebrates. *Neurosci. Biobehav. Rev.* **24**:391–416.

Städler E (2002) Plant chemical cues are important for egg deposition by herbivorous insects. In: *Chemoecology of Insect Eggs and Egg Deposition,* Hilker M, Meiners T, Eds., Blackwell Science, Berlin, pp. 171–204.

Städler E, Renwick JAA, Radke CD, Sachdev-Gupta K (1995) Tarsal contact chemoreceptor response to glucosinolates and cardenolides mediating oviposition in *Pieris rapae. Physiol. Entomol.* **20**:175–187.

Szentesi A, Bernays EA (1984) A study of behavioural habituation to a feeding deterrent in nymphs of *Schistocerca gregaria. Physiol. Entomol.* **9**:329–340.

Thompson KJ (1986a) Oviposition digging in the grasshopper. I. Functional anatomy and the motor programme. *J. Exp. Biol.* **122**:387–411.

Thompson KJ (1986b) Oviposition digging in the grasshopper. II. Descending neural control. *J. Exp. Biol.* **122**:413–425.

Tousson E, Hustert R (2000) Central projections from contact chemoreceptors of the locust ovipositor and adjacent cuticle. *Cell Tiss. Res.* **302**:285–294.

Uvarov B (1977) *Grasshoppers and Locusts: A Handbook of General Acridology. Vol. 2. Behaviour, Ecology, Biogeography, Population Dynamics.* Centre for Overseas Pest Research, Cambridge University Press, Great Britain.

Varkevisser B, Peterson D, Ogura T, Kinnamon SC (2001) Neural networks distinguish between taste qualities based on receptor cell population responses. *Chem. Senses* **26**:499–505.

White PR, Chapman RF (1990) Tarsal chemoreception in the polyphagous grasshopper *Schistocerca americana*: behavioural assays, sensilla distributions and electrophysiology. *Physiol. Entomol.* **15**:105–121.

Woodrow DF (1965) The responses of the African migratory locust *Locusta migratoria migratorioides* R and F to the chemical composition of the soil at oviposition. *Anim. Behav.* **13**:348–356.

Yates PI, Newland PL (2003) Contact chemoreception and its role in the oviposition behaviour of the desert locust. *Comp. Biochem. Physiol.* **134**:S67.

Zacharuk RY (1985) Antennae and sensilla. In: *Comprehensive Insect Physiology, Biochemistry, and Pharmacology*, Vol. 6, Kerkut GA, Gilbert LI, Eds., Pergamon, Oxford, U.K., pp. 1–69.

Zill SS, Moran DT (1981) The exoskeleton and insect proprioception. I. Responses of tibial campaniform sensilla to external and muscle-generated forces in the American cockroach, *Periplaneta americana. J. Exp. Biol.* **91**:1–24.

12 Insect Olfactory Neuroethology — An Electrophysiological Perspective

Rickard Ignell and Bill Hansson

CONTENTS

0-8493-2024-0/05/$0.00+$1.50

12.1 INTRODUCTION

To a large extent, insects rely on olfactory stimuli for their reproduction and survival. The male often uses female-emitted sexual pheromones to locate a mate and, once in close contact, often emits his own pheromone bouquet to coerce the female into mating. Similarly, females may use host-emitted cues to determine where to lay eggs or where to find a blood meal. Odor cues thus play an extremely important role in shaping insect behavior, and studies of the olfactory system, its structure, and its function become paramount to our understanding of insect neuroethology. This fact was realized early, in the mid-1950s, when Professor Dietrich Schneider performed the first electrophysiological recordings from the peripheral olfactory system of insects. He and his colleagues developed both the electroantennogram (EAG) and the single-sensillum recording (SSR) technique (see Section 12.2). The work of Schneider and his disciples allowed studies of insect odor detection at an unprecedented scale, and a considerable volume of scientific literature has been published on the subject (reviewed by, e.g., Hansson, 1995, 1999).

With an increased understanding of the peripheral olfactory system, a natural question was to examine what happens at the next level. The primary olfactory center of the insect brain is the *antennal lobe* (AL). Here, olfactory receptor neurons (ORNs), arising from sensory hairs called *sensilla* (singular: sensillum) first synapse onto a variety of antennal lobe interneurons (see also Chapter 13 and Chapter 14). Electrophysiological recordings from AL neurons were first performed extracellularly by Professor Jürgen Boeckh in the mid-1970s, but the major breakthrough occurred when intracellular methods that allowed both physiological recording and morphological staining came into use (Matsumoto and Hildebrand, 1981; review: Christensen and Hildebrand, 2002a). Again, these methodological developments have given rise to a large number of studies of AL function in a number of insects (for recent reviews, see Hansson and Christensen, 1999; Christensen and White, 2000).

The work of the pioneers mentioned previously has formed the foundation for the remarkable progress that has taken place in the area of insect olfaction during the last 20 years. In this chapter, we will discuss how different electrophysiological methods have contributed to our understanding of how odors govern the behavior of insects.

12.2 PROBING THE PERIPHERAL OLFACTORY SYSTEM

As mentioned above, electrophysiological techniques have been used extensively for the analysis of peripheral olfactory organs for almost half a century. Despite this, the essentials of a typical electrophysiological station have not changed dramatically and still consist of a microscope (with long working-distance objectives), a vibration-isolation table, and a Faraday cage (screen) to reduce unwanted electrical noise. A recent technological advancement has been the introduction of higher-power compound microscope[12,19,21], which allow the tiniest insect sensilla to be resolved clearly (de Bruyne et al., 1999; Shields and Hildebrand, 2001). Electrodes[34], micromanipulators[18,27,34], and a specialized amplifier[6,28,34] are also key ingredients to successful electrophysiological data recordings. The following section is a short description of the most common techniques currently used in the field; for a more detailed description see, for example, Kaissling (1995) and Millar (1992).

12.2.1 The Electroantennogram

Ever since Schneider made the first electroantennogram (EAG) recording from the *Bombyx mori* antenna in 1957, this method has been used as a convenient screening procedure for odors that could be of biological significance to the insect. Although most work has been on filamentous antennae, EAGs have also been performed on other types of antennae (Figure 12.1a). Data from the latter type are, however, more difficult to interpret. In filamentous antennae, EAGs are presumably formed by the summation of receptor potentials of many ORNs, where the amplitude of the EAG is proportional to the number of ORNs between the electrodes (Nagai, 1981; White, 1991). Due to the odor-induced depolarization of the dendrites, each ORN is thought to become a small dipole upon stimulation (Kaissling, 1971). As a result of this serial arrangement, the receptor potentials of many ORNs between the tip and the base of the antenna are summed, and the tip will become negative with respect to the base. In nonfilamentous antennae like the globular antennae of flies (see Chapter 10), the EAG is probably formed in a different way (Crnjar et al., 1989; Kelling, 2001). Unlike filamentous antennae, globular antennae of flies have a low internal resistance due to a large internal cavity of the antenna filled with conducting haemolymph. This prevents insertion of both electrodes into the hemolymph as this leads to a short-circuit the electrodes (Den Otter and Saini, 1985). The cuticle, however, has a high resistance, which allows for a clear potential difference when a current flows. As a consequence, the EAG is probably formed by the summated potentials over the cuticle of ORNs near the electrode (Crnjar et al., 1989; Kelling, 2001). Thus, when measuring EAGs from a nonfilamentous antenna, possible differences in sensitivity of regionalized fields along the antenna must be considered (Ayer and Carlson, 1992; de Bruyne et al., 2001).

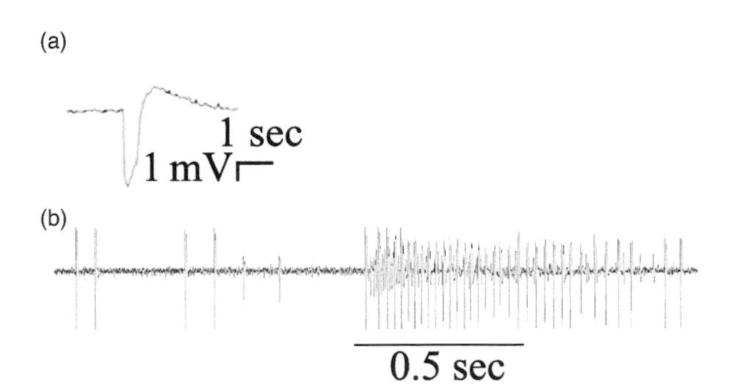

FIGURE 12.1 Electroantennogram (EAG) and single-sensillum recording (SSR) in the fruit fly *Drosophila melanogaster*. **a**: An EAG recording showing a depolarization in response to a common green leaf odor, 1-hexanol. **b**: Recording from a single sensillum containing two olfactory receptor neurons (differentiated by different amplitudes). Both neurons showed a response to 2,3-dibutanone (horizontal marker).

EAGs are usually recorded with glass capillary electrodes on intact animals, detached heads, or detached antennae with the indifferent electrode placed on, at, or near the base of the antenna. Several custom-made holders have also been used to obtain successful EAGs (Leal et al., 1991; Sauer et al., 1992; Hillbur, 2001). The reference and recording electrodes are connected to a high-impedance differential amplifier with a low-pass filter (~0.01 to 10 Hz) and the slow change in DC potential is visualized on an oscilloscope or a computer screen (Figure 12.1a).

12.2.2 Extracellular Single-Sensillum Recording Techniques

The limitations set by the EAG technique (i.e., in sensitivity and specificity) are partly overcome by single-sensillum recording (SSR) techniques, which monitor the electrical events elicited in ORNs when stimulated by different odors. The SSR techniques allow for analysis of the cellular basis of odor coding in some detail. The electrical events arise due to currents in the medium surrounding the ORNs, and these are most likely generated by receptor potentials across the cell membranes of the ORNs (Stengl et al., 1992; Kaissling, 1995). These currents, in turn, generate voltages in the medium, which may be detected as extracellular potentials, referred to as *spikes* or *sensillar potentials* (Figure 12.1b).

12.2.2.1 Types of Electrodes

Tungsten microelectrodes have been widely used to analyze the response properties of insect olfactory sensilla ever since the first impulses were recorded by Boeckh (1962) and Schneider et al. (1964). These electrodes are usually made from 0.01 to 0.1 mm tungsten wire, which is electrolytically sharpened by repeated dipping in a ~10% $NaNO_2$ or KNO_2 solution, while passing a 0.1 to 1 mA current through the solution until the desired shape of the electrode has been obtained. Difficulties in obtaining reproducible electrodes are considered by some to be a major disadvantage of this type of electrode. The ability to obtain very fine microelectrodes, however, has probably led to the widespread use of this technique because it allows recordings from very small sensilla. Tungsten microelectrodes are used to penetrate either the antennal cuticle at the base of the sensillum or the shaft of the sensillum (see also Chapter 10). As the variance of the membrane potential is very small along the sensory dendrite (Vermeulen and Rospars, 2001), the measured extracellular potential is largely independent of the placement of the electrode.

Glass capillary electrodes may also be used to penetrate individual sensilla (e.g., Dobritsa et al., 2003), but the use of glass electrodes is usually restricted to larger sensilla. This type of electrode, a borosilicate glass capillary with an internal filament that is pulled to produce a tip of about 1 μm, is generally filled with sensillum lymph ringer (Kaissling and Thorson, 1980) or Beadle and Ephrussi solution (Ashburner, 1989) and slipped over an AgCl-coated silver wire.

Glass capillary electrodes for single-sensillum recordings were introduced by Kaissling (1974). By cutting the tip of a sensillum with a microscopic glass knife, he was able to gain access to the inside of the sensillum and the outer dendritic segments of the ORNs. A main advantage of using glass electrodes is that it allows the recording of both the receptor potential and the action potential of ORNs. Direct access to the sensillum lymph also allows for administration of neuronal tracers (Hansson et al., 1992) and other chemicals, which allow experimental changes of the ORN environment (e.g., Pophof et al., 2000; Pophof, 2002a,b).

In order to acquire an electrophysiological signal, both the recording and the reference electrodes should be connected to a high-impedance, high-gain, differential amplifier. We use the data acquisition system (IDAC) from Syntech[(28)] and Iso-DAM preamplifier from WPI[(34)]. Amplification of the input signal ensures that the signal is brought out of the noise level of the amplifier and also allows recording of the low-amplitude spikes. Most amplifiers are also equipped with a set of filters that allow the operator to remove both low- and high-frequency signals outside the spike bandwidth (~100 Hz to 3 kHz). When receptor potentials of ORNs are to be recorded, a filter that passes DC signals must be used. The signal is then typically recorded on tape or digitized on a computer and visualized on an oscilloscope or directly imported, via an analog-to-digital converter, to data acquisition software.

12.2.2.2 Spike Classification

Discrimination of the recorded spikes is required in order to assess the activity of multiple ORNs that are usually compartmentalized in the same sensillum. The shape and amplitude of the recorded spikes are known to be affected by the size of the ORNs and the physical environment in which they are located (De Kramer et al., 1984; Hansson et al., 1994; Vermeulen and Rospars, 2001); these characteristics have been used repeatedly for spike classification and enumerating spike frequency (see Chapter 14). In cases in which the shape and amplitude of cocompartmentalized ORNs are indistinguishable, however, these methods are not always useful. The reliability of these methods has been questioned because they rely on visual classification. The increased use of statistical sorting algorithms (Lewicki, 1998) may result in a more unbiased classification and aid in discriminating spikes during periods of high activity. Analysis using principal components analysis of ORNs has shown the potential of this type of spike classification methodology (Mackenzie, Pearce, and Hansson, unpublished data). An estimate of the spiking frequency of each individual ORN can also be achieved through interspike-interval assessment (Meunier et al., 2003). A more detailed discussion of extracellular spike-sorting methods can be found in Chapter 14.

12.3 FUNCTIONAL CHARACTERISTICS OF THE PERIPHERAL OLFACTORY SYSTEM

The olfactory stimuli reaching an insect antenna are variable in qualitative, quantitative, and temporal aspects, and the insect's behavior depends strongly on all three of these variables. For example, a pheromone plume must contain the right constituents, often at a certain concentration and in a specific temporal pattern, to elicit upwind flight in a male moth (e.g., Baker et al., 1985; Baker, 1990). To understand how insects are able to discriminate different odors, how they distinguish the concentration of these, and how well temporal fluctuations are resolved, the function of the olfactory system — from periphery to the brain to premotor centers — must be studied. In this section, we discuss how the peripheral olfactory system deals with these various processes.

12.3.1 Selectivity of Olfactory Receptor Neurons

The level of ORN specificity, or *tuning*, is generally determined by the threshold at which an ORN responds to a spectrum of odorants. Recent studies have shown that this tuning is highly dependent on the membrane-bound receptor site at which the odorant molecule binds (Malnic et al., 1999; Dobritsa et al., 2003; Elmore et al., 2003), as well as the perireceptor environment housing the ORN dendrites (Prestwich and Du, 1997; Vogt 2003). These studies are germane to previous work performed on sex pheromone–specific ORNs (Liljefors et al., 1984, 1985, 1987; Bengtsson et al., 1990). These authors made structural modifications to one of the three sex pheromone components of *Agrotis segetum* and tested their activity on ORNs tuned to the original component. The activity shown by each analog was found to be directly proportional to the energy needed to fold the molecule, calculated using molecular mechanics, implying that certain important structural molecular characteristics fall into the same or a similar spatial location as they would in the lowest energy form of the original molecule. Optimal tuning, in this case, was found to be highly dependent on the molecular shape, chirality, and electron distribution. This is similar to that found in other insects (Priesner et al., 1975; Priesner, 1979; Bestmann, 1981; Bestmann and Vostrowsky, 1982; Wojtasek et al., 1998; Larsson et al., 1999).

The studies presented previously assumed that only a single receptor site is present on each ORN (for a discussion on this topic regarding mammals, see Mombaerts, 2004). Validation of this assumption in insects has traditionally been performed by differential adaptation of ORNs, which in most cases has rejected the hypothesis of multiple-receptor ORNs (Payne and Dickens, 1976). More recently, this hypothesis has been strengthened by receptor expression studies in *Drosophila melanogaster* (Clyne et al., 1999; Gao et al., 1999; Vosshall et al., 1999, 2000). Functional analysis of members of the recently identified family of seven-transmembrane domain proteins in *D. melanogaster* has, furthermore, allowed an integration of the molecular and the cellular definition of the basis of odor selectivity (Dobritsa et al., 2003; Elmore et al., 2003). Analysis of ORNs across species, such as the one performed by Stensmyr et al. (2003) on members of the *D. melanogaster* subgroup, may also contribute to understanding receptor–ligand interactions at the molecular level. The ability to express olfactory receptors ectopically in mutant ORNs of *D. melanogaster* (Dobritsa et al., 2003), as well as the ability to introduce point mutations in these receptors, offers several novel opportunities to study the molecular basis of selectivity at the receptor level (see also Chapter 9 and Chapter 10).

The specificity of single ORNs is also regulated by the perireceptor environment, specifically by odorant binding proteins (OBPs) or by pheromone binding proteins (PBPs) (see reviews in Blomquist and Vogt, 2003). These binding proteins are present in the sensillum lymph and have been shown to form a complex with the odorant ligand (Du and Prestwich, 1995; Breer, 1997; Vogt, 2003). This implies that PBPs and OBPs may play a role as filters, which bind and consequently translocate only physiologically relevant ligands. Until recently, OBPs and PBPs have been defined at least partly due to their ability to bind odorants, but recent publications indicate that members of the OBP family of genes presumably have no function in chemodetection (Vogt, 2003). Experiments performed by Wojtasek et al. (1998) also indicate limitations in the odor-discrimination capabilities of PBPs. Further work is thus needed to elucidate the role of OBPs and PBPs in olfactory processing.

Concerning selectivity at the cellular level, the previous hypothesis of broadly tuned ORNs formulated in early investigations (e.g., Behan and Schoonhoven, 1978; Hansson et al., 1989) has now repeatedly been attributed to lack of a well-defined key stimulus and the use of systematically high concentrations of odorants. ORNs previously identified as broadly tuned, such as ORNs responding to plant-derived volatiles, however, exhibit relatively high selectivity and sensitivity when stimulated with an appropriate signal (e.g., Priesner, 1979; Dickens, 1990; Anderson et al., 1995; Heinbockel and Kaissling, 1996; Hansson et al., 1999; Larsson et al., 2001; Shields and Hildebrand, 2001; Skiri et al., 2004). In order to provide the central nervous system with sufficient information for odor discrimination, more generally tuned ORNs are required because they display overlapping tuning curves, which provide a combinatorial coding scheme for encoding odor identity (Boeckh and Ernst, 1983; Todd and Baker, 1999). (In our opinion, "generally" tuned ORNs differ from "broadly" tuned ORNs in that the former display selectivity to particular molecular features [or odotopes], such as functional group, location of double bonds, and chirality.) Such coding schemes have recently been illustrated by de Bruyne et al. (2001) and Shields and Hildebrand (2001), showing that a single ORN type selectively responds to multiple odorants and that a single odorant elicits a response in different types of ORNs. Furthermore, different combinations of odorants elicit responses in different combinations of ORNs, but at different thresholds. Resolution of chemical identity and concentration of a complex odor thus seem to arise from an across-fiber pattern of ORN activity (Dethier, 1976; Christensen and White, 2000). In contrast, information carried by ORNs that are highly tuned to a single odorant, with sex-pheromone component–specific ORNs being a classic example (Hansson, 1995; Hansson and Christensen, 1999), seem to be represented as a labeled-line pattern at the peripheral level. The selectivity of this type of ORN thus provides sufficient intrinsic discrimination capability for correct identification of the signal.

12.3.2 Sensitivity of the Peripheral Olfactory System

Selectivity and sensitivity are inextricably linked in olfaction, implying that the identity of a particular odorant cannot be discriminated unless the sensitivity of a particular ORN to all possible odorants is defined. Differential tuning, in which various ORN types within an across-fiber pattern increase and decrease their firing rates in response to changes in the flux of a particular odorant, is thus a prerequisite for odor discrimination. This requires that cocompartmentalized ORNs display independent tuning curves because mixture interactions at this level would alter the content of information before reaching the central nervous system (reviewed by Todd and Baker, 1999).

Odor-induced behavioral (Kaissling and Priesner, 1970; Kaissling, 1971) and cardiac responses (Angioy et al., 2003) have shown that insects are able to respond to extremely low concentrations of odorants. Single-sensillum recordings generally indicate a response threshold of sex pheromone–specific ORNs in the range of 10^{-7} to 10^{-12} g (Boeckh and Selsam, 1984; Christensen and Hildebrand, 1987a; Hartlieb et al., 1997). Plant-volatile responding ORNs generally seem to have a much higher threshold: 10^{-6} g and higher. Highly sensitive plant volatile–specific ORNs responding to stimulus loads down to 10^{-12} g have, however, been found (Hansson et al., 1999; Jönsson and Anderson, 1999; Larsson et al., 2001). These thresholds are, however, arbitrary. By applying more rigorous analysis of ORN spike trains, however (e.g., raster plots, peristimulus histograms, and inter-spike interval superposition), lower thresholds may be found because these methods allow for an unbiased analysis. When making comparative studies of ORN thresholds, special attention should be paid to differences in stimuli volatility, airflow speed, stimulation time, etc., because these factors greatly affect the response of ORNs.

12.3.3 Encoding Stimulus Frequency in Olfactory Receptor Neurons

Accumulating data over the years has shown that temporal variation in the odor stimulus in many cases has a dramatic effect on the orientation behavior of insects (Kennedy et al., 1981; Baker et al., 1985; Baker, 1990; Mafra-Neto and Cardé, 1994; Vickers and Baker, 1994). An intermittent odor stimulus has been shown to be extremely important for evoking behaviors such as sustained flight or walking to an upwind odor source. The ability of insects to resolve a flickering stimulus at the antennal level was tested by Bau et al. (2002), who showed that EAGs of three moth species were able to resolve the temporal structure of stimulus pulses delivered (amazingly) up to 33 Hz. Although some moth species have been shown to orient in plumes pulsed at up to 25 Hz (Justus et al., 2002), signal fusion at even lower rates seems to occur in other species and has been shown to elicit behavioral inhibition (Kennedy et al., 1981; Baker et al., 1985). At the neuronal level, individual ORNs have also been shown to follow stimuli mimicking the temporal patterns in natural odor plumes (Kaissling, 1986; Baker et al., 1988, 1989; Marion-Poll and Tobin, 1992). In the moths *Antheraea pernyi, Agrotis segetum, Grapholita molesta,* and *Manduca sexta,* populations of ORNs have been found that follow pulses of up to 3 Hz, whereas some ORNs in *Helicoverpa zea* are able to follow 10 to 12 Hz pulses (Kaissling, 1986; Baker et al., 1988; Almaas et al., 1991; Marion-Poll and Tobin, 1992). At this resolution, ORNs would be able to encode a fluctuating signal, which naturally occurs at around 2 to 15 Hz (Murlis and Jones, 1981; Mafra-Neto and Cardé, 1998).

The ability of ORNs to follow pulsed stimuli is a function of their relative rates of adaptation and disadaptation (Kaissling, 1987; Borroni and O'Connell, 1992; Marion-Poll and Tobin, 1992; Kodadova, 1996). Rapidly disadapting neurons generally display a phasic temporal response pattern with a rapid decay that is better suited for tracking stimuli pulsed at high frequencies. Slowly disadapting ORNs, however, display a more tonic temporal profile that generally prevents their response to fast consecutive pulses of stimuli (Rumbo 1983; Rumbo and Kaissling, 1989). Although limited in their ability to follow pulsed stimuli, tonic and phasic–tonic responding ORNs may, however, serve as a "peripheral memory" that keeps the insect in an alert state (Almaas and Mustaparta, 1991; Den Otter and Van der Goes van Naters, 1993). These types of temporal patterns may also serve to drive long-lasting behavioral responses, such as casting or looping in clean air following odor loss. It is important to note, however, that adaptation and disadaptation characteristics of ORNs may be concentration- and time-dependent (Kaissling, 1987; Almaas et al., 1991; Heinbockel and Kaissling, 1996).

Clear differences in temporal response patterns within the antenna, and even among ORNs of one type, have been reported. For example, in *Antheraea polyphemus*, *A. pernyi,* and *Epiphyas postvittana*, separate ORN types have been found that are tuned to different pheromone components and that display different temporal response patterns (Zack, 1979; Kaissling, 1987; Meng et al., 1989; Rumbo and Kaissling, 1989). Variability of temporal response characteristics within ORNs responding to the same odorant has also been reported. In *H. virescens,* ORNs housed in different sensilla show phasic and phasic–tonic responses to the major pheromone component (Berg et al., 1995). In *H. zea* and in *B. mori*, ORNs housed in morphologically indistinguishable trichoid sensilla seem to span a continuum of temporal response patterns (Almaas et al., 1991; Heinbockel and Kaissling, 1996).

Raster plot analysis is a useful tool for defining the range and variability of such intricate temporal patterns. Raster plots of pheromone-sensitive ORNs in trichoid sensilla of *S. littoralis* (Mackenzie, Pearce, and Hansson, unpublished data) responding to repeated stimulations of the major pheromone component (*Z*9, *E*11–14:Ac), indicated a wide variation in temporal response patterns. These patterns ranged from highly tonic to highly phasic, with a variety of phasic–tonic response profiles, which suggests the existence of several physiological subtypes of specific ORNs. Based on these results, one may speculate that the differential ability of ORNs to encode and report temporal information to higher olfactory centers has developed to fine-tune the behavioral response of the insect.

12.4 PROBING THE CENTRAL NERVOUS SYSTEM

Olfactory receptor neurons detect specific odors with a varying degree of specificity and sensitivity. Information gathered by several hundred thousand peripheral neurons converges onto the primary olfactory centers of the insect brain: the antennal lobes (ALs). Each physiological type of ORN targets a specific AL glomerulus, the spherical neuropil structure typically found in the primary olfactory centers of all higher organisms. This is the first level in the brain where interactions between different types of odor input take place (see also Chapter 13 and Chapter 14). Thus, it is at this level where computations regarding mixture interactions, contrast enhancement, amplification, and other integrative events are performed.

In the AL, three main types of neurons have been investigated: local interneurons (LNs), projection neurons (PNs), and centrifugal neurons (CNs). Among these, the

LNs are generally considered to be highly integrative elements, allowing extensive lateral interactions between AL glomeruli (Christensen et al., 1993; Hansson and Christensen, 1999; Laurent, 2002; Wilson et al., 2004). Processing in the AL is also influenced by centrifugal input via CNs from higher brain areas. The final output signal of the AL is transferred through PNs to higher olfactory centers, such as the mushroom bodies and the lateral protocerebrum. In the protocerebrum, the signal is distributed over a number of different brain areas and becomes integrated with other sensory input. An output signal is finally supplied to premotor centers in the thoracic ganglia (Kanzaki et al., 1991a,b; Lei et al., 2001). This whole system, from AL to premotor centers, is open to be probed by different methods measuring neural activity.

12.4.1 Intracellular Recording Techniques

Substantial progress in understanding the function of the insect central nervous system has been gained since the introduction of intracellular recordings with sharp microelectrodes. This technique allows characterization of physiological and morphological properties of individual neurons in relatively unperturbed *in vivo* preparations. In order to obtain good-quality recordings, similar requirements of the basic electrophysiological setup described earlier should be met. The major difference is the use of a specialized intracellular amplifier, which should be equipped with current injection, capacity compensation, and an oscillation or "buzz" circuit (which facilitates penetration of the neuron cell membrane)[(3,6)]. An important recent technical development has been the introduction of new and various intracellular markers with favorable properties (e.g., low toxicity and rapid, complete distribution in the recorded neuron), ranging from Lucifer yellow, dextrans, biocytin (and its derivatives such as neurobiotin; Figure 12.2) to fluorescent hydrazide derivatives (Alexa Fluor dyes)[(14)]. A general method for preparation of insects for intracellular recordings, as well as specifics for obtaining intracellular recordings from the insect central nervous system, has been described in detail by Christensen and Hildebrand (1987b). Several recent reviews contain additional detailed information on intracellular recording techniques (Silinsky, 1992; Christensen and Hildebrand, 2002b).

12.4.2 Whole-Cell Patch Clamp Technique

Intracellular recordings with sharp microelectrodes are limited in that it is not always possible to impale smaller neurons, mainly due to difficulties in maintaining long-lasting contacts with tiny dendritic and axonal processes. As a result, large populations of small neurons in the central olfactory pathway (e.g., Kenyon cells in the paired mushroom bodies) have been studied only occasionally and often not as thoroughly as other neuron types (Laurent and Naraghi, 1994; Perez-Orive et al., 2002). By using a different approach, patch-clamp recording in the whole-cell configuration (which is in many ways equivalent to intracellular recordings with sharp microelectrodes), electrophysiological recordings can be made from cell bodies that are considerably larger than dendrites, thus allowing recordings from a variety of neurons, regardless of their size. One drawback is that recordings from

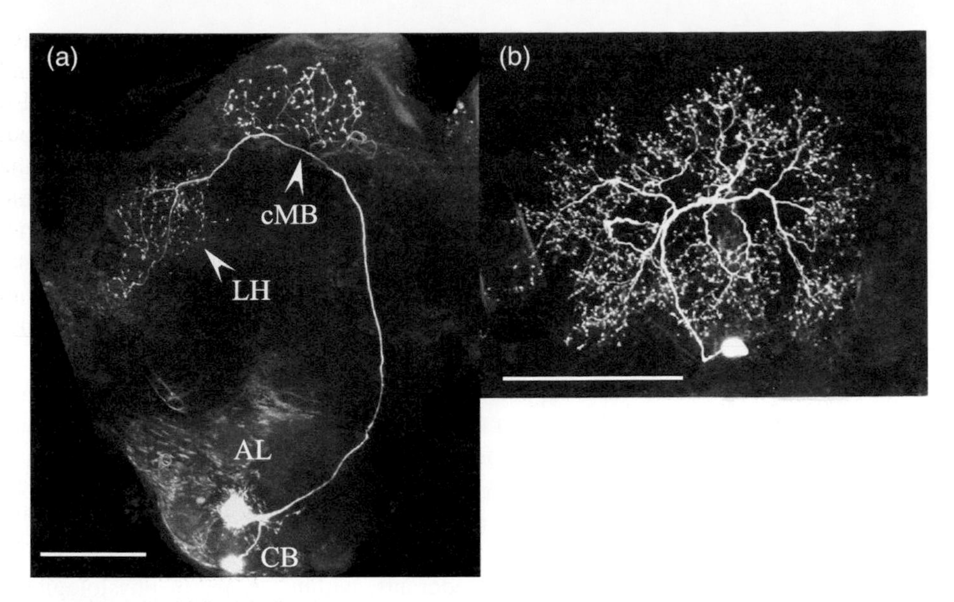

FIGURE 12.2 Intracellular staining of different antennal lobe neurons with neurobiotin in *Spodoptera littoralis*. **a**: A projection neuron displaying uniglomerular dendritic arborization in the antennal lobe (AL) and a projection to higher olfactory centers, the calyces of the mushroom body (cMB), and the lateral horn (LH) of the protocerebrum. **b**: Wide-field arborization of a local interneuron in the AL. CB: cell body. Scale bars: 100 μm.

the soma may not be fully representative of potentials at the dendrite; this may be particularly true for widely branching cells like olfactory LNs, but it has not been a problem in our studies of multiglomerular PNs in locusts. This technique has only recently been pursued for analysis of electrophysiological properties of insect neurons in an *in vivo* preparation (Kloppenburg et al., 1999a,b; Wilson et al., 2004; Schmidt et al., submitted). An in-depth description of the patch-clamp technique is beyond the scope of this chapter, so we refer instead to Sakmann and Neher (1995), which provides a practical guide to the patch-clamp setup and the experimental procedure.

12.5 PHYSIOLOGICAL CHARACTERISTICS OF ANTENNAL LOBE NEURONS

In the AL, all vital characteristics of the olfactory signal must be extracted from the information arriving via the ORNs. The main features extracted are information regarding quality, intensity, and fluctuations over time. All of these characters have been shown to be highly important as determinants for odor-induced behavior.

12.5.1 Encoding Odor Quality

A primary task of the AL is to receive olfactory information from the peripheral olfactory epithelium and to organize this information into a neural representation of odor quality, which may be read by higher olfactory centers. In most cases, this has been shown to involve progressive transformations of the olfactory information between layers of pre- and postsynaptic neurons in a hierarchical fashion. In insects, the first order of transformation seems to be a reorganization of the seemingly heterogeneous peripheral olfactory epithelium (e.g., Lee and Strausfeld, 1990; Shanbhag et al., 1999; Shields and Hildebrand, 2001) into a highly organized chemotopic map in the AL glomeruli (Galizia and Menzel, 2000; Gao et al., 2000; Vosshall et al., 2000; see Chapter 13 and Chapter 14). In *D. melanogaster*, this chemotopic map constitutes a more or less spatially invariant glomerular array, where individual glomeruli receive afferent input from ORNs expressing the same odorant receptor (Vosshall et al., 1999, 2000; Gao et al., 2000; Scott et al., 2001; Dobritsa et al., 2003).

These newer studies fully support previous physiological and morphological staining studies of functionally distinct ORNs in other insect species (e.g., Hansson et al., 1992; Christensen et al., 1995; Ochieng' et al., 1995; Todd et al., 1995). The functional significance of a chemotopic map in odor quality processing has more recently been studied by calcium imaging of ORN ensembles (Joerges et al., 1997; Galizia et al., 1999a,b, 2000; Carlsson et al., 2002; Wang et al., 2002; Ng et al., 2003). These studies all emphasize that odorants, within a range of physiological odor concentrations (see Section 12.5.3), elicit a distinct and relatively sparse pattern of glomerular activity that is stereotyped in different individuals of the same species. In addition, more detailed studies have showed that the chemical identity of an odor is spatially represented within the glomerular array of the AL (Sachse et al., 1999; Meijerink et al., 2003).

The next question is whether this spatial representation is transformed at higher levels in the olfactory pathway. Results from intracellular and neural-ensemble recordings, as well as imaging of PN activity, provide support for different degrees of signal transformation in the AL. As we shall see, these differences may arise (at least partially) as a result of the different methods used to measure and quantify odor-evoked activity. At one extreme, no transformation may occur, suggesting a faithful transmission of sensory information to higher brain centers (Ng et al., 2002; Wang et al., 2003) (Figure 12.3a). This finding, obtained through fluorescence imaging in genetically engineered *Drosophila*, implies that downstream decoders must, therefore, be able to recognize and discriminate the different molecular features relayed through labeled lines that, in combination, represent a given odor.

These results, however, stand in contrast to a variety of other studies arguing that AL interneurons actively shape the responses of PNs. First, a recent electrophysiological study in *Drosophila* concluded that odors generally elicit responses in large, distributed ensembles of AL neurons and that PNs are more broadly tuned than ORNs (Wilson et al., 2004) (Figure 12.3b). Second, using neural-ensemble recording methods (see Chapter 14), Christensen and colleagues found evidence in the moth *Manduca* for dynamic network interactions both within and between identified AL glomeruli (Christensen et al., 2000) (Figure 12.3c). Earlier electrophysiological studies in the locust and honeybee also showed that synaptic interactions within the AL can impose odor-specific slow temporal patterns on the AL network (Laurent et al., 1996a,b; Laurent, 2002; Figure 12.3d) (see Section 12.5.3). Lateral inhibitory interactions within the AL, conveyed by GABAergic LNs, have

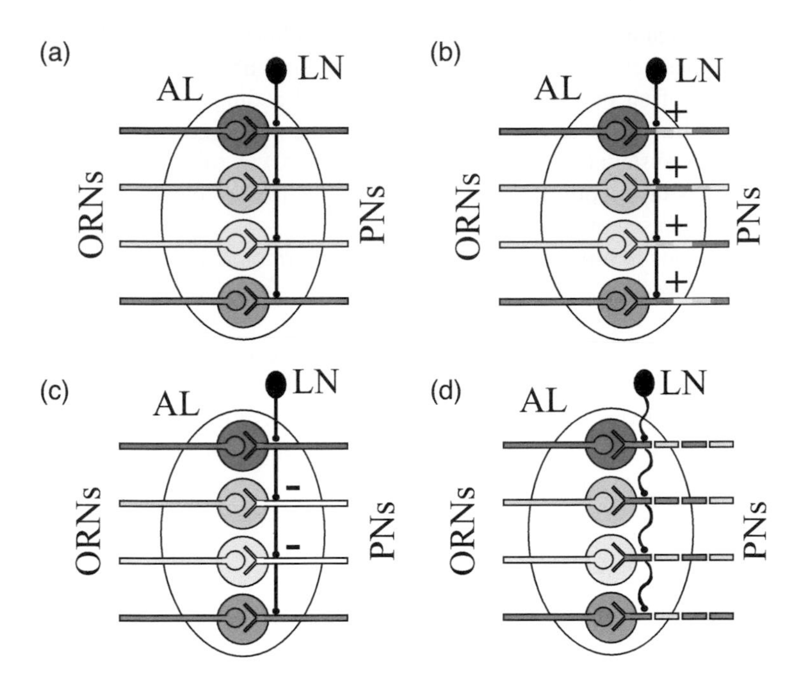

FIGURE 12.3 (See color insert following page 202.) Models for transformation of olfactory information in the antennal lobe. **a**: In the simplest model, no transformation of the olfactory information from the olfactory receptor neurons (ORNs) to the projection neurons (PNs) occurs, implying that local interneurons (LNs) are silent in this aspect of olfactory coding. **b**: Lateral interactions via LNs may, however, serve to broaden the response of the primary (ORN) to the secondary (PN) neuronal level, thus increasing the discrimination capability of the olfactory system by enhancing the combinatorial code. **c**: Lateral interactions may also sharpen the tuning curve of each PN through lateral inhibition, thus restricting the spatial pattern of output from the odor-activated glomeruli. **d**: A dynamic pattern of synchronized PNs, driven by an oscillatory neural network of LNs, may also drastically transform the original olfactory information

been shown to modulate glomerular output activity (Waldrop et al., 1987; Christensen et al., 1998; Lei et al., 2002; Sachse and Galizia, 2002; Wilson et al., 2004). These lateral interactions may sharpen the tuning curve of each glomerular channel through lateral inhibition (Waldrop et al., 1987; Christensen et al., 1998; Sachse and Galizia, 2002), thus restricting the spatial pattern of output from the odor-activated glomeruli. Alternatively, lateral interactions may facilitate decoding of the representation by separating odors in stimulus space (Wilson et al., 2004; see also Ng et al., 2002). Irrespective of the transformation mechanism, there is a growing consensus among olfactory physiologists regarding some basic concepts:

Olfactory coding as a combinatorial process — Discrimination of odors depends on a combinatorial coding process, involving the ensemble activity of a diverse array of PNs in the AL (reviewed by Hansson and Christensen, 1999; Christensen and White, 2000; Christensen and Hildebrand, 2002b).

One implication of this is that neither labeled lines nor across-fiber patterns by themselves are able to encode information about different molecular features of an odor blend. Instead, a hybrid coding strategy, across-label coding, has been suggested to better describe the processing scheme in the olfactory system (Christensen and White, 2000). A full explanation of how an across-label-coding strategy may function in odor blend coding is given by Christensen and White (2000).

Synchronous firing among PNs — Another principle emerging from olfactory processing studies is that synchronous firing among PNs is a key mechanism for strengthening the neural representations of odors in the brain. There remains some disagreement, however, over whether synchronous firing is (Laurent, 2002) or is not (Christensen et al., 2003) functionally linked to network-level oscillations. For reviews of these concepts and their relevance to information coding, see Christensen and White (2000) and Laurent (2002) and Section 12.5.3.

12.5.2 Coding of Odor Intensity

12.5.2.1 Sensitivity of Antennal Lobe Neurons

Unlike the peripheral olfactory system, less is known about the sensitivity of individual central olfactory neurons, primarily due to technical difficulties related to limited recording periods. Nonetheless, the threshold value for response (i.e., the concentration at which the action potential frequency statistically differs from that elicited by a control stimulus) and dose-response curves of single PNs have been constructed.

In pheromone- and plant-odor responding PNs of *S. gregaria* (Ignell et al., 1998, 1999; Ochieng' and Hansson, 1999) and in plant-odor responding PNs of *P. americana* (Boeckh and Ernst, 1987), PN sensitivity is more or less identical to that reported for ORNs. This is in contrast to what has been found for sex-pheromone responding PNs of both moths (Hansson et al., 1991; Christensen and Hildebrand, 1994; Hartlieb et al., 1997) and cockroaches (Boeckh and Ernst, 1987). In these cases, a tenfold to 10,000-fold increase in sensitivity of single PNs, compared with that of ORNs, has been observed. This is partly explained by spatial summation due to a large convergence of ORNs onto PNs (reviewed by Hansson and Christensen, 1999). Sensitivity enhancement seems, however, also to be centrally regulated because the increase in sensitivity from the periphery to the AL is not uniform. As an example, two components of the sex pheromone of *M. sexta* are detected by two colocalized ORNs, implying approximately similar abundance of these ORNs. However, PNs that respond to these two pheromone components display threshold sensitivities that differ by a factor of 10^2 (Hansson et al., 1991). In this case, one cannot rule out the possibility that there exist single ORNs with extremely high sensitivity. Perhaps more likely, dendritic processing (possibly through the release of neuromodulators) may affect response threshold and PN sensitivity, as has been suggested in other neural systems (Magee, 2000). Direct and indirect roles of neuromodulators on the sensitivity of AL neurons have, in fact, been shown, for example, in *Agrotis ipsilon* (Anton and Gadenne, 1999) and in *M. sexta* (Kloppenburg et al., 1999a). However, these cases probably constitute slow-acting modulatory systems, and further work is needed to elucidate the effect of other neuromodulators in the olfactory system.

12.5.2.2 Intensity vs. Identity Coding

Psychophysical studies have shown that *Apis mellifera* and *Periplaneta americana* are able to discriminate not only between different odorants but also between varying concentrations of the same odorant (Kramer, 1976; Getz and Smith, 1991; Sakura et al., 2002). This is not surprising because both the odor identity and its concentration are likely to contain valuable information about the odor source. In insects that have to recognize odorants occurring in plumes of varying concentration, it is generally believed (although not proven) that the central olfactory system more or less independently extracts these two parameters (stimulus identity and intensity) in order to optimize odor coding. Such a mechanism would correct for decreases in odor specificity with increasing odor concentration, as has been observed at the peripheral olfactory level (discussed earlier).

In the honeybee *Apis mellifera*, Sachse and Galizia (2003) proposed that encoding of odor information is performed by different parallel neuronal pathways resulting in an intensity-coding (concentration- or odor quantity–coding) channel and a concentration-invariant (odor-quality) channel. Odor intensity was suggested to be processed by a distinct neuronal subsystem of PNs (Abel et al., 2001; Sachse and Galizia, 2003), which, due to its homogeneous multiglomerular innervation, was suggested to encode odor intensity by measuring the strength of glomerular activity overall (Sachse and Galizia, 2003). In line with this proposal, multiglomerular PNs of other insects (Anton and Homberg, 1999) might play a similar role, but in cases in which multiglomerular PNs have been functionally and morphologically characterized (Hansson et al., 1994; Anton and Hansson, 1995, 1999; Abel et al., 2001), intensity coding has not been thoroughly investigated. Further analysis of this type of PN is required in order to draw further conclusions concerning its role as genuine odor-intensity encoders.

Several strategies for a concentration-invariant code have been presented over the last few years, based primarily on results from functional imaging of odor-evoked calcium changes in ORNs and PNs. By imaging ensembles of ORNs, it has been shown in several different species that different odors evoke activity in different subsets of glomeruli, and the particular subset of glomeruli that is activated by a given odor is relatively consistent across stimulus concentrations (Galizia et al., 2000; Carlsson and Hansson, 2003; Meijerink et al., 2003; Sachse and Galizia, 2003; Wang et al., 2003). The dynamic range of optical signals is usually 2 to 4 orders of magnitude, which is in accordance with extracellular recordings from single ORNs (Ljungberg et al., 1993; Anderson et al., 1995; Berg et al., 1995; de Bruyne et al., 2001). Similar to single ORNs, odor-evoked patterns of activity across glomeruli, presumably representing the activity of the ORN assembly, generally display sigmoidal dose-response curves with resulting saturation at high concentrations. At these higher concentrations, however, several "promiscuous" glomeruli may appear, probably reflecting the recruitment of ORNs with lower affinity. Whether the olfactory system is able to discriminate odors at these concentrations is still a matter of debate (Ng et al., 2002; Carlsson and Hansson, 2003; Stopfer et al., 2003; Wang et al., 2003). If an activity pattern of ORNs is to be read by downstream brain regions, a "promiscuous" code implies that similar odors should be more difficult to discriminate (Carlsson and Hansson, 2003; Wang et al., 2003). Alternatively, one may argue that an increasing number of "promiscuous" glomeruli serves to expand the discrimination capability of the olfactory system by enhancing the combinatorial code (Ng et al., 2002; Wilson et al., 2004). Interestingly, behavioral data from *A. mellifera* indicate that the discriminatory ability increases at high concentrations (Bhagavan and Smith, 1997). In either case, there is now abundant evidence from insects showing that the spatial pattern of glomerular activity plays a critical role in odor coding in the AL (but see Section 12.5.3 for a discussion of how time in the nervous system is also an important coding parameter).

Although the subset of activated glomeruli is generally consistent across physiological concentrations, the relative contributions of the members of this subset may be very different at different stimulus intensities. How do higher centers read the AL output? One strategy could be to read out the glomerular array in terms of relative response intensity. This concept has been around for many years (e.g., Boeckh and Ernst, 1987; Hansson et al., 1991; Christensen and Hildebrand, 1994; Hartlieb et al., 1997) and is supported by more recent functional imaging studies in *A. mellifera* and *D. melanogaster* (Sachse and Galizia, 2003; Wang et al., 2003). It is important to note, however, that this strategy applies only in cases where the response function of PNs is shifted by two or more decadic steps relative to the ORNs. A relative-intensity strategy would not work in cases in which the dose-response functions of ORNs are similar to those of PNs (Anton et al., 1997; Ignell et al., 1998, 1999; Ochieng' and Hansson, 1999). PNs nevertheless respond reliably to ORN input (Anton et al., 1997; Ignell et al., 1998, 1999), suggesting that the excitation ratio between the most responsive glomeruli could also be used as a relevant code (Ng et al., 2002; Sachse and Galizia, 2003). This would require an across-fiber pattern with differentially tuned PNs. Such a pattern has been suggested by functional imaging of PN assemblies (Carlsson, 2003; Sachse and Galizia, 2003; Wang et al., 2003) and corroborated by electrophysiological recordings of single PNs (Sadek et al., 2002; Greiner et al., 2003). All of these hypotheses, however, presuppose that neural decoders downstream of the PN ensemble can evaluate the combinatorial depth of the PN signal. At present, few studies have examined this question, and little is known about higher processing stages (Perez-Orive et al., 2002).

Contrast in the odor representation may be elevated through lateral inhibition of PNs by inhibitory LNs that function as adaptive gain control elements in the AL (Christensen et al., 1993; Lei et al., 2002; Ng et al., 2002; Sachse and Galizia, 2002). Recently, evidence for concentration-invariant, odor-specific contrast enhancement has been found through functional imaging studies (Ng et al., 2002; Sachse and Galizia, 2003). The specific neural interactions underlying this effect, however, need to be studied in more detail.

How else might odor contrast be enhanced in a concentration-invariant manner? One idea is that odor identity in insects and vertebrates can be encoded by specific spatiotemporal activity patterns of PN assemblies over time (Laurent et al., 1996a,b; Friedrich and Laurent, 2001; Laurent, 2002). In this model, an odor representation may be considered a multidimensional vector of PN states that evolve over the duration of a stimulus. A recent study in *Schistocerca americana* showed that the spatiotemporal pattern of PN activity changed relatively continuously over a range of concentrations for each odorant (Stopfer et al., 2003). Thus, the high-dimensional state of the PN assembly over time is more or less concentration invariant. A problem with this model is that it requires that the stimulus be present over a relatively long period of time, a condition that is not often found in nature (see Section 12.5.3). In *M. sexta*, a different type of spatiotemporal model has been suggested, in which different odor concentrations are represented by specific temporal relationships between combinations of PNs. These relationships do not evolve but are transient and dictated by the temporal pattern of the stimulus (Christensen et al., 2000; Lei et al., 2002). The following section addresses these concepts further.

12.5.3 Time as a Factor in Antennal Lobe Processing

Odors do not occur as a homogeneous fog in space. Rather, they flow through space as filaments of higher concentration interspersed with odor-free air. This means that odor molecules will hit the ORNs as clusters, interspersed with "silent" periods. The structure of the odor plume has been shown to be imperative for a number of insects, especially moths, to exhibit odor-driven behavior (see Chapter 2). It therefore stands to reason that there would be neural mechanisms in the insect brain to preserve this critical information.

12.5.3.1 Encoding Stimulus Frequency in Antennal Lobe Neurons

The ability of AL neurons to follow intermittent pulsed stimuli has been studied extensively in the sex-pheromone processing subsystem, the macroglomerular complex, of *M. sexta* (Christensen and Hildebrand, 1988; Heinbockel et al., 1999) and in *A. segetum* (Lei and Hansson, 1999), and in PNs innervating glomeruli that process plant-odor information in female *M. sexta* (King et al., 2000). Lemon and Getz (1998), furthermore, studied stimulus tracking of PNs to a general odorant, 1-hexanol, in *P. americana*. In all species, PNs were found to be differentially able to follow pulses at frequencies between 1 and 10 Hz. In *M. sexta* and *A. segetum*, this pulse-tracking capability was found to be related to a triphasic (-/+/-) and biphasic (+/-) response pattern, respectively. In *M. sexta*, the amplitude of the initial short-latency inhibitory postsynaptic potential (IPSP) was found to be highly correlated with the rate of pulsing a neuron could resolve (Christensen and Hildebrand, 1997). Furthermore, the IPSP was found to be blocked by bicuculline or picrotoxin, but this treatment did not affect the excitatory or secondary inhibition (Waldrop et al., 1987; Christensen et al., 1998). In *A. segetum*, the inhibitory period of the biphasic response was found to be the key factor regulating the punctuation of the depolarized state of a PN, enabling it to track fast consecutive pulses (Lei and Hansson, 1999).

In *M. sexta,* most specialist and generalist PNs are able to follow pheromone pulses up to 3 to 4 Hz (Christensen and Hildebrand, 1988; Heinbockel et al., 1999; King et al., 2000), equivalent to those found at the peripheral level (Marion-Poll and Tobin, 1992). As an exception, sex-pheromone blend-specific PNs follow pulses of up to 10 Hz (Christensen and Hildebrand, 1988). Recently, CO_2-sensing PNs in *Manduca* also were shown to track 10 Hz pulses (P. Guerenstein, unpublished). In *A. segetum*, on the other hand, no discrepancy is found between PN types, and both pheromone-component- and blend-specific PNs are able to follow pulses of up to 10 Hz (Lei and Hansson, 1999); this should be compared with 3 Hz at the ORN level (Baker et al., 1988). The ability to follow pulsed stimuli at higher frequency than individual ORNs may be related to the spatial arrangement of inhibitory inputs relative to the spike-initiation zone in the PNs. This arrangement could provide a more precise regulation of spiking input that closely reflects the intermittent input, allowing PNs to extract more information about intermittently pulsed stimuli.

In *M. sexta,* PNs, either depolarized or hyperpolarized by the two sex-pheromone components, show significantly higher synchronization in response to the blend of these odorants (Lei et al., 2002). This finding suggests that interglomerular inhibitory input may influence the output from a given PN, thus differentially increasing the encoding ability of certain key features of the odor stimulus. Whether a similar system may explain the general increase in frequency encoding of both component-specific and blend-specific PNs remains to be studied.

12.5.3.2 Temporal Representation of Odors

12.5.3.2.1 Rate Encoders vs. Temporal Encoders

Theoretical and experimental investigations (Softky, 1995; Theunissen and Miller, 1995) have indicated that sensory information may be represented either as a temporal code or as a rate code. Information carried within a temporal code has a higher density because the information is encoded in the timing of the spikes, compared with a rate code that carries information in the average rate of spikes. At points of convergence in the nervous system, as in olfactory glomeruli in the AL, information carried as a rate code may be translated into a temporal code employed by a single neuron or a few neurons only (Lemon and Getz, 2000). Information transmission utilizing such a strategy will occupy less neural space but retain the high information content of the original signal. According to the criterion for rate encoders, the encoded information should be correlated with the average number of spikes within the encoding window (i.e., the shortest time period of neural activity in which the stimulus can be presented) (Theunissen and Miller, 1995). As discussed previously, the activity of some PNs carries information concerning the temporal structure of the stimulus, similar to that found at the peripheral olfactory level, which implies that at least some PNs and ORNs may act as rate encoders. The criterion for temporal encoders, on the other hand, is that the encoded information must be correlated with some aspect of the temporal pattern of spikes within the encoding window (Theunissen and Miller, 1995).

Decomposition of the responses of PNs of *P. americana* has shown that stimulus identity is correlated with higher-order principal components of the activity of PNs, consistent with a temporal encoding scheme (Theunissen and Miller, 1995; Lemon and Getz, 2000). These results are consistent with those found in the locust, *S. americana* (Laurent and Davidowitz, 1994; Laurent et al., 1996a,b), where each responsive PN produces an odorant-specific, temporally complex response, often not timed to odor delivery, which typically outlasts the stimulus. This is in contrast to PN responses in moths, which typically do not outlast the stimulus, are locked to stimulus onset, and generally display stereotypic multiphasic responses (Christensen et al., 2000). However, in both *P. americana* (Lemon and Getz, 1998) and *M. sexta* (Christensen et al., 2000), PNs (recorded at different sites and responding to different odors) are found that exhibit the same temporal dynamic pattern. The significance of these findings is that temporal synchronization of firing between neurons with the same response dynamics may be involved in interglomerular integration of odor information. Upon stimulation with a single puff of odor, individual PNs of *S. americana* generate a dynamic, odor-dependent pattern that unfolds over time (Laurent et al., 1996a,b). This temporal summation mechanism is thought to bind together multiple signals into a more coherent representation of an odor blend, which reinforces the PN population code and increases the probability of activating target neurons in higher olfactory centers (Laurent, 2002). Ensemble recordings from PNs in *M. sexta* also support this interglomerular temporal binding of odor blends but indicate that the expression of binding is highly context dependent (Christensen et al., 2000; Lei et al., 2002).

Dual intracellular recordings from PNs innervating the same glomerulus and tuned to the same odorant have, furthermore, indicated a tight intraglomerular synchrony of PN activity (Lei et al., 2002). Lei et al. (2002) and Christensen et al. (2000) suggested that the function of such synchronous firing might be to integrate information signals from a selected subset of the functionally diverse population of PNs that arise from a single glomerulus and, thus, reinforce odorant-specific signals. By superimposing distinct temporal dynamic patterns on spatially segregated glomeruli, other features of the odorant signal may thus be extracted. For example, different subsets of PNs have been shown to synchronize at different stimulus concentrations, which suggests functional partitioning within glomeruli (Christensen et al., 2000 Lei et al., 2002).

12.5.3.2.2 Spatiotemporal Patterns and Odor Discrimination

Temporal patterns are linked to the spatial representation of an odorant across the glomerular array (Hildebrand and Shepherd, 1997; Hansson and Christensen, 1999; Stopfer et al., 1999; Christensen and White, 2000), thus creating a spatiotemporal pattern of neuronal activity. The added dimensions resulting from a high number of possible spatiotemporal combinations have been proposed to allow for a large coding space that facilitates stimulus identification. Although it is generally agreed that key features, such as odor intensity and the dynamics and quality of odor blends, may be encoded by these spatiotemporal patterns, their role in odor discrimination is debated (Stopfer et al., 1999; Christensen et al., 2000, 2003; Laurent, 2002; Lei et al., 2002).

Recordings from honeybees and locusts have indicated that responses of PNs are highly dynamic and characterized by epochs of increased and decreased firing that are both neuron and odor specific (Laurent et al., 1996a,b; Wehr and Laurent, 1996; Stopfer et al., 1999; Laurent, 2002). As a result, PN representations will evolve in a stimulus-specific manner over time, generating an oscillating local field potential (LFP) to which the synchronized firing and updating of participating PNs is linked (Laurent, 1999; Stopfer et al., 1999). This dynamic evolution seems, however, to be contingent on the applied stimulus paradigm. Using brief, intermittent stimulus pulses (as opposed to pulses of seconds in locusts and honeybees), Christensen et al. (2003) were unable to find any correlation between the temporal patterns of PN firing and the LFP in the moth *Manduca*. From a behavioral point of view, the time course of stimulation is an important topic to address in future experiments because brief intermittent stimulation is critical for odor discrimination in many insects.

12.6 CONCLUSIONS AND OUTLOOK

We are approaching the 50th birthday of insect olfactory neurophysiology. Immense progress has been made during this period, and our knowledge of the system has improved considerably. Still, what we know does not make up more than a fraction of what we do not know about the system. Our understanding of the molecular processes shaping the antennal signal is today quite exhaustive, but the processes according to which a molecule is *de facto* identified by the receptor protein are still unclear. The olfactory map among antennal lobe glomeruli has been shown using optical imaging, but the neural elements shaping the map and how they interact are definitely not understood. When approaching the higher brain areas, our understanding of the system gets even more rudimentary. Excellent studies have demonstrated the architecture of, for example, the mushroom bodies, but what these structures indeed do is still under debate.

To keep pushing the limit of our capabilities when investigating the insect olfactory system, we need to continue developing new methodology and refining practices already in use. The pioneers in the field had a huge impact by adapting electrophysiology to the system we work on, and still quantum leaps appear every now and then when a method is developed or adapted. By posing new questions and by developing the techniques to answer them, our understanding of the different parts of the system will increase. Over time, this will allow us to weave a more and more complete picture of the neural web that enables insects to smell.

ACKNOWLEDGMENTS

The authors would like to thank Sylvia Anton, Teun Dekker, and Jocelijn Meijerink, as well as other members of the Division of Chemical Ecology in Alnarp. This work was supported by the Scientific Council of Sweden.

REFERENCES

Abel R, Rybak J, Menzel R (2001) Structure and response patterns of olfactory interneurons in the honeybee, *Apis mellifera*. *J. Comp. Neurol.* **437**:363–383.

Almaas TJ, Mustaparta H (1991) *Heliothis virescens*: response characteristics of receptor neurons in *sensilla trichodea* Type 1 and Type 2. *J. Chem. Ecol.* **17**:953–972.

Almaas TJ, Christensen TA, Mustaparta H (1991) Chemical communication in heliothine moths. I. Antennal receptor neurons encode several features of intra- and interspecific odorants in the male corn earworm moth *Helicoverpa zea*. *J. Comp. Physiol. A* **169**:249–258.

Anderson P, Hansson BS, Löfkvist J (1995) Plant-odour-specific receptor neurones on the antennae of female and male *Spodoptera littoralis*. *Physiol. Entomol.* **20**:189–198.

Angioy AM, Desogus A, Barbarossa IT, Anderson P, Hansson BS (2003) Extreme sensitivity in an olfactory system. *Chem. Senses* **28**:279–284.

Anton S, Gadenne C (1999) Effect of juvenile hormone on the central nervous processing of sex pheromone in an insect. *Proc. Natl. Acad. Sci. USA* **96**:5764–5767.

Anton S, Hansson BS (1995) Sex pheromone and plant-associated odour processing in antennal lobe interneurons of male *Spodoptera littoralis* (Lepidoptera: Noctuidae). *J. Comp. Physiol. A* **176**:773–789.

Anton S, Hansson BS (1999) Physiological mismatching between neurons innervating olfactory glomeruli in a moth. *Proc. Roy. Soc. Lond. B* **266**:1813–1820.

Anton S, Homberg U (1999) Antennal lobe structure. In: *Insect Olfaction,* Hansson BS, Ed., Springer, Berlin, pp. 97–124.

Anton S, Löfstedt C, Hansson BS (1997) Central nervous processing of sex pheromones in two strains of the European corn borer *Ostrinia nubilalis* (Lepidoptera: Pyralidae). *J. Exp. Biol.* **200**:1073–1087.

Ashburner M (1989) *Drosophila: A Laboratory Handbook*. Cold Spring Harbor Press, Cold Spring Harbor, NY.

Ayer RK, Carlson JR (1992) Olfactory physiology in the *Drosophila* antenna and maxillary palp: acj6 distinguishes two classes of odorant pathways. *J. Neurobiol.* **23**:965–982.

Baker TC (1990) Upwind flight and casting flight: complementary phasic and tonic systems used for location of sex pheromone sources by male moths. In: *ISOT X, Proc. 10th Int. Symp. Olfaction Taste,* Doving K, Ed., GCS/AS, Oslo, pp. 18–25.

Baker TC, Willis MA, Haynes KF, Phelan PL (1985) A pulsed cloud of sex pheromone elicits upwind flight in male moths. *Physiol. Entomol.* **10**:257–265.

Baker TC, Hansson BS, Löfstedt C, Löfkvist J (1988) Adaptation of antennal neurons in moths is associated with cessation of pheromone mediated upwind flight. *Proc. Natl. Acad. Sci. USA* **85**:9826–9830.

Baker TC, Hansson, BS, Löfstedt C, Löfkvist J (1989) Adaptation of male moth antennal neurons in a pheromone plume is associated with cessation of pheromone-mediated flight. *Chem. Senses* **14**:439–448.

Bau J, Justus KA, Cardé RT (2002) Antennal resolution of pulsed pheromone plumes in three moth species. *J. Insect Physiol.* **48**:433–442.

Behan M, Schoonhoven LM (1978) Chemoreception of an oviposition deterrent associated with eggs in *Pieris brassicae*. *Entomol. Exp. Appl.* **24**:163–179.

Bengtsson M, Liljefors T, Hansson BS, Löfstedt C, Copaja SV (1990) Structure-activity relationships for chain-shortened analogs of (Z)-5-decenyl acetate, a pheromone component of the turnip moth, *Agrotis segetum*. *J. Chem. Ecol.* **16**:667–684.

Berg BG, Tumlinson JH, Mustaparta H (1995) Chemical communication in heliothine moths IV. Receptor neuron responses to pheromone compounds and formate analogues in the male tobacco budworm moth *Heliothis virescens*. *J. Comp. Physiol. A* **177**:527–534.

Bestmann HJ (1981) Pheromon Rezeptor-Wechsilwirkung bei Insekten. *Mitt. dt. Ges. Allg. Angew. Ent.* **2**:242–247.

Bestmann HJ, Vostrowsky O (1982) Peripheral aspects of olfacto-endocrine interactions. Structure-activity. In: *Olfaction and Endocrine Regulation,* Breipohl W, Ed., IRL Press, London, pp. 253–265.

Bhagavan S, Smith BH (1997) Olfactory conditioning in the honeybee, *Apis mellifera*: effects of odor intensity. *Physiol. Behav.* **61**:107–117.

Blomquist GJ, Vogt RG (2003) *Insect Pheromone Biochemistry and Molecular Biology.* Elsevier, Amsterdam.

Boeckh J (1962) Elektrophysiologishe Untersuchungen an einzelnen Geruchsrezeptoren auf den Antennen des Totengräbers (Necrophorus, Coleoptera). *Z. Vergl Physiol* **46**:212–248.

Boeckh J, Ernst KD (1983) Olfactory food and mate recognition. In: *Neuroethology and Behavioral Physiology,* Hyber F, Markl H, Eds., Springer, Berlin, pp. 78–94.

Boeckh J, Ernst KD (1987) Contribution of single unit analysis in insects to an understanding of olfactory function. *J. Comp. Physiol. A* **161**:549–565.

Boeckh J, Selsam P (1984) Quantitative investigation of the odour specificity of central olfactory neurones in the American cockroach. *Chem. Senses* **9**:369–380.

Borroni PF, O'Connell RJ (1992) Temporal analysis of adaptation in moth (*Trichoplusia ni*) pheromone receptor neurons. *J. Comp. Physiol. A* **170**:691–700.

Breer H. (1997) Sense of smell: signal recognition and transduction in olfactory receptor neurons. In: *Handbook of Biosensors and Electronic Noses: Medicine, Food and the Environment,* Kress-Rogers E, Ed., CRC Press, Boca Raton, FL, pp. 521–532.

Carlsson MA (2003) A sensory map of the odour world in the moth brain, Ph.D. thesis, Agraria, Alnarp, Sweden.

Carlsson MA, Hansson BS (2003) Dose-response characteristics of glomerular activity in the moth antennal lobe. *Chem. Senses* **28**:269–278.

Carlsson MA, Galizia G, Hansson BS (2002) Spatial representation of odours in the antennal lobe of the moth *Spodoptera littoralis* (Lepidoptera: Noctuidae). *Chem. Senses* **27**:231–244.

Christensen TA, Hildebrand JG (1987a) Functions, organization, and physiology of the olfactory pathways in the lepidopteran brain. In: *Arthropod Brain: Its Evolution, Development, Structure and Functions,* Gupta AP, Ed., Wiley, Chichester, pp. 457–484.

Christensen TA, Hildebrand JG (1987b) Male-specific, sex pheromone-selective projection neurons in the antennal lobes of the moth *Manduca sexta*. *J. Comp. Physiol. A* **160**:553–569.

Christensen TA, Hildebrand JG (1988) Frequency coding by central olfactory neurons in the sphinx moth *Manduca sexta*. *Chem. Senses* **13**:123–130.

Christensen TA, Hildebrand JG (1994) Neuroethology of sexual attraction and inhibition in heliothine moths. In: *Neural Basis of Behavioural Adaptations. Progress in Zoology,* vol. 39, Schildberger K, Elsner N, Eds., Fischer, Stuttgart, pp. 37–46.

Christensen TA, Hildebrand JG (1997) Coincident stimulation with pheromone components improves temporal pattern resolution in central olfactory neurons. *J. Neurophysiol.* **77**:775–781.

Christensen TA, Hildebrand JG (2002a) Pheromonal and host-odor processing in the insect antennal lobe: how different? *Curr. Opin. Neurobiol.* **12**:393–399.

Christensen TA, Hildebrand JG (2002b) Electrophysiological analysis of olfactory coding in the CNS. In: *Methods and Frontiers in Chemosensory Research*, Simon SA, Nicolelis MAI, Eds., CRC Press, Boca Raton, FL, pp. 325–337.

Christensen TA, White J (2000) Representation of olfactory information in the brain. In: *The Neurobiology of Taste and Smell,* Finger TE, Silver WL, Restrepo D, Eds., Wiley-Liss Inc., New York, pp. 201–232.

Christensen TA, Waldrop BR, Harrow ID, Hildebrand JG (1993) Local interneurons and information processing in the olfactory glomeruli of the moth *Manduca sexta. J. Comp. Physiol. A* **173**:385–399.

Christensen TA, Harrow ID, Cuzzocrea C, Randolph PW, Hildebrand JG (1995) Distinct projections of two populations of olfactory receptor axons in the antennal lobe of the sphinx moth *Manduca sexta. Chem. Senses* **20**:313–323.

Christensen TA, Waldrop B, Hildebrand JG (1998) GABAergic mechanisms that shape the temporal response to odors in moth olfactory projection neurons. *Ann. N.Y. Acad. Sci.* **855**:475–481.

Christensen TA, Pawlowski VM, Lei H, Hildebrand JG (2000) Multi-unit recordings reveal context dependent modulation of synchrony in odor-specific neural ensembles. *Nat. Neurosci.* **3**:927–931.

Christensen TA, Lei H, Hildebrand JG (2003) Coordination of central odor representations through transient, non-oscillatory synchronization of glomerular output neurons. *Proc. Natl. Acad. Sci. USA* **100**:11076–11081.

Clyne PJ, Warr CG, Freeman MR, Lessing D, Kim J, Carlson JR (1999) A novel family of divergent seven-transmembrane proteins: candidate odorant receptors in *Drosophila. Neuron* **22**:327–338.

Crnjar R, Scalera G, Liscia A, Angioy AM, Bigiani A, Pietra P, Barbarossa IT (1989) Morphology and EAG mapping of the antennal receptors in *Dacus oleae. Entomol. Exp. Appl.* **51**:77–85.

de Bruyne M, Clyne PJ, Carlson JR (1999) Odor coding in a model olfactory organ: The *Drosophila* maxillary palp. *J. Neurosci.* **19**:4520–4532.

de Bruyne M, Foster K, Carlson JR (2001) Odor coding in the *Drosophila* Antenna. *Neuron* **30**:537–552.

De Kramer JJ, Kaissling K-E, Keil T (1984) Passive electrical properties of insect olfactory sensilla may produce the biphasic shape of spikes. *Chem. Senses.* **8**:289–295.

Den Otter CJ, Saini RK (1985) Pheromone perception in the tsetse fly *Glossina morsitans morsitans. Entomol. Exp. Appl.* **39**:155–161.

Den Otter CJ, van der Goes van Naters W (1993) Responses of individual antennal olfactory cells of tsetse flies (*Glossina m. morsitans*) to phenols from cattle urine. *Physiol. Entomol.* **18**:43–49.

Dethier VG (1976) *The Hungry Fly*, Harvard University Press, Cambridge, MA.

Dickens JC (1990) Specialized receptor neurons for pheromone and host plant odors in the boll weevil, *Anthonomus grandis* Boh. (Coleoptera: Curculionidae). *Chem. Senses.* **15**:311–331.

Dobritsa A, van der Goes van Naters W, Warr CG, Steinbrecht RA, Carlson JR (2003) Integrating the molecular and cellular basis of odor coding in the *Drosophila* antenna. *Neuron* **37**:827–841.

Du G, Prestwich GD (1995) Protein structure encodes the ligand binding specificity in pheromone binding proteins. *Biochemistry* **34**:8726–8732.

Elmore T, Ignell R, Carlson JR, Smith DP (2003) Targeted mutation of a *Drosophila* odor receptor defines receptor requirement in a novel class of sensillum. *J. Neurosci.* **23**:9906–9912.

Friedrich RW, Laurent G (2001) Dynamic optimization of odor representations by slow temporal patterning of mitral cell activity. *Science* **291**:889–894.

Galizia CG, Menzel R (2000) Odour perception in honeybees: coding information in glomerular patterns. *Curr. Opin. Neurobiol.* **10**:504–510.

Galizia CG, Menzel R, Hölldobler B (1999a) Optical imaging of odor-evoked glomerular activity patterns in the antennal lobes of the ant *Camponotus rufipes. Naturwissenschaften* **86**:533–537.

Galizia CG, Sachse S, Rappert A, Menzel R (1999b) The glomerular code for odor representation is species specific in the honey bee, *Apis mellifera. Nature* **2**:473–478.

Galizia CG, Sachse S, Mustaparta H (2000) Calcium responses to pheromones and plant odours in the antennal lobe of male and female moth *Heliothis virescens. J. Comp. Physiol. A* **186**:1049–1063.

Gao Q, Yuan B, Chess A (1999) Convergent projections of *Drosophila* olfactory neurons to specific glomeruli in the antennal lobe. *Nat. Neurosci.* **3**:780–785.

Getz WM, Smith KB (1991) Olfactory perception in honeybees: concatenated and mixed odorant stimuli, concentration and exposure effects. *J. Comp. Physiol. A* **169**:215–230.

Greiner B, Gadenne C, Anton S (2003) A 3D antennal lobe atlas of a male moth (*Agrotis ipsilon*) and glomerular representation of projection neurons responding to plant volatiles. In: *Proceedings of 8th ESITO,* July 2–7, Harstad, Norway. p. 63.

Hansson BS (1995) Olfaction in Lepidoptera. *Experientia* **51**:1003–1027.

Hansson BS, Ed. (1999) *Insect Olfaction*, Springer-Verlag, Berlin.

Hansson BS, Christensen TA (1999) Functional characteristics of the antennal lobe. In: *Insect Olfaction*, Hansson BS, Ed., Springer-Verlag, Berlin, pp. 125–161.

Hansson BS, Löfstedt C, Foster SP (1989) Z-linked inheritance of male olfactory response to sex pheromone components in two species of tortricid moths, *Ctenopseustis obliquana* and *Ctenopseustis sp. Entomol. Exp. Appl.* **53**:137–145.

Hansson BS, Christensen TA, Hildebrand JG (1991) Functionally distinct subdivisions of the macroglomerular complex in the antennal lobe of the male sphinx moth *Manduca sexta. J. Comp. Neurol.* **312**:264–278.

Hansson BS, Ljungberg, H, Hallberg E, Löfstedt C (1992) Functional specialization of olfactory glomeruli in a moth. *Science* **256**:1313–1315.

Hansson BS, Hallberg E, Löfstedt C, Steinbrecht RA (1994) Correlation between dendrite diameter and action potential amplitude in sex pheromone specific receptor neurons in male *Ostrinia nubilalis* (Lepidoptera: Pyralidae). *Cell Tissue Res.* **26**:503–512.

Hansson BS, Larsson M, Leal WS (1999) Green leaf volatile-detecting olfactory receptor neurones display very high sensitivity and specificity in a scarab beetle. *Physiol. Entomol.* **24**:121–126.

Hartlieb E, Anton S, Hansson BS (1997) Dose-dependent response characteristics of antennal lobe neurons in the male moth *Agrotis segetum* (Lepidoptera: Noctuidae). *J. Comp. Physiol. A* **181**:469–476.

Heinbockel T, Kaissling K-E (1996) Variability of olfactory receptor neuron responses of female silkmoths (*Bombyx mori* L.) to benzoic acid and (±)-linalool. *J. Insect Physiol.* **42**:565–578.

Heinbockel T, Christensen T, Hildebrand JG (1999) Temporal tuning of odor responses in pheromone-responsive projection neurons in the brain of the sphinx moth *Manduca sexta. J. Comp. Neurol.* **409**:1–12.

Hildebrand JG, Shepherd GM (1997) Mechanisms of olfactory discrimination: converging evidence for common principles across phyla. *Ann. Rev. Neurosci.* **20**:595–631.

Hillbur Y (2001) *Tracking the tiny: identification of the sex pheromone of the pea midge as a prerequisite for pheromone-based monitoring*, Ph.D. thesis, Agraria, Alnarp, Sweden.

Ignell R, Anton S, Hansson BS (1998) Central nervous processing of behaviourally relevant odours in solitary and gregarious fifth instar locusts, *Schistocerca gregaria. J. Comp. Physiol. A* **183**:453–465.

Ignell R, Anton S, Hansson BS (1999) Integration of behaviourally relevant odours at the central nervous level in solitary and gregarious third instar locusts, *Schistocerca gregaria. J. Insect Physiol.* **45**:993–1000.

Joerges J, Küttner A, Galizia GC, Menzel R (1997) Representations of odours and odour mixtures visualized in the honeybee brain. *Nature* **387**:285–288.

Jönsson M, Anderson P (1999) Electrophysiological response to herbivore-induced host plant volatiles in the moth *Spodoptera littoralis. Physiol. Entomol.* **24**:377–385.

Justus KA, Schofield SW, Murlis J, Cardé RT (2002) Flight behaviour of *Cadra cautella* males in rapidly pulsed pheromone plumes. *Physiol. Entomol.* **27**:58–66.

Kaissling K-E (1971) Insect olfaction. In: *Handbook of Sensory Physiology, Vol. IV, Chemical Senses*, Beidler LM, Ed., Springer-Verlag, Berlin, pp. 351–431.

Kaissling K-E (1974) Sensory transduction in insect olfactory receptors. In: *Biochemistry of Sensory Functions,* Jaenicke D, Ed., Springer, Berlin, pp. 243–273.

Kaissling K-E (1986) Chemo-electrical transduction in insect olfactory receptors. *Ann. Rev. Neurosci.* **9**:121–145.

Kaissling K-E (1987) Temporal characteristics of pheromone receptor cell responses in relation to orientation behaviour of moths. In: *Wright Lectures in Insect Olfaction,* Colbow K, Ed., Simon Fraser University, Burnaby, BC, Canada, pp. 1–190.

Kaissling K-E (1995) Single unit and electroantennogram recordings in insect olfactory organs. In: *Experimental Cell Biology of Taste and Olfaction*, Spielman AI, Brand JG, Eds., CRC Press, Boca Raton, FL, pp. 361–377.

Kaissling K-E, Priesner E (1970) Die Riechschwelle des Seidenspinners. *Naturwissenschaften* **57**:23–28.

Kaissling K-E, Thorson J (1980) Insect olfactory sensilla: structural, chemical and electrical aspects of the functional organization. In: *Receptors for Neurotransmitters, Hormones and Pheromones in Insects,* Sattelle DB, Hall LM, Hildebrand JG, Eds., Elsevier, Amsterdam, pp. 261–282.

Kanzaki R, Arbas EA, Hildebrand JG. (1991a) Physiology and morphology of protocerebral olfactory neurons in the male moth *Manduca sexta. J. Comp. Physiol. A* **168**:281–298.

Kanzaki R, Arbas EA, Hildebrand JG. (1991b) Physiology and morphology of descending neurons in pheromone-processing olfactory pathways in the male moth *Manduca sexta. J. Comp. Physiol. A* **169**:1–14.

Kelling FJ (2001) *Olfaction in houseflies: morphology and electrophysiology*, Ph.D. thesis, University of Groningen, the Netherlands.

Kennedy JS, Ludlow AR, Sanders CJ (1981) Guidance of flying male moths by wind-borne sex pheromone. *Physiol. Entomol.* **6**:395–412.

King JR, Christensen TA, Hildebrand JG (2000) Response characteristics of an identified, sexually dimorphic olfactory glomervlus, *J. Neurosci.* **20**:2391–2399.

Kloppenburg P, Ferns D, Mercer AR (1999a) Serotonin enhances central olfactory neuron responses to female sex pheromone in the male sphinx moth *Manduca sexta. J. Neurosci.* **19**:8172–8181.

Kloppenburg P, Kirchhof BS, Mercer AR (1999b) Voltage-activated currents from adult honeybee (*Apis mellifera*) antennal motor neurons recorded *in vitro* and *in situ. J. Neurophysiol.* **81**:39–48.

Kodadova B (1996) Resolution of pheromone pulses in receptor cells of *Antherea polyphemus* at different temperatures. *J. Comp. Physiol. A* **179**:301–310.

Kramer E (1976) The orientation of walking honeybees in odour fields with small concentration gradients. *Physiol. Entomol.* **1**:27–37.

Larsson MC, Leal WS, Hansson BS (1999) Olfactory receptor neurons specific to chiral sex pheromone components in male and female *Anomala cuprea* beatles (Coleoptera: Scarabaeidae). *J. Comp. Physiol. A* **184**:353–359.

Larsson MC, Leal WS, Hansson BS (2001) Olfactory receptor neurons detecting plant odours and male volatiles in *Anomala cuprea* beetles (Coleoptera: Scarabaeidae). *J. Insect Physiol.* **47**:1065–1076.

Laurent G (1999) A systems perspective on early olfactory coding. *Science* **286**:723–728.

Laurent G (2002) Olfactory network dynamics and the coding of multidimensional signals. *Nat. Rev. Neurosci.* **3**:885–895.

Laurent G, Davidowitz H (1994) Encoding of olfactory information with oscillating neural assemblies. *Science* **265**:1872–1875.

Laurent G, Naraghi M (1994) Odorant-induced oscillations in the mushroom bodies of the locust. *J. Neurosci.* **14**:2993–3004.

Laurent G, Wehr M, Davidowitz H (1996a) Temporal representations of odors in an olfactory network. *J. Neurosci.* **16**:3837–3847.

Laurent G, Wehr M, Macleod K, Stopfer M, Leitch B, Davidowitz H (1996b) Dynamic encoding of odors with oscillating neuronal assemblies in the locust brain. *Biol. Bull.* **191**:470–475.

Leal WS, Mochizuki F, Wakamura S, Yasuda, T (1991) Electroantennographic detection of *Anomala cuprea* Hope (Coleoptera: Scarabaeidae) sex pheromone. *Appl. Entomol. Zool.* **27**:289–291.

Lee J-K, Strausfeld NJ (1990) Structure, distribution and number of surface sensilla and their receptor cells on the olfactory appendage of the male moth *Manduca sexta*. *J. Neurocytol.* **19**:519–538.

Lei H, Hansson BS (1999) Central processing of pulsed pheromone signals by antennal lobe neurons in the male moth *Agrotis segetum*. *J. Neurophysiol.* **81**:1113–1122.

Lei H, Anton S, Hansson BS (2001) Olfactory protocerebral pathways processing sex pheromone and plant odor information in the male moth *Agrotis segetum*. *J. Comp. Neurol.* **432:**356–370.

Lei H, Christensen TA, Hildebrand JG (2002) Local inhibition modulates odor evoked synchronization of glomerulus-specific output neurons. *Nat. Neurosci.* **5**:557–565.

Lemon WC, Getz WM (1998) Responses of cockroach antennal lobe projection neurons to pulsatile stimuli. *Ann. N.Y. Acad. Sci.* **855**:517–520.

Lemon WC, Getz WM (2000) Rate code input produces temporal code output from cockroach antennal lobes. *BioSystems* **58**:151–158.

Lewicki MS (1998) A review of methods for spike sorting: the detection and classification of neural action potentials. *Network: Comp. Neur. Syst.* **9**:R53–R78.

Liljefors T, Thelin B, van der Pers JNC (1984) Structure–activity relationships between stimulus molecules and response of a pheromone receptor cell in turnip moth, *Agrotis segetum*: modifications of the acetate group. *J. Chem. Ecol.* **10**:1661–1675.

Liljefors T, Thelin, B, van der Pers JNC, Löfstedt C (1985) Chain-elongated analogues of a pheromone component of the turnip moth, *Agrotis segetum*. A structure-activity study using molecular mechanics. *J. Chem. Soc. Perkin. Trans. II:* 1957–1962.

Liljefors T, Bengtsson M, Hansson BS (1987) Effects of double-bond configuration on interaction between a moth sex pheromone component and its receptor: a receptor-interaction model based on molecular mechanics. *J. Chem. Ecol.* **13**:2023–2040.

Ljungberg H, Anderson P, Hansson BS (1993) Physiology and morphology of pheromone-specific sensilla on the antennae of male and female *Spodoptera littoralis* (Lepidoptera: Noctuidae). *J. Insect Physiol.* **39**:253–260.

Mafra-Neto A, Cardé RT (1994) Fine-scale structure of pheromone plumes modulates upwind orientation of flying moths. *Nature* **369:**142–144.

Mafra-Neto A, Cardé RT (1998) Rate of realized interception of pheromone pulses in different wind speeds modulates almond moth orientation. *J. Comp. Physiol. A* **182**:563–572.

Magee JC (2000) Dendritic integration of excitatory synaptic input. *Nat. Rev. Neurosci.* **1**:181–190.

Malnic B, Hirono J, Sato T, Buck LB (1999) Combinatorial receptor codes for odors. *Cell* **96**:713–723.

Marion-Poll F, Tobin TR (1992) Temporal coding of pheromone pulses and trains in *Manduca sexta. J. Comp. Physiol. A* **171**:505–512.

Matsumoto SG, Hildebrand JG (1981) Olfactory mechanisms in the moth *Manduca sexta*: response characteristics and morphology of central neurons in the antennal lobes. *Proc. Roy. Soc. Lond. B* **213**:249–277.

Meijerink J, Carlsson MA, Hansson BS (2003) Spatial representation of odorant structure in the moth antennal lobe: a study of structure–response relationships at low doses. *J. Comp. Neurol.* **467**:11–21.

Meng LZ, Wu CH, Wicklein M, Kaissling KE, Bestmann HJ (1989) Number and sensitivity of three types of pheromone receptor cells in *Antherea pernyi* and *A. polyphemus. J. Comp. Physiol. A* **165**:139–146.

Meunier N, Marion-Poll F, Lansky P, Rospars JP (2003) Estimation of the individual firing frequencies of two neurons recorded with a single electrode. *Chem. Senses.* **28**:671–679.

Millar J (1992) Extracellular single and multiple unit recording with microelectrodes. In: *Monitoring Neuronal Activity: A Practical Approach,* Stamford JA, Ed., IRL Press, Oxford, pp. 1–27.

Mombaerts P (2004) Odorant receptor gene choice in olfactory sensory neurons: the one receptor–one neuron hypothesis revisited. *Curr. Opin. Neurobiol.* **14**:31–36.

Murlis J, Jones CD (1981) Fine-scale structure of odour plumes in relation to distant pheromone and other attractant sources. *Physiol. Entomol.* **6**:71–86.

Nagai T (1981) Electroantennogram response gradient on the antenna of the European corn borer, *Ostrinia nubilalis. J. Insect Physiol.* **27**:889–894.

Ng M, Roorda RD, Lima SQ, Zemelman BV, Morcillo P, Miesenböck G (2002) Transmission of olfactory information between three populations of neurons in the antennal lobe of the fly. *Neuron* **36**:463–474.

Ochieng' SA, Hansson BS (1999) Responses of olfactory neurones to behaviourally important odours in the adult gregarious and solitarious desert locust, *Schistocerca gregaria. Physiol. Entomol.* **24**:28–36.

Ochieng' SA, Anderson P, Hansson BS (1995) Antennal lobe projection patterns of olfactory receptor neurons involved in sex pheromone detection in *Spodoptera littoralis* (Lepidoptera: Noctuidae). *Tiss. Cell* **27**:221–232.

Payne TL, Dickens JC (1976) Adaptation to determine receptor system specificity in insect olfactory communication. *J. Insect Physiol.* **22**:1569–1572.

Perez-Orive J, Mazor O, Turner GC, Cassenaer S, Wilson RI, Laurent G. (2002) Oscillations and sparsening of odor representations in the mushroom Body. *Science* **297**:359–365.

Pophof B (2002a) Moth pheromone binding proteins contribute to the excitation of olfactory receptor cells. *Naturwissenschaften* **89**:515–518.

Pophof B (2002b) Octopamine enhances moth olfactory responses to pheromones, but not those to general odorants. *J. Comp. Physiol. A* **188**:659–662.

Pophof B, Gebauer T, Ziegelberger G (2000) Decyl-thio-trifluoropropanone, a competitive inhibitor of moth pheromone receptors. *J. Comp. Physiol. A* **186**:315–323.

Prestwich G, Du G (1997) Pheromone-binding proteins, pheromone recognition, and signal transduction in moth olfaction. In: *Insect Pheromone Research: New Directions*, Cardé RT, Minks AK, Eds., Chapman & Hall, New York, pp. 131–143.

Priesner E (1979) Progress in the analysis of pheromone receptor systems. *Ann. Zool. Ecol. Anim.* **11**:533–546.

Priesner E, Jacobson M, Bestmann HJ (1975) Structure–response relationships in noctuid sex pheromone reception. *Z. Naturf.* **30:**283–293.

Rumbo ER (1983) Differences between single cell responses to different components of the sex pheromone in males of the light brown apple moth (*Epiphyas postvittana*). *Physiol. Entomol.* **8**:195–201.

Rumbo ER, Kaissling K-E (1989) Temporal resolution of odour pulses by three types of pheromone receptor cells in *Antherea polyphemus. J. Comp. Physiol. A* **165**:281–291.

Sachse S, Galizia CG (2002) Role of inhibition for temporal and spatial odor representation in olfactory output neurons: a calcium imaging study. *J. Neurophysiol.* **87**:1106–1117.

Sachse S, Galizia CG (2003) The coding of odour-intensity in the honey bee antennal lobe: local computation optimizes odour representation. *Eur. J. Neurosci.* **18**:2119–2132.

Sachse S, Rappert A, Galizia CG (1999) The spatial representation of chemical structures in the antennal lobe of honeybees: steps towards the olfactory code. *Eur. J. Neurosci.* **11**:3970–3982.

Sadek MM, Hansson BS, Rospars JP, Anton S. (2002) Glomerular representation of plant volatiles and sex pheromone components in the antennal lobe of the female *Spodoptera littoralis. J. Exp. Biol.* **205**:1363–1372.

Sakmann B, Neher E (1995) *Single-Channel Recording*, Plenum Press, New York.

Sakura M, Okada R, Mizunami M (2002) Olfactory discrimination of structurally similar alcohols by cockroaches. *J. Comp. Physiol. A* **188**:787–797.

Sauer AE, Karg G, Koch UT, De Kramer JJ, Milli R (1992) A portable EAG system for the measurement of pheromone concentrations in the field. *Chem. Senses* **17**:543–553.

Schmidt M, Ignell R, Sibbe M (submitted) An isolated brain preparation to study the electrophysiology and morphology of Kenyon cells, lamina monopolar cells, and other small neurons of the adult insect brain with the patch-clamp technique *in situ. J. Neurosci.*

Schneider D (1957) Elektrophysiologische untersuchen con Chemo- und Mechanorezeptoren der Antenne des Seidenspinners *Bombyx mori. Z. Vergl. Physiol.* **40**:8–41.

Schneider D, Lacher V, Kaissling K-E (1964) Die Reaktionsweise und das Reaktionsspektrum von Riechzellen bei *Antherea pernyi* (Lepidoptera, Saturniidae). *Z. Vergl. Physiol.* **48**:632–662.

Scott K, Brady R, Cravchik A, Morozov P, Rzhetsky A, Zuker C, Axel R (2001) A chemosensory gene family encoding candidate gustatory and olfactory receptors in *Drosophila. Cell* **104**:661–673.

Shanbhag SR, Muller B, Steinbrecht RA (1999) Atlas of olfactory organs of *Drosophila melanogaster* 1. Types, external, organization, innervation and distribution of olfactory sensilla. *Int. J. Insect Morphol. Embryol.* **28**:377–397.

Shields VDC, Hildebrand JG. (2001) Recent advances in insect olfaction, specifically regarding the morphology and sensory physiology of antennal sensilla of the female sphinx moth *Manduca sexta. Micros. Res. Techniq.* **55**:307–329.

Silinsky EM (1992) Intracellular recording methods for neurons. In: *Monitoring Neuronal Activity: A Practical Approach,* Stamford JA, Ed., IRL Press, Oxford, pp. 29–58.

Skiri HT, Galizia CG, Mustaparta H (2004) Representation of primary plant odorants in the antennal lobe of the moth *Heliothis virescens* using calcium imaging. *Chem. Senses* **29**:253–267.

Softky WR (1995) Simple codes versus efficient codes. *Curr. Opin. Neurobiol.* **5**:239–247.

Stengl M, Hatt H, Breer H (1992) Peripheral processes in insect olfaction. *Ann. Rev. Entomol.* **54**:665-681.

Stensmyr MC, Giordano E, Balloi A, Angioy A-M, Hansson BS (2003) Novel natural ligands for *Drosophila* olfactory receptor neurones. *J. Exp. Biol.* **206**:715–724.

Stopfer M, Wehr M, MacLeod K, Laurent G (1999) Neural dynamics, oscillatory synchronization, and odour codes. In: *Insect Olfaction,* Hansson, BS, Ed., Springer, Berlin, pp. 163–180.

Stopfer M, Jayaraman V, Laurent G (2003) Intensity versus identity coding in an olfactory system. *Neuron* **39**:991–1004.

Theunissen FE, Miller JP (1995) Temporal encoding in nervous systems: a rigorous definition. *J. Comput. Neurosci.* **2**:149–162.

Todd JL, Baker TC (1999) Function of peripheral olfactory organs. In: *Insect Olfaction*, Hansson BS, Ed., Springer-Verlag, Berlin, pp. 67–96.

Todd JL, Anton S, Hansson BS, Baker TC (1995) Functional organization of the macroglomerular complex related to behaviorally expressed olfactory redundancy in male cabbage looper moths. *Physiol. Entomol.* **20**:349–361.

Vermeulen A, Rospars JP (2001) Electrical circuitry of an insect olfactory sensillum. *Neurocomputing* **38–40**:1011–1017.

Vickers NJ, Baker TC (1994) Reiterative responses to single strands of odor promote sustained upwind flight and odor source location by moths. *Proc. Natl. Acad. Sci. USA* **91**:5756–5760.

Vogt RG (2003) Biochemical diversity of odor detection: OBPs, ODEs and SNMPs. In: *Insect Pheromone Biochemistry and Molecular Biology*, Blomquist GJ, Vogt RG, Eds., Elsevier, Amsterdam, pp. 391–445.

Vosshall LB, Amrein H, Morozov PS, Rzhetsky A, Axel R (1999) A spatial map of olfactory receptor expression in the *Drosophila* antenna. *Cell* **96**:725–736.

Vosshall LB, Wong AM, Axel R (2000) An olfactory sensory map in the fly brain. *Cell* **102:**147–159.

Waldrop B, Christensen TA, Hildebrand JG (1987) GABA-mediated synaptic inhibition of projection neurons in the antennal lobes of the sphinx moth, *Manduca sexta. J. Comp. Physiol. A* **161**:23–32.

Wang JW, Wong AM, Flores J, Vosshall LB, Axel R (2003) Two-photon calcium imaging reveals an odor-evoked map of activity in the fly brain. *Cell* **112**:271–282.

Wehr M, Laurent G (1996) Odour encoding by temporal sequences of firing in oscillating neural assemblies. *Nature* **384**:162–166.

White PR (1991) The electroantennogram response: effects of varying sensillum numbers and recording electrode position in a clubbed antenna. *J. Insect Physiol.* **37**:145–152.

Wilson RI, Turner GC, Laurent G (2004) Transformation of olfactory representations in the *Drosophila* antennal lobe. *Science* **303**:366–370.

Wojtasek H, Hansson BS, Leal WS (1998) Attracted or repelled? A matter of two neurons, one pheromone binding protein, and a chiral center. *Biochem. Biophys. Res. Commun.* **250**:217–222.

Zack C (1979) Sensory Adaptation in the Sex Pheromone Receptor Cell of Saturniid Moth. Ph.D. thesis, Ludwig Maximilians University, Munich.

13 Optical Methods for Analyzing Odor-Evoked Activity in the Insect Brain

C. Giovanni Galizia and Richard S. Vetter

CONTENTS

0-8493-2024-0/05/$0.00+$1.50

13.1 INTRODUCTION

A number of new and exciting methods for studying insect neurophysiology have been developed over the last few decades. Optical imaging techniques measure neuronal activity using light, generally by introducing activity-dependent dyes into the neurons. Ever since the development of dyes that change their optical properties as a function of some metabolic variable (e.g., the concentration of certain ions or molecules or a change in membrane potential), the field of optical imaging has flourished. Many excellent reviews and edited books have been written about the underlying physics and methods (Cohen et al., 1978; Yuste et al., 1999; Frostig, 2002; Toga and Mazziotta, 2002). In this chapter, we will limit the overlap with the existing literature by focusing on those aspects that are peculiar to *in vivo* recordings of insect brains when stimulated with natural stimuli, using the olfactory system as an example. We will cover details about the dissection, the loading of the dye (including genetically engineered reporter proteins), recording, and data analysis:

depending on the species, the dye used, and the particular question asked, many aspects will differ. We assume that most readers have only limited prior knowledge about imaging, so some of the descriptions may seem basic to seasoned imagers. Our hope is that the detailed descriptions will allow more researchers to master these fascinating techniques. The chapter starts with a brief overview of the olfactory system because all examples are taken from studies about insect olfaction.

13.1.1 The Insect Olfactory System

In insects, the axons from olfactory sensory neurons (OSNs) on the antennae (and in some species, the maxillary palps) converge in the *antennal lobe* (AL), the first area in the brain that processes olfactory information. Its functional units are the olfactory *glomeruli*, which are small, spheroidal structures (10 to 50 μm in diameter, with great size differences for different species). Most synapses in the AL reside within the glomeruli (Gascuel and Masson, 1991). OSNs expressing a given olfactory receptor converge to one (or a few) glomeruli, so that each glomerulus is a collection site of a functionally homogeneous family of OSN axons. Local neurons (LNs) synapse among glomeruli, and projection neurons (PNs) project to other areas in the brain, notably the mushroom bodies and the lateral protocerebrum (see also Chapter 1, Chapter 12, and Chapter 14). Finally, a few neurons, generally with widespread branching patterns, innervate the ALs and other brain areas. These neurons provide modulation and feedback to the AL network (Hammer, 1997).

Glomeruli are the functional units of the AL. Their numbers differ widely among insects: fruit flies have about 43, moths between 60 and 70, and honeybees about 160. Glomeruli also differ in their shape and size, but within a species, their spatial arrangement is reproducible, allowing for consistent and unique assignment of nomenclature (Rospars, 1988). Thus, morphological atlases have been created for ALs in several species, including *Drosophila* (Laissue et al., 1999), moths (Berg et al., 2002), and honeybees (Flanagan and Mercer, 1989; Galizia et al., 1999a). Several atlases are digital and can be accessed via the Internet (e.g., http://galizia.ucr.edu/honeybeeALatlas). These atlases can be used to identify a glomerulus in an insect preparation if its shape, neighboring glomeruli, and some landmarks are clearly visible. This approach has made it possible to compare physiological responses to odors across individuals, showing that their odor-response patterns are predictable within a species (Galizia and Menzel, 2001). Such studies form the basis for creating functional atlases of the ALs in order to understand the olfactory code. Wasps and locusts have in the range of 1000 "mini"-glomeruli with different cytoarchitecture and innervation patterns, suggesting that in those species glomeruli may not form the functional units. Creating an atlas for these species will be extremely difficult, at best.

13.1.2 Overview of Imaging Methods

Imaging is a valuable technique for investigating the olfactory code because it allows simultaneous measurement of neural activity in different places, and thus a combinatorial pattern of olfactory glomeruli activated by a given odor (see also Chapter 12) can be imaged with each stimulus. In recent years, various methods have been developed, all with their advantages and respective drawbacks. We are using the word *method* as a broad term, including the preparation, data acquisition, and data analysis. We will give a very brief discussion of some methods and then give the protocols in detail in Section 13.3.

13.1.2.1 Bath-Applied Calcium-Sensitive Dye

This approach gives a signal with limited temporal resolution but good spatial resolution. Signal-to-noise ratio is also good. The output is a compound signal that comes from many different cells. It is dominated by OSN activity in bees, but less so in moths. The preparation is reliable.

13.1.2.2 Bath-Applied Voltage-Sensitive Dye

This approach gives good temporal and spatial resolution, but signal-to-noise ratio of voltage-sensitive dyes is poor. Bath application leads to a compound signal that summates signals from many cells at the same time. The preparation is reliable.

13.1.2.3 Selective Mass Staining of PNs with Calcium-Sensitive Dye

Spatial and temporal resolution are both excellent, probably mostly limited by cellular time constants. For calcium concentration measurements, signal decay can be considerable, but signal-to-noise ratio is excellent. Within each glomerulus, individual PNs cannot be separated. The preparation is stable but technically challenging.

13.1.2.4 Staining of Individual Neurons with Calcium-Sensitive Dyes

Although network properties cannot be measured, survival time of these preparations is long, and the spatial extent of activity within a single neuron can be analyzed. The technical difficulty is equal to that of electrophysiological single-cell recordings.

13.1.2.5 Labeling with Genetically Engineered Probes

With the increase of available probes, which are no longer limited to measuring intracellular calcium, this approach is very promising. The main difficulty is presented by the small size of *Drosophila* (which still is the only animal available for this approach) and the spurious signals from cells that ectopically express the reporter.

13.1.3 Abbreviations

ACT: antenno-cerebralis tract
AL: antennal lobe
AM: acetoxymethylester

ECFP: enhanced cyan fluorescent protein
EYFP: enhanced yellow fluorescent protein
fMRI: functional magnetic resonance imaging
FRET: fluorescence resonance energy transfer
GFP: green fluorescent protein
lACT: lateral antenno-cerebralis tract
LN: local neuron
LP: long-pass optical filter
LPL: lateral protocerebral lobe
mACT: medial antenno-cerebralis tract
MB: mushroom body
mlACT: mediolateral antenno-cerebralis tract
NA: numerical aperture
OSN: olfactory sensory neuron
PN: projection neuron
VSD: voltage-sensitive dye

13.2 THE *IN VIVO* PREPARATION

13.2.1 General Considerations

In general, the techniques involved in this research have undergone many evolutions over the years, with each experimenter developing his or her own technique and procedure. Some aspects are critical for proper results; other aspects can be varied without much overall effect. Decisions about which techniques to follow will depend on the individual performing the dissections and the subsequent success.

13.2.2 Materials

13.2.2.1 Saline Solution

For preparations in which the nerve sheath (perineurium and neural lamella) is kept intact, the saline solution can be adjusted to mimic the insect's hemolymph. The situation becomes more critical if the blood–brain barrier has to be damaged during the preparation procedure. The saline solution that we use in the honeybee is (in mM): 130 NaCl, 6 KCl, 4 $MgCl_2$, 5 $CaCl_2$, 160 sucrose, 25 glucose, 10 HEPES, pH 6.7, 500 mOsmol. The composition is not based on a hemolymph analysis, but rather on a growing medium used for isolated honeybee neurons *in vitro* (Goldberg et al., 1999).

For the moth *Heliothis virescens* we use (in mM): 150 NaCl, 3 KCl, 3 $CaCl_2$, 10 TES buffer, 25 sucrose, pH 6.9. For *Drosophila* we use (in mM): 130 NaCl, 5 KCl, 2 $MgCl_2$, 2 $CaCl_2$, 36 sucrose, 5 HEPES, pH 7.3. A discussion and recipes for *Drosophila* larval saline solution have been published (Macleod et al., 2002).

13.2.2.2 Plastic Stage and Dissecting Equipment

We always fix the animals in a custom-built Plexiglas stage (Figure 13.1a), the design of which will vary for different species, the research goals to be achieved, and the microscope objectives used. The main considerations are the following:

- The animal should be fixed (movement artifacts are to be avoided), but at the same time air should reach the spiracles for sufficient oxygenation.
- The antennae (and maxillary palps, if applicable) need to be freely exposed to the air in order to stimulate them with odors.
- The brain area to be imaged needs to be optically accessible.

This last point may strongly influence the stage design. For example, due to the position of the AL in the head capsule, the angle for viewing moth ALs (Galizia et al., 2000b) is different from that for the bee (Joerges et al., 1997; Galizia et al., 1998; Sachse and Galizia, 2002).

Standard microdissecting equipment is used. Great importance is given to the blades, which should be as sharp as possible (blade breaker and breakable razor blades, in some instances sapphire blades), and forceps must be extremely fine as well (DuMont's Biologie 5 or finer)[9,34].

13.2.2.3 Stabilizing the Head: Waxes, Glues, etc.

A variety of materials can be used to fix insect cuticle to the stage. Waxes work very well but cause damage when too hot. Most glues do not attach well to the waxy cuticle. The best material must be tested for each preparation.

13.2.2.3.1 Waxes

The best method for stabilizing the head is to use waxes. Beeswax is mechanically suitable but has a strong odor. There are a variety of hard dental waxes that do not smell strong and have low melting points (around 40 to 45°C). This is sufficiently low to not cause any damage when applied to the head capsule of bees, but it is sufficiently high to completely abolish odor responses if applied to the antennae. These waxes offer excellent adhesion to both the cuticle and the plastic stage, and they are not brittle. The one we use is yellow hard wax (Deiberit 502) (Galizia et al., 1997). For melting, small soldering tips with a variable voltage supply can be used.

An alternative is n-eicosane[27]. The melting point at 37°C is so low that it can also be used to fix antennae without causing damage. However, the material is fairly brittle and of limited mechanical resistance. Some dental waxes are very malleable and can be used after warming and working them between the fingers. They are excellent for forming molds and fixing the animals when applied mechanically, but they do not adhere to the cuticle. Therefore, they do not form a seal against saline percolation. The wax we use for bees is Kerr Utility Wax Rods, #09731, available from dentists' supply stores. There are softer and harder variants.

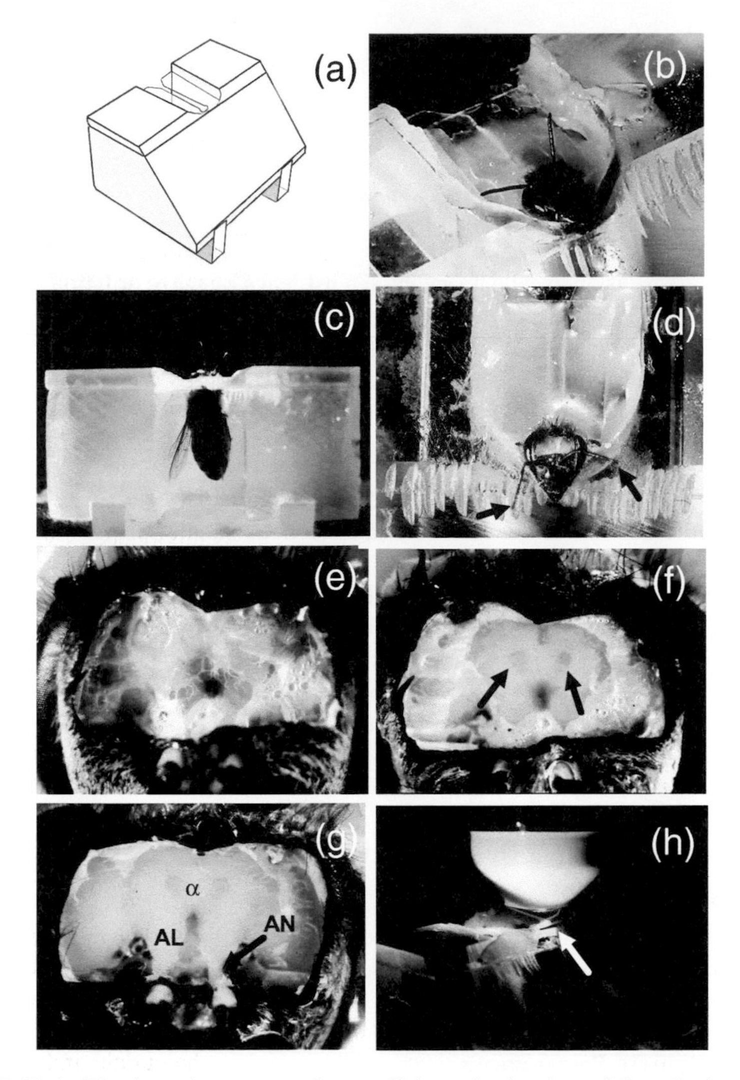

FIGURE 13.1 The honeybee preparation. **a**: Schematic drawing of the plastic stage for restraining the honeybee. Note the angles at the back of the slot, which allow for easier manipulation of the bee's head into the slot. **b**: Honeybee secured in the stage. Note the plastic square behind the bee's head, secured with wax, which holds the bee firmly in position and prevents movement. **c**: Rear view of the bee in the stage. **d**: Top view of the restrained bee after incisions are made and the window in the head is removed, exposing the bee's brain. The antennae are not yet restrained (arrows). **e**: Exposed brain of the bee covered with tracheae. **f**: Exposed brain of the bee with tracheae partially removed. The α-lobes appear as paired medial spots (arrows). **g**: Bee brain with most of the tracheae removed. α: right α-lobe; AL: antennal lobe; AN: antennal nerve. **h**: View of the preparation under the water-immersion objective of the microscope. Note the saline bubble under the objective and the antennae exposed to air (arrow).

13.2.2.3.2 Silicone

A fast-curing, two-component silicone such as Kwik-Sil[34] is excellent for fixing antennae in place. The material is easily applied. Even though it does not bond to the cuticle, it creates a tight seal. However, it will easily detach. In fact, we have used this property to cover antennae temporarily with silicone in order to prevent stimulation, and then freed the antenna again.

13.2.2.3.3 Petroleum Jelly

Stopcock grease is an excellent sealant against leaks of saline solution but gives no mechanical stability. Furthermore, most petroleum jelly products form a greasy film that travels on the cuticle. We found that 30 minutes after applying petroleum jelly to the neck of a fruit fly, the antennae were already covered with a thin greasy film, changing (if not totally occluding) olfactory responses.

13.2.2.3.4 Glue and Cement

Superglue is based on cyanoacrylate and cures rapidly. It has a strong odor, which vanishes after a short time. However, adhesion to cuticle is of limited strength. Furthermore, cyanoacrylate binds water when curing, which in small animals such as *Drosophila* can easily dessicate the entire head capsule. Even a minute drop close to *Drosophila*'s antennae prevented any odor responses in our preps.

Multiple-component glues, such as two- or three-component dental cements, are laborious, but work very well. They bond well to the cuticle and offer excellent stability. We have used these products to fix foils onto *Drosophila* heads (see Section 13.2.3.3).

We have had less success with single-component dental cements that are cured with UV light, because even after curing they are covered with a thin fluid film, and this affected chemosensory responses if it was too close to the antennae.

13.2.2.3.5 Rosin

Excellent results for gluing cuticle can be obtained with rosin. However, in order to be applied, rosin must be dissolved in solvents that invariably have a strong odor. Furthermore, time to desiccation can be fairly long.

13.2.3 Animal Preparations

The preparation method one chooses depends on the species used. The main decision is whether to use an isolated brain preparation or an *in vivo* approach. The worst detriment of any imaging measurement is movement of the preparation. Therefore, the isolated brain preparation has a strong appeal, due to the possibility of completely removing any muscles that could cause movement artifacts. Several optical imaging studies in the insect olfactory system using an isolated brain preparation have been published (Galizia et al., 1997; Ng et al., 2002; Wang et al., 2003). However, an isolated brain preparation invariably causes more damage to the brain than an *in vivo* preparation and may cut feedback connections, with as-yet unknown effects for the measured signals. Furthermore, in our hands, the survival time and the quality of the signals are higher with the *in vivo* approach. We therefore focus on an *in vivo* preparation here.

In the following three sections, we review in detail the methods that we use in the preparation of three insects: bees, moths, and flies. Most readers of this chapter are likely to be looking for instructions that they can apply to their research animal of choice. Therefore, each section includes both basic information that can be used across species and specialized information that is particularly useful for certain insects.

13.2.3.1 Bees

13.2.3.1.1 Collecting and Anesthetizing

Bees are collected at a beehive and transferred into a plastic vial with a perforated lid. Bees do not survive long in small vials and need to be provided with honey or sugar water for survival. We anesthetize them by placing them on ice for a few minutes to cool them down sufficiently for handling. Once fixed in the plastic stand, bees can survive for several days without problems, if fed.

Every researcher will know best how to treat his or her species of interest. Remember that some anesthetics affect the insects' physiology. For example, CO_2 anesthesia in fruit flies changes chemosensory responses and should be avoided.

13.2.3.1.2 Placing Bees in the Plastic Stage

The insect should be fixed stably at the neck, so that movements of legs, wings, and abdomen do not transmit to the head. The stage should be custom-made for the objective used, in order to have optical access to the brain, and it should have a base that allows it to be placed at and removed from the microscope quickly. It should allow saline supply to the brain, avoid any leakage of the saline solution, and provide for the antennae to be kept dry and accessible to odor stimulation.

The plastic stage we use for securing honeybees is shown in Figure 13.1a. It is a block of solid plastic in which the bee can be positioned by sliding its neck through a slit (Figure 13.1b,c). A piece of plastic cover slip (7 × 7 mm) is slid underneath the back of the head to immobilize the bee (Figure 13.1b,d). This step helps in standardizing the angle at which bees are fixed. Hot wax may be used for fixation but should not exceed 40°C (see Section 13.2.2.3.1). This temperature does not cause harm because honeybees can heat up above 40°C during flight (Heinrich, 1974). Alternatively, the animal maybe fixed with soft wax. It is useful to align the thorax and abdomen so that the bee is not contorted. Additionally, a small piece of foam may be inserted into the cavity to immobilize the thorax and abdomen and to prevent the researcher from being stung.

13.2.3.1.3 Fixing the Antennae

This is among the more delicate steps, because mistreatment of the antennae invariably affects the results. The main purpose is to protect the antennae, restrain their movement, make them accessible to odor stimulation, and keep them away from saline solution during the experiment. Once the head is secure, minuten pins are used to secure the antennae into the desired position, with the bee projecting parallel and forward. The pins are anchored into the wax on the side of the head. The antennae should not be forcibly manipulated or physically stretched. The flagella of the antennae should be exposed to the air and should never be touching other surfaces because mechanoreceptors and contact chemoreceptors will cause sensory input that might interfere with the investigated brain activity. A two-component silicone (see Section 13.2.2.3.2) is used to secure the antenna in place and provide a fluid-proof seal around the bee head. The two components are mixed for about 10 to 15 sec, and then a glob of silicone is taken from one side of the stage to the other side, laying down a strand of silicone across the scape of the antenna. The remainder of the silicone should be spread around the bee's head, providing a fluid-tight seal.

An alternative approach uses soft dental wax instead of silicone (see Section 13.2.2.3.1). A very thin strip of dental wax is placed across the mandibles, and two small grooves are made in the wax near the antennae. The antennae are placed in these grooves, and another small strip of wax is placed on top to hold the antennae in place if necessary.

13.2.3.1.4 Making the Incisions

When cutting a hole into the head cuticle to expose the brain, there are two important things to consider. First, trivially, do not damage the antennal nerves or the brain. Second, make the hole as large as possible above the structure to be imaged, so that no shadow is cast during imaging. Remember that objectives with a high numerical aperture collect light from fairly shallow angles (see Section 13.4.2).

In the bee, the first cut should be as close to the toruli (antennal joint) as possible. The second and third cuts are made vertically just medial to the eyes. The last cut is made horizontally from eye to eye such that it occurs anterior to the ocelli, and the cuticular flap is then removed (Figure 13.1d–g). A drop of saline solution should be placed on the bee's brain immediately if dry.

13.2.3.1.5 Exposing the Brain

There are three goals:

1. Expose the area of interest to make it optically accessible.
2. Remove any structures that will disturb measurements, in particular, those structures that cause movement artifacts. These include muscles and trachea that connect the ALs to muscles.
3. Remove structures that are strongly fluorescent. Bath-applied dyes, for example, strongly stain muscles and glands, and these create fluorescent light that affects the measurements.

It is important to keep the brain soaked in saline solution. The yellowish hypopharyngeal glands are removed first, followed by the white tracheae that surround the brain, the ALs, and the antennal nerves (Figure 13.1e). Removal of tracheae is a delicate step and is probably among the most critical steps because injury can render a preparation useless. Never pull tracheae perpendicular to the direction of the antennal nerves, to prevent stretching them. The brain cavity should be rinsed frequently with saline solution to remove cellular enzymes (which might have been released in removal of the tissue) because they can "digest" the brain. The process from cooling the bee to final dissection should take around 10 minutes. The dye is then bath applied or injected (see Section 13.3 for details of staining procedures). The bee is stored in a cool, dark, insulated box with high humidity, and care is taken to make sure that the preparation does not leak during the staining process (see Section 13.3.1.1).

13.2.3.1.6 How to Control Movement

There are several sources of movement in an *in vivo* insect preparation, and several techniques to reduce them. Movements that directly displace the brain can be avoided only by severing the muscles that produce them and the ligaments and trachea that transmit that movement to the imaged area. This can be tedious because some muscles are either very close to the brain (making it difficult to remove them without causing damage to the brain), or they may be hidden in areas that are not directly accessible.

We have also tried to reduce movements pharmacologically, by applying the wasp venom *philanthotoxin,* which blocks the glutamate receptors of the insect neuromuscular junction. Although this approach was highly effective, we discontinued it after discovering that *philanthotoxin* also blocks GABA-induced currents in cultivated Kenyon cells of the mushroom bodies (A. Wersing and B. Grünewald, personal communication), with the consequent danger that using such a pharmacological approach might influence the neural activity to be measured.

Another approach that works very well is to embed the brain in low-melting agarose. We use Sigma A2576[27], which melts above 60°C and gels below 17°C, at a concentration of 1 to 1.5%. After the agarose has been dissolved and melted in saline solution, it can be kept liquid at room temperature or slightly above it. After covering the brain with agarose, the insect has to be cooled down to 10 to 13°C for several minutes and can then be brought back to room temperature. We found that this procedure significantly reduces movement artifacts. However, with very small insects such as *Drosophila,* it is difficult to get the agar to the back of the brain, where it is most needed. The agar does not significantly reduce nutrient and oxygen flow to the brain; indeed, pharmacological experiments with TTX (tetrodotoxin) showed no difference in the time needed for the toxin to reach the honeybee AL between bees embedded in agar and the standard preparation.

Many movements do not originate in the head capsule but are caused by respiratory movements in the abdomen that pump hemolymph into the head capsule. Applying a thin thread of melted wax along the abdominal segments will block or at least reduce these movements. Alternatively, a foam block may be squeezed against the bee's body to reduce movement, or the abdomen can be cut off shortly before measurement begins. In bees, this cutting should be done away from the measurement setup, because the severed abdomen will release alarm pheromone into the air.

Other strong movement sources are the mouthparts and the esophagus. Removing these structures will invariably remove gustatory input to the brain. Although we found no differences in odor responses between animals with and without the mouthparts, this is clearly a step which moves the preparation in the direction of an *in vitro,* rather than an *in vivo,* situation. In bees, a small square is cut in the bee's head between the antennae and the mouthparts. Fine forceps then grab and pull the U-shaped sclerotized structure, which also removes the esophagus. Alternatively, the esophagus and the mouthparts can be directly removed from between the ALs with fine forceps.

13.2.3.2 Moths

A similar preparation has been used for several moth species: *Heliothis virescens* (Galizia et al., 2000b), *Spodoptera littoralis* (Carlsson et al., 2002; Carlsson and Hansson, 2003; Meijerink et al., 2003), and *Manduca sexta* (Hansson et al., 2003). Each species has its own geometrical peculiarities, which result in different solutions: fixing the antennae, fixing the animal, getting optical access to the ALs. In the preceding text we have tried to justify every step taken in the bee preparation. With the appropriate modifications, therefore, these techniques can easily be transferred. In moths, the ALs face sidewise rather than to the front, and the anterior angle is best for viewing, after the mouthparts have been removed. Another observation is that using melted wax always resulted in reduced signals, suggesting that the moth brain may be more sensitive to heat than the bee brain. Differences in the staining success of bath-applied AM dyes are reported Section 13.3.1.3.

13.2.3.3 Flies

The fruitfly *Drosophila melanogaster* is an ideal model system for investigating the molecular mechanisms of olfactory coding (see also Chapter 10). In recent years, the availability of genetically engineered activity-sensitive proteins has made the fly even more attractive for physiological studies. The preparations used by different groups have varied considerably. The main decisions to be made are, first, how much of the fly should remain intact, and second, at what angle should the brain be viewed. When studying odor-evoked activity, the antennae have to remain dry and accessible for stimulation, and the antennal nerve intact. Because of the small size of the fly antennae and their position right on the head capsule, this creates the main difficulty in the preparation. When one is looking at the ALs directly, the antennae themselves must be pulled slightly forward. This corresponds to a natural position of the fly when it is extending the mouthparts and is achieved by building up internal pressure. However, when preparing the fly, pressure cannot be created within the head capsule because it is cut open. Therefore, the antennae must be pulled forward with a thin string. Alternative approaches include looking at the ALs through the entire brain from the back of the head. This is possible by exploiting the great penetration power of a two-photon microscope, but it is not feasible with traditional epifluorescent excitation. When the mushroom bodies are the focus of interest, rather than the ALs, the situation becomes geometrically less challenging.

Optical imaging studies of odor-evoked activity in the *Drosophila* AL have been published for an *in vivo* preparation (Fiala et al., 2002) and for isolated brain preparations (Ng et al., 2002; Wang et al., 2003). A preparation published for electrophysiological studies is also suitable for optical imaging (Wilson et al., 2004).

Here is the procedure used in our lab (Fiala et al., 2002; Fiala and Spall, 2003). Flies are immobilized on ice and then fixed to a Plexiglas stage. When recording from the ALs, flies are fixed at their neck, and the antennae are pulled forward with a fine metal wire. The head is covered with a polyethylene film, which is sealed against the cuticle with silicone. The purpose of the film is to be able to submerge the brain in saline solution without wetting the antennae. Then, a hole is cut through film and cuticle, allowing optical access to the ALs. As compared with the bee or the moth preparation described previously, cutting the fly cuticle is more difficult, mainly because of its soft and elastic consistency. Using a very sharp dissection knife is helpful (e.g., sapphire blades). The brain is immediately bathed with saline solution, tracheal air sacs are removed from the head capsule, and the preparation is placed under the microscope. For mushroom body measurements, it is not necessary to pull the antennae forward.

13.3 STAINING PROTOCOLS

Calcium-sensitive dyes, such as Fura dyes or Calcium Green[14], are derived from calcium chelators and are large polar molecules. They are not membrane permeable. Consequently, when they are used to measure intracellular calcium concentrations, the very high calcium levels that are present in the intercellular space do not affect them. However, this property makes loading the cells more difficult. Furthermore, great care should be taken not to have any residual dye in the extracellular space. Four approaches are presented here: bath-applied dyes (Sections 13.3.1 through 13.3.3, and 13.3.5), backfilling with dextrans (Sections 13.3.4 and 13.3.5), loading single cells (Section 13.3.6), and using genetically engineered probest (Section 13.7).

13.3.1 Bath-Applied Dyes — AM Dyes

Calcium-sensitive dyes bound to acetoxymethylester (AM) are used for bath-applying the dye to the tissue. In the AM-bound form, the dye is membrane permeable and does not interact with calcium. Intracellular esterases split the AM group and release the calcium-sensitive dye, which is now membrane impermeable. This approach is a standard technique for cultured cells. In the *in vivo* situation, two peculiarities appear that must be considered. First, before reaching the target cells, other cells (such as the perineurium and neural lamella) must be passed. Second, the resulting calcium-concentration measurements will come from several cells, which will result in a more complex interpretation of the data and a reduced signal-to-noise ratio. This latter point is shared by all methods that measure physiological responses using bath-applied dyes or intrinsic signals.

13.3.1.1 Staining Technique

For the insect AL, we have used Calcium Green AM with much success. To make the dye, 50 μg Calcium Green AM dye (Molecular Probes, Inc.[14]) are dissolved in 50 μL Pluronic F-127 (20% in dimethylsulfoxide, DMSO, also from Molecular Probes, vortex for at least one minute) and then diluted in 800 to 950 μL saline solution (again, vortex for at least one minute, and then sonicate for 5 minutes). The brain is stained by bath-applying 50 μL of dye to the open brain cavity. This step will further dilute the dye. If no staining occurs, dilute the dye further, because this will improve the staining, even if this may appear counterintuitive. The animal is then placed in a moist and darkened place for 1 hour at a temperature of about 13°C. We had no success with staining times below 45 minutes but no problems with 2 hours The low temperature probably inhibits intracellular esterase activity, therefore, allowing the dye to cross several cell layers and reach deeply into the brain. As with all fluorescent dyes, excessive light exposure should be avoided. Use a yellow filter for illumination under the dissection microscope.

After staining, the bees are placed at room temperature in a dark box. This will allow the intracellular esterases to become active and, thus, to cleave the AM from the dye. After about 15 minutes, the head capsule must be thoroughly rinsed with fresh saline solution several times. The bee is now ready for imaging.

13.3.1.2 Other Brain Areas and Other Species

Calcium Green dye has also been used for other areas in the bee brain, notably for the mushroom bodies and the α-lobe (Faber and Menzel, 2001). The same technique has also been used in other insects, such as the ant *Camponotus rufipes* (Galizia et al., 1999b) and various moth species (Galizia et al., 2000b; Carlsson et al., 2002; Carlsson and Hansson, 2003; Hansson et al., 2003; Meijerink et al., 2003). Interestingly, the staining as described previously cannot always be applied to other species without modification. In our work with the ant, penetration of the dye into the brain proved very difficult and only succeeded after we mechanically removed or damaged the sheath. With *Drosophila melanogaster*, we never achieved successful Calcium Green AM staining. In several moth species, however, the technique described in the preceding protocol works well.

13.3.1.3 Other Dyes

We have tested a variety of dyes in honeybees, with varying degrees of success. The following is a list of the dyes that gave good odor-evoked signals when applied in their AM form: Calcium Green AM (1, 2, and 5N), Oregon Green AM, Fluo-3 AM, and Fura 2 AM (the last with limited success). With the following dyes we were not successful either in loading the brain or in recording olfactory signals: Calcium Crimson 1 AM and rhod 2 (A. Delorenzi, unpublished observations). As a general rule, we found that the optimal excitation and emission wavelengths were shifted toward shorter wavelengths with respect to an *in vitro* situation. For example, the dye-manufacturers handbook gives 506 nm/531 nm as excitation/emission wavelengths for Calcium Green 1, whereas the best values in our *in vivo* application are 480 nm/LP505. We did not further investigate the reasons for the differences between dyes and for the spectral shift. Such a study has been performed in cell culture (Thomas et al., 2000).

Recently, the NO-sensitive AM dye DAF-FM DA has been successfully applied to this antennal lobe preparation to measure odor-evoked NO production (Collmann et al., 2004). As the development of new dyes progresses it will become possible to investigate even more cellular signalling mechanisms.

13.3.1.4 Which Cells Are Stained?

An important question when interpreting the results from bath-applied AM dyes concerns the origin of the signals. At least four mechanisms contribute to this question:

1. The bath-applied dye does not reach all cells equally.
2. All cells do not take up the dye equally.
3. Cells differ in their response magnitudes and other relevant properties, such as resting calcium levels, total cytosol volume, and intracellular calcium distribution, which all contribute to the signal.
4. The geometrical position of cells in the light-path differ, with some cells shielding others, thus biasing their contribution to the measurements.

The first and second points lead to unequal loading of different cells. For example, glial cells surround olfactory glomeruli and are therefore more exposed to the dye than neurons. They may, in fact, take up more dye than neurons. Indeed, confocal analysis of Calcium Green AM loading in the AL of the moth *Manduca sexta* has suggested such a bias toward glial cells (Christian Lohr, personal communication). Furthermore, cells have a variety of selective transport mechanisms that expel dyes from them, and cells differ in their transporter complement. For example, *Xenopus* OSNs possess a variety of transport mechanisms that efficiently expel several calcium-sensitive dyes (Manzini and Schild, 2003). It should be noted that the fluorescence that can be seen under resting conditions is not a function of the dye loading alone but also of the resting calcium concentration. When equally loaded with dye, cells with higher resting calcium levels will be more fluorescent than those with lower concentrations.

This leads also to the third point: changes in calcium fluorescence depend on the magnitude of calcium change and the dose-response curve of the dye. Several dyes are available with different affinities (e.g., Calcium Green 1, 2, and 5A, with dissociation constant $K_d[Ca^{2+}]$ of 0.19 μM, 0.55 μM, and 14 μM, respectively), in order to choose the best response range. Dyes with different K_d values can also be combined in order to extend the dynamic range. However, this also leads to a shallower dose-response curve, with the consequence of reduced changes in fluorescence (i.e., a decrease in signal-to-noise ratio). It is an important trade-off: either cover a wide concentration range with limited signal-to-noise or cover a small range with high signal-to-noise, with the consequence of being blind above and below the optimal range (see Section 13.5.3.6, and Figure 13.8).

The *fourth* aspect leads to cells that are deeper in the tissue, contributing less to the signal when they are shielded by more peripheral cells. For example, OSN terminals strongly innervate the outer shell of honeybee olfactory glomeruli and possibly shield the core, that is mostly innervated by LNs and PNs.

With all these uncertainties, how can we interpret the results that we get? As a first approximation, we cannot attribute the signal to any given cell population. Therefore, we are in a situation akin to the measurement of intrinsic signals (i.e., changes in optical properties of the tissue without using a dye) or fMRI data, both of which reflect metabolic rather than neural activity but are interpreted as neural activity because the two magnitudes are strongly correlated. A strong increase in intracellular calcium concentration in a glomerulus means that stimulating the animal with that odor activates at least some, if not all, neural processes of that glomerulus. However, we can use additional information: in bees, for example, we see neither spontaneous activities nor off-responses (i.e., responses at the end of the stimulus) in AM-stained preparations. Insect LNs and PNs display a strong spontaneous background activity and off-responses to odors (Hansson and Christensen, 1999; Abel et al., 2001; Müller et al., 2002), which is also visible in calcium imaging experiments (Sachse and Galizia, 2002, 2003; Galizia and Kimmerle, 2004). We can, therefore, exclude a relevant contribution of these two cell types to the signals, leaving OSNs and glial cells as possible candidates. If glial cells contribute to the signal, they are likely to mirror activity of neurons. In honeybees, they would reflect the OSN signal because otherwise we would expect the spontaneous activity and off-responses of LNs and PNs to be apparent. The close apposition of the glial cells with the numerically dominant OSNs would favor such an interpretation. Therefore, response magnitude can be taken to be proportional to OSN input.

This same argument does not apply to the moth, *Heliothis virescens*. In this species, we do observe spontaneous activity and off-responses, suggesting that for unknown reasons, LNs or PNs or both must also contribute to the signals in this species (Galizia et al., 2000b).

13.3.2 Afterstaining for Glomerular Mapping

Because the glomerular structure is not visible during calcium imaging when bath-applied Calcium Green AM is used, glomeruli must be counterstained after functional imaging for glomerular identification. The neural sheath is digested with Protease Type XIV(27) for 5 minutes, and then the AL is stained with the lipophilic dye RH795(14). Other proteases may be used or are even necessary for different species. The purpose of the protease is to increase penetration of the dye. After 30 to 45 minutes, the glomerular structure becomes visible in epifluorescent light. Ideally, the preparation is not moved during this process, so that the AL can be imaged under the microscope in the same position as when the odor-evoked activity was measured. This approach also allows taking images of the RH795 staining every 5 minutes, such that the moment of strongest contrast is not missed. At this point, the brain is washed with saline solution, which increases the contrast by removing dye molecules still floating in the fluid. Fluorescent photographs are then taken at a series of different focal planes.

The images are sharpened with image-processing software (e.g., Adobe Photoshop) in order to reconstruct the glomerular borderlines for each animal measured (Figure 13.3a). Afterstaining images must be registered with respect to the activity images using prominent landmarks, such as trachea or borderlines of brain structures. This is necessary even if the preparation was not moved because the protease may have digested some ligaments, in which case the AL may have moved. Other groups have used other dyes for counterstaining ALs, such as Lucifer Yellow (Okada and Kanzaki, 2001).

13.3.3 Bath-Applied Voltage-Sensitive Dyes

Voltage-sensitive dyes can be easily bath applied. For bees and moths, we have used RH795[(14)]. The dye is first dissolved in 70% ethanol and then diluted adding saline solution to reach a final concentration of 20 μM. A drop of dye is added to the head capsule and incubated for 5 to 10 minutes. Excessive dye is quickly rinsed off, and the bee is ready for imaging.

Odor-evoked olfactory activity has been measured with voltage-sensitive dyes in bees and bumblebees, (Galizia et al., 2000a; Okada and Kanzaki, 2001). In species with stronger sheath, it may be necessary to remove the sheath mechanically or to digest it with a protease.

13.3.4 Staining Projection Neurons with Dextrans

Staining populations of neurons by backfilling them has become a popular method in a variety of systems (Gelperin and Flores, 1997; Delaney et al., 2001; Macleod et al., 2002, 2003; Sachse and Galizia, 2002, 2003). By placing a high concentration of dye into the axonal tract, the dye is forced into the cells and retrogradely transported to the cell bodies and dendrites, where it can be measured. Injection site and measurement site must be sufficiently distant from each other because the fluorescence from the dye bolus at the injection site will create a strong light halo if illuminated, making any imaging impossible. The mechanisms for forcing axons to take up the dye may differ, including mechanical damage or osmotic pressure. We have used mechanical damage with success: a glass microelectrode is coated with dye crystals and stabbed into the axonal tract, damaging the axons (Figure 13.2b). We have never observed odor-evoked activity in the AL in preparations where the dye has been applied but the PNs were not stained, showing that unspecific loading does not occur.

Dye is injected into the brain with glass micropipettes that have been pulled into fine-pointed electrodes in a pipette puller. Glass size and type are not critical. The drawn electrode will have a flexible tip, which needs to be truncated slightly (thin enough to effect a clean injection, not so flexible as to bend during loading with dye or during injection). The dye solution is made by taking two to three crystals of Fura dextran 10,000 MW[(14)] dissolved in a 1 μl drop of 2% bovine serum albumin (BSA) solution on a glass slide. BSA is added to create a pasty consistency, which prevents the electrostatic liberation of small dye crystals from the electrode prior to brain contact. The dissolved dye should be slightly yellow in color. The dextran is used to accelerate intracellular transport along the axons. Generally, smaller dextrans travel faster. Place 0.5 μl of filtered water on the dye and load electrodes at the water–dye interface by rolling their tips into the gelatinous dye. The best loaded electrode will have an elongated globule of dye near its tip. Loaded electrodes can be stored in the freezer. Dye should be protected from strong light at all times. Using a yellow filter for the light source under the dissecting microscope is advisable.

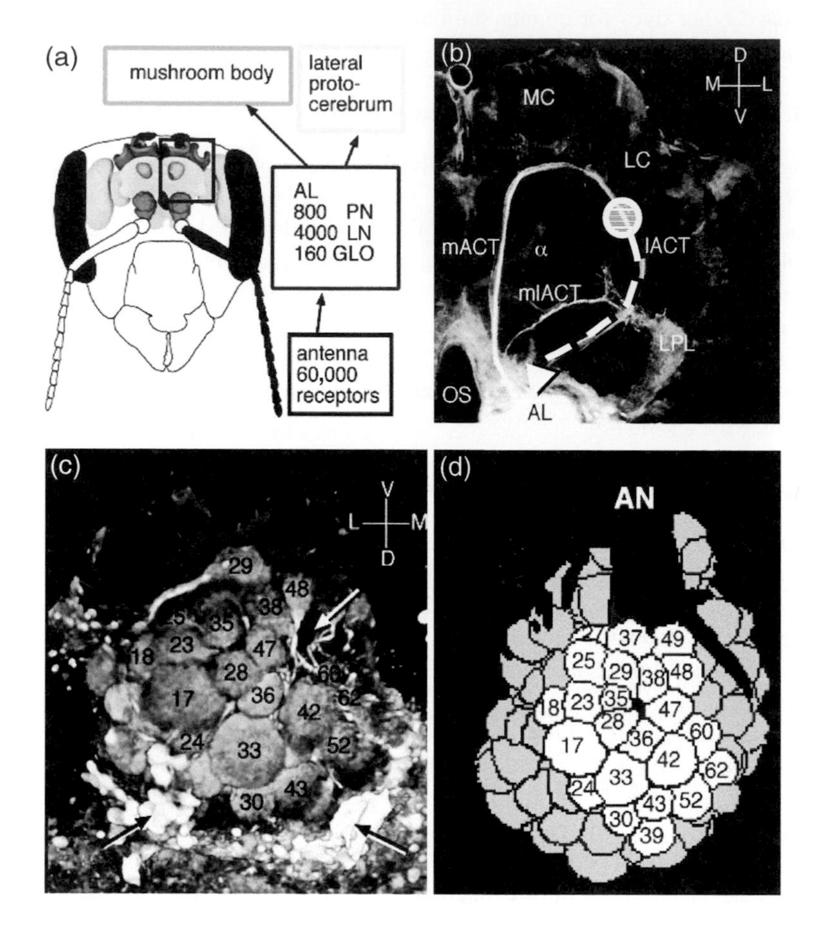

FIGURE 13.2 Staining of honeybee PNs. **a**: Schematic view of the honeybee olfactory system. All 60,000 OSNs are on the antenna. Their axons project to the antennal lobe (AL), which contains 160 glomeruli (GLO) and about 4,000 local interneurons (LN). Eight hundred projection neurons (PN) leave the AL toward the mushroom body and the lateral protocerebrum. **b**: View of a confocal reconstruction of the honeybee brain. The imaged area is outlined in **a**. PNs have been labeled in the AL with the fluorescent dye rhodamine dextran, and their axons are visible as tracts (mACT, mlACT, lACT) that invade the mushroom body calyces (MC, LC) and the lateral protocerebral lobe (LPL). By injecting dye into the axonal tract (circle), PN dendrites in the AL can be filled in a retrograde fashion (curved arrow). OS: oesophagus **c**: Confocal extended-focus view of the AL with PNs backfilled with rhodamine dextran. Note the stained PN somata (arrows). Glomeruli are clearly visible and many have been numbered: all of these glomeruli belong to a specific subset called T1-glomeruli. Only axons in the lACT were labeled, as seen by the gap between glomeruli 48 and 60 (white arrow), where a glomerulus innervated by the mACT is not visible because it is not stained (glomerulus T3-45). **d**: Schematic view of the honeybee AL, with some of the glomeruli innervated by T1 identified. (Adapted from Sachse S, Galizia CG, *J. Neurophysiol.* **87**:1106–1117, 2002. With permission.)

The dye is injected by jabbing the loaded electrode into the brain with a short quick movement, then keeping the electrode in place until the dye has dissolved, and finally removing it slowly. The aim is to hit (and damage) the ACT tract, which carries the PN axons from the ALs to the mushroom bodies (Figure 13.2b). The injection spot is extremely critical and can be mastered only with lots of experience. Initially, a neophyte may hit the tract only once in about 50 bees, if that often. The head capsule must be rinsed thoroughly after an injection in order to remove dye traces that may have spilled. The bee is then stored in a cool, dark place for at least four hours as the dye travels to the AL.

13.3.5 Double Staining with AM Dyes and Dextrans

Selective staining of PNs and bath-applied dye can be combined. In the honeybee, where the bath application gives a good estimate of the OSN input to the AL, this can be used to observe the input–output function of olfactory glomeruli within the same animal (Sachse and Galizia, 2003). The combination is straightforward because the two approaches do not directly interfere with each other (Figure 13.3b,c).

The emission spectrum of the two dyes overlaps strongly. Fura has a broad emission spectrum and can be reliably measured with a long-pass filter above 500 nm. Therefore, both Fura and Calcium Green can be measured with the same emission filters. However, Calcium Green is not excited at 340 nm and 380 nm (the wavelengths for Fura), and Fura is not excited at 475 nm (the wavelength for Calcium Green), so switching different excitation wavelengths allows the measurement of the two dyes within the same preparation simultaneously, with interleaved frames. Technically, the most important aspect to consider is the choice of the dichroic mirror: common dichroics at 500 nm reflect 475 nm, but not UV light (340 nm and 380 nm), and therefore are not suitable for triple-wavelength measurements. Special dichroics that also reflect in the UV spectrum have been developed for applications involving UV light for uncaging, where flashes of UV light are used to activate physiologically active substances. These dichroics are perfectly suited for combined measurements as described here. The main drawback of this double-labeling is that apparently Calcium Green is not very stable under UV light: although it can be measured for at least 1.5 hours in a stand-alone measurement, generally there are no signals left after 30 minutes when measurements are interleaved with UV exposure for Fura measurements. So the advantage of double-labeling within a preparation comes at a cost of shorter measurement time.

13.3.6 Single-Cell Staining

Single-cell staining is a standard technique in electrophysiological studies, where cells that are recorded intracellularly are then loaded with a dye by applying an iontophoretic current (see Chapter 12, this volume). This dye can be a calcium-sensitive dye, such as Fura or Calcium Green, which are both negatively charged ions. Combining electrophysiology with single-cell calcium imaging has been developed and successfully adopted in studies of the optical system in blowflies (Egelhaaf and Borst, 1995; Haag and Borst, 2000; Kurtz et al., 2000; see also Chapter 7).

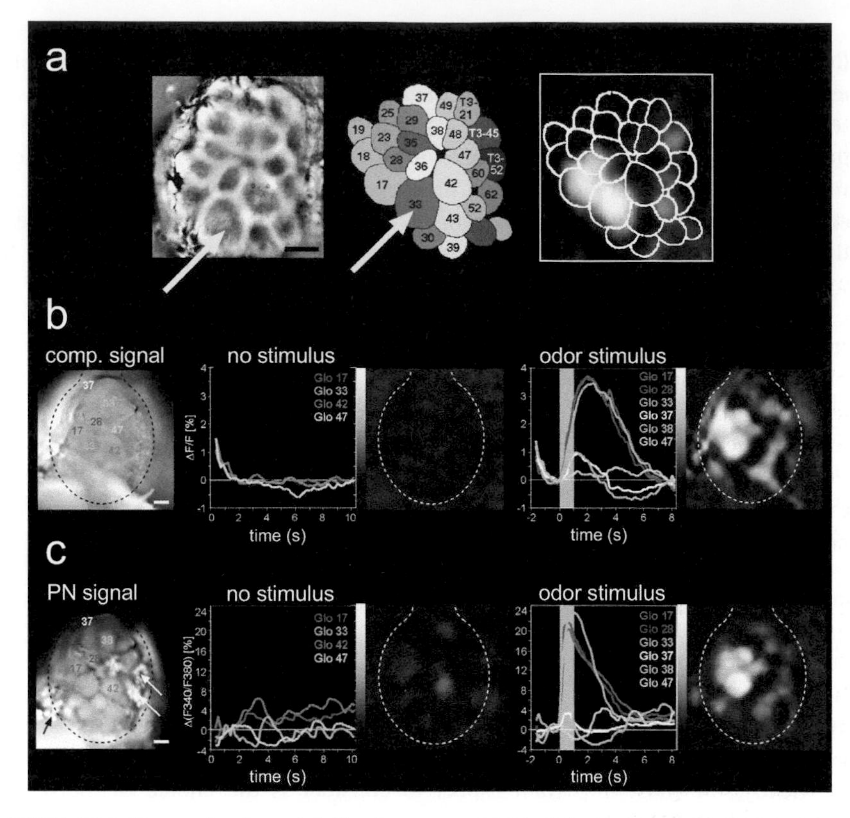

FIGURE 13.3 (See color insert following page 202.) Odor responses in the honeybee. **a**: Method for identifying AL glomeruli. Left to right: after calcium imaging, the AL was stained with the membrane-selective dye RH795, making glomerular borders clearly visible. From these images, a map of the glomeruli was created, and glomerular identity was determined by comparison with the standard morphological atlas, on the basis of glomerular size, shape, and relative position. Finally, the glomerular lattice was superimposed onto odor-evoked activity images (here showing the response to 1-nonanol). **b,c**: Simultaneous measurements of two calcium-sensitive dyes in one animal stained with bath application of Calcium Green-2 AM (**b**), and with selective staining of PNs with Fura-Dextran (**c**). Scale bar = 50 μm. For clarity, fewer traces are shown in the "no stimulus" panels. **b,c**: Left: raw video images of the AL in one animal obtained with filter settings for Calcium Green (**b**) or Fura (**c**). The approximate AL border is marked with a dotted line; antennal nerve is at the top. A subset of glomeruli was identified and numbered using the digital AL atlas (http://galizia.ucr.edu/honeybeeALatlas). Middle column: false-color spatial activity pattern (right) and time course (left) of calcium signal without odor stimulation. Right column: spatial and temporal patterns measured in different glomeruli in response to an odor stimulus, a mixture of hexanol, octanol, and nonanol. **b**: Measurement of the compound (comp.) signal obtained with bath application of dye. **c**: Measurement of the PN signal obtained by specific staining. Arrows indicate the somata of the lateral ACT neurons. The glomerular structure is visible, which allows identification of the glomeruli. In contrast to the compound signal in **b**, the PNs are spontaneously active. Note that PNs can also be inhibited during olfactory stimulation (e.g., glomerulus 47). (Images adapted from the following: **a**: Galizia, C.G. et al., *Nat. Neurosci.* **2**:473–478, 1999; **b,c**: Sachse, S., Galizia, C.G., *Eur. J. Neurosci.* **18**:2119–2132, 2003. With permission.)

In our studies, we injected Fura-2 as a calcium indicator and coinjected a fluorescent dye that has a different fluorescence spectrum, is insensitive to calcium, and is fixable (Alexa 568)[(14)], in order to reconstruct and, where possible, identify the cells in confocal sections that were recorded after fixing and embedding the brain. In our experiments, we first did the electrophysiological analysis, injected the cells, and then transferred the preparation to the optical imaging setup. There is no reason why the first two steps should not be combined into one setup, thus allowing the fast measurement of electrophysiological responses and the spatial extent of calcium activity simultaneously. The main limitation may be posed by the stability of the electrophysiological experiment: if the time for which a cell can be held is too short, the dye may not yet have diffused in the entire cell during the electrophysiological measurement, thus limiting the amount of information gathered simultaneously about the spatial extent of the activity and the electrical membrane potential. For the calcium imaging alone, time appears to be less of a problem: we could measure calcium responses from individually loaded cells for several hours. An example is given in Figure 13.4. Our method has been described in detail (Galizia and Kimmerle, 2004).

13.3.7 Genetically Engineered Dyes

The availability of genetically engineered activity-dependent proteins has rapidly increased over the last few years. Today, several different calcium-sensitive proteins are available, as well as probes for other parameters of cell function (Guerrero and Isacoff, 2001; Miyawaki et al., 2003). Probes for cAMP, cGMP, glucose, membrane voltage, and a score of other molecules important for intracellular communication are available or are being developed in several labs around the world, and this review is bound to be out of date by the time it has reached readers. Therefore, we will not try to give a comprehensive overview of the available probes.

Most of the probes are based on modified GFP (green fluorescent protein) and often exploit proteins or parts of proteins that in natural cell physiology interact with the molecule of interest. For example, the calcium-sensitive reporter cameleon uses a section of the calmodulin sequence of its target peptide M13 (Miyawaki et al., 1997, 1999). Upon binding with calcium, the conformation of the molecule changes, which leads to a change in its fluorescent properties. Some reporters, such as cameleon, use fluorescence resonance energy transfer (FRET) between two fluorophores (see also Figure 13.5b). When excited at 440 nm, the emission light is strongest at about 480 nm (ECFP fluorescence) at low calcium levels, but at high calcium levels, fluorescence at this wavelength decreases whereas emission around 540 nm increases (EYFP fluorescence) (Miyawaki et al., 1999). Both wavelengths should be measured simultaneously. Calcium concentration changes are then calculated as the ratio of both wavelengths (see Section 13.5.3.2). An example is given in Figure 13.5.

These reporter genes can, in principle, be expressed in any cell of interest in any species if the necessary molecular tools are available. In *Drosophila*, the UAS-GAL4 technique has been extremely powerful for this purpose (Brand and Perrimon,

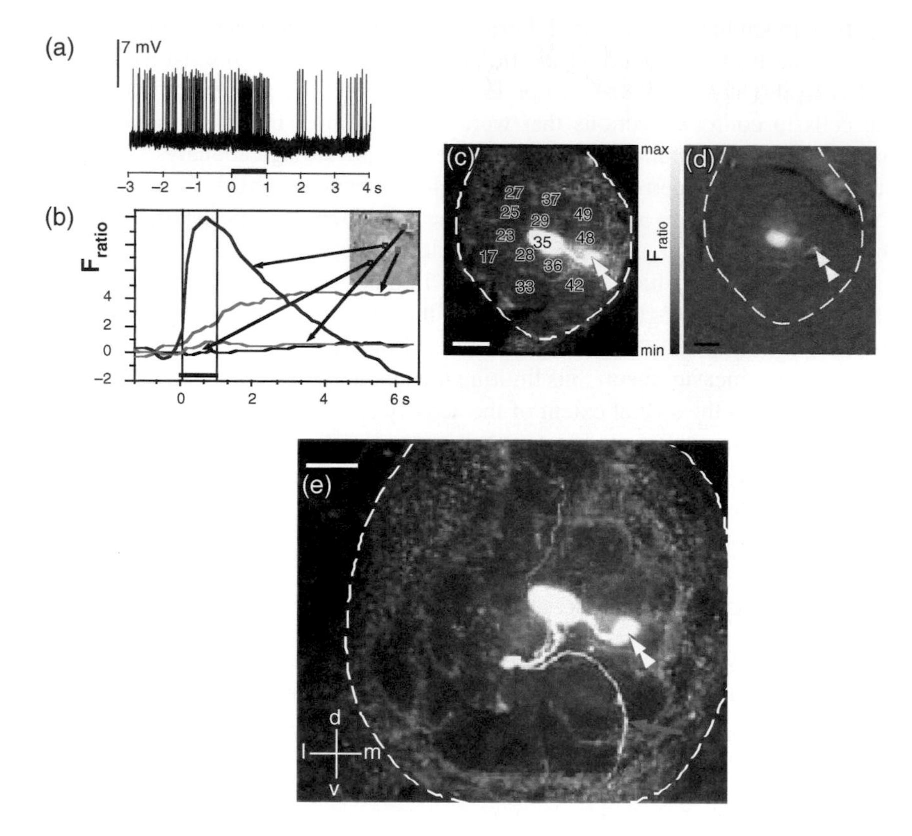

FIGURE 13.4 (See color insert following page 202.) Single-cell imaging in the honeybee of a uniglomerular PN filled with Fura by iontophoretic injection. **a**: Electrophysiological response to clove oil. An increase in action potential frequency occurred upon stimulation (black bar), followed by an inhibitory period during which the membrane potential was hyperpolarized and no spiking occurred. **b**: Calcium responses to clove oil recorded in the glomerular dendrites (upper trace, dark blue), the soma (red), and two reference areas within the AL (same cell as shown in **a**). Stimulus time is given by the bar and vertical lines. The intracellular calcium concentration in the glomerulus rapidly increases upon stimulation, decreases after stimulation, and finally falls below baseline. Calcium in the soma also increases, but with a delay and more slowly, and it does not decrease during the sampled period. No activity is apparent in the remainder of the AL (green and black lines). **c**: CCD video image of the same specimen. The glomerular borders were clearly visible in this preparation, allowing the identification of glomerulus T1-35 as the innervated glomerulus. Note the large soma (double arrowhead) and the strong labeling in the glomerulus due to dense dendritic branching. The approximate AL borders are indicated with dashed lines (also in **d** and **e**). Scale bar = 50 μm. **d**: False-color coded image of the calcium response to clove oil (compare with **b**). Note the strong signal in the glomerulus, the weaker signal in the soma, and a very weak signal in the primary neurite connecting soma and glomerular dendrites. Also note the absence of signals in the remainder of the AL. Scale bar = 50 μm. **e**: Confocal extended focus view of this neuron (compare with **c**). Note the staining of the axon leaving the AL toward the protocerebrum (arrow). Scale bar = 50 μm. (Adapted from Galizia, C.G., Kimmerle, B., *J. Comp. Physiol. A* **190**:21–38, 2004. With permission.)

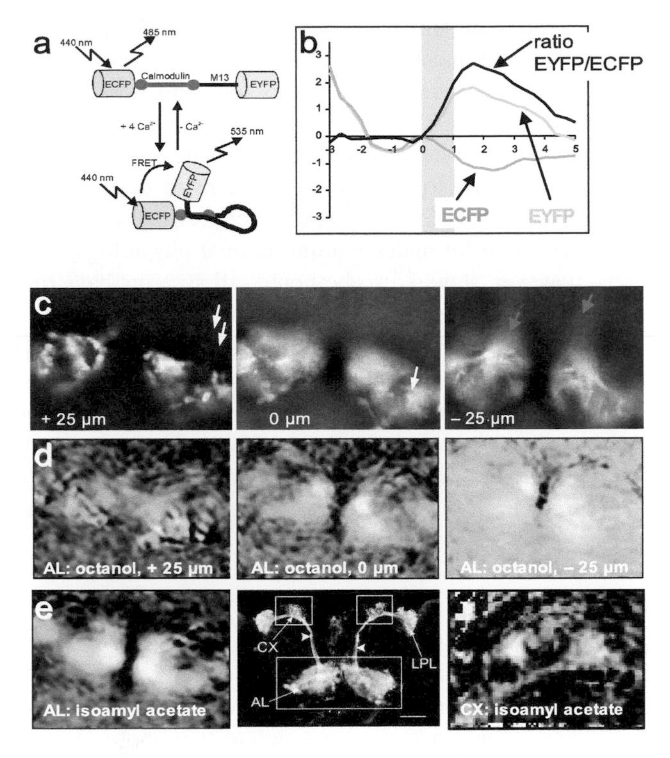

FIGURE 13.5 (See color insert following page 202.) Odor responses recorded in *Drosophila* using a genetically engineered calcium reporter, cameleon. **a**: Cameleon consists of two fluorescent proteins (ECFP and EYFP), linked by the calcium-sensitive calmodulin-M13 complex. Calcium binding causes calmodulin to wrap around the M13 domain, bringing the two fluorescent proteins closer together, with the resulting fluorescence of ECFP exciting the EYFP molecule in a process known as *fluorescence resonance energy transfer* (FRET). **b**: Time course of an odor-driven response in *Drosophila* AL (stimulus indicated by gray bar). Upon binding calcium, EYFP fluorescence increases (yellow curve), while ECFP fluorescence decreases (cyan curve). The curves show bleaching of both proteins before the odor is presented. Taking the ratio yields a flat curve before the odor response (black line). **c**: Video images of a *Drosophila* AL at different focal depths, obtained in fluorescent mode with a CCD camera. PNs were fluorescently labeled with cameleon at reduced magnification (GH146-GAL4 crossed with UAS-cameleon). Note the cell bodies of PNs (white arrows) and the tract leading to the mushroom bodies (red arrow). Antennal nerve faces down and is not visible because no marker is expressed in OSNs. **d**: False-color coded images of the responses to octanol, at the same focal depths as in **c**. Glomerular responses are visible only at the intermediate focal level, and blurred at +25 µm. The three focal depths were recorded during a single stimulation, by rapidly moving the focal depth and acquiring interleaved images. **e**: Examples of odor responses to isoamyl acetate in the AL (left) and in the mushroom body cortex (CX, right). The central image shows the labeling in the brain (GH146-GAL4). Approximate areas for images of the AL and the CX are shown with white boxes. Note the axonal tract connecting the AL with the CX (arrowheads) and the LPL (lateral protocerebral lobe). (Figures adapted from Miyawaki, A. et al., *Nature* **388**:882–887, 1997, and Fiala, A. et al., *Curr. Biol.* **12**:1877–1884, 2002. With permission.)

1993). Once the reporter protein has been engineered, fly lines are created that express the reporter protein under the control of a UAS promoter (see also Chapter 9 and Chapter 10). Then, these lines are crossed with flies that express GAL4 specifically in the cells of interest. For many cell populations, such lines are available; for specific studies, they must be created. Such crossings yield exquisitely selective labeling. Together, the ability to express the proteins in selective cell populations and the increasing availability of probes for different cellular mechanisms will prove an extremely powerful tool for understanding animal physiology.

The obvious, however, should also be noted: all systems have their limitations. Calcium-sensitive proteins share most properties with calcium-sensitive dyes: they measure calcium within a given concentration range, with a given dynamics and a given signal-to-noise ratio. And the GAL4-system is leaky: not only the strongly visible cells express the protein, but there is often a population of other cells that express the protein, if only weakly (Ito et al., 2003). This can also be the case in lines where very specific promotors have been used (Bhalerao et al., 2003). Furthermore, a suitable promoter may not be known for some cells of interest, or possibly may not even exist at all.

13.4 ACQUISITION OF IMAGING DATA

An optical imaging setup consists of a light source, a microscope with appropriate optics, and a sensor (e.g., a video camera CCD chip). We will first review these points and then address some special cases (such as two-photon measurements) and describe the system that we use as an example. Because good literature is available about these subjects and there is little to add that would be specific to insect physiology, we will keep this section brief.

13.4.1 THE LIGHT SOURCE

The light needed depends on the dye used. Fura is excited by ultraviolet light and requires an arc lamp. Many other dyes, including Calcium Green and all GFP-based probes, are excited with visible light, so an incandescent source is sufficient. Great attention must be given to the low noise level of the lamp, because any ripples in the light will result in increased noise in the measurements. We have had good results with a halogen lamp as light source (150 W) driven by a stabilized power supply with the dye Calcium Green (filter: BP 448-495 for excitation).

The use of a monochromator is not essential. However, we found it very useful in developing new preparations because often the optimal excitation wavelength *in vivo* was not the same as what is measured in the *in vitro* calibration systems. A monochromator allows making measurements across a spectrum of wavelengths and then taking the best wavelength for the experiments. For example, the best excitation wavelength for Calcium Green is shifted to a shorter wavelength with respect to the values given in the handbook from Molecular Probes[14]. A technical treatment of arc lamps and monochromators has been published (Uhl, 1999).

13.4.2 The Optics

The most important part of the optical path is the microscope objective. Light consists of photons that follow a Poisson distribution. This statistical nature of light causes what is called *shot noise* (i.e., fluctuations in the signal that are entirely due to physical characteristics of light). Therefore, among the main tasks in optical imaging is measuring as many photons as is feasible. The higher the numerical aperture of an objective, the more light it collects. Also, *dip objectives* (i.e., water immersion objectives for physiology) are ideally suited to physiological experiments because there is no cover slip to deteriorate the optical path and they generally have long working distances.

One drawback of objectives with a very high numerical aperture is that their light cone has a wide angle, which can cast shadows in the image, with a consequent deterioration of the signals. In such a situation, either the hole in the cuticle must be cut larger or the whole preparation must be tilted. For example, in the honeybee the ALs are about 400 μm below the cuticle. If the cuticle is excised in a way that one border of the cut is just above the border of the area to be imaged, the AL will be crisp and clear with a low-numerical aperture objective (e.g., a 10× water immersion objective with $NA = 0.3$). However, with an objective with a higher numerical aperture, the rim of the cuticle will cast a shadow (e.g., a 20× water-immersion objective with $NA = 0.95$). With a numerical aperture of $NA = 0.95$ in water, the angular aperture is almost 47°, and at a distance of 400 μm from the focal plane, the area necessary for the light to pass will have a diameter of almost 860 μm. This compares with less than 200 μm for a numerical aperture of $NA = 0.3$. As a reminder, the formula for the numerical aperture is $NA = n * \sin(\alpha)$, where α is the angular aperture and n is the refractive index of the medium (i.e., $n \approx 1$ for water).

With an objective of high numerical aperture, the focal sectioning becomes more accurate. There are situations, however, in which this is not desirable. For example, in the honeybee AL, the glomeruli form almost a monolayer around the central core of the AL. With a very high numerical aperture and focusing slightly into the AL, only a ring of glomeruli contributes to the signal: neither the central glomeruli above the focal plane nor the lateral glomeruli below the focal plane will appear in the signal (see Figure 13.5c, where images of *Drosophila* ALs at different focal depths are shown). With a lower numerical aperture, the signal will be weaker, but more glomeruli can be measured simultaneously. However, on the lateral edge of the AL, the signal may then consist of a superposition of several glomeruli. There are good reviews about optics and microscopes relevant to imaging studies (e.g., Lanni and Keller, 1999).

13.4.3 The Detector

A CCD chip used as the detector should have good quantum efficiency, low noise, and at least a 12-bit dynamic range. (Twelve-bit gives 4095 gray values. When using voltage-sensitive dyes, 14-bit or 16-bit is better.) Generally, spatial and temporal resolution share an inverse relationship: with higher spatial resolution comes lower temporal resolution. This is due to two concomitant effects:

1. Technically, the chip needs more time to be read and the data need more time to be transferred to the computer.
2. Physically, reducing either the size of each pixel or its integration time reduces the amount of light that is measured.

In imaging, individual information points consist of a count of photons, and this count should be large enough in order to get signals that are not dominated by the light's shot noise. Photon events follow a Poisson distribution. The variance of a Poisson distribution is proportional to the square root of the counts, which means that the shot noise grows with the square root of the number of photons. Therefore, the more photons that are measured in one bin, the smaller the *relative* noise (i.e., the higher the signal-to-noise ratio). CCD chips can be binned on chip; that is, the spatial resolution can be lowered, thus gaining in speed or allowing a reduction in light exposure times. Photodiode arrays are faster than CCD chips but have less spatial resolution and higher noise. If the entire image is not measured at once — that is, if imaging is done in scanning mode (confocal imaging, two-photon imaging) — the detector is generally a photomultiplier or a photodiode. See Christenson (1999) and Zochowski et al. (2002) for excellent reviews.

13.4.4 The System We Use

As an example of a complete system, we list the equipment that we use and the function of each part. Most, if not all, pieces can also be found from other suppliers — the fact that this system works well for us does not mean that other companies do not build products that are just as good or possibly better. The purpose of this list is to present a sample, to be modified as necessary by each researcher. Most parts have been bought as a package from a single supplier as a turnkey system.

Monochromator (TILL-Photonics)[(30)] — With a tunable excitation light and very fast switching between wavelengths, a monochromator is valuable for dual-excitation applications (e.g., Fura) and when different dyes are used or new dye protocols are being developed.

Shutter (built into the monochromator) — This is not standard but is highly advisable. The monochromator is based on a moving grating and is switched "off" by setting its wavelength to 200 nm, a range where the system has no light. However, lateral peaks of the diffraction pattern of the grating cause a limited but still visible spurious light, even in the off position. Adding a shutter that will totally shut off the light between measurements helps to reduce exposure of the dyes to light.

Light guide and condenser (TILL-Photonics)[(30)] — The light guide scatters the light from the monochromator, and the condenser must be specifically adjusted to each microscope model in order to optimize light efficiency.

Filters — The filters to be used depend on the dye. With very delicate dyes excited in the visible light range, it may be useful to use an excitation filter, even if a monochromator is employed, because the movement of the grating creates a very fast sweep through the UV range that may affect some dyes. This may be relevant for some of the GFP-based dyes. Table 13.1 lists the filter settings that we use with different dyes. These are given as examples; changes are possible and should be tested.

Microscope (Olympus BX-50WI)[21] — The microscope needs to be an upright model with fixed stage. The fixed stage allows the animal to be positioned and the focusing accomplished by moving the objective rather than the animal. In this way, movement of the animal is avoided, and — more importantly — relative movement between the animal and the peripheral equipment, such as the stimulus device, is avoided. Because the animal in an *in vivo* situation is not transparent, we have removed all optics and equipment related to transmission microscopy, creating space beneath the preparation.

Objectives — The magnification depends on the animal used. For moths and bees, we use a 20× water-immersion physiology objective with a long working distance and a numerical aperture of 0.95 (Olympus)[21]. This particular objective has a very good quality and high numerical aperture but an unusual thread and size, requiring a modification of the microscope. It is not compatible with other objectives. For flies, we either use the same objective, with an additional magnification in the optical pathway, or change to a 40× or 60× water objective.

In early experiments, we covered the animal with a cover slip and used an air objective. In these experiments, it was crucial to perfuse the brain constantly with oxygenated saline. When using a water-immersion objective, we found no improvement in survival time or signal strength when perfusing with saline, as opposed to placing a single drop at room temperature. Apparently, the oxygen available to the brain through diffusion is sufficient. This observation does not hold true for our *Drosophila* preparation: here, a constant supply of Ringer's solution is necessary; alternatively, saline can be exchanged at regular intervals. Whether this is due to the smaller brain in a relatively larger drop and, therefore, a reduced diffusion efficient for oxygen that has to cover relatively long distances, remains to be established.

TABLE 13.1
Filter Settings Used with Different Dyes

Dye	Excitation	Dichroic	Emission
RH795	546 nm	580 nm	LP 590
Fura	340/380 nm	410 nm	LP 440
Calcium Green	475 nm	500 nm	LP 515
cameleon	440 nm	470/520 nm	BP 473–494/ BP 530–565
gCamP	480 nm,	490 nm	BP 506–547 nm

Microfocusing piezo device for positioning objective (PIFOC; www.physikinstrumente.com) — This device allows images to be taken at different focal depths at the end of a recording (standardized z-stacks). Such images can be used for morphological reconstructions. For example, if after a measurement with Calcium Green the borders of olfactory glomeruli are fluorescently labeled with another dye, images can be taken at different focal depths, in order to reconstruct their positions and identify them. Because the PIFOC can change focal plane within 1 to 2 msec, interleaved measurements at two (or more) focal depths can be taken (e.g., if two glomeruli at different focal depths are to be monitored). The range of the PIFOC is 100 μm.

CCD camera (Till-Photonics)[30] — Binning, exposure time, and frame rate can be varied, so that photodamage (caused by the energy of strong light exposure) can be traded off against temporal and spatial resolution.

Micro-Imager (www.optical-insights.com) — This device splits the emitted light into two parallel paths using a dichroic mirror and two emission filters. The two images, which depict the same area but different spectral ranges, are projected onto two halves of the same CCD chip. This allows simultaneous measurement of two wavelengths, which is advisable for FRET-based dyes. The equipment is user friendly and, in its newer version, allows exchanging the filter block quickly. Unfortunately, the dichroic filter used is not a standard filter, which means that for each wavelength pair used in the lab, a dedicated filter set must be purchased. An alternative design for measuring two wavelengths simultaneously is achieved by using two CCD cameras with a dichroic mirror. The components needed for this design are standard microscope parts, but the cost of a second CCD camera with its dedicated control unit and software is higher than a Micro-Imager.

xyz-Microscope specimen stage (www.luigs-neumann.com) — An xyz-table allows the performance of complex multisite measurements. It also affords the possibility to record z-stacks over a wider range than the 100 μm limitation of the PIFOC.

Olfactometer (custom-built or Syntech)[29] — We use a variety of olfactometers, depending on the goal of the experiment. All olfactometers have their strengths: some are best for complex temporal patterns, others for measuring many odors, still others for precise control of stimulus concentration. A description of olfactometers is beyond the scope of this chapter. An important feature of any stimulus device is that it should be controllable from the imaging computer.

Vibration isolation table — Vibrations, like movements, are detrimental for imaging experiments. The faster the measurements, the more they will be affected by vibrations.

Darkened room — The light level in the room should be very low. When very fast measurements are taken, the 50 or 60 Hz power fluctuation in artificial light sources may affect the results. In this case, even the weakest light powered by power supply must be avoided and replaced with DC-powered lamps. Alternatively, or additionally, the microscope may be shielded by a black curtain (be sure to use lint-free cloth to protect the optical equipment).

Imaging system, general considerations — An important aspect in these experiments is that the different components integrate into each other and are controllable all from the same computer. That is, the core system must allow a flexible array of digital and analog input and output channels for experimental control, which can be integrated in the software commands.

Data analysis software — Most imaging systems have their own data analysis software. The Till Photonics system[30] also comes with software, TillVision, which allows control of the imaging parameters (wavelengths, recording parameters, sophisticated sequences of recordings), the periphery (stimulus controller, Ringer's perfusion, stage control, etc.), and data analysis. Nevertheless, for complex and more fine-tuned analysis, we found it more convenient to develop custom-built data analysis software. Languages such as MATLAB, IDL, or LabView are appropriate — we use IDL (Research Systems, Inc.). The initial effort required is greatly outweighed by the increase in flexibility and precision. Some of the features that we have implemented in our software are described Section 13.5.

13.4.5 Two-Photon Imaging

It is not the purpose of this chapter to describe two-photon imaging. Good reference chapters are available (Denk, 1999; Tsai et al., 2002). We would like, however, to list the most important advantages and disadvantages of two-photon imaging vs. a CCD-camera system.

The main advantage of a two-photon system is its superior spatial resolution. This is further increased by the fact that the infrared light used for excitation penetrates the tissue better than visible light, making it possible to measure deeply into the tissue. For *in vivo* situations, where invariably a 3D resolution is desirable, this is an unbeatable advantage.

Principally, a two-photon imaging system is a scanning device; that is, different points in space are imaged at different times. This can create artifacts if activity spots move spatially within the field of view at a high rate. At every single pixel, the image consists of only a minute fraction of the time attributed to that image: such a short sampling time with respect to the temporal resolution does lead to the danger of aliasing effects (i.e., the creation of wrong low-frequency fluctuations when in fact there are only high-frequency data). A major advantage of CCD cameras is that all pixels are recorded at exactly the same time.

CCD cameras have less noise when measuring photons and a linear dynamic range, making quantifications very reliable. They can integrate over long times, so that slow measurements in unfavorable light conditions are possible. One important advantage of a CCD system is that it can be bought as a turnkey system and used without any special skills, including by students without previous imaging experience. CCD-based imaging systems need little, if any, maintenance over time. Two-photon systems are technically challenging and require the operator to have a good background in physics. They are an order of magnitude more expensive and require constant care and costly maintenance.

13.5 ANALYSIS OF IMAGING DATA

Many aspects of data analysis for *in vivo* imaging in insects follow standard techniques used in imaging applications. There are some aspects, however, that are peculiar, and we will emphasize these. Naturally, these techniques are in continuous evolution as technology develops.

13.5.1 Pretreatment of the Data

13.5.1.1 Movement Correction

When movement leads to a change in focus, the effects are substantial and difficult — if not impossible — to correct. Such measurements must be discarded altogether. If they occur frequently, the preparation must be modified in order to obtain more stable measurements. However, if movement leads to a limited lateral shift, it is easy to adjust the effect offline by shifting each image accordingly. This should be done on the raw data, using morphological structures as reference points: trachea and borders of brain areas are particularly good references. The treated signals (i.e., $\Delta F/F$ or ratio images) should not contain morphological detail and, therefore, are not suitable as reference images for movement corrections.

A correlation-based algorithm can estimate the best movement correction, but it is always necessary to check the result visually. This can be done easily by playing the corrected image sequence as a fast-forward movie: it should look like a movie of a steady object. Unless pixel values are interpolated, movement correction cannot correct for movements that are less than a pixel size in magnitude. Visual inspection is particularly important for measurements in which there is only limited contrast, or where there are contrast-rich areas outside the brain that spoil the image, such as some tracheal sacs or muscles. An unsupervised algorithm may adjust changes based on these regions, with detrimental effects for the area of interest.

13.5.1.2 Structure Identification

Among the greatest advantages in using insect models is the relative ease of identifying homologous neurons and structures. For example, in the AL, individual glomeruli can be identified by their shape and relative position if their layout is visible. A comparison of their physiological properties then becomes more useful because a circular argument is avoided about their identity, on one hand, and their physiological tuning on the other. Nevertheless, in some studies the physiological properties may also have to be considered in order to identify an area. Areas of interest must also be chosen in studies in which the units to be considered are compartments such as the dendritic area vs. the soma of a neuron or a group of neurons.

Two steps must be kept logically distinct:

1. Selecting a set of areas of interests
2. Naming or identifying the structures in these areas

13.5.2 Data Filtering and Correction

13.5.2.1 Filters

Filters that reduce noise in the data are low-pass filters or median filters, applied either within each image (i.e., in the spatial domain) or across images (i.e., in the temporal domain). These filters are often instrumental in finding activity areas, but they can also create the illusion of active areas where none exist and, therefore, must be used with due care. In particular, all low-pass filters smear data. In the temporal domain, this means that a response onset or offset may be estimated with limited confidence; in the spatial domain, it means that an inactive area that is close to a very active spot may have activity erroneously attributed to it.

The use of a spatial median filter on data that originate from CCD cameras is highly recommended because in these data there are always individual "bad pixels" (i.e., technically motivated outliers). A median filter removes these pixels, as long as they are isolated, without affecting the remainder of the image. A median filter is often employed in image analysis to remove "salt-and-pepper" noise (i.e., noise consisting of individual and isolated unproportionally bright or dark pixels).

13.5.2.2 Scattered Light Correction

All microscope images have a limited lateral resolution. *In vivo* measurements, however, are particularly affected, because of the poor optical properties in the aqueous medium and because of the thickness of the preparation. The ideal lateral resolution that is possible with embedded objects under the microscope is not achievable. Fluorescence light from a point may scatter in all directions and be reflected back into the objective (Figure 13.6a). This light cannot be distinguished from fluorescent light that may have originated in that location. As a consequence, any bright fluorescent light spot will create an extended halo of reflected light around itself, limiting the spatial resolution. This will also affect fluorescence changes due to activity in one area, which may lead to an artificial lateral correlation in activity patterns. In the olfactory system, for example, we found that some neighboring glomeruli are correlated in their physiological responses, but others are not. If scattered light created artificial correlations between neighboring glomeruli, the question about neighborhood correlation may erroneously be given a positive response in all instances for purely physical reasons, without biological evidence. Finding ways to correct for scattered light is, therefore, an important task in the data analysis process.

We have developed a method to reduce the effect of scattered light with an offline treatment of the data. The technique is borrowed from an image-processing

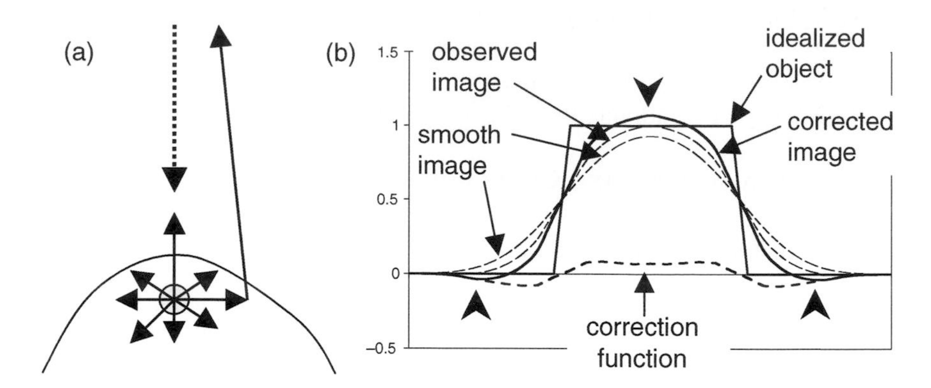

FIGURE 13.6 The principle of scattered light correction. **a**: Any source of fluorescence emits light in all directions. In an *in vivo* situation, where the object is thick, this creates a 3D halo around the image. In this schematic view, excitation light is shown as a dashed arrow, emitted fluorescent light as solid arrows radiating from the source, and the tissue border as a curved line. **b**: An idealized finite object with sharp edges will appear as a blurred image (observed image). By calculating an artificially blurred image (smoothed image) and subtracting the difference between the smoothed and the observed image (correction function) from it, we obtain the "corrected image" (solid curve), which represents a closer approximation of the original idealized object than the observed image. Note the overshoot (top of curve) and undershoots (ends of curve) of the corrected image with respect to the idealized object (arrowheads). When neighboring pixels are averaged, e.g., when calculating the response of the surface belonging to a particular glomerulus, the overshoot averages out with the remaining blur in the image.

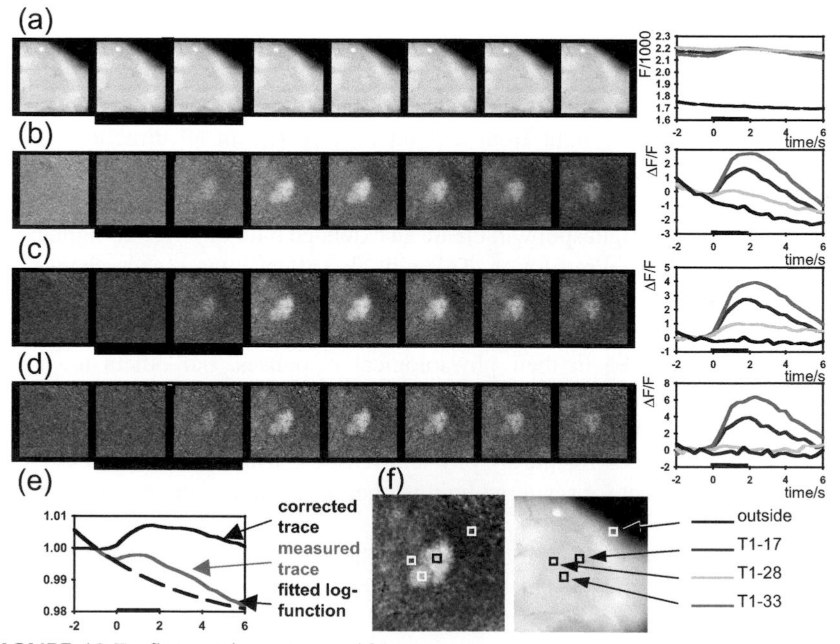

FIGURE 13.7 See caption on page 381.

FIGURE 13.7 (See color insert following page 202.) Scattered light and bleach correction. The AL was stained with the calcium-sensitive dye Oregon Green 488 BAPTA-5N AM by bath application. The stimulus (2 s) was 1-nonanol (red bar in images, black bar in time-traces). For this demonstration, the AL is deliberately positioned laterally in the image. In **a–d**, image frames are shown to the left, and time traces for four selected regions (black, blue, green, red) to the right (**f** shows a spatial map of the four regions). (Data courtesy of A. Delorenzi, unpublished.) **a**: Left: sequence of recorded images, scaled in black and white. Time between frames is 1 sec. Note the trachea on the AL and the lack of apparent change in the pictures upon stimulation, due to the small changes relative to the high contrast in the images. Time traces (right) give fluorescent values as derived from the A/D converter of the CCD camera divided by 1000. **b**: Sequence of false-color coded images of ΔF/F of the measurement in **a**. Odor stimulation leads to focal increase in the calcium signal. Note the overall decay in brightness over time, which is also evident in the time traces. **c**: Same data as in **b**, after bleach correction as described in the text. Note the lack in brightness decay over time, as apparent in the images and the time traces. Compare with **e**. **d**: Same data as in **c**, but after additional light-scattering correction. Note that in the false-color coded sequence the difference is hardly noticeable. However, in the traces, the apparent response in glomerulus T1-28 is now missing: the green curve (T1-28) and the black curve (area outside the AL) overlap, indicating that the apparent response in T1-28 in **c** resulted from light scattering. Notice also that the response in glomerulus T1-33 is enhanced. **e**: Overall brightness function as calculated by averaging the fluorescence over the entire image (red continuous curve), and the fitted log-function (black dashed line). The difference of the two lines is also shown (corrected trace). Note that the odor response is so strong that it is also apparent in the averaged brightness. Therefore, this portion of the fluorescence decay should not be used to fit the correction curve. **f**: Left: false-color coded image of the response to 1-nonanol in this preparation (ΔF/F at maximum intensity minus ΔF/F just before stimulus), showing the four areas selected for the time traces in **a–d**. Right: video image of the AL (compare with **a**), with the superimposed areas chosen in the time traces. The individual glomeruli are labeled. In this preparation, their identity has been estimated from the response (1-nonanol elicits activity in the glomeruli T1-17 and T1-33); therefore, the position of T1-28 is tentative. The reference area is clearly outside the AL.

technique generally called *unsharp masking*. It consists of subtracting a smoothed image from each image of raw data. The formula is as follows (Figure 13.6; an example is given in Figure 13.7d): Let a measurement consist of n fluorescent images, F_1, F_2 ... F_n. For each image, F_i, calculate a spatially low-pass filtered image, sm(F_i). Because scattered light is caused in the tissue, the radius used should be given in micrometers at the object and empirically determined. In our studies of the AL, we use a value of 50 μm as low-pass filter kernel. We take the difference between the smoothed image and the observed image as a correction factor (i.e., we subtract this difference from the original image). We therefore calculate

$$F_i' = F_i - \left[sm(F_i) - F_i\right] = 2 * F_i - sm(F_i) .$$

The images F_i' are the scattered-light corrected images. Note that this procedure does not change the measured values in areas where there is no spatial contrast. In these areas, the low-pass filtered image has the same value as the original image, and consequently, $F_i' = F_i$.

13.5.2.3 "Bleach" Correction

In vivo measurements in the insect AL are often dominated by an overall, fast decay in fluorescence. This decay is particularly strong in measurements that use excitation wavelengths in the range of 450 to 500 nm. Its strength depends on the exposure time: when very short exposure times are used with longer interframe intervals, the decay is weaker because the fluorescence recovers during the pauses. Similarly, intervals during short measurements lead to a recovery in the fluorescence values. The fast recovery argues that the origin of the fluorescence decay is not a bleaching phenomenon, but rather due to photoisomerization. Unlike with bleaching, the fluorescent molecule is not destroyed when it photoisomerizes, and it recovers in the dark. Part of the intensity decay is independent of the dye used and reflects photoisomerization of tissue autofluorescence. The net result remains: the response to a given stimulus is overlain by a strong, exponential decay of the response.

The most common technique for removing such artifacts is to run a "blank" measurement and subtract it from the measurement of interest. For example, a control (e.g., a measurement when stimulating with clean air) response is subtracted from a stimulus (e.g., an odor stimulation) response. The result contains the response component exclusively due to the stimulus alone. This is a good approach, but it has two drawbacks. First, it assumes that "bleaching" is identical in different trials. However, we found that the "bleaching" slope declines over time, leading to a slightly rising or falling tail in measurements corrected by subtracting a reference measurement. Second, it assumes that responses are additive. This is of no concern if the control measurement does not contain a response. However, this technique may also be used to remove unwanted response components, such as mechanosensory stimulation responses. Although appealing, the approach is problematic. Additivity of responses is only warranted if the responses come from different cells. Within cells, combination of stimuli is governed by a variety of mechanisms, from logarithmic dose-response curves at transduction to the binding properties of the activity-dependent dye itself. Therefore, if the "artifact" originates in the same cells as the signal, more appropriate methods than subtraction should be used.

For "bleach" correction we have developed a method that is very robust and does not make assumptions other than the shape of bleaching (logarithmic, and equal for the entire region imaged). This method needs a series of mathematical steps. In short, the steps are the following:

1. Calculate relative data (akin to $\Delta F/F$).
2. Fit a log-function.
3. Subtract the fitted log-function.
4. Transform back to fluorescence data (see Figure 13.7).

In more detail, let a measurement consist of n fluorescent images, F_1, F_2 ... F_n. Let F be a reference image, as is used for the calculation of $\Delta F/F$ values. Such an image can be obtained by averaging a few images at the beginning of the measurements, for example, $F = \frac{1}{3}\sum_{i=2}^{4} F_i$. We exclude the very first images because bleaching in these images does not follow the exponential shape. Furthermore, we should not include any of the images that contain the stimulus response. Theoretically, taking a single image would also work, but averaging a few images reduces noise. We now calculate the *relative fluorescence* for each image; that is, for each F_i we get an image that we denote with S_i, for signals: $S_i = \frac{F_i}{F}$. This is akin to calculating $\Delta F/F$; in fact, the calculated value is identical to $\Delta F/F$ if 1 is subtracted (see Section 13.5.3.1). The images S_i contain the changes in fluorescence relative to the reference images: in the example, the images 2, 3, and 4. Let v_i denote the average relative fluorescence pixel brightness of the image S_i. The value v_i should not be the average of the entire image, but only the average of the brain area being investigated (e.g., the AL, because the decrease in fluorescence of, for example, Ringer's solution is of no interest). This gives a vector $\mathbf{v} = (v_i, v_2, \ldots v_n)$. Generally, the values will be equidistant in time, but in more complex protocols, time intervals between frames may differ, making it necessary to associate with the vector $\mathbf{v}$ a vector of time-stamps for each frame.

It should be noted, however, that varying time intervals between frames leads to an infringement of the logarithmic shape of the "bleaching" function and is, therefore, not recommended unless the decay function is modified accordingly. We define a logarithmic decay function as $w(t) = a_0{*}e^{a_1{*}t} + a_2$. Using a least-squares algorithm, we can fit this decay function to the vector $\mathbf{v}$, obtain the best-fitting parameters a_0, a_1, and a_2, and calculate a vector of fitting decay values $\mathbf{w} = (w_i, w_2, \ldots w_n)$. The values that correspond to the stimulus response should not be used for estimating the decay parameters, leaving a series of values at the beginning and the end of the decay, which is generally sufficient for a good fit. These values are subtracted from each frame (i.e., the same value w_i is subtracted from all pixels of each frame S_i), but a different value is used for each frame, giving S' with $S_i' = S_i - w_i$. S' contains bleach-corrected data, but relative data, and therefore must be converted back into fluorescence data, inverting the initial procedure: $F_i' = S_i' * F$.

In the following, images F' are treated just as if they were the original images, with the advantage that the overall decay in fluorescence has been removed.

The calculation must go through the step of calculating relative signals because the parameters for the logarithmic function differ widely for places in the image view that have different brightness. Because, in an *in vivo* situation, brightness will

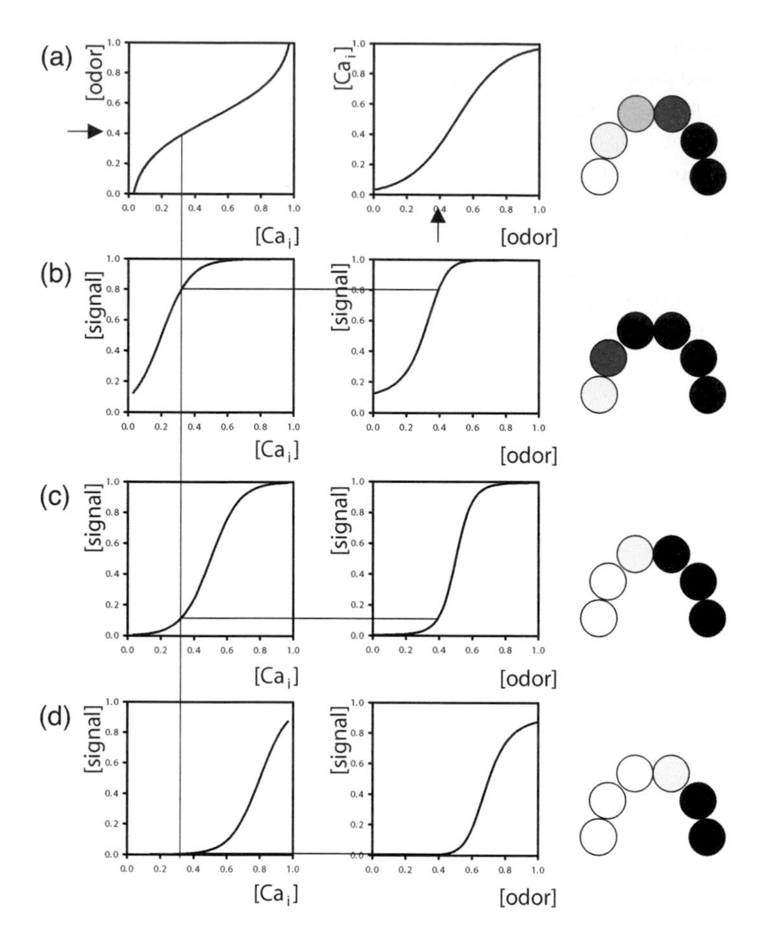

FIGURE 13.8 Calcium affinity and the binding properties of calcium dyes. **a**: Assume that an odor stimulus varies over a range of concentrations (0 to 1 arbitrary units), leading to an intracellular calcium signal (Ca_i) in the cells (0 to 1 arbitrary units) that follows a standard sigmoidal response curve (center column of figure). The graph in the left column is shown with inverted axes; the arrows indicate the exemplary odor concentration (0.4). The rightmost column depicts six schematic glomeruli showing graded responses over the entire concentration range. **b–d**: Left: plot of the relationship between intracellular calcium and fluorescence (signal). Center: plot of the relationship between odor concentration and fluorescence. Lines comparing the responses for the exemplary odor concentration 0.4 are shown. Right: the apparent odor response in an array of glomeruli is shown. **b**: If this calcium response is measured with a high-affinity dye, only the lower range can be resolved; that is, most glomeruli appear active, and almost to the same degree (right). **c**: With a dye of intermediate affinity, the central range of the spectrum is well covered, but lower and higher ranges cannot be resolved dynamically. **d**: With a low-affinity dye, only very strong responses are apparent.

vary substantially for different areas in the brain, no exponential function will be able to correct the fluorescence decay equally well at different positions. However, with relative data that is possible: all areas have comparable values. If, for example, a single image is used as reference image *F*, that image will have the value 1 in all its pixels in the corresponding signal image *S*. Under these circumstances, a single logarithmic function fits all pixels well, and the corresponding value can be subtracted with no harm to the data and, in particular, no change to any spatial component of the response.

In our data, this procedure gives excellent results. The single exponential curve does not, however, fit the first one to two frames of the measurements; here the decay is steeper. Rather than fitting a double exponential function (and thus increasing the parameters to be estimated in each measurement), we accept that these two frames are not perfectly corrected. For quality purposes, it is always good practice to inspect a graphical display of the brightness time course with the superimposed fitted function for each measurement.

13.5.3 Data Analysis

13.5.3.1 Calculation of ΔF/F

Single-wavelength data must be transformed into relative fluorescence data for quantitative analysis, because the absolute change in brightness is less informative than the relative change. In other words, if in an area with brightness values of, for example, 2000, a calcium increase may lead to levels of, for example, 2020, the same increase will change an area with brightness of 1000 to 1010 — that is, a 1% increase in both places (with a 12-bit CCD camera, measured values are between 0 and 4095). The calculation of relative signals is straightforward: the values are divided by a reference image, and 1 is subtracted. Let a measurement consist of n fluorescent images, F_1, F_2 ... F_n. Let F be a reference image; we then can calculate *relative fluorescences* for each image, S_i (S for signals) as follows:

$$S_i = \frac{F_i}{F} - 1 = \frac{F_i - F}{F} = \frac{\Delta F}{F}$$

The question is: what is the best reference image F? In order to reduce noise, this should be the average of more than one image. In order to be relevant to the stimulus response, it should be close to the stimulus. And, in order not to spoil the responses themselves, it should not contain any frame during the stimulus response (including those frames would lead to a slight underestimation of the responses). We therefore generally take a few (three to six) frames just before stimulus onset. For example, for a measurement with stimulus onset at frame 6, we may take,

$$F = \frac{1}{4}\sum_{i=2}^{5} F_i$$

Note, however, that this procedure needs particular attention when looking at the data afterwards. Let us consider an extreme case: the reference image is frame 5 (no averaging), and stimulus onset is at frame 6. All pixels will have the value of 0 at frame 5 — and even without any response but only background activity, some areas will increase and some will decrease in fluorescence at frame 6 (i.e., at stimulus onset). In measurements with strong random activity, as in measurements of olfactory PNs, false responses may be interpreted from these curves.

13.5.3.2 Calculation of Ratios

The calculation of ratios of two wavelengths is equally straightforward. These data arise from double excitation measurements, such as for Fura, with image pairs excited at 340 nm and 380 nm, respectively, or from double emission measurements, such as for the FRET-based indicator cameleon. Given a pair of fluorescent images, FA_i and FB_i, the ratio is calculated as follows:

$$S_i = \frac{FA_i}{FB_i}$$

Generally, both wavelengths have similar bleaching behaviors, making a prior bleaching correction unnecessary. If, however, the bleaching is unequal, each wavelength should be corrected independently following the procedure described previously, and the ratio taken thereafter.

In cell culture, taking the ratio for dyes such as Fura, can be applied to calculating absolute calcium levels within the cells, for example, by calibrating the system with a series of calcium buffers and a calcium ionophore after the measurements. This approach is not possible for *in vivo* measurements if more than one cell is loaded with the dye and their images superimpose, because the ratios are a combination of the intracellular calcium concentrations of all cells involved. Two superimposed cells with intermediate calcium concentration may give the same ratio as one cell with high and one with low concentration, when combined.

Taking the ratio results in different basal levels at different places within the field of view, due to differences in background fluorescence, dye loading, or resting calcium concentration. Therefore, a background ratio image must be subtracted. Following the argument for the background fluorescence image used when calculating relative fluorescences above, it is best to calculate an average from a few images before stimulation. The same caveat as before follows: if not scrutinized critically, this procedure may create the impression of stimulus-correlated activity.

13.5.3.3 False-Color Coded Images

Moving images (i.e., films) can best show spatial activity in time. For documentation on paper, however, single frames are more useful. These are often shown in false colors (e.g., with colors ranging from blue [no activity] to green to yellow to red [maximal activity]). Other color scales may be used, or even black-and-white scaling. An advantage of false colors is that activity spots appear more prominent and are easier to identify. A disadvantage is given by the properties of our visual system, which creates categories for individual colors. So, even though the passage of colors is smooth along the false-color scale, to our eyes a change from yellow to red is more prominent than a change of equal size within the yellow range. Black-and-white scaling avoids this problem, but small activity spots, which are still visible as, for example, green spots against a blue background in color, may vanish in black-and-white coding. Another disadvantage of the most common false-color scales is that part of their range is difficult to interpret for colorblind readers (see http://jfly.iam.u-tokyo.ac.jp/html/color_blind).

Another issue to be considered is the scaling range, which can dramatically affect the image. For example, when studying the response to different odor concentrations within the AL, we found that the response pattern is remarkably similar over several orders of magnitude (Sachse and Galizia, 2003). However, when odor concentration increases, the total activity size also increases. In order to show the increase in activity with concentration, the responses at different concentrations must be shown with a common scaling, resulting in blue images at low concentration and red images at high concentration. However, in order to show the similarity of the response pattern, the same data must be shown scaled to each frame maximum and minimum. So, different scaling procedures result in different images and messages from the same data.

The most important aspect to be considered when presenting single false-color coded images as responses is what information to calculate for each pixel: the maximum activity size after stimulation, the average activity over a particular time window (which window?), the difference between the activity at two points in time (e.g., just before and just after stimulus time), or the value of a fitted function. With data that have relatively slow time courses, such as the data from bath-applied Calcium Green measurements, these different possibilities are mainly a question of signal-to-noise ratio. However, with more dynamic data, such as the temporally complex responses to odors found in selectively labeled PNs, a peak response image may look totally different from an average response image because in some places of the AL there is a small, but persistent response, whereas other areas give short, but very strong responses. Furthermore, the time of peak activity differs between glomeruli, so if the peak activity value is taken, different pixels of the image reflect activities at different times. This may be the best representation for some analyses. There is no rule about what function is the best — the choice depends on the data and the study.

13.5.3.4 Time Course

Time-course plots give the signals over time for particular positions. These are particularly useful for temporally complex responses, such as odor responses in PNs. For statistical analyses, however, the time courses should derive from a comparable area in different animals (for the AL, this would mean that identifying the glomerulus of origin should be a high priority), and a characteristic feature of the curve should be extracted quantitatively.

13.5.3.5 Fitting Response Functions

When the shape of the response is known, fitting an appropriate function can be used to quantify the magnitude of response parameters. This approach gives the best signal-to-noise ratio possible. For example, when the maximum response is estimated from the data curve directly, this value is effectively estimated from the reading of a single frame (i.e., the one with the maximal response). However, when a function is fitted to the entire time course, the value is estimated from all frames, with a substantial increase in signal-to-noise. A function can be fitted to estimate several parameters simultaneously, such as stimulus onset time and stimulus size. A description of this approach for bath applied Calcium Green data in the honeybee AL has been published (Stetter et al., 2001).

13.5.3.6 Data Interpretation

Interpreting the biological meaning of imaging data is beyond the scope of this chapter. But we would like to mention one aspect, common to all measurements, but particularly important in imaging experiments: a biochemical probe (the dye) is used to report about a physiological parameter of the cells investigated (e.g., calcium). This probe has particular binding properties, so every result should be interpreted considering the limited window that the particular dye opens on cellular processes. Other parameters, such as dye concentration, dissociation constants, number of collectible photons, and dynamic range of the experiment, all affect the measured responses and their interpretation.

An example with three calcium dyes of different binding affinities should clarify the point (Figure 13.8). Assume a stimulus has been given at different concentrations, and the response, in terms of intracellular calcium increase, increases with stimulus concentration (Figure 13.8a). Increased calcium concentration will lead to stronger fluorescence of the dye (Figure 13.8b–d). However, high-affinity (Figure 13.8b), intermediate-affinity (Figure 13.8c), and low-affinity dyes (Figure 13.8d) will each emphasize a different spectrum of the response and ignore another part. As a result, very graded responses, as indicated in the right column of Figure 13.8, may be measured as almost binary responses, as in Figure 13.8c.

13.6 CONCLUSIONS AND OUTLOOK

Optical imaging techniques are powerful tools for analyzing insect physiology. The combination of spatial analysis, identified units, and diversity of physiological mechanisms analyzed in an *in vivo* situation is particularly revealing. The next few years will mark a further leap in this technology. New optical methods will be available for more efficient measurements. For example, two-photon microscopy with spectral and temporal analysis of the emitted photons will allow simultaneous recording of several probes and will improve both temporal and spatial resolution. New probes are being developed to visualize diverse aspects of cell physiology. And new techniques for introducing probes into specific cells (notably in *Drosophila*) or for use in different species will be available. These techniques are ideally suited for insects. As compared with mammalian studies, *in vivo* physiology is easy: the number of neurons is reduced, and identified neurons can be targeted. Furthermore, the great diversity of solutions that evolution provided within the insects offers sublime insight into nature's toolbox.

ACKNOWLEDGMENTS

Many thanks to Mathias Ditzen, Philipp Peele, Daniela Pelz, Silke Sachse, and Ana Silbering for excellent discussions and valuable comments.

REFERENCES

Abel R, Rybak J, Menzel R (2001) Structure and response patterns of olfactory interneurons in the honeybee, *Apis mellifera*. *J. Comp. Neurol.* **437**:363–383.

Berg BG, Galizia CG, Brandt R, Mustaparta H (2002) Digital atlases of the antennal lobe in two species of tobacco budworm moths, the Oriental *Helicoverpa assulta* (male) and the American *Heliothis virescens* (male and female). *J. Comp. Neurol.* **446**:123–134.

Bhalerao S, Sen A, Stocker R, Rodrigues V (2003) Olfactory neurons expressing identified receptor genes project to subsets of glomeruli within the antennal lobe of *Drosophila melanogaster*. *J. Neurobiol.* **54**:577–592.

Brand AH, Perrimon N (1993) Targeted gene expression as a means of altering cell fates and generating dominant phenotypes. *Development* **118**:401–415.

Carlsson MA, Hansson BS (2003) Dose-response characteristics of glomerular activity in the moth antennal lobe. *Chem. Senses* **28**:269–278.

Carlsson MA, Galizia CG, Hansson BS (2002) Spatial representation of odours in the antennal lobe of the moth *Spodoptera littoralis* (Lepidoptera: Noctuidae). *Chem. Senses* **27**:231–244.

Christenson M (1999) The application of scientific-grade CCD cameras to biological imaging. In: *Imaging Neurons: A Laboratory Manual*. Yuste R, Lanni F, Konnerth A, Eds., Cold Spring Harbor Press, Cold Spring Harbor, NY, pp. 6.1–6.14.

Cohen LB, Salzberg BM, Grinvald A (1978) Optical methods for monitoring neuron activity. *Annu. Rev. Neurosci.* **1**:171–182.

Collmann C, Carlsson MA, Hansson BS, Nighorn A (2004) Odorant-evoked nitric oxide signals in the antennal lobe of *Manduca sexta*. *J. Neurosci.* **24**:6070–6077.

Delaney K, Davison I, Denk W (2001) Odour-evoked [Ca^{2+}] transients in mitral cell dendrites of frog olfactory glomeruli. *Eur. J. Neurosci.* **13**:1658–1672.

Denk W (1999) Principles of multiphoton-excitation fluorescence microscopy. In: *Imaging Neurons: A Laboratory Manual*. Yuste R, Lanni F, Konnerth A, Eds., Cold Spring Harbor Press, Cold Spring Harbor, NY, pp. 17.11–17.18.

Egelhaaf M, Borst A (1995) Calcium accumulation in visual interneurons of the fly: stimulus dependence and relationship to membrane potential. *J. Neurophysiol.* **73**:2540–2552.

Faber T, Menzel R (2001) Visualizing mushroom body response to a conditioned odor in honeybees. *Naturwissenschaften* **88**:472–476.

Fiala A, Spall T (2003) *In vivo* calcium imaging of brain activity in *Drosophila* by transgenic cameleon expression. *Sci. STKE* **2003**:PL6.

Fiala A, Spall T, Diegelmann S, Eisermann B, Sachse S, Devaud JM, Buchner E, Galizia CG (2002) Genetically expressed cameleon in *Drosophila melanogaster* is used to visualize olfactory information in projection neurons. *Curr. Biol.* **12**:1877–1884.

Flanagan D, Mercer AR (1989) An atlas and 3-D reconstruction of the antennal lobes in the worker honey bee, *Apis mellifera* L. (Hymenoptera: Apidae). *Int. J. Insect Morphol. Embryol.* **18**:145–159.

Frostig R (2002) In Vivo *Optical Imaging of Brain Function*. CRC Press, Boca Raton, FL.

Galizia CG, Kimmerle B (2004) Physiological and morphological characterization of honeybee olfactory neurons combining electrophysiology, calcium imaging and confocal microscopy. *J. Comp. Physiol. A* **190**:21–38.

Galizia CG, Menzel R (2001) The role of glomeruli in the neural representation of odours: results from optical recording studies. *J. Insect Physiol.* **47**:115–130.

Galizia CG, Joerges J, Kuttner A, Faber T, Menzel R (1997) A semi-*in-vivo* preparation for optical recording of the insect brain. *J. Neurosci. Methods* **76**:61–69.

Galizia CG, Nägler K, Hölldobler B, Menzel R (1998) Odour coding is bilaterally symmetrical in the antennal lobes of honeybees (*Apis mellifera*). *Eur. J. Neurosci.* **10**:2964–2974.

Galizia CG, McIlwrath SL, Menzel R (1999a) A digital three-dimensional atlas of the honeybee antennal lobe based on optical sections acquired by confocal microscopy. *Cell Tissue Res.* **295**:383–394.

Galizia CG, Menzel R, Hölldobler B (1999b) Optical imaging of odor-evoked glomerular activity patterns in the antennal lobes of the ant *Camponotus rufipes*. *Naturwissenschaften* **86**:533–537.

Galizia CG, Sachse S, Rappert A, Menzel R (1999c) The glomerular code for odor representation is species specific in the honeybee *Apis mellifera*. *Nat. Neurosci.* **2**:473–478.

Galizia CG, Küttner A, Joerges J, Menzel R (2000a) Odour representation in honeybee olfactory glomeruli shows slow temporal dynamics: an optical recording study using a voltage-sensitive dye. *J. Insect Physiol.* **46**:877–886.

Galizia CG, Sachse S, Mustaparta H (2000b) Calcium responses to pheromones and plant odours in the antennal lobe of the male and female moth *Heliothis virescens*. *J. Comp. Physiol. A* **186**:1049–1063.

Gascuel J, Masson C (1991) A quantitative ultrastructural study of the honeybee antennal lobe. *Tissue Cell* **23**:341–355.

Gelperin A, Flores J (1997) Vital staining from dye-coated microprobes identifies new olfactory interneurons for optical and electrical recording. *J. Neurosci. Meth.* **72**:97–108.

Goldberg F, Grunewald B, Rosenboom H, Menzel R (1999). Nicotinic acetylcholine currents of cultured Kenyon cells from the mushroom bodies of the honey bee *Apis mellifera*. *J. Physiol.* **514**:759–768.

Guerrero G, Isacoff EY (2001) Genetically encoded optical sensors of neuronal activity and cellular function. *Curr. Opin. Neurobiol.* **11**:601–607.

Haag J, Borst A (2000) Spatial distribution and characteristics of voltage-gated calcium signals within visual interneurons. *J. Neurophysiol.* **83**:1039–1051.

Hammer M (1997) The neural basis of associative reward learning in honeybees. *TINS* **20**:245–252.

Hansson BS, Christensen TA (1999) Functional characteristics of the antennal lobe. In: *Insect Olfaction*, Hansson BS, Ed., Springer, Berlin, pp. 125–161.

Hansson BS, Carlsson MA, Kalinova B (2003) Olfactory activation patterns in the antennal lobe of the sphinx moth, *Manduca sexta*. *J. Comp. Physiol. A* **189**:301–308.

Heinrich B (1974) Thermoregulation in endothermic insects. *Science* **185**:747–756.

Ito K, Okada R, Tanaka NK, Awasaki T (2003) Cautionary observations on preparing and interpreting brain images using molecular biology-based staining techniques. *Microsc. Res. Tech.* **62**:170–186.

Joerges J, Küttner A, Galizia CG, Menzel R (1997) Representations of odours and odour mixtures visualized in the honeybee brain. *Nature* **387**:285–288.

Kurtz R, Durr V, Egelhaaf M (2000) Dendritic calcium accumulation associated with direction-selective adaptation in visual motion-sensitive neurons *in vivo*. *J. Neurophysiol.* **84**:1914–1923.

Laissue PP, Reiter C, Hiesinger PR, Halter S, Fischbach KF, Stocker RF (1999) Three-dimensional reconstruction of the antennal lobe in *Drosophila melanogaster. J. Comp. Neurol.* **405**:543–552.

Lanni F, Keller HE (1999) Microscopy and microscope optical systems. In: *Imaging Neurons: A Laboratory Manual.* Yuste R, Lanni F, Konnerth A, Eds., Cold Spring Harbor Press, Cold Spring Harbor, NY, pp. 1.1–1.72.

Macleod GT, Hegstrom-Wojtowicz M, Charlton MP, Atwood HL (2002) Fast calcium signals in *Drosophila* motor neuron terminals. *J. Neurophysiol.* **88**:2659–2663.

Macleod GT, Suster ML, Charlton MP, Atwood HL (2003) Single neuron activity in the *Drosophila* larval CNS detected with calcium indicators. *J. Neurosci. Methods* **127**:167–178.

Manzini I, Schild D (2003) Multidrug resistance transporters in the olfactory receptor neurons of *Xenopus laevis* tadpoles. *J. Physiol.* **546**:375–385.

Meijerink J, Carlsson MA, Hansson BS (2003) Spatial representation of odorant structure in the moth antennal lobe: a study of structure-response relationships at low doses. *J. Comp. Neurol.* **467**:11–21.

Miyawaki A, Llopis J, Heim R, McCaffery JM, Adams JA, Ikura M, Tsien RY (1997) Fluorescent indicators for Ca^{2+} based on green fluorescent proteins and calmodulin. *Nature* **388**:882–887.

Miyawaki A, Griesbeck O, Heim R, Tsien RY (1999) Dynamic and quantitative Ca^{2+} measurements using improved cameleons. *Proc. Natl. Acad. Sci. USA* **96**:2135–2140.

Miyawaki A, Sawano A, Kogure T (2003) Lighting up cells: labelling proteins with fluorophores. *Nat. Cell Biol.* **Suppl**:S1–S7.

Müller D, Abel R, Brandt R, Zockler M, Menzel R (2002) Differential parallel processing of olfactory information in the honeybee, *Apis mellifera* L. *J. Comp. Physiol. A* **188**:359–370.

Ng M, Roorda RD, Lima SQ, Zemelman BV, Morcillo P, Miesenbock G (2002) Transmission of olfactory information between three populations of neurons in the antennal lobe of the fly. *Neuron* **36**:463–474.

Okada K, Kanzaki R (2001) Localization of odor-induced oscillations in the bumblebee antennal lobe. *Neurosci. Lett.* **316**:133–136.

Rospars JP (1988) Structure and development of the insect antennodeutocerebral system. *Int. J. Insect Morphol. Embryol.* **17**:243–294.

Sachse S, Galizia CG (2002) Role of inhibition for temporal and spatial odor representation in olfactory output neurons: a calcium imaging study. *J. Neurophysiol.* **87**:1106–1117.

Sachse S, Galizia CG (2003) The coding of odour-intensity in the honeybee antennal lobe: local computation optimizes odour representation. *Eur. J. Neurosci.* **18**:2119-2132.

Stetter M, Greve H, Galizia CG, Obermayer K (2001) Analysis of calcium imaging signals from the honeybee brain by nonlinear models. *Neuroimage* **13**:119–128.

Thomas D, Tovey SC, Collins TJ, Bootman MD, Berridge MJ, Lipp P (2000) A comparison of fluorescent Ca^{2+} indicator properties and their use in measuring elementary and global Ca^{2+} signals. *Cell Calcium* **28**:213–223.

Toga AW, Mazziotta JC (2002) *Brain Mapping: The Methods.* Academic Press, New York.

Tsai PS, Nishimura N, Yoder EJ, Dolnick EM, White GA, Kleinfeld D (2002) Principles, design, and construction of a two-photon laser-scanning microscope for *in vitro* and *in vivo* brain imaging. In: In Vivo *Optical Imaging of Brain Function.* Frostig R, Ed., CRC Press, Boca Raton, FL.

Uhl R (1999) Arc lamps and monochromators for fluorescence microscopy. In: *Imaging Neurons: A Laboratory Manual.* Yuste R, Lanni F, Konnerth A, Eds., Cold Spring Harbor Press, Cold Spring Harbor, NY, pp. 2.1–2.8.

Wang JW, Wong AM, Flores J, Vosshall LB, Axel R (2003) Two-photon calcium imaging reveals an odor-evoked map of activity in the fly brain. *Cell* **112**:271–282.

Wilson RI, Turner GC, Laurent G (2004) Transformation of olfactory representations in the *Drosophila* antennal lobe. *Science* **303**:366–370.

Yuste R, Lanni F, Konnerth A (1999) *Imaging Neurons: A Laboratory Manual.* Cold Spring Harbor Press, Cold Spring Harbor, NY.

Zochowski M, Cohen LB, Falk CX, Wachowiak M (2002) Voltage-sensitive and calcium-sensitive dye imaging of activity: examples from the olfactory bulb. In: In Vivo *Optical Imaging of Brain Function.* Frostig R, Ed., CRC Press, Boca Raton, FL.

14 A Primer on Multichannel Neural Ensemble Recording in Insects

Vincent M. Pawlowski, Thomas A. Christensen, Hong Lei, and John G. Hildebrand

CONTENTS

0-8493-2024-0/05/$0.00+$1.50

14.1 INTRODUCTION

A fundamental tenet of neurobiology is that complex signals in the brain are often represented across distributed populations of often highly interconnected neurons (Sanger, 2003; Averbeck and Lee, 2004). In order to understand how information is encoded in an intricate network of neurons, it would be advantageous to have a method that allows the experimenter to record simultaneously from many neurons across the brain. Fortunately for neurobiologists, recent rapid advances in microfabrication techniques have led to the development of an ever growing variety of multichannel microelectrode arrays that are proving to be indispensable for probing neural function and information processing in populations of neurons. A fortunate consequence of these exciting technological advances is that the microminiaturization of these devices makes them quite suitable for work in the tiny brains of many of the larger insects (Christensen et al., 2000; Galizia and Menzel, 2000). The first section of this chapter reviews the relevant background literature and introduces the state of the art in *multichannel neural ensemble recording* (MNER). Later sections cover current and advanced technologies and techniques that are especially relevant to insect sensory neurobiology. Many more ideas and techniques are being developed than we have space to address in this chapter, and we are certain that new ones will have surfaced by the time the book is published. Our intent, therefore, is to provide a broad overview of ensemble recording methods, focusing on applications that are particularly relevant to the neurobiology of insects. We hope that increased awareness of this technology will act as an impetus to its further development for use in insect neurobiology.

14.1.1 In the Beginning ...

Using simple wire electrodes inserted into the cortex, the first recordings of brain activity were obtained by Lord Adrian and colleagues in the 1920s (Adrian and Zotterman, 1926; review: Schmidt, 1999). It soon became apparent that the more wires added to the recording apparatus, the greater the number of neural processes that could be monitored. Each wire electrode usually records the activity of several neurons, some of which may have very similar wave shapes, making them at times difficult to discriminate as single events (see Figure 14.1b). This problem was solved by twisting two (stereotrode) or four (tetrode) electrodes together, thus obtaining simultaneous recordings of the waveforms from slightly different positions in the tissue (McNaughton et al., 1983; Gray et al., 1995). Today, hundreds or even thousands of these microwires are fashioned into arrays that are being used to study the functions of large-scale neural ensemble activity in larger animals. At the other end of the spectrum, however, micromachined silicon or ceramic-based microelectrodes (e.g., Wise et al., 1969; Ji et al., 1990) allow small brain areas to be monitored in greater detail. Since their introduction, these electrode arrays have been improved through the use of different electrode and insulation materials, new configurations to help improve the separation of signals from different neurons, and enhanced methods for implanting and stabilizing the arrays (Nicolelis, 1999). The use of this technology in insect sensory, neurobiology research is a relatively new development (Christensen et al., 2000), but its popularity is steadily growing. Originally developed for use in the mammalian brain, this method is now used routinely in our laboratory to study odor-evoked activity patterns in the insect olfactory system.

14.1.2 Silicon-Based MNER Arrays Suited for Insect Use

It soon became clear that the ability to monitor and characterize the responses of *ensembles* of neurons, as opposed to one or two, revealed information about cellular interactions in these circuits that could not be resolved with conventional single-unit recording methods (see Chapter 5, Chapter 7, Chapter 11, and Chapter 12). In the olfactory system of the moth *Manduca sexta*, we use multichannel silicon microprobes based on thin-film microfabrication methods developed at the Center for Neural Communication Technology (CNCT) at the University of Michigan (www.engin.umich.edu/center/cnct). We use a 16-channel array of iridium-plated electrodes that fits the dimensions of the moth's antennal lobe, allowing us to record neural ensemble activity both within and across the olfactory glomeruli (Figure 14.1a). Results from experiments using this technology have quickly led us to the conclusion that the neural computations performed in olfactory circuits at the first stage of processing are considerably more complex than previously thought. For example, we now have evidence that ensembles of glomerular output neurons use multiple and overlapping coding strategies to process olfactory information, and these strategies are matched to the particular context surrounding odor presentation (Christensen et al., 2000, 2003; Lei et al., 2002; see also Chapter 12 and Chapter 13).

14.2 ACQUISITION OF SPIKE DATA

The technology that makes multichannel neural ensemble recording possible is a combination of special-purpose electronics along with computer hardware and software. As experimental complexity and the number of electrodes continue to increase, one of the biggest hurdles we face is the handling and analysis of huge volumes of data that result from even the simplest experiments. Throughout the history of MNER, advances in recording methods have kept pace with improvements in computer technology, new analytical software algorithms, and advanced data visualization tools. This section, therefore, will begin with a review of the basic techniques, providing a foundation for later discussion of more recent advances.

In a recent review, Sameshima and Baccal (1999) provide an excellent introduction to the basic technology that goes into MNER instrumentation. The digital *acquisition* of analog action potentials can be accomplished in one of three fundamental ways:

1. As a continuous, filtered, high-rate data stream
2. As discrete events (spikes, field potentials, etc.)
3. As event times only

Before purchasing one of the growing number of commercially available MNER packages (see Section 14.4.2), it is important to know that all three of these recording

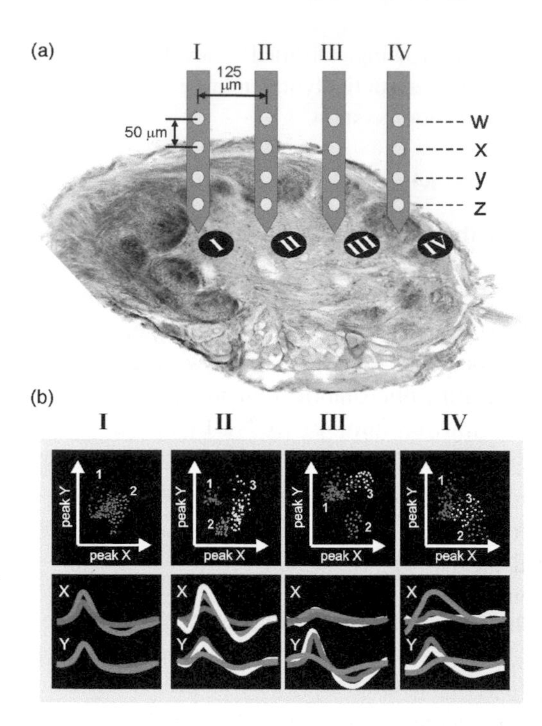

FIGURE 14.1 (See color insert following page 202.) Diagram of the silicon multichannel recording probes (www.engin.umich.edu/center/cnct) we currently use in the characterization of olfactory networks in the moth *Manduca*. **a**: Placement of the four-shank multielectrode array in the moth's antennal lobe. Histological examination and confocal microscopy following recording are used to reveal the placement of the four shanks (I–IV) in the lobe. **b**: Cluster analysis was performed on each shank of the array. Comparison of a given feature (in this simple example, peak amplitude) extracted from waveforms recorded on adjacent sites (x, y) results in a two-dimensional scatter plot for each shank (upper cluster diagrams), illustrating the separation of different spike waveforms into different clusters. After separation, the average waveforms recorded on sites x and y (lower) show slight differences in shape, and amplitude differences provide useful information about the origin of the signals.

methods may or may not be available in all systems, so it is important to decide before you make your purchase which of these recording methods will best suit the needs of your experiment. Although the hardware for acquiring digital signals is well developed and readily available, some prefer to write their own software. Nevertheless, the selection of a complete commercial package can save years of costly development time in the absence of an experienced, in-house software development team.

It is important to note that the details of the experimental paradigm can dictate the choice of the most appropriate spike acquisition system, so we will begin with the fundamentals of the three data collection systems (continuous acquisition, event recording, and time-stamped data) as they pertain to different experimental situations. Obviously, continuous recording is the most suitable method if the analysis of slow waveform changes (oscillations, EPSPs, IPSPs) is required, but other situations may also call for this method. When the experiments are complex and difficult (or expensive) to repeat, and the cellular constituents of the brain area are still relatively unknown, we recommend digitizing all data at a high decimation rate and saving it to disk (see Section 14.4.2.2). One should also bear in mind, however, that these continuous acquisition systems carry the highest price tags and generate huge and unwieldy datasets. A viable alternative is event recording. If the experiments are easy to set up and replicate and the results require only well-defined spike data, then the more cost-effective method is event recording, where only a brief window of data surrounding each waveform is saved to disk (see the discussion of windowed acquisition in Section 14.2.1). Waveforms that are difficult to separate can then be teased apart offline, sorting errors can be checked, and the data can be reanalyzed. Finally, in cases in which only a single, well-isolated spike waveform per channel is obtained (usually facilitated by independently positionable electrodes) and the neural signal types are well defined, only the time of occurrence or *timestamp* of the waveform may be needed. In other applications, the experimental goal may be to close a neural-feedback loop, for example, to drive a neural prosthesis (Bossetti et al., 2004) or to provide feedback to regulate an insect's flight maneuvers (Gray et al., 2002; see also Chapter 4). In these cases, the neural events need not be recorded permanently even as timestamps, but can instead provide immediate input to the calculation of a statistical feedback variable that can then be used to initiate or stimulate motor activity. As these real-time neurophysiology tasks become more common, the need to fuse data inputs simultaneously from several sensory modalities will continue to drive increases in the speed of data acquisition and analysis. In sum, each experimental application determines the type of data acquisition system required, and thus no particular system may be best for all needs.

14.2.1 Analog-to-Digital Conversion and Spike Detection

After amplification and band-pass filtering, an analog waveform recorded from an electrode must be converted to a digital signal that the computer can interpret. In the process of data acquisition, the computer decimates, or divides, the waveform into a series of discrete voltage points acquired at a fixed time interval set by the user. The inverse of this interval is the *sampling rate*, and it is critical that the acquisition system be fast enough to provide an accurate digitized representation of every waveform on every channel, timestamp the waveforms, and save them all to disk. As mentioned earlier, some commercial systems record data continuously (RC Electronics[(24)], Plexon's MAP system[(23)]), but this strategy can result in unmanageably huge datasets because the system records the spike data and everything in between. Another strategy is to select and record only the data of interest (e.g., spikes) in discrete time windows, thus saving only the windowed data and discarding the rest. *Windowed acquisition* is a very useful tool as long as you have determined without doubt that the data points thrown away in the process are unnecessary. Windowed acquisition is useful in cases when recordings last for long periods of time (hours to days) and it is not practical to save continuous data streams on multiple channels recorded at a high sampling rate. For most of our work, we use a sampling rate of 25 kHz on each of 16 channels, which translates to 32 sample points in a 1.28 msec data window (Figure 14.2). Each spike is saved in its own window, and

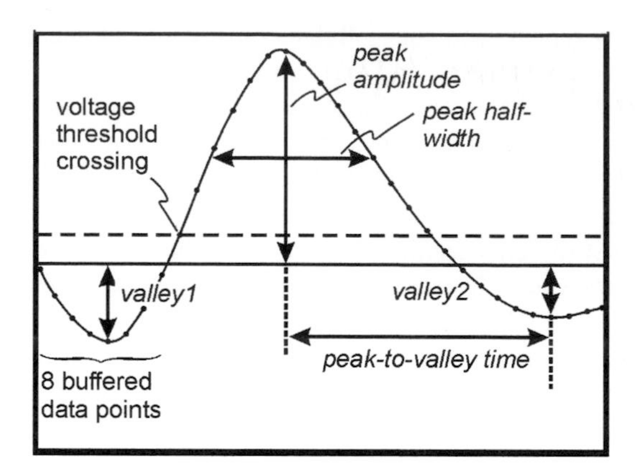

FIGURE 14.2 Windowed data acquisition and waveform analysis. A typical acquisition window consisting of 32 data points for a total window width of 1.8 msec is shown. In the method of simple thresholding, when the voltage threshold (horizontal dashed line) set by the user is crossed, this instructs the A/D converter to begin acquiring data. We recommend using a system that continuously buffers some data points (eight in this example) in order to retain some prethreshold data (see text). Once the 32-point waveform is saved to disk, the process of data extraction can begin. Five simple parameters (italics) are extracted in this example, but the number of possible parameters that one can extract is much larger and can include linear (peak or valley amplitude, peak half-width, etc.), reduced (principal components), or transformed measures (wavelet reduction).

each window is timestamped (typically at a resolution of 1 to 10 msec), so that the precise timing relationships between events are preserved. Because the average extracellular spike lasts less than 1 msec, the window provides us with the complete waveform, as well as some prespike and postspike information. This latter information can be useful for sorting similar waveforms), as long as the window length is set carefully. For example, when more than one unit is recorded on a single channel, the postspike information can often alert the user to a superposition of several near-simultaneous spikes from different neurons in the same window (Figure 14.3a). Because the refractory period of most neurons is several times the window duration, one can assume that a single neuron cannot spike more than once during this interval, so any doublet or triplet waveforms likely originate from multiple sources. The superposition or near-superposition of different neural events is a key determinate in understanding the role of synchrony, or coincident firing, in sensory signal decoding (see Chapter 5 and Chapter 12). Windowing spike data may not be the best approach if these are important considerations. However, the window method of acquisition is very efficient in that it preserves all of the spike data, including all timing relationships between spikes, while dramatically reducing the volume of data collected. As we will see, this volume reduction becomes a distinct advantage when sorting the data.

14.2.2 Thresholding

If the data are not recorded continuously, how can all the neural events be recorded with any accuracy? In the case of windowed acquisition, each channel's voltage is

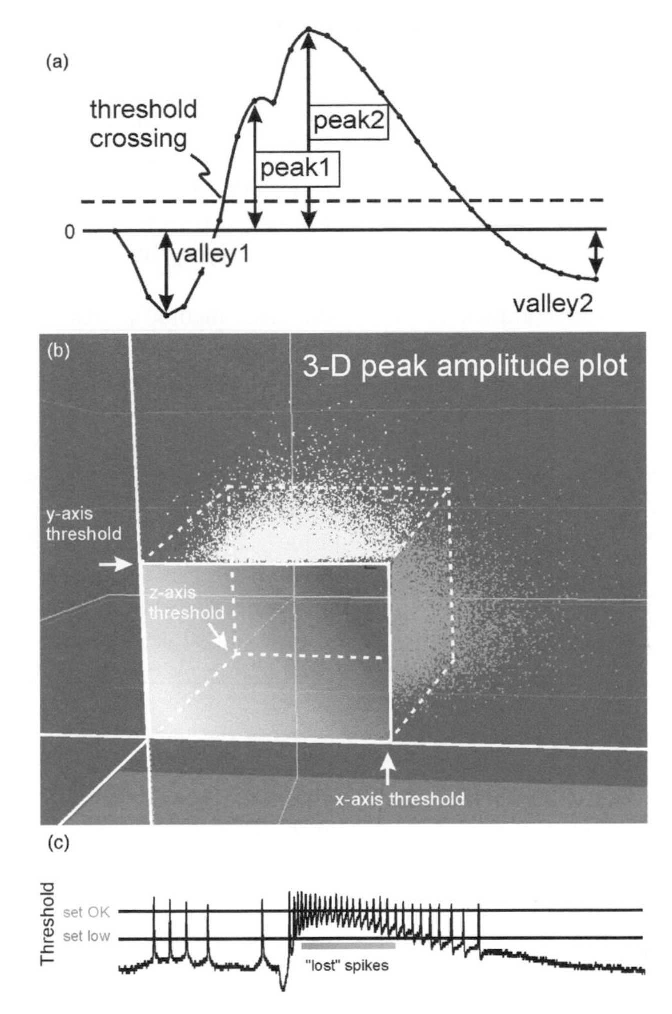

FIGURE 14.3 (See color insert following page 202.) Some pitfalls of simple fixed-voltage thresholding. **a**: Superposition of two near-simultaneous spikes results in a summated waveform that may be interpreted as a third spike and, thus, a third unit. **b**: What you might see if your thresholds are set too high. When applied to stereotrodes, the projection of the two channels' peak amplitude values along the axes of a scatterplot leaves a rectangle in which no spikes are visible because they fall below threshold and are discarded. In this figure, a 3D plot illustrates the rectangular box that results from the projection of clusters based on parameter values from three electrodes. When applied to tetrodes, the subthreshold rectangle becomes a hypercube, and so on. This method is known as *linear thresholding* (see text) and can sometimes result in the undersampling of spikes in a given cluster (two well-separated but clipped clusters are shown). **c**: Spikes do not always maintain the same amplitude during a recording, which can sometimes compromise correct classification and sorting. This intracellular recording illustrates how spike amplitude in a single neuron can attenuate over time. If the single fixed threshold is set too low (lower horizontal line), the A/D system will fail to collect a complete dataset, resulting in "lost" spikes.

monitored nearly simultaneously, and the system searches and temporarily buffers a few (typically eight) sample points until a minimum voltage *threshold* (set by the user) is crossed. Then, the next millisecond or two of data, plus the previously buffered data samples, are saved (Figure 14.2). With voltage thresholding, only waveforms that are stable over time and are 3.5 to 4 standard deviations above the baseline noise level are sufficient for correct sorting (Quian Quiroga et al., 2004). This simple method is popular because it does not represent a major burden on a computer processor already busy buffering, timestamping, and saving many channels of data.

A slightly more complex version of the basic method is *ganged thresholding*. In this recording configuration, two or four channels (stereotrodes or tetrodes) are monitored together (ganged). Data are acquired simultaneously for all ganged channels at the moment that any one of them exceeds the threshold. This method has some advantages over basic single-channel voltage thresholding. For example, by comparing the amplitude of spikes on each of the multiple channels, units that originate from outside the recording area can be isolated by triangulation (Jog et al., 2002). A recent improvement on multiple-ganged thresholding (or hypercubic thresholding) is called *hyperellipsoid thresholding* (Rebrik et al., 1999), as illustrated in the three-dimensional scatter plot in Figure 14.3b. Rather than setting the threshold on each axis as a constant, hyperellipsoid thresholding takes into account the aggregate voltage values at all four channels of the tetrode. When the voltage values for two or more channels are simultaneously close to the threshold, the thresholds are adjusted to a lower value, which can help avoid the underestimation of spikes in each cluster, as illustrated in Figure 14.3b.

Historically, thresholding to detect a spike was achieved with analog electronics, and now digital acquisition has followed this technique. Sometimes, dual or multiple thresholds were used to sort multiple channels in real time, but anything more complex was difficult with a PC-based system. Most major vendors of MNER systems base their hardware architecture on two distinct technologies. The older format uses a data acquisition card controlled by an acquisition computer, usually a powerful PC. Some newer technology uses a digital signal processor (DSP) to control data acquisition and initial processing, freeing the PC to do other tasks. This allows for the implementation of more complex algorithms for spike detection. These improvements in technology, along with the growing need to address ways to study neural tissues with small neurons, are fueling a burst of new algorithms for spike detection. For more details on DSPs in spike acquisition, see Section 14.4.

One of the advantages of DSP is that it allows alternative methods of spike thresholding. One common problem is illustrated in Figure 14.3c: how does one set a fixed threshold in situations when a neuron fires in a bursting pattern that exhibits a change in spike amplitude over time? If the threshold is set to capture the beginning of the burst, it can easily miss many of the attenuated spikes that occur later in the burst (Figure 14.3c). Detection algorithms using waveform spike detection schemes, among others, have been proposed to remedy this problem (Sim et al., 1998).

As greater computing power is harnessed, methods that detect inflection points on the waveform can be implemented, and other, more complex methods, such as wavelet detection (Hulata et al., 2002) and firing rate changes (Ritov et al., 2002), are also being investigated. The ability to implement these complex methods may grow in importance as researchers begin looking at more finely structured neural tissues with smaller and more complex spike patterns. When using only a few independently movable electrodes, voltage thresholding is a sufficient method, but with the development of larger and denser array patterns on silicon-based probes, the detection algorithm will have an increasingly important impact on the power and utility of the sorting method employed. This is currently a very active area of research.

14.3 SPIKE SORTING AND PROCESSING

From an engineering viewpoint, spike sorting is essentially a problem in pattern recognition. To separate the waveforms adequately from a multineuron recording that shows different spike shapes, a set of salient waveform features must first be extracted from each spike in the dataset. Whether the extracted waveform parameters are director linear (peak or valley amplitude, peak half-width, etc.), reduced or integrated (principal components), or transformed measures (Fourier transforms or wavelet-reduction) (Letelier and Weber, 2000), a comparison of any two spike features results in a two-dimensional scatter plot, or *cluster diagram,* in which the different spike waveforms are distinguished as separate clusters (Figure 14.1b). Likewise, using multidimensional sorting algorithms, three spike parameters may be compared in a 3D matrix (Figure 14.3b), and *n* features can be mapped into *n*-dimensional space. When multidimensional sorting is employed, great care must be taken in choosing the most appropriate parameters to extract from the recorded data on each channel. Thorough visualization of an assortment of parameters is a necessary step in sorting spike clusters. The method of principal components can be helpful because it emphasizes the parameters with the greatest variation (Wheeler, 1999).

Another advantage of multielectrodes such as tetrodes is that the *ratio* of the parameters recorded between electrodes is much more likely to remain constant from one spike to the next. Therefore, these ratio values are more likely to lead to a clearer separation between clusters. However, the accuracy of spike separation by microwire tetrodes is not always assured (Harris et al., 2000).

The signals recorded with extracellular electrodes arise from currents generated by nearby neurons. Extracellular spike waveforms may appear very different from intracellular action potentials. In some cases, at least during the rising phase of the action potential, the extracellular waveform resembles the derivative of the intracellular waveform (Brooks and Eccles, 1947). Several more recent studies have tried to derive a precise relationship between intracellular and extracellular waveforms. Henze et al. (2000), working on the rat hippocampus, used wire tetrodes and silicon microprobes along with simultaneous intracellular recordings to show that extracellular spikes started approximately 0.01 msec before and ended 0.02 msec earlier than corresponding intracellular waveforms. However, *in vitro* research with invertebrate cells recorded intracellularly while lying on top of unique field effect transistors shows a variety of corresponding extracellular waveforms that have more variable relationships with the intracellular waveform (Fromherz, 2003).

The voltage that we see in an ensemble recording depends on the inverse of the distance between the source and the electrode, which implies that the strongest signals come from those neurons closest to the electrode (Sameshima and Baccal, 1999). Once the waveforms that are most alike are separated and sorted (or "cut") into clusters, each cluster is generally thought to represent the activity of a separate neuron. There is still much debate, however, about whether multiple spike waveforms may arise from a single neuron. In fact, there is evidence from intracellular recordings that neurons with multiple spike initiating zones are common in the olfactory networks of some insects (Christensen et al., 1993). We, therefore, prefer not to equate a sorted set of spikes with the term *neuron*; instead, we use the term *unit* to describe their activity. The olfactory responses of two units extracted from a 15-unit ensemble in the *Manduca* antennal lobe are illustrated in Figure 14.4. For more on the analytical methods used to study the spatial and temporal patterns of ensemble activity after the unit data have been extracted, see Christensen et al., (2000, 2003) and Lei et al., (2002).

The separation and sorting of unit waveforms from multiple channels is seldom a straightforward process, and thus many groups have put a lot of energy into developing accurate and efficient sorting algorithms and toolsets. Most of these are automated to varying degrees. Purely manual sorting tools have lost their popularity,

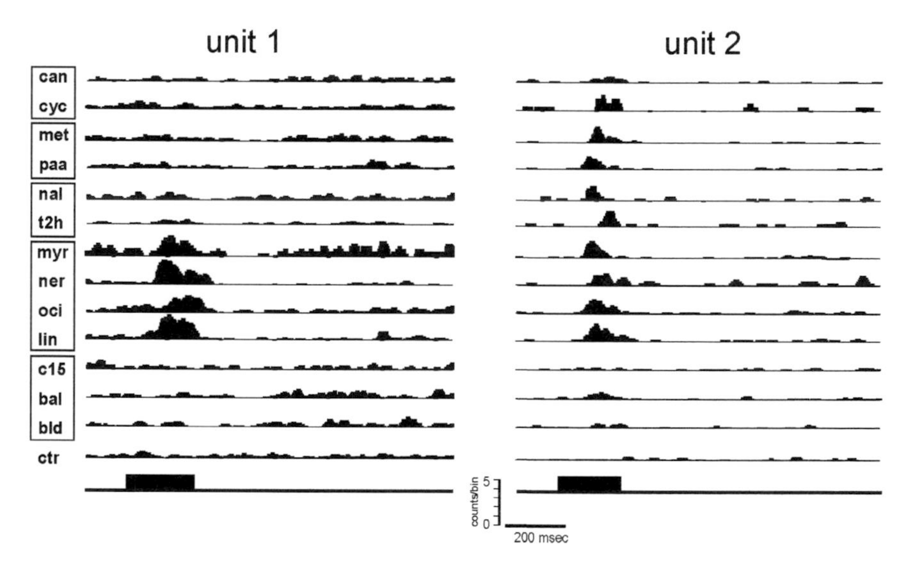

FIGURE 14.4 Response histograms from two well-separated units illustrating dramatic differences in their odor-tuning characteristics. Histograms were generated with Neuroexplorer (Nex Technologies)[17]. Our olfactory studies employ a wide range of different odorants, shown here grouped by category: simple organic solvents — cyclohexane (can) and cyclohexanone (cyc); aromatics — methyl salicylate (met) and phenylacetaldehyde (paa); aliphatics — nonanal (nal) and trans-2-hexenal (t2h); monoterpenoids — myrcene (myr), nerol (ner), ocimene (oci), linalool (lin); and pheromones — C15, bal, and a blend of the two (bld); control air (ctr). **a**: Unit 1 showed selective, stimulus-locked responses to the monoterpenoids. **b**: Unit 2 was broadly responsive across all odorants except pheromones.

and most of the recent spike-sorting software releases include at least a minimal level of unsupervised or semiautomated sorting. In the sections that follow, we discuss the most common methods used in manual or semiautomated sorting: *K-means clustering* and *template matching.*

14.3.1 K-Means Clustering

This clustering method is a widely used iterative technique for automatic classification and separation of spikes (Wheeler, 1999). To begin the analysis, one must first specify the number of expected clusters (K) or manually "seed" the algorithm with the K cluster means. Each data point (representing the n-dimensional vector from each spike) is then assigned to its best-fit cluster. The new mean calculated from the assigned points becomes the new mean for the cluster. The process is repeated until there is no longer a change in the assignments, usually resulting in a progressively more distinct separation between individual clusters. Each cluster boundary is described by a hyperellipsoid in the n-dimensional cluster space, with n associated centroids, means, and variances. Separation between cluster centroids can then be verified statistically using a multivariate ANOVA.

Most unsupervised K-means algorithms have little trouble with this step, as long as an accurate number of seed values is generated for gathering the data points into a cluster. The step of selecting the centroids and means to start the algorithm is not always straightforward, so several sorting packages provide the user with help in predetermining the number of clusters or their centroids. On a practical level, the choice of these cluster seeds can have tremendous impact on the results, and this problem becomes nontrivial in the case of overlapping cluster distributions. Typically, most manual or semiautomated packages allow the operator to select cluster centers, change the acceptable variance, limit the linear distance or Mahalanobis distance between nearby or overlapping clusters, and merge, delete, or split clusters. Some laboratories may choose to write their own spike-sorting software, but few of these in-house packages will be complete or very flexible. They usually rely on one sorting method that works well for a particular dataset. When used in the manual mode, the same sorting algorithm can produce variable results (Wood et al., 2003).

14.3.2 Template Matching and Other Sorting Methods

Another widely used sorting method is template matching. As the name implies, a set of manually selected "typical" waveforms is used to generate a template, which is then compared with the entire spike data set (Stitt et al., 1997). Cluster cutting methods, such as template matching and the method of principal components, have been shown to be mathematical variations of a Bayes classifier (Stitt et al., in prep). Due to the recent application of clustering to problems in other fields, such as data mining, web searching, and genomics, there has been a substantial increase in the number of papers on new clustering methods; an in-depth analysis of all these would simply not be practical here. We will instead concentrate on several of the more popular methods of spike sorting. It should be pointed out, however, that various statistical and transformation tools can all be linked in a general purpose program such as MClust (www.cbc.umn.edu/~redish/mclust) as separate "cutting engines," thereby allowing side-by-side comparison of the results of each method.

Importantly, new data-handling and open software standards, such as the Neuroshare initiative (www.neuroshare.org) and NeuroMAX (www.neuromax.org), are now evolving to help with compatibility issues among the different data formats used by the various commercial and custom data acquisition systems. The ability to compare different datasets will undoubtedly be crucial in facilitating objective comparisons of results obtained with the growing number of different sorting algorithms. Standard simulated data sets based on neural models will also be helpful in performing comparative benchmark tests with the many new sorters.

14.3.3 Can Cluster Cutting Be Fully Automated?

Is there any way to fully automate the process of spike sorting? Whether a human operator or a HUMANN (hybrid unsupervised multilayer artificial neural network) algorithm (Garcia et al., 1998) can more accurately sort spikes remains debatable. Many reason that because these "unsupervised" methods eliminate human error, one avoids the subjectivity of the fallible human operator. Although this may be true in some circumstances, we find that there will inevitably be cases that necessitate user intervention and manual editing. Regardless of the type of algorithm you select for cluster cutting, we believe that it is important to use one that allows the cluster boundaries to be edited. By displaying a three-dimensional plot of all spike data separated into clusters, the strength of human pattern recognition, along with the user's *a priori* knowledge of the dataset, can greatly help to clarify ambiguous clusters.

14.3.3.1 Noise

Perhaps the single biggest factor favoring manual editing is non-Gaussian *noise* — especially unpredictable noise that can arise from multiple sources, both intrinsic and extrinsic to the preparation. Electromagnetic interference, movement artifacts, or even intermittent potentials from distant neurons can all negatively impact a clean cluster analysis. Any clusters that are contaminated with noise must be excluded, so it is important that convenient editing tools be part of a comprehensive spike-sorting package. These tools should provide the user with the ability to select specific subranges of data and to alter the setup parameters for recalculating unsupervised results.

14.3.3.2 Other Sources of Error

It is also important for anyone new to MNER to realize that there are potential sources of error in the recordings that may render some multiunit spike data essentially unsortable (Gruner and Johnson, 2001). One source of error arises from the design of the electronic filters commonly used to reduce noise. Low-pass filters can produce an artifact that results in waveform deflections that can be superimposed on spike waveforms. When spikes are of particularly low amplitude (e.g., < 30 to $40\ \mu V$), these artifacts can result in erroneous waveform separation (Fiore et al.,

1996). Even if these minute spikes are detectable in the multiunit environment, they may not be discriminated from noise. At other times, these "dirty" spikes may be sorted, but their responses will be undifferentiated from their "spontaneous" activity.

Other sources of noise that may get past unsupervised sorters include sources of transient biological and electronic noise, or the occasional temporal summation of several events that have the appearance of a single larger waveform. When large numbers of high-amplitude units are firing together, the biological background frequently increases, resulting in many of these small, difficult-to-sort clusters. In mammalian research, where data sampling is from large structures with well-defined neural architecture and large numbers of independently movable electrodes, these results are simply discarded. In contrast, in tiny zones of finely differentiated insect tissue with few neurons, this is often not an option, and new sorting methods specifically designed for insect sensory neurophysiology applications are needed.

Another typical source of clustering error arises from changes in unit amplitude over time. The most common sources can arise both within and outside the preparation. One internal source is the complex bursting behavior in some neurons, which often leads to variations in spike amplitude (Figure 14.3c). A rapidly firing neuron with amplitude-modulated spikes occurring over the burst duration can easily split into two clusters, resulting in the erroneous conclusion that there are two or more different units with identical responses. Unsupervised algorithms tend to "overcut" clusters in this manner, separating spike incidences that originate from the same source. A common external source of error is from physical movement of the electrodes slow swelling or shifting of tissue over time, or even tissue recoil after electrode penetration. It is no easy task to account fully for these mechanical and biological causes of drift. Some clustering software allows a drift time constant to be included (DataWave's AutoSort[(7)]). A source of error to which some sorting algorithms may be particularly susceptible is a spike alignment error (Lewicki, 1998). This misalignment appears as a slight jitter in the peak time of the spike, which arises from the decimation process. For all the preceding reasons, we believe that it will be some time before cluster analysis is fully automated.

Theoretically, it may not even be possible for any unsupervised algorithm to sort all spike events. In particular, when recording from two-dimensional microelectrode arrays, two identical neurons, equally spaced on opposite sides of the electrode plane or below the leading edge of the probe shanks, may not be distinguishable. Neither the voltage parameters themselves nor their derivatives contain sufficient information to separate these waveforms. Unless time-domain information on the firing rates of the individual spike sources can be determined, even tetrode recording cannot triangulate on a spike source in this case.

14.3.3.3 Checking for Errors

What does all this mean to the user of MNER technology? Simply put, no sorting method based on waveform information is perfect under all circumstances. This situation requires that postcutting tests must also be performed carefully. We recommend that after a preliminary sorting, an interspike interval histogram should be calculated for each putative unit. Any unit that displays an interspike interval less than the minimum expected (based on known activity levels from intracellular recording, for example) is assumed to be contaminated and must be resorted or eliminated from further ensemble analysis. Importantly, however, this does not mean that unsortable multiunit activity is useless. For qualitative purposes, if the temporal response of the putative multiunit is strongly correlated with some experimental parameter, much may be learned from knowing that several spike sources in a given space are responsive in a similar way.

Dubiously separated clusters may, therefore, still have some utility, depending on the purpose of the analysis. For example, if several hard-to-sort multiunit clusters exhibit similar, but not identical, responses to other well-sorted single units and only qualitative results are needed for gross mapping studies, then the suspect clusters may be included in the analysis. On the other hand, if the purpose of the study is to determine the fine temporal relationships between nearby substructures (such as the olfactory glomeruli in our studies), then these multiunits must be viewed with suspicion. Whenever new methods are being developed or tested, conclusions must be drawn from several data sets: some that are easily sorted and others that produce more challenging results. In a previously unexamined neural structure with unusual spike waveforms, it may be necessary to perform simultaneous intracellular and multichannel extracellular recordings (Henze et al., 2000).

In summary, as useful as unsupervised or automatic sorting algorithms may be, they present many users with a dilemma because they are forced to give up much or all control over the sorting process. At present, most investigators are of the mindset that, until we can be absolutely certain that we have unequivocally tied each waveform to a specific unit, editing the sorted data will be a necessity. Adapting an automated algorithm in this manner is a significant programming task, but much progress is being made, as is detailed in the following sections.

14.4 REVIEW OF SOME CURRENT MNER SYSTEMS

Many software developers now offer automated sorting algorithms with editing capability (MClust, DataWave's AutoSort, Plexon's OffLine Sorter). Several spike-sorting packages provide a wide range of cutting algorithms and editing features. Manufacturers such as Alpha Omega(1) and Bionic Technologies(4) have sorting software that is limited to their systems. Neuralynx relies on the open MClust, with additional proprietary extensions. Plexon's OffLine Sorter is one commercial package that provides data input capabilities in many external formats. MClust, OffLine Sorter, and Autosort all provide a broad range of sorting and editing capabilities, as discussed in the following section.

14.4.1 Cluster-Cutting Suites

It is possible to compare unsupervised cutting algorithms on particular real or synthesized data sets (Menne et al., 2002). However, with the exception of MClust and a few others, most open sorting packages from academic labs do not include enough editing tools to be of general use without an in-house programmer or at least some customization for use by new users. Most manufacturers of commercial MNER systems use proprietary sorting software or an interface to a more general package.

DataWave's previous software, Autocut 3, one of the earliest commercial autosorters, provided flexible editing capability along with an automated sorter. The software used a Gaussian hyperellipsoid cutting algorithm with probabilistic assignment of outlying data points. The user was able to visualize clusters in 28 dimensions and weight each separately, even though not all were displayed. Its weakness was that it was a DOS-based package that is no longer supported. The weighting feature could be very useful, but it is not provided in any of the current automated sorters we have tested.

OffLine Sorter is relatively new, and although still undergoing development, it appears to have stabilized with the latest release. The package includes cluster selection in principal components space, K-means sorting, and template matching, and it comes with some flexibility in setting sort parameters. The manual cluster-drawing capabilities are limited to two-dimensional polygons, but flexible editing capabilities make up for the lack of ability to cut manually in more dimensions. It also includes the ability to select a range of sorting parameters for manual sorting, including principal components and nonlinear energy, as well as two automatic sorting algorithms (Shoham et al., 2003). Offline Sorter and other packages allow users to add or remove individual or groups of spikes in a cluster and swap or merge clusters.

MClust is an open spike-sorting package based on MATLAB(13) which allows for new "cutting engines" to be added to a flexible cluster editor. MClust includes a polygon cluster–outlining tool and editing features that make it a flexible tool for general use. With the addition of various tools from several academic sources and MATLAB, all but the newest cutting algorithms are generally available and readily integrated. MClust 3.3 includes two external sorting algorithms: BubbleClust and KlustaKwik. Other examples of academic sorters include Spiker, Spike-O-Matic, and MCLUST (unrelated to MClust). We recommend that the new user closely examine these options, with the cautionary note that some of these sorters' Web sites show little current development.

Although new sorting algorithms are being developed reglularly, few make the transition from demonstration or in-house use to becoming generally available and accepted tools. Whichever tool one chooses for spike sorting, a novice user must spend many hours practicing with the cutting tool in order to learn all its features and develop confidence in appraising the accuracy of the results. Comparisons of several different manual or semiautomated algorithms with different operators show that this is a complex process with many pitfalls and personal preferences (Wehr et al., 1999).

14.4.2 Review of MNER Hardware

Most major vendors of MNER systems base their hardware architecture on one of two distinct technologies:

1. Older systems use a data acquisition card controlled by an acquisition computer, usually a powerful PC.
2. Newer technology uses a digital signal processor (DSP) to control data acquisition and initial processing.

These systems and major commercial vendors are compared in the following sections.

14.4.2.1 PC-Based Acquisition

DataWave Technologies (formerly Brainwave Systems) originally developed and marketed a popular DOS-based package called Discovery, which in the last few years has been reborn as a Windows-based package called SciWorks(7). This software package, coupled with Lynx-8 Amplifiers (Neuralynx)(16), provides up to 32 A/D channels, whereas Neuralynx's own Cheetah systems provide up to 256-channel capability. Both MNER systems are based on high-speed, PC-controlled, bus-based data acquisition cards from Data Translation(8). A number of other systems use similar architectures but are designed for a smaller number of recording channels and may be too limited for most MNER applications. They illustrate the problems inherent in burdening a PC with the task of acquiring, processing, and storing large amounts of data. With most of these lower-end products, the limitations are not lessened with more powerful PCs, because the bottlenecks are in the CPU-dependent internal boards, legacy operating system, and PCI standard buses connecting them to the PC.

Evolving from a 160-channel system developed for NASA, new Cheetah 32- and 16-channel technology has been developed by Neuralynx. These systems use the Lynx-8 eight-channel analog amplifier with software-controlled gain and filter settings, a laboratory standard for the past decade. The Cheetah 16/32 systems use the DT3010 A/D card from Data Translation. As with the other systems mentioned previously, the Cheetah system includes a headstage amplifier for the University of Michigan probes. It also includes Neuralab analysis software and a proprietary version of the BubbleClust spike sorting software.

DataWave Technologies' SciWorks and Experimenter software package is a system based outside the realm of neural-ensemble recording, so it may be of additional interest in a multipurpose setting. The system package combines Neuralynx Lynx-8 amplifiers and headstages with Data Translation's DT3010 A/D card. SciWorks offers the advantage of being compatible with the previous generation of Experimenter's Workbench and Discovery software for users familiar with that standard. The Experimenter plug-in for SciWorks includes a unique N-Probe multisite electrode spike detection feature that can be especially handy for odd probe array site configurations, such as CNCT's 2-by-6 array or the Acreo "on-demand" probes referred to Section 14.5.1.

National Instruments' RT series(15) of data acquisition boards with embedded 486 processors help offload some tasks from the PC's main processor. Also, new embedded microcontrollers combine the capabilities of RISC and DSP architectures, along with A/D converters and high speed serial ports. Additionally, there are a few vendors, such as RC Electronics(24), that provide custom, high-speed, computerized acquisition systems that are capable of recording many channels continuously. In the past, the cost of high-speed data storage limited these systems to special applications. As a result, a legacy of limited hardware engendered a paradigm of online data compression and parameter extraction. This technology is becoming the most popular among cost-conscious users. These changes are blurring the distinction between the PC-based and DSP-type systems (described next).

14.4.2.2 DSP-Based Acquisition

Systems offered by Plexon[(23)], Tucker-Davis Technologies (TDT)[(31)], and Alpha Omega[(1)] use DSP-based acquisition hardware to minimize the workload of the PC, freeing it for other tasks, such as experiment control. These systems also provide the capability for data-processing functions that are not practical with a typical PC. When combined with fourth-generation graphical programming languages, they also provide a degree of flexibility that is currently unattainable with PC card–based systems.

The DSP (or several) may be located in a separate module or rack, or on a bus-based card within the main computer cabinet. The major advantage of the DSP-based systems is that the acquisition computer is not loaded down with all the tasks of controlling the experiment while also processing and storing the huge amounts of data typically generated in an MNER experiment. Historically, however, programming a DSP has been a difficult, specialist's task.

The TDT System III is a new generation of hardware updated significantly from TDT's previous technology, which consisted of the four-channel System 2 with BrainWare software, used for many years in auditory neurophysiology. The new System III is a DSP-based package that includes a 16-channel headstage, designed specifically for use with multichannel probes, and the Medusa 16-channel preamp with low-noise fiber optic output. The new OpenEx software is a turnkey multichannel neurophysiology package with a graphical block-diagram programming language called RPVds. This interfaces to MATLAB for both instrument control and data handling, if desired. The new Pentusa system extends this capability to 32 channels.

Plexon has offered the MAP (Multichannel Acquisition Processor) DSP-based, high channel-count system for many years. This system is in the higher price range, along with the systems from Neuralynx, Alpha Omega, and RC Electronics. Recently, though, Plexon has begun offering a smaller system, more suited to insect work, that is based on National Instruments' E-series PCI A/D cards. A complete system is available with the Recorder acquisition software, 16-channel headstage amplifier, and 16/32-channel preamplifier technology adapted from the MAP system.

14.5 NEW TECHNOLOGIES AND APPLICATIONS

14.5.1 New Probe Designs

New developments since wire tetrodes include ceramics (Xu et al., 2002), flexible polyimide multitrodes (Rousche et al., 2000; Takeuchi et al., 2004), and microelectrode arrays with heating elements (Chen et al., 1997). Due partly to their ready and long-term availability, silicon probes from the Center for Neural Communication Technology (CNCT) at the University of Michigan have quickly gained in popularity for use in insects (Christensen et al., 2000; Gray et al., 2002). Among their practical advantages are the precise replication of dimensions from probe to probe (which facilitates reproducible placement in the tissue), and their rigidity, reusability, and relatively low noise levels. The availability of these probes is likely to increase, because CNCT has announced the commercialization of their distribution. As is typical in the semiconductor industry, standardization and wider usage could lead to a significant decrease in production costs. However, this would apply only to the standard probe designs, which are aimed at the largest user base (namely, labs that study mammals and other vertebrates). We expect that insect research will benefit more and more from this technology as news of its usefulness spreads. At least four companies — DataWave Technologies, Neuralynx, Plexon, and Tucker-Davis Technologies — offer commercial MNER systems designed for use with the University of Michigan's silicon microprobes.

Within the next 5 to 10 years, multichannel probes will be getting smaller even as the number of recording sites increases, and these designs will increasingly be aimed at use in insects. Already, both the NeuroNexus Technologies (www.neuronexus.com) and Acreo (www.acreo.com) probe catalogs offer at least a few probes that are useful for smaller tissues. The CNCT catalog offers three 16-channel probe designs that we have used extensively in our studies: the 4 × 4 3mm50, the 4 × 1 tetrodes, and the 2 × 2 tetrodes. All of these probes have 3.0 or 3.75 mm shank lengths. The 4 × 4 array (Figure 14.1a) is available in different shank and site spacings from 125 by 50 μm to 200 by 200 μm. Both the 2 × 2 and 4 × 1 "tetrodes" have a 2D diamond arrangement of 177 μm^2 sites spaced 50 μm on a diagonal. The tetrode center-to-center distances are all 150 μm. Because the probes are two dimensional, they differ from wire tetrodes in that they are guaranteed coplanar with a defined spacing. This could allow accurate triangulation calculations to determine the spike source distance (Jog et al., 2002).

Surface treatments of existing probe designs can improve their performance dramatically (Paik and Cho, 2002). The University of Michigan recommends an activation procedure described on its Web site (www.engin.umich.edu/center/cnct/activ.html). Using just the first part of the procedure, a cleaning process consisting of a quick alternation of positive and negative 2 to 3 V DC pulses, electrode impedances can be lowered by as much as 50%.

The Acreo probes were originally designed as part of a European Union–funded consortium project called VSAMUEL (Hofmann et al., 2000). Some probe designs with a 250 μm minimum intershaft spacing are too large and sparse for most insect use (Norlin et al., 2002). However, this group also makes denser probes. In particular, Acreo's 32-channel U2 and 64-channel U4 probes, with 50 μm spacing from site to site, 60 μm intershaft spacing, and 25 μm shaft width (at the working end), should prove to be very useful for insect applications (www.acreo.se/upload/Partners/acr007826-r3.pdf).

An on-demand manufacturing process recently proposed by the Acreo team promises to make variable electrode spacing probes possible (Kindlundh et al., 2004). The sites are laid out on a grid pattern, which allows one to select electrode sites at specified locations. These custom electrode arrays could be especially useful for recording from highly structured neural tissues, such as the insect antennal lobe (Figure 14.1a).

14.5.2 Integrated "On-Chip" Processing

In an article entitled "A Unified Framework for Advancing Array Signal Processing Technology of Multichannel Microprobe Recording Devices" (Oweiss, 2002), the author proposes to put a complete MNER system on a single integrated circuit chip. Given the demand for human applications in cochlear implants, for instance, combining the recording probes, amplifiers, signal processing, and telemetry on integrated circuit substrates will be a coming trend (see, e.g., Gadicke and Albus, 1997; Kim and Kim, 2000). Although most of the initial effort in this area will be geared toward human neuroprosthetic applications, some laboratories are focusing integration efforts on telemetric systems designed to be carried by an untethered insect (Noriyasu et al., 2002; Takeuchi and Shimoyama, 2004). Even though most of the new amplifier designs are aimed at mammalian applications and may at first have only a minor impact on insect research, newer designs that incorporate the specific requirements of insect experimentation (e.g., smaller, low-noise amplifier designs) will likely debut in the next few years.

14.6 CONCLUSIONS AND OUTLOOK

The rapid integration of advanced electronic and computer technology with micromachined electrode fabrication techniques has rapidly advanced the art of MNER to new heights (Bai and Wise, 2001; Harrison and Charles, 2003). Higher levels of system integration that were developed for special applications, such as deep brain stimulation and neuroprosthesis control, will make the technology much more powerful and progressively smaller (Nicolelis, 2002; Katz and Grinvald, 2002). This is particularly good news for insect electrophysiological research (Galizia and Menzel, 2000; Christensen et al., 2000; Gray et al., 2002). Because the smallest wire multielectrodes are too large for some insect research, micromachined electrodes are rapidly being adopted in many areas of sensory and motor neurophysiology. The spread of new technologies for MNER, exemplified by new probe designs, lower-cost multichannel acquisition systems, and advanced data analysis tools, promises to make MNER a valuable and ultimately common tool in insect neurophysiology research. The increasing power and system integration of ensemble recording packages is being combined with other neural-recording, imaging, data analysis, and visualization techniques and is changing the way extracellular spike data are being used. These new combinations of diverse technologies are creating entirely new applications and fundamentally unique experimental approaches (Bragin et al., 1997; Gray et al., 2002; Hanein et al., 2002).

What does all this mean for the future of insect neurophysiology, and where does a newcomer to MNER begin? In short, there are many more ideas and techniques being developed than there are opportunities to use them. Ideas are no longer limited by the available technology, and new probe designs in particular show how much can be accomplished with a little imagination and a basic familiarity with semiconductor fabrication methods. Commercialization of MNER systems is now in full swing, and these advances will increasingly allow scientists to focus on science rather than on engineering and system design. Microprobes designed specifically for insect electrophysiology are now being designed (Spence et al., 2001, 2003), and these will undoubtedly continue to become more common. The recent explosion of interest in MNER technology for neuroscience research has spawned a number of new options for recording and analyzing spike data. Simultaneously, the developing need for better, faster, semiautomated clustering methods in applications such as image segmentation and data mining has led to new methods for data sorting (for a general review of classical methods, see Jain and Dubes, 1988). Several recent reviews provide comparisons of different spike-clustering methods using quantitative statistical tests (Stitt et al., 1997; Sim et al., 1998; Menne et al., 2002). This chapter has outlined some of the oldest and most popular packages. Contact the vendors, get to know them, and ask them your questions. You may even be able to take a system for a "test drive" before making a substantial purchase.

Regardless of the package ultimately chosen, the user must learn to make objective decisions about the validity of the sorting results. Spike clusters with dubious statistical separation must be either merged or discarded, but which option to choose is not always clear. A frequent complaint about unsupervised sorting methods is a tendency to "overcut" the data. At some point, therefore, the *a priori* knowledge of the human sorter may have to come into play. To use a more familiar analogy, if apples are round and red and pears are pear-shaped and green or yellow, where do round Asian pears and green Granny Smith apples fall on a multidimensional scatter plot? With this simple example, we hope to leave the reader with the realization that sorting spikes from ensemble data is not yet an exact science. Spike sorting is an engineering optimization problem that must be tailored to the specific needs of the research question. As such, the tools we use to make these sorting decisions become an integral part of our results, and we must, therefore, choose these tools wisely. Our wish is that this chapter will serve as a good place to begin.

ACKNOWLEDGMENTS

We wish to thank the many talented individuals in our lab who have contributed to this work over the past several years, especially Andrew Dacks and Heather Stein. This work is supported by NIH grants DC-05652 to TAC and DC-02751 to JGH.

REFERENCES

Adrian ED, Zotterman Y (1926) The impulses produced by sensory nerve endings. Part II. The response of a single end organ. *J. Physiol.* **61**:151–171.

Averbeck BB, Lee D (2004) Coding and transmission of information by neural ensembles. *Trends Neurosci.* **27**:225–230.

Bai Q, Wise KD (2001) Single-unit neural recording with active microelectrode arrays. *IEEE Trans. Biomed. Eng.* **48**: 911–920.

Bossetti CA, Carmena JM, Nicolelis MAL, Wolf PD (2004) Transmission latencies in a telemetry-linked brain-machine interface. *IEEE Trans. Biomed. Eng.* **51**:919–924.

Bragin A, Csicsvari J, Penttonen M, Buzsaki G (1997) Epileptic afterdischarge in the hippocampo–entorhinal system: current source density and unit studies. *Neuroscience* **76**:1187–1203.

Brooks CM, Eccles JC (1947) Electrical investigations of the monosynaptic pathway through the spinal cord. *J. Neurophysiol.* **10**:251--274.

Chen JK, Wise KD, Hetke JF, Bledsoe SC (1997) A multichannel neural probe for selective chemical delivery at the cellular level. *IEEE Trans. Biomed. Eng.* **44**:760–769.

Christensen TA, Waldrop B, Harrow ID, Hildebrand JG (1993) Local interneurons and information processing in the olfactory glomeruli of the moth *Manduca sexta*. *J. Comp. Physiol. A* **173**:385–399.

Christensen TA, Pawlowski VM, Lei H, Hildebrand JG (2000) Multi-unit recordings reveal context-dependent modulation of synchrony in odor-specific neural ensembles. *Nature Neurosci.* **3**:927–931.

Christensen TA, Lei H, Hildebrand JG (2003) Coordination of central odor representations through transient, non-oscillatory synchronization of glomerular output neurons. *Proc. Natl. Acad. Sci. USA* **100**:11076–11081.

Fiore L, Corsini G, Geppetti L (1996) Application of non linear filters based on the median filter to experimental and simulated multiunit neural recordings. *J. Neurosci. Meth.* **70**:177–184

Fromherz P (2003) Neuroelectronic interfacing: semiconductor chips with ion channels, nerve cells, and brain. In: *Nanoelectronics and Information Technology.* Waser R, Ed., Wiley-VCH, Berlin, 781–810.

Gadicke R, Albus K (1997) Performance of real time separation of multineuron recordings with a DSP32C microprocessor. *J. Neurosci. Meth.* **75**:187–192.

Galizia CG, Menzel R (2000) Probing the olfactory code. *Nature Neurosci.* **3**:853–854.

Garcia P, Suarez CP, Rodriguez J, Rodriguez M (1998) Unsupervised classification of neural spikes with a hybrid multilayer artificial neural network. *J. Neurosci. Meth.* **82**:59–73.

Gray CM, Maldonado PE, Wilson M, McNaughton B (1995) Tetrodes markedly improve the reliability and yield of multiple single-unit isolation from multi-unit recordings in cat striate cortex. *J. Neurosci. Meth.* **63**:43–54.

Gray JR, Pawlowski V, Willus MA (2002) A method for recording behavior and multineuronal CNS activity from tethered insects flying in virtual space. *J. Neurosci. Meth.* **120**:211–223.

Gruner CM, Johnson DH (2001) Comparison of optimal and suboptimal spike sorting algorithms to theoretical limits. *Neurocomputing* **38–40**:1663–1669.

Hanein Y, Bohringer KF, Wyeth RC, Willows AOD (2002) Towards MEMS probes for intracellular recording. *Sensors Update* **10**:1–29.

Harris K, Henze DA, Hirase H, Csicsvari J, Buzsáki G (2000) The accuracy of tetrode spike separation as determined by simultaneous intracellular and extracellular measurements. *J. Neurophysiol.* **84**:401–414.

Harrison RR, Charles C (2003) A low-power low-noise CMOS amplifier for neural recording applications. *IEEE J. Solid-State Circ.* **38**:958–965.

Henze DA, Borhegyi Z, Csicsvari J, Mamiya A, Harris KD, Buzsaki G (2000) Intracellular features predicted by extracellular recordings in the hippocampus *in vivo*. *J. Neurophysiol.* **84**:390–400.

Hofmann UG, Folkers A, Malina T, Biella G, De Curtis M, De Schutter E, Yoshida K, Thomas U, Höhl D, Norlin P (2000) Towards a versatile system for advanced neuronal recordings using silicon multisite microelectrodes. *Biomed. Technik.* **45**:169–170.

Hulata E, Segev R, Ben-Jacob E (2002) A method for spike sorting and detection based on wavelet packets and Shannon's mutual information. *J. Neurosci. Meth.* **117**:1–12.

Jain AK, Dubes RC (1988) *Algorithms for Clustering Data.* Prentice Hall, Englewood Cliffs, NJ.

Ji J, Najafi K, Wise KD (1990) A scaled electronically-configurable multichannel recording array. *Sens. Actuat.* **A22**:589–591.

Jog MS, Connolly CI, Kubota Y, Iyengar DR, Garrido L, Harlan R, Graybiel AM (2002) Tetrode technology: advances in implantable hardware, neuroimaging, and data analysis techniques. *J. Neurosci. Meth.* **117**:141–152.

Katz LC, Grinvald A (2002) New technologies: molecular probes, microarrays, microelectrodes, microscopes and MRI. *Curr. Opin. Neurobiol.* **12**:551–553.

Kim KH, Kim SJ (2000) Noise performance design of CMOS preamplifier for the active semiconductor neural probe. *IEEE Trans. Biomed. Eng.* **47**:1097–1105.

Kindlundh M, Norlin P, Hofmann UG (2004) A neural probe process enabling variable electrode configurations. *Senso. Actuat. B*: *Chem.* **102**:51–58.

Lei H, Christensen TA, Hildebrand JG (2002) Local inhibition modulates odor-evoked synchronization of glomerulus-specific output neurons. *Nat. Neurosci.* **5**:557–565.

Letelier JC, Weber PP (2000) Spike sorting based on discrete wavelet transform coefficients. *J. Neurosci. Meth.* **101**:93–106.

Lewicki, MS (1998) A review of methods for spike sorting: the detection and classification of neural action potentials. *Network: Computation in Neural Systems.* **9**:R53–R78.

McNaughton BL, O'Keefe J, Barnes CA (1983) The stereotrode: a new technique for simultaneous isolation of several single units in the central nervous system from multiple unit records. *J. Neurosci. Meth.* **8**:391–397.

Menne KML, Folkers A, Malina T, Maex R, Hofmann UG (2002) Test of spike-sorting algorithms on the basis of simulated network data. *Neurocomputing* **44–46:**1119–1126.

Nicolelis MAL, Ed. (1999) *Methods for Neural Ensemble Recordings.* CRC Press, Boca Raton, FL.

Nicolelis MAL (2002) Multielectrode recordings: the next steps. *Curr. Opin. Neurobiol.* **12**:602–606.

Noriyasu A, Shimoyama I, Kanzaki R (2002) A dual-channel FM transmitter for acquisition of flight muscle activities from the freely flying hawkmoth, *Agrius convolvuli. J. Neurosci. Meth.* **115**:181–187.

Norlin P, Kindlundh M, Mouroux A, Yoshida K, Hofmann UG (2002) A 32-site neural recording probe fabricated by DRIE of SOI substrates. *J. Micromech. Microeng.* **12**:414–419.

Oweiss KG (2002) A unified framework for advancing array signal processing technology of multichannel microprobe recording devices. *Proc. 2nd IEEE Conf. Microtech. Med. Biol.,* Madison, WI, pp. 245–250.

Paik S-J, Cho D (2002) A novel surface modification process for bioelectrodes with a low interface impedance for high sensitivity. *Pacific Rim Workshop on Transducers and Micro/Nano Technologies*, ATIP/Xiamen, P.R. China, pp. 757–760.

Quian Quiroga R, Nadasdy Z, Ben-Shaul Y (2004) Unsupervised spike detection and sorting with wavelets and super-paramagnetic clustering. *Neur. Comput.* **16**:1661–1687.

Rebrik SP, Wright BD, Emondi AA, Miller KD (1999) Cross-channel correlations in tetrode recordings: implications for spike-sorting. *Neurocomputing* **26**:1033–1038.

Ritov Y, Raz A, Bergman H (2002) Detection of onset of neuronal activity by allowing for heterogeneity in the change points. *J. Neurosci. Meth.* **122**:25–42.

Rousche PJ, Pellinen D, Pivin DP, Williams JC, Vetter R, Kipke DR (2000) Flexible polyimide-based intracortical electrode arrays with bioactive capability. *IEEE Trans. Biomed. Eng.* **4**:361–372.

Sameshima K, Baccal, LA (1999) Trends in multichannel neural ensemble recording instrumentation. In: *Methods for Neural Ensemble Recordings.* Nicolelis MAL, Ed., CRC Press, Boca Raton, FL, pp. 47–60.

Sanger TD (2003) Neural population codes. *Curr. Opin. Neurobiol.* **13**:238–249.

Schmidt EM (1999) Electrodes for many single neuron recordings. In: *Methods for Neural Ensemble Recordings.* Nicolelis MAL, Ed., CRC Press, Boca Raton, FL, pp. 1–23.

Shoham S, Fellows MR, Normann RA (2003) Robust, automatic spike sorting using mixtures of multivariate *t*-distributions. *J. Neurosci. Meth.* **127**:111–122.

Sim AWK, Jin CT, Chan LW, Leong PHW (1998) A comparison of methods for clustering of electrophysiological multineuron recordings. *Proc. Int. Conf. IEEE Eng. Med. Bio. Soc.* **20**:1381–1384.

Spence AJ, Buschbeck CJ, Hoy RR, Isaacson MS, Craighead H (2001) Studies of invertebrate neural networks using custom microfabricated electrode arrays. *Proc. Soc. Neurosci.*, San Diego, CA.

Spence AJ, Hoy RR, Isaacson MS (2003) A micromachined silicon multielectrode for multiunit recording. *J. Neurosci. Meth.* **126**:119–126.

Stitt JP, Gaumond RP, Frazier JL, Hanson FE (1997) A comparison of neural spike classification techniques (caterpillar taste organs application). *Proc. 19th Int. IEEE Eng. Med. Bio. Sec. Conf.*, **3**:1092–1094.

Stitt JP, Gaumond RP, Frazier JL, Hanson FE (in preparation) Multiunit neural spike sorting using a discrete Gaussian approximation of a Bayes classifier. *IEEE Trans. Biomed. Eng.*.

Takeuchi S, Shimoyama I (2004) A radio-telemetry system with a shape memory alloy microelectrode for neural recording of freely moving insects. *IEEE Trans. Biomed. Eng.* **51**:133–137.

Takeuchi S, Takafumi S, Kunihiko M, Hiroyuki F (2004) 3D flexible multichannel neural probe array. *J. Micromech. Microeng.* **14**:104–107.

Wehr M, Pezaris JS, Sahani M (1999) Simultaneous paired intracellular and tetrode recordings for evaluating the performance of spike sorting algorithms. *Neurocomputing* **26–27:**1061–1068.

Wheeler BC (1999) Automatic discrimination of single units. In: *Methods for Neural Ensemble Recordings.* Nicolelis MAL, Ed., CRC Press, Boca Raton, FL, pp. 61–77.

Wise K, Angell J, Starr A (1969) An integrated circuit approach to extracellular microelectrodes. *Proc. 8th Int. Conf. Med. Biol. Eng.* **1**:14–15.

Wood F, Black MJ, Vargas-Irwin C, Fellows M, Donoghue JP (2004) On the variability of manual spike sorting. Preprint to *IEEE Trans. Biomed. Eng.*, **51**:912–918.

Xu C, Lemon W, Liu C (2002) Design and fabrication of a high-density metal microelectrode array for neural recording. *Sens. Actuat. A* **96**:78–85.

Appendix

This appendix contains an alphabetized list of vendors for the specialized equipment, tools, and supplies used in the specific research methods described in this book.

1. **Alpha Omega** (Chapter 14)
   Biomedical and clinical neurorecording equipment
   P.O. Box 810
   Nazareth Illit 17105 Israel
   Tel: 972-4-6563-327; fax: 972-4-6574-075
   www.alphaomega-eng.com

2. **Aurora Scientific Co.** (Chapter 3)
   Acoustic research; lever systems (300B series) and force transducers
   360 Industrial Parkway S., Unit 4
   Aurora, Ontario, Canada L4G 3V7
   Tel: 877-878-4784; fax: 905-713-6882
   www.aurorascientific.com

3. **Axon Instruments, Inc. (Molecular Devices Corp.)** (Chapters 12, 14)
   Microelectrode amplifiers; data acquisition
   3280 Whipple Road
   Union City, CA 94587 U.S.A.
   Tel: 510-675-6200; fax: 510-675-6300
   www.axon.com

4. **Bionic Technologies (Cyberkinetics, Inc.)** (Chapter 14)
   Brain–computer interfaces; data acquisition
   100 Foxborough Boulevard, Suite 240
   Foxborough, MA 02035 USA
   Tel: 508-549-9981; fax: 508-549-9985
   www.cyberkineticsinc.com

5. **Brüel & Kjær North America, Inc.** (Chapter 3)
   Sound and vibration equipment; microphones
   2815-A Colonnades Court
   Norcross, GA 30071-1588 U.S.A.
   Tel: 800-332-2040; fax: 800-236-8351
   www.bkhome.com

6. **Dagan Corp.** (Chapter 12)
   Microelectrode amplifiers; microscopy
   2855 Park Avenue
   Minneapolis, MN 55407 U.S.A.
   Tel: 612-827-5959; fax: 612-827-6535
   www.dagan.com

7. **DataWave Technologies** (Chapter 14)
   Data acquisition and analysis
   Tel: 800-736-9283; fax: 970-532-4171
   www.dwavetech.com

8. **Data Translation, Inc.** (Chapter 14)
   A/D data acquisition boards
   100 Locke Drive
   Marlboro, MA 01752-1192 U.S.A.
   Tel: 800-525-8528; fax: 508-481-8620
   www.datatranslation.com

9. **Fine Science Tools (USA), Inc.** (Chapter 13)
   Fine dissecting equipment
   373-G Vintage Park Drive
   Foster City, CA 94404-1139 U.S.A.
   Tel: 800-521-2109; fax: 800-523-2109
   www.finescience.com

10. **Invitrogen Corp.** (Chapter 9)
    Molecular and genetic products and services
    1600 Faraday Avenue
    P.O. Box 6482
    Carlsbad, CA 92008 U.S.A.
    Tel: 760-603-7200; fax:760-602-6500
    www.invitrogen.com

11. **LDS-Nicolet** (Chapter 3)
    Vibration test equipment; minishakers
    8551 Research Way, M/S 140
    Middleton, WI 53562 U.S.A.
    Tel: 608-821-6600; fax: 608-821-6691
    www.lds-group.com

12. Leica Microsystems (Chapters 6, 12)
Microscopes, optics, lenses
Ernst-Leitz-Strasse 17-37
Wetzlar 35578 Germany
Tel: +49 6441 29-0; fax: +49 6441 29-2590
www.leica-microsystems.com

13. The MathWorks (Chapters 4, 14)
MATLAB — technical computing software
Worldwide Headquarters
3 Apple Hill Drive
Natick, MA 01760-2098 U.S.A.
Tel: 508-647-7000; fax: 508-647-7001
www.mathworks.com

14. Molecular Probes, Inc. (Chapters 3, 6, 12, 13)
Fluorescent cell markers and calcium-sensitive dyes
29851 Willow Creek Road
Eugene, OR 97402 U.S.A.
Tel: 541-335-0338; fax: 541-335-0305
www.probes.com

15. National Instruments (Chapter 14)
Test, measurement, and automation equipment
11500 N. Mopac Expressway
Austin, TX 78759-3504 U.S.A.
Tel: 888-280-7645; fax: 512-683-8411
www.ni.com

16. Neuralynx (Chapter 14)
Data acquisition and analysis equipment
2434 North Pantano Road
Tucson, AZ 85715 U.S.A.
Tel: 520-722-8144; fax: 520-722-8163
www.neuralynx.com

17. Nex Technologies (Chapter 14)
Neuroexplorer data analysis package
31 Surrey Road
Littleton, MA 01460 U.S.A.
www.neuroexplorer.com

18. Newport Corp. (Chapter 12)
Photonics, optics, mechanics, vibration isolation
1791 Deere Avenue
Irvine, CA 92606 U.S.A.
Tel: 800-222-6440; fax: 949-253-1680
www.newport.com

19. Nikon USA (Chapter 12)
Microscopes, cameras, imaging supplies
1300 Walt Whitman Road
Melville, NY 11747 U.S.A.
Fax: 631-547-4025
www.nikonusa.com

20. Olivier Carmona (Chapter 8)
Khepera robots
K-Team S.A.
Chemin de Vuasset, CP 111
1028 Preverenges, Switzerland
Tel:+41 21 802 5472; fax:+41 21 802 5471
carmona@k-team.com

21. Olympus America Inc. (Chapters 12, 13)
Microscopes, cameras, imaging supplies
2 Corporate Center Drive
Melville, NY 11747 U.S.A.
Tel: 800-622-6372
www.olympusamerica.com

22. Piezo Systems, Inc. (Chapter 3)
Mechanosensory research; bending transducers
186 Massachusetts Avenue
Cambridge, MA 02139 U.S.A.
Tel: 617-547-1777; fax: 617-354-2200
www.piezo.com

23. Plexon Inc. (Chapter 14)
Multichannel microelectrode systems
6500 Greenville Avenue, Suite 730
Dallas, TX 75206 U.S.A.
Tel: 214-369-4957; fax: 214-369-1775
www.plexoninc.com

24. **RC Electronics** (Chapter 14)
Data acquisition systems and amplifiers
6464 Hollister Avenue
Santa Barbara, CA 93117 U.S.A.
Tel: 805-685-7770; fax: 805-685-5853
www.rcelectronics.com

25. **Research Systems, Inc.** (Chapter 13)
Data visualization and image analysis
4990 Pearl East Circle
Boulder, CO 80301 U.S.A.
Tel: 303-786-9900; fax: 303-786-9909
www.rsinc.com

26. **RUN Technologies** (Chapter 14)
Data acquisition equipment
22702 Via Santa Maria
Mission Viejo, CA 92691 U.S.A.
Tel: 949-348-1234; fax: 949-305-9588
www.runtech.com

27. **Sigma-Aldrich** (Chapter 7)
Research chemicals
3050 Spruce Street
St. Louis, MO 63103 U.S.A.
Tel: 800-325-3010
www.sigma-aldrich.com

28. **Sutter Instrument Corp.** (Chapter 12)
Pullers and manipulators for glass microelectrodes
51 Digital Drive
Novato, CA 94949 U.S.A.
Tel: 415-883-0128; fax: 415-883-0572
www.sutter.com

29. **Syntech** (Chapters 2, 9, 12, 14)
Data acquisition software and hardware
PO Box 1547
1200 BM Hilversum, The Netherlands
Tel: +31 (0) 35 6219760; fax: +31 (0) 35 6218377
www.syntech.nl

30. **Till Photonics** (Chapter 13)
Equipment and software for fluorescence imaging
411 Trebbiano Place
Pleasanton CA 94566 U.S.A.
Tel: 925-600-1871; fax: 925-397 3386
www.till-photonics.com

31. **Tucker-Davis Technologies** (Chapter 14)
MNER, signal processing hardware and software
11930 Research Circle
Alachua, FL 32615 U.S.A.
Tel: 386-462-9622; fax: 386-462-5365
www.tdt.com

32. **UDT Sensors, Inc.** (Chapter 4)
Position sensors
12525 Chadron Avenue
Hawthorne, CA 90250 U.S.A.
Tel: 310-978-0516; fax: 310-644-1727
www.udt.com/positionsensitivedetectors.htm

33. **Vossloh-Schwabe Wustlich & Co.** (Chapter 7)
LEDs, optoelectronic supplies
Carl-Friedrich-Gauβ-Straβe 3
D-47475 Kamp-Lintfort, Germany
Tel: +49 (0) 2842 980-0; fax: +49 (0) 2842 980-299
www.vs-optoelectronic.com

34. **World Precision Instruments** (Chapters 12, 13, 14)
Electrophysiological and lab equipment
International Trade Center
175 Sarasota Center Boulevard
Sarasota, FL 34240-9258 U.S.A.
www.wpiinc.com

Index

B

C

D

E

F

G

H

I

J

K

N

O

P

R

S

T

U

V

W

T - #0191 - 251019 - C12 - 234/156/21 - PB - 9780367393465